T0155886

Lecture Notes in Computer Science 13973

Founding Editors

Gerhard Goos
Juris Hartmanis

Editorial Board Members

Elisa Bertino, *Purdue University, West Lafayette, IN, USA*
Wen Gao, *Peking University, Beijing, China*
Bernhard Steffen, *TU Dortmund University, Dortmund, Germany*
Moti Yung, *Columbia University, New York, NY, USA*

The series Lecture Notes in Computer Science (LNCS), including its subseries Lecture Notes in Artificial Intelligence (LNAI) and Lecture Notes in Bioinformatics (LNBI), has established itself as a medium for the publication of new developments in computer science and information technology research, teaching, and education.

LNCS enjoys close cooperation with the computer science R & D community, the series counts many renowned academics among its volume editors and paper authors, and collaborates with prestigious societies. Its mission is to serve this international community by providing an invaluable service, mainly focused on the publication of conference and workshop proceedings and postproceedings. LNCS commenced publication in 1973.

Chun-Cheng Lin · Bertrand M. T. Lin ·
Giuseppe Liotta
Editors

WALCOM: Algorithms and Computation

17th International Conference and Workshops, WALCOM 2023
Hsinchu, Taiwan, March 22–24, 2023
Proceedings

Springer

Editors
Chun-Cheng Lin ⓘ
National Yang Ming Chiao Tung University
Hsinchu, Taiwan

Bertrand M. T. Lin ⓘ
National Yang Ming Chiao Tung University
Hsinchu, Taiwan

Giuseppe Liotta ⓘ
University of Perugia
Perugia, Italy

ISSN 0302-9743   ISSN 1611-3349 (electronic)
Lecture Notes in Computer Science
ISBN 978-3-031-27050-5   ISBN 978-3-031-27051-2 (eBook)
https://doi.org/10.1007/978-3-031-27051-2

This Springer imprint is published by the registered company Springer Nature Switzerland AG
The registered company address is: Gewerbestrasse 11, 6330 Cham, Switzerland

Preface

WALCOM 2023, the 17th International Conference and Workshops on Algorithms and Computation, was held during March 22–24, 2023 at National Yang Ming Chiao Tung University, Hsinchu, Taiwan. The workshop covered diverse areas of algorithms and computation, namely, approximation algorithms, computational complexity, computational geometry, graph algorithms, graph drawing, visualization, online algorithms, parameterized complexity, and property testing.

The quality of the workshop was ensured by a Program Committee comprising 27 researchers of international reputation from Australia, Bangladesh, Belarus, Brazil, Canada, Germany, Greece, India, Ireland, Israel, Italy, Japan, the Netherlands, Russia, Taiwan, and USA. This proceedings volume contains 30 contributed papers and two invited papers presented at WALCOM 2023. The Program Committee thoroughly reviewed each of the 75 submissions from 32 countries and accepted 30 of them for presentation at the conference after elaborate discussions on 223 review reports prepared by Program Committee members together with 95 external reviewers. The image of the workshop was highly enhanced by the two invited talks of eminent and well-known researchers Prof. Jan Kratochvil, Charles University, Prague, Czech Republic, and Prof. Michael Kaufmann, Universität Tübingen, Germany.

As editors of this proceedings, we would like to thank all the authors who submitted their papers to WALCOM 2023. We also thank the members of the Program Committee and external reviewers for their hard work in reviewing the manuscripts. Our sincere appreciation goes to the invited speakers for delivering wonderful talks from which researchers of this field benefited immensely. We acknowledge the continuous encouragements of the advisory board members Prof. M. Kaykobad and Prof. C. Pandu Rangan. The Steering Committee members of WALCOM always supported us with their valuable suggestions. We sincerely thank the Organizing Committee led by Prof. Chun-Cheng Lin for his excellent services that made the workshop a grand success. We would like to thank Springer for publishing this proceedings in their prestigious LNCS series. Finally, we acknowledge the EasyChair conference management system for providing a beautiful platform for conference administration.

March 2023

Chun-Cheng Lin
Bertrand M. T. Lin
Giuseppe Liotta

Organization

WALCOM Steering Committee

Tamal Dey	Purdue University, USA
Seok-Hee Hong	University of Sydney, Australia
Costas S. Iliopoulos	King College London (KCL), UK
Giuseppe Liotta	University of Perugia, Italy
Petra Mutzel	Technische Universität Dortmund, Germany
Shin-ichi Nakano	Gunma University, Japan
Subhas Chandra Nandy	Indian Statistical Institute, Kolkata, India
Md. Saidur Rahman	Bangladesh University of Engineering and Technology (BUET), Bangladesh
Ryuhei Uehara	Japan Advanced Institute of Science and Technology, Japan

Organizing Institution

National Yang Ming Chiao Tung University, Taiwan

Program Committee

Aritra Banik	National Institute of Science Education and Research, India
Tiziana Calamoneri	Sapienza University of Rome, Italy
William Evans	University of British Columbia, Canada
Martin Fuerer	Pennsylvania State University, USA
Patrick Healy	University of Limerick, Ireland
Alexander Kononov	Russian Academy of Sciences, Russia
Mikhail Y. Kovalyov	National Academy of Sciences of Belarus, Belarus
Bertrand M. T. Lin (Co-Chair)	National Yang Ming Chiao Tung University, Taiwan
Chun-Cheng Lin (Co-Chair)	National Yang Ming Chiao Tung University, Taiwan
Giuseppe Liotta (Co-Chair)	University of Perugia, Italy
Tamara Mchedlidze	Utrecht University, The Netherlands

Debajyoti Mondal	University of Saskatchewan, Canada
Krishnendu Mukhopadhyaya	Indian Statistical Institute, India
Shin-ichi Nakano	Gunma University, Japan
Rahnuma Islam Nishat	University of British Columbia, Canada
Yoshio Okamoto	The University of Electro-Communications, Japan
Chrysanthi N. Raftopoulou	National Technical University of Athens, Greece
Md. Saidur Rahman	BUET, Bangladesh
Ignaz Rutter	University of Passau, Germany
Saket Saurabh	The Institute of Mathematical Sciences, Chennai, India
Uéverton Souza	Universidade Federal Fluminense, Brazil
Ioannis Tollis	University of Crete, Greece
Ryuhei Uehara	Japan Advanced Institute of Science and Technology, Japan
Sue Whitesides	University of Victoria, Canada
Hsu-Chun Yen	National Taiwan University, Taiwan
Meirav Zehavi	Ben-Gurion University, Israel
Yakov Zinder	University Technology Sydney, Australia

Organizing Committee Chair

Chun-Cheng Lin	National Yang Ming Chiao Tung University, Taiwan

Technical Co-sponsors

Information Processing Society of Japan (IPSJ), Japan; The Institute of Electronics, Information and Communication Engineers (IEICE), Japan; Japan Chapter of the European Association of Theoretical Computer Science (EATCS Japan), Japan; Operations Research Society of Taiwan (ORSTW); Chinese Institute of Industrial Engineers (CIIE), Taiwan.

External Reviewers

Ageev, Alexander	Araki, Tetsuya
Ahmed, Abu Reyan	Bandopadhyay, Susobhan
Ahmed, Shareef	Bayzid, Md. Shamsuzzoha
Ahn, Taehoon	Bekos, Michael
Alam, Md. Jawaherul	Bhagat, Subhash

Bhore, Sujoy
Bhyravarapu, Sriram
Biniaz, Ahmad
Biswas, Arindam
Brakensiek, Joshua
Bredereck, Robert
Chatterjee, Abhranil
Corò, Federico
Cunha, Luis
Das, Gautam K
de Castro Mendes Gomes, Guilherme
Dósa, György
Eidenbenz, Stephan
Eisenstat, David
Epstein, Leah
Espenant, Jared
Fink, Simon D.
Fujii, Kaito
Förster, Henry
Gorain, Barun
Habib, Mursalin
Haeusler, Hermann
Hakim, Sheikh Azizul
Harrigan, Martin
Horiyama, Takashi
Ibiapina, Allen
Imai, Hiroshi
Jain, Pallavi
Jelínek, Vít
Ju, Andrew
Kakoulis, Konstantinos
Kanesh, Lawqueen
Kare, Anjeneya Swami
Kasthurirangan, Prahlad Narasimhan
Kawahara, Jun
Khachay, Michael
Khandeev, Vladimir
Kiyomi, Masashi
Kryven, Myroslav
Lionakis, Panagiotis
Liotta, Giuseppe
Lokshtanov, Daniel
Lucarelli, Giorgio

Madireddy, Raghunath Reddy
Manea, Florin
Mann, Kevin
Marcilon, Thiago
Mieno, Takuya
Miltzow, Till
Mondal, Kaushik
Mukhopadhyaya, Srabani
Münch, Miriam
Nascimento, Julliano
Ortali, Giacomo
Otachi, Yota
Papan, Bishal Basak
Parvez, Mohammad Tanvir
Pedrosa, Lehilton L. C.
Pfretzschner, Matthias
Pokorski, Karol
Sahu, Abhishek
Salvo, Ivano
Sampaio, Rudini
Satti, Srinivasa Rao
Schnider, Patrick
Schweitzer, Pascal
Sen, Sagnik
Sinaimeri, Blerina
Skiena, Steven
Stumpf, Peter
Suzuki, Akira
Symvonis, Antonios
Tabatabaee, Seyed Ali
Tappini, Alessandra
Tsakalidis, Konstantinos
Tsidulko, Oxana
Uchizawa, Kei
Verbeek, Kevin
Verma, Shaily
Viglietta, Giovanni
Wang, Haitao
Wasa, Kunihiro
Watrigant, Rémi
Xu, Chao
Xue, Jie
Yamanaka, Katsuhisa

Contents

String Algorithm

Optimization

Graph Algorithm

Approximation Algorithm

Parameterized Complexity

Invited Talks

Graph Covers: Where Topology Meets Computer Science, and Simple Means Difficult

Jan Kratochvíl[(✉)] [ID]

Department of Applied Mathematics, Faculty of Mathematics and Physics,
Charles University, Prague, Czech Republic
honza@kam.mff.cuni.cz

Abstract. We survey old and recent results on the computational complexity of graph covers, also known as locally bijective graph homomorphisms. This notion opens doors to interesting connections. The motivation itself comes from the classical notion of covering spaces in general topology, graph covers find computer science applications as a model of local computation, and in combinatorics they are used for constructing large highly symmetric graphs.

More than 30 years ago, Abello et al. [1] asked for a complete characterization of the computational complexity of deciding if an input graph covers a fixed one, and until this day only isolated results are known. We look at this question from several different angles of view – covers as locally constrained graph homomorphisms, covers of multigraphs, covers of graphs with semi-edges, or the list variant of the graph covering question. We also mention several open problems, including the Strong Dichotomy Conjecture for graph covers of Bok et al. [6], stating that for every target multigraph H, the H-COVER problem is either polynomial time solvable for arbitrary input graphs, or NP-complete for simple graphs on input. We justify this conjecture for several infinite classes of target (multi)graphs.

Keywords: Graph · Graph cover · Graph homomorphism · Multigraph · Computational complexity

1 Definitions

A *simple graph* is a pair $G = (V, E)$ where $V = V(G)$ is a set of vertices and $E = E(G)$ is a set of un-ordered pairs of vertices, called edges. We will only consider finite graphs, i.e., graphs whose vertex sets are finite. Two vertices are called *adjacent* if they are connected by an edge. The *(open) neighborhood* $N_G(u)$ of a vertex u is the set of vertices u is adjacent to. The *degree* of a vertex is the number of vertices it is adjacent to, i.e., $\deg_G(u) = |N_G(u)|$. A graph is k-*regular* if the degree of every vertex is k, a 3-regular graph is called *cubic*. A *path* in the graph is a sequence of distinct vertices, every two consecutive ones being

C.-C. Lin et al. (Eds.): WALCOM 2023, LNCS 13973, pp. 3–11, 2023.
https://doi.org/10.1007/978-3-031-27051-2_1

adjacent. The length of a path is the number if its edges, i.e., the number of its vertices minus 1. The graph is *connected* if any two if its vertices are connected by a path in the graph. A *cycle* in the graph is a path of length at least 2 whose end-vertices are adjacent.

In a more general setting, we allow pairs of vertices to be connected by several parallel edges, so called *multiple edges*, and we allow edges that are incident with a single vertex only. The latter are *loops* or *semi-edges*, their difference lying in how they contribute to the degrees of their vertices (a loop contributes 2, a semi-edge contributes 1). In other words, the degree of a vertex is the number of edges it is incident to, loops being counted twice. (Multi)graphs with parallel edges and loops are studied from the early days of graph theory, while the semi-edges are being considered only recently, mainly because of the applications in mathematical physics and topological graph theory. In the sequel, we allow graphs to have multiple edges, loops and/or semi-edges. We call a graph *simple* if it contains no loops, no semi-edges and no parallel edges. A vertex of a graph is called *semi-simple* if it is incident with no loops, no multiple edges and to at most one semi-edge.

Now we are ready to introduce the main character of this paper, the notion of *graph cover*.

Definition 1. *Let G and H be simple connected graphs. A* covering projection *from G to H is a mapping $f : V(G) \to V(H)$ such that for every vertex $u \in V(G)$, the neighborhood $N_G(u)$ is mapped by f bijectively onto the neighborhood $N_H(f(u))$. We say that G covers H, and write $G \to H$, if a covering projection from G to H exists.*

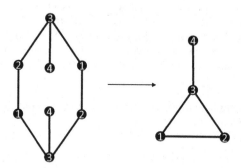

Fig. 1. Illustration to the definition of covers of simple graphs. The covering projection is visualized by displaying the names of the images of the vertices of the covering graph.

Informally speaking, if we imagine an agent moving on the vertices of a graph and being able to see the name of the vertex he/she is in and the names of its neighbors, the agent cannot determine whether he/she is moving through the graph H or through its cover G. This has been exploited by Angluin [2] for establishing models of local computation. For more recent results in this direction, cf. [7,9]. This computer science connection led several researchers to exploring the question of computational complexity of deciding if one graph covers another one. Bodlaender [3] considered the case when both graphs are part of the input and showed that the problem is NP-complete. Abello et al. [1] considered the target graph H fixed and asked about the complexity of the problem

H-COVER

Input: A graph G.

Question: Does G cover H?

parameterized by the target graph H. They asked for a complete characterization, and showed first examples of graphs H for which the problem is NP-complete. In that paper the authors already consider multigraphs, in fact, they prove their NP-hardness result for the so called dumbbell graph (a 2-vertex graph with a loop incident to each of its vertices and a single normal edge connecting them). At this point we are ready for the full definition of graph covers, even when semi-edges are allowed.

Definition 2. *Let G and H be graphs. A covering projection from G to H is a pair of mappings $f = (f_V, f_E)$ such that*

- *f_V maps vertices of G onto vertices of H,*
- *f_E maps edges of G onto edges of H,*
- *f_V and f_E are incidence preserving,*
- *the preimage of a loop of H incident with a vertex u is a disjoint union of cycles spanning the subgraph of G induced by the preimage $f_V^{-1}(u)$ of u,*
- *the preimage of a semi-edge of H incident with a vertex u is a disjoint union of single edges and semi-edges spanning the subgraph of G induced by the preimage $f_V^{-1}(u)$ of u, and*
- *the preimage of a normal edge of H incident with distinct vertices u and v is a matching spanning the subgraph of G induced by $f_V^{-1}(u) \cup f_V^{-1}(v)$.*

Kratochvíl et al. [23] showed that in order to characterize the complexity of H-COVER for all simple graphs H, one has to be able to characterize it for all graphs that allow multiple edges and loops. The presence of semi-edges provides a connection to edge-coloring problems, cf. the following example. Let $F(1,1)$ denote the 1-vertex graph with a loop and a semi-edge, and let $F(3,0)$ denote the 1-vertex graph with 3 semi-edges.

Proposition 1. *A simple cubic graph covers $F(1,1)$ if and only if it has a perfect matching. Hence $F(1,1)$-COVER is solvable in polynomial time. On the other hand, a simple cubic graph covers $F(3,0)$ if and only if it is 3-edge-colorable, and hence $F(3,0)$-COVER is NP-complete.*

The complexity of the H-COVER problem for graphs with semi-edges has been studied only recently by Bok et al. [4–6]. It is immediately clear, already from the

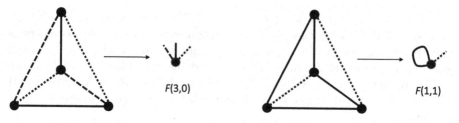

Fig. 2. Illustration to the definition of covers of (multi)graphs with semi-edges. In the case of 1-vertex graphs, the vertex mapping is uniquely defined. The edge part of the covering projection is demonstrated by dashed lines of different patterns.

example in Proposition 1, that the presence of semi-edges makes the covering problem much harder. To explain the oxymoron from the title, we observe that proving the NP-hardness of H-COVER is easier when the input graph is allowed to have loops, multiple edges and semi-edges. However, in all cases when the problem is known to be NP-complete, it remains NP-complete for simple graphs on input as well. This phenomenon has been conjectured to hold true in general, and was named the Strong Dichotomy Conjecture on graph covers in [6]. Attempts to prove this conjecture for large classes of graphs have led to introduction of a notion generalizing the concept of snarks known from and well studied in the theory of edge-colorings of graphs. We will comment on the results and open problems in this recently developing area of research in Sects. 5 and 7. But first we reveal several interesting connections of graph covers to other topics.

2   Negami's Conjecture

Planar graphs belong to the most popular and most studied special classes of graphs. In 1988, Negami [24] posed the following beautiful and still unresolved conjecture.

Conjecture 1. A connected simple graph has a finite planar cover if and only if it is projective planar.

Note here that every graph is covered by an infinite tree, called the universal cover. Thus the requirement "finite" is vital for the non-triviality of this conjecture. It is well known from the topology that the Euclidean plane is a double-cover of the projective plane. It follows that every projective planar graph does have a finite planar cover. The opposite implication is far less trivial (if true at all). The conjecture was formulated in the time when the Robertson-Seymour theory of graph minors was being developed, and it was soon observed that both the class of projective planar graphs, and the class of graphs admitting a finite planar cover are closed in the minor order. Luckily, the minimal forbidden minors for the projective planar graphs were already described. What remained was to check one by one that none of the connected ones admits a finite planar cover.

By a frontal attack of several researchers, 28 out of 32 of these graphs have been directly resolved, and, assuming the conjecture holds true, it suffices to prove the last one - the complete 4-partite graph K2,2,2,1. For an accessible survey of the most recent progress in this question see [19].

3   Locally Constrained Graph Homomorphisms

For simple graphs, a graph homomorphism is an adjacency preserving vertex mapping between two graphs. If $f : V(G) \rightarrow V(H)$ is such a mapping, then for any vertex $u \in V(G)$, $f(N_G(u)) \subseteq N_H(f(u))$. We have already seen that f is a covering projection when the restricted mapping $f|_{N_G(u)}$ is a bijection of $N_G(u)$ onto $N_H(f(u))$. In this sense, graph covering projections are also referred to as *locally bijective homomorphisms*. The following definition is a natural relaxation of the bijectivity restriction.

Definition 3. *A homomorphism* $f : G \rightarrow H$ *is called* locally surjective *if for every vertex* $u \in V(G)$, *the restricted mapping* $f|_{N_G(u)}$ *maps* $N_G(u)$ *surjectively onto* $N_H(f(u))$. *And it is called* locally injective *if for every vertex* $u \in V(G)$, *the restricted mapping* $f|_{N_G(u)}$ *maps* $N_G(u)$ *injectively into* $N_H(f(u))$.

Locally surjective homomorphisms are also called *role assignments* and they find applications in the social network theory. Fiala and Paulusma [15] gave a complete characterization of its complexity. They proved that for any connected graph H with at least 3 vertices, deciding if an input graph admits a locally surjective homomorphism onto H is NP-complete. Locally injective homomorphisms are also called *partial covers* because a graph admits a locally injective homomorphism into a graph H if and only if it is an induced subgraph of a graph that covers H. Partial covers are closely related to the so called Frequency Assignment Problem, motivated by the practical task of assigning frequencies in mobile networks. One particular subproblem is $L(2, 1)$-labeling of graphs (for a graph G, $L_{2,1}(G)$ is the smallest integer λ such that the vertices of G can be labeled by integers from the range $0, \dots, \lambda$ so that the labels of adjacent vertices differ by at least 2 and labels of vertices with a common neighbor are different), cf. [8,12,16,17]. It can be easily seen that $L_{2,1}(G) \leq \lambda$ if and only if G is a partial cover of the complement of the path of length λ. For every fixed $\lambda \geq 4$, deciding $L_{2,1}(G) \leq \lambda$ is NP-complete, and so is the partial covering of complements of paths. However, the catalog of known results on the complexity of partial covers is far from being complete.

4   List Covering

Many graph theory problems are also studied in their list versions, in which the colors (or labels or images) of vertices are restricted to be assigned values from lists of admissible ones. For every problem, its list version is at least as difficult as the plain version. If a problem is parameterized (like the H-COVER problem), this means that the class of parameters that define polynomially solvable instances is narrower. This may (or may not) imply that it is easier to describe. Compare the situation for graph homomorphisms for simple graphs:

Theorem 1. [18] *Deciding if a simple input graph allows a homomorphism into a simple graph H is polynomial time solvable when H is bipartite and NP-complete otherwise.*

Theorem 2. [10] *The* LIST-*H*-HOMOMORPHISM *problem is solvable in polynomial time for bipartite graphs whose complement is a circular arc graph, and NP-complete otherwise.*

On the other hand, for the locally injective homomorphisms, the catalog of complexity is unknown and presumably hard to achieve, while for the list variant it has been determined:

Theorem 3. [13] *Let H be a connected simple graph. Then* LIST LOCALLY INJECTIVE *H*-HOMOMORPHISM *is solvable in polynomial time if H has at most one cycle, and NP-complete otherwise.*

For locally bijective homomorphisms, i.e., for graph covers, the lists are helpful as well. It is known that for simple regular graphs of valency greater than 2, the *H*-COVER problem is NP-complete (this was proven first for k-regular k-edge-colorable graphs in [22] and then for general k-regular graphs by Fiala [11], cf. also [14]). Though the intuition says that covering graphs with loops, multiple edges and semi-edges should be at least as difficult as covering simple graphs, the arguments used in the NP-hardness reduction of the aforementioned result breaks down in the presence of multiple edges. It is known that *H*-COVER is NP-complete if all vertices of *H* are semi-simple [4], but when only some vertices are semi-simple, lists come to help. The following result is proven in [6].

Theorem 4. *If a k-regular graph H, $k \geq 3$, contains a semi-simple vertex, then the* LIST *H*-COVER *problem is NP-complete, even for simple input graphs.*

In this theorem, *H* may contain multiple edges, loops, and semi-edges. The problem remains NP-complete even if the lists are restrictive only for the vertex mapping, the lists for the edge mapping being full.

5   Strong Dichotomy

Note that the NP-hardness result of Theorem 4 is stated for simple input graphs. This follows the urge of proving the results in their strongest form. Construction of gadgets for the NP-hardness reductions would be (sometimes much) easier if multiple edges/loops/semi-edges were allowed. It is not even granted that a problem NP-complete for (multi)graphs would be NP-complete also for simple input graphs. However, in case of graph covering problems this has so far always been the case. Bok et al. [6] have formulated the following Strong Dichotomy Conjecture for graph covers:

Conjecture 2. For every graph *H*, the *H*-COVER problem is either polynomial time solvable for arbitrary input graphs, or NP-complete even for simple graphs on the input.

6 Disconnected Graphs

In most situations one can freely say that we are only interested in connected graphs, since the problem can be solved for each component separately otherwise. And this have been done in several papers on graph covers. Only recently, Bok et al. [5] noted, that this is not that obvious for graph covers, at least when multiple edges, loops or semi-edges are present. They have argued that the following definition of covers of disconnected graphs is the right one.

Definition 4. *Let G and H be disconnected graphs, the components of G being $G_1, G_2, \ldots, G_p$ and the components of H being $H_1, H_2, \ldots, H_q$. A mapping $f : G \to H$ is a covering projection of G onto H if*

- *for each $i = 1, 2, \ldots, p$, there is a j such that $f|_{G_i}$ is a covering projection from G_i onto H_j, and*
- *the preimage of any vertex of H has size $\frac{|V(G)|}{|V(H)|}$.*

Even with this most restrictive definition the following holds true.

Theorem 5. [5] *If every component H_i of H defines a polynomial time solvable problem H_i-COVER, then H-COVER is polynomial time solvable. On the other hand, if H_i-COVER is NP-complete for some component H_i of H for simple input graphs, then H-COVER is NP-complete for simple input graphs.*

The proof of the NP-hardness part of this theorem, i.e., the proof of H_i-COVER $\propto$ H-COVER, is non-constructive in the following sense. For two components H_j and H_i of H we use a simple graph G_j (as a component of an input graph G that we construct) such that G_j covers H_j and G_j does not cover H_i, if such a graph G_j exists (otherwise we use an arbitrary simple cover of H_j). Since H (and each of its components) are fixed graphs for the reduction, this is a legal move, though we do not know how to decide if such a graph G_j exists or not. This somewhat unexpected twist has led to introduction of a new relation between connected graphs which will be the topic of the concluding section.

7 Look Who is Stronger, and Mind Generalized Snarks

Insisting on proving the NP-hardness results for simple input graphs leads to the following definition which we find interesting in its own. Thus insisting in simple graphs does not only make the proofs more difficult, as the title of the article promised, but also brings in a new concept with intriguing open problems.

Definition 5. *Let A and B be graphs. We say that A is stronger than B, and write $A \triangleright B$ if it holds true that every simple graph that covers A also covers B.*

It is straightforward to observe that if A covers B, then A is stronger than B. The converse is, however, not true. Consider the graphs $F(1,1)$ and $F(3,0)$ from Proposition 1. Every simple graph that covers $F(3,0)$ contains a perfect matching (the edges that map onto the same semi-edge of $F(3,0)$) and hence it covers $F(1,1)$. Thus $F(3,0) \triangleright F(1,1)$. But obviously, $F(3,0)$ does not cover $F(1,1)$.

Another easy observation states that if A is a simple graph, then A is stronger than B if and only if A covers B. In the Open Problem Session at GROW 2022 [20], we have conjectured that the impact of semi-edges is vital for the existence of non-trivial pairs of graphs in the $\triangleright$ relation.

Conjecture 3. If A has no semi-edges, then $A \triangleright B$ if and only if $A \to B$.

In [21], we have confirmed Conjecture 3 for all pairs A, B such that $B = F(3,0)$ or $B = F(1,1)$. In particular, we proved that $A \triangleright F(3,0)$ if and only if A is 3-edge-colorable, i.e., if and only if $A \to F(3,0)$. Showing that a graph A is not stronger than $F(3,0)$ requires constructing a non-3-edge-colorable cubic graph that covers A. Two-connected non-3-edge-colorable graphs are called snarks, and thus a snark that covers A is a witness that $A \not\triangleright F(3,0)$. Snarks have been hunted for decades, but not from the point of view which graphs they cover. In our proofs, 2-connectedness is not necessary, but being so close to the Wonderland, we cannot resist the temptation to pose the last open problem.

Problem 4. Is it true that for every non-3-edge-colorable cubic graph A, there exists a 2-connected non-3-edge-colorable graph that covers A?

Given a pair of graphs A and B, one way of proving that $A \not\triangleright B$ is to construct a witness, i.e., a simple graph H such that $H \to A$ and $H \not\to B$. We call such witnesses *generalized snarks*. Hunting for generalized snarks has the best chance to become quite an adrenaline sport, but it can hardly be avoided if one wants to prove Conjecture 3 for larger infinite classes of graphs. Got interested? Join us in the game!

Acknowledgements. Supported by Research grant GAČR 20-15576S of the Czech Science Foundation.

References

1. Abello, J., Fellows, M.R., Stillwell, J.C.: On the complexity and combinatorics of covering finite complexes. Aust. J. Comb. **4**, 103–112 (1991)
2. Angluin, D.: Local and global properties in networks of processors. In: Proceedings of the 12th ACM Symposium on Theory of Computing, pp. 82–93. ACM, New York (1980)
3. Bodlaender, H.L.: The classification of coverings of processor networks. J. Parallel Distrib. Comput. **6**, 166–182 (1989)
4. Bok, J., Fiala, J., Hliněný, P., Jedličková, N., Kratochvíl, J.: Computational complexity of covering multigraphs with semi-edges: small cases. In: Bonchi, F., Puglisi, S. J. (eds.), MFCS 2021, LIPIcs, vol. 202, pp. 21:1–21:15. Schloss Dagstuhl - Leibniz-Zentrum für Informatik (2021)
5. Bok, J., Fiala, J., Jedličková, N., Kratochvíl, J., Seifrtová, M.: Computational complexity of covering disconnected multigraphs. In: Bampis, E., Pagourtzis, A. (eds.) FCT 2021. LNCS, vol. 12867, pp. 85–99. Springer, Cham (2021). https://doi.org/10.1007/978-3-030-86593-1_6

6. Bok, J., Fiala, J., Jedličková, N., Kratochvíl, J., Rzazewski, P.: List covering of regular multigraphs. In: Bazgan, C., Fernau, H. (eds.) Combinatorial Algorithms. LNCS, vol. 13270, pp. 228–242. Springer, Heidelberg (2022). https://doi.org/10.1007/978-3-031-06678-8_17

7. Chalopin, J., Paulusma, D.: Graph labelings derived from models in distributed computing: a complete complexity classification. Networks **58**(3), 207–231 (2011)

8. Chang, G.J., Kuo, D.: The $L(2,1)$-labelling problem on graphs. SIAM J. Discrete Math. **9**, 309–316 (1996)

9. Courcelle, B., Métivier, Y.: Coverings and minors: applications to local computations in graphs. Eur. J. Comb. **15**, 127–138 (1994)

10. Feder, F., Hell, P., Huang, J.: List homomorphisms and circular arc graphs. Combinatorica **19**(4), 487–505 (1999)

11. Fiala, J.: Locally injective homomorphisms, Ph.D. thesis. Charles University, Prague (2000)

12. Fiala, J., Kloks, T., Kratochvíl, J.: Fixed-parameter complexity of lambda-labelings. Discret. Appl. Math. **113**(1), 59–72 (2001)

13. Fiala, J., Kratochvíl, J.: Locally injective graph homomorphism: lists guarantee dichotomy. In: Fomin, F.V. (ed.) WG 2006. LNCS, vol. 4271, pp. 15–26. Springer, Heidelberg (2006). https://doi.org/10.1007/11917496_2

14. Fiala, J., Kratochvíl, J.: Locally constrained graph homomorphisms - structure, complexity, and applications. Comp. Sci. Rev. **2**(2), 97–111 (2008)

15. Fiala, J., Paulusma, D.: A complete complexity classification of the role assignment problem. Theor. Comput. Sci. **349**(1), 67–81 (2005)

16. Griggs, J.R., Yeh, R.K.: Labelling graphs with a condition at distance 2. SIAM J. Discret. Math. **5**(4), 586–595 (1992)

17. Hasunuma, T., Ishii, T., Ono, H., Uno, Y.: A linear time algorithm for $L(2,1)$-labeling of trees. Algorithmica **66**(3), 654–681 (2013). https://doi.org/10.1007/s00453-012-9657-z

18. Hell, P., Nešetřil, J.: On the complexity of H-coloring. J. Comb. Theory Ser. B **48**(1), 92–110 (1990)

19. Hliněný, P.: 20 years of negami's planar cover conjecture. Graphs Comb. **26**, 525–536 (2010). https://doi.org/10.1007/s00373-010-0934-9

20. Kratochvíl, J: Towards strong dichotomy of graph covers. In: S. Cabello, M. Milanič (eds.), GROW 2022 - Book of Open Problems, pp. 10 (2022). https://grow.famnit.upr.si/GROW-BOP.pdf

21. Kratochvíl, J., Nedela, R.: Covers and semi-covers: who is stronger? In: Preparation (2023)

22. Kratochvíl, J., Proskurowski, A., Telle, J.A.: Covering regular graphs. J. Comb. Theory Ser. B **71**(1), 1–16 (1997)

23. Kratochvíl, J., Proskurowski, A., Telle, J.A.: Complexity of colored graph covers I. colored directed multigraphs. In: Möhring, R.H. (ed.) WG 1997. LNCS, vol. 1335, pp. 242–257. Springer, Heidelberg (1997). https://doi.org/10.1007/BFb0024502

24. Negami, S.: Graphs which have no planar covering. Bull. Inst. Math. Acad. Sinica **16**(4), 377–384 (1988)

The Family of Fan-Planar Graphs

Michael Kaufmann[✉][iD]

Wilhelm-Schickard-Institute, Tübingen, Germany
`michael.kaufmann@uni-tuebingen.de`

1 The Origins

Beyond-planarity [24,28] has been developed in the areas of graph drawing and topological graph theory as a core topic. Planar graphs have been a key class here since several decades, although most of the graphs in practical applications are not planar at all. Nevertheless, most of the models and layout algorithms are based on the concept of planarity and aim for crossing-minimization, since too many edge crossings may lead to clutter and visual errors.

The very first class in the landscape of beyond-planarity are 1-planar graphs, i.e. graphs that have a drawing where edges have at most one crossing. This generalization has been introduced by Ringel in 1965 [34] in the context of graph coloring planar and near-planar graphs, see also [15]. Later 1-planar graphs of maximum number of edges, called maximum density, has been characterized in [14], and later extended to larger crossing numbers per edge by Pach and Tóth [33], where first bounds for the maximum density has been given. Meanwhile, many combinatorial and algorithmic aspects in particular for 1-planar and partially for 2-planar graphs have been considered, like recognition [7,27], layout algorithms [4], generation [37]. In particular, many subclasses like outer-1-planar graphs, IC-planar and NIC-planar graphs etc have been considered, and numerous results on the structure of such classes, the recognition problem as well as efficient layout algorithms have been found. Surveys on 1-planar graphs can be found in [32] and [28].

In follow-up works of [33], a complete characterization of 2-planar graphs of maximum density and tight bounds for 3-planar graphs have been given in [10].

A more advanced model are quasiplanar graphs, i.e. graphs that have drawings without 3 mutually crossing edges. These graphs have been introduced in the early days of beyond-planarity as well. In a series of publications, the bounds for maximum densities have been [3,35,36] gradually improved. In particular, for k-quasiplanar graphs, where any k edges are forbidden to mutually cross, there have been remarkable achievements [1,2,26]. We mention only the work in [5] where it is shown that any simple k-planar drawing of a graph can be transformed into a simple $(k + 1)$-quasiplanar drawing.

Another branch that had been developed so far are the class of RAC-drawable graphs, i.e. graphs that have drawings with right-angle crossings [22,23]. In principle, there are no limitations on the number of crossings along a single edge,

This work has been supported by DFG grant KA812-18/2.

on the other hand, in the straight-line drawing, two edges incident to the same vertex may not cross any other edge. Hence, fan-crossings are forbidden in this scenario.

Overall, mostly the density questions and the inclusion relations between the graph classes beyond planarity have been studied [9]. This was the state of the landscape, when we considered the new graph class of fan-planar graphs. We aimed for a class in the beyond-planarity landscape that reflects some practical needs: First of all, sometimes an unlimited number of crossings should be allowed. Note that the basic models of 1-, 2-, 3–planarity etc. seem a simple but artificial restriction which is often not useful in practice. A second practical aspect is the model of bundling edges. Bundling means edges that are routed similarly are bundled into one single route and the question is how to count possible crossings. This should be taken into account and supported by the model. Third aspect is of course the necessity to develop a counterpart to the RAC model, where fan-crossings are impossible for straight-line drawings. We want to explicitly allow such fan crossings.

2 The First Generation

In the original paper [29], which appeared in a journal only recently [30], we formally introduced the class of fan-planar graphs: In this graph class, graphs have drawings where no two independent edges cross the same edge, i.e. an edge can be crossed by several other edges as long as they have a common vertex (Configuration I) (Fig. 1). Unfortunately, this condition is not enough as we had to exclude the case that two edges which are incident to the same vertex cross another edge coming from two different sides (Configuration II). Graphs with drawings avoiding Configurations I and II will be called weakly fan-planar following [21]. Much later, Klemz et al. [31] pointed out that our proof works when we further generalize configuration II and give a new forbidden configuration III. We call such graphs strongly fan-planar graphs. In [29], we give a combinatorial proof for a bound of $5n - 10$ on the density of simple strongly fan-planar graphs of n vertices. In this proof, we assume right from the beginning that we consider only simple graphs, i.e. graphs without self-loops, parallel edges and non-intersecting incident edges, that have maximum density, and further that the number of uncrossed edges is also maximal. We partition the edges into several subsets and count the cardinality of these subsets. E.g. one subset comprises the uncrossed edges. In a fan-planar drawing, those edges define faces, i.e. (not necessarily simple) connected regions in the plane surrounded by the uncrossed edges. The edges that are being crossed are contained in those faces. We describe the ways that those edges might be drawn and hence we are able to bound the number of such edges, depending on the length and the properties of the faces. We can then characterize the shape of the faces such that the total number of edges is maximized and finally achieve the claimed result.

In [29], we additionally describe several fan-planar graphs with different structural properties that achieve the claimed upper bound of $5n - 10$ on the edge

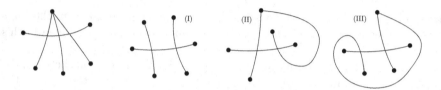

Fig. 1. A fan-crossing and the three forbidden Configurations I, II and III

density, we shortly discuss the option to remove configurations II and III for straight-line fan-planar drawings, and show what might happen when allowing incident edges that might intersect each other.

In follow-up papers [16,17], Brandenburg discusses the variant where Configuration II is allowed but only Configuration I is forbidden. Among others, he interestingly showed how to express this graph class in first-order logic. Further, he claimed that graphs only obey forbidden Configuration I have the same density bound as the fan-planar graphs as defined above.

As for almost all the graphs beyond-planarity, the recognition problem plays a prominent role, but the results are the same: For the general setting, Binucci et al. in the general setting [13] showed NP-hardness using a reduction from 1-planarity testing. And even if the rotation system of the input graph, i.e. the cyclic order of the incident edges for every vertex is being prescribed, Bekos et al. [8] show NP-hardness as well using a reduction from 3-Partition. More feasible variants with respect to the recognition problem are the maximal outer fan-planar graphs and the 2-layer fan-planar graphs. In the 2-layer variant, usually the input graph is bipartite, and the vertices of the two partitions are placed onto two parallel (horizontal) lines and the edges are drawn in between the two lines. In [12] the authors were able to completely characterize 2-layer fan-planar graphs. Efficient recognition algorithms as well as tight density bounds could be found. For the variant of outer-fan-planar graphs, all the vertices should be adjacent to the outer face in addition to the general requirements of fan-planar graphs. Bekos et al. [8] gave efficient recognition algorithms for the maximal variant, where no extra edge can be inserted without violating one of the requirements of the model. Additionally, the authors derived important properties in the respective drawings.

3   The Relatives

A seemingly counterpart of the fan-planarity model, where only fan crossings are allowed, is the fan-crossing free model, where only independent edges can be crossed by the same edge. In an early work, Cheong et al. [20] explored this graph class and gave an upper bound on such graphs of size $4n - 8$, which is the same as for 1-planar graphs, and hence remarkably low. Brandenburg [18] has put the fan-crossing free graphs in relation to several other classes beyond planarity.

Remarkably, the same density bound as the fan-planar graphs is obtained by 2-planar graphs. Although 2-planar graphs of maximum density are also fan-planar, there are 2-planar graphs which are not fan-planar [13] (independent edges cross the same edge), and there are fan-planar graphs which are not 2-planar, and not even k-planar for bounded k. Hence the two classes are incomparable.

A more recent approach exploits the bundling abilities of fans: In a 1-sided fan-bundle-planar drawing, the edges of a fan are grouped into a bundle such that a crossing of such a fan-bundle with another edge counts as one crossing [6]. The crossing rule is that each bundle is crossed only once, and the bundling rule is that an edge can be bundled only at one of its endpoints. The two rules imply that graphs with such properties are fan-planar. For graphs with such a 1-sided fan-bundle-planar drawings Angelini et al. [6] also showed a tight bound on the density of that graph class. They obtained $(13n - 36)/3$, which then implies that there are fan-planar graphs, i.e. those of maximum density, that are not 1-fbp.

A last relative that we mention here is by Biedl et al. [11]. They consider a generalization of 2-layer fan-planar drawings to multi-layer drawings, and here several variants depending on the route of edges between neighboring layers or even connecting non-neighboring layers. Due to the restricted setting by the horizontal layers, we were able to develop FPT algorithms for the recognition problem w.r.t. the number of layers, if the embedding is asssumed to be fixed. Other properties like a bounded treewidth and drawability as a bar-1-visible drawing have been obtained as well.

4   The Second Generation

We will mention here some recent results, even partially unpublished, which show the attractiveness and the active research in this area.

4.1   Thickness

Graph thickness is a well-known graph parameter that describes in how many planar graphs a graph can be decomposed. It is closely related to arboricity, which denotes the number of edge-disjoint forests a graph can be decomposed. For k-planar graphs, the arboricity, and also the thickness is $3.8\sqrt{k}$ which follows from the density bound of $3.8\sqrt{k}n$ for general k planar graphs, while for 2- and 3-planar graphs, a thickness of 5 and 6 follow from the density bounds respectively. Analogously, from the $5n - 10$ bound from the density, an upper bound for the thickness of 5 holds for fan-planar graphs.

Cheong, Pfister and Schlipf improved this simple bound and showed that any fan-planar drawings that obey the three forbidden configurations can be partitioned into 3 non-crossing set of edges [21]. The main technique is to consider odd cycles in the intersection graph, which then imply odd cycles in the original graph. Dependencies between odd cycles can be resolved and finally the result is achieved. Due to the absence of odd cycles in the intersection graph, the bounds for the thickness is even two, for bipartite fan-planar graphs.

The main task left open for the future here is to find a fan-planar graph, which has thickness exactly 3.

4.2 Non-simple Fan-Planar Graphs

Klemz, Knorr, Reddy and Schröder [31] made a thorough research about non-simple fan-planar graphs, they identified some flaws in the original paper by Kaufmann and Ueckerdt [29] and introduced the new configuration III (ref. to the definitions), see also [30]. They showed how to make non-simple fan-planar drawings simple without introducing any new crossings. Rerouting some edges to simplify the drawing is particularly tricky as naive approaches might destroy fan-planarity.

4.3 Insights on Configuration III

New progress has also been made concerning Configuration III [19]. As a first result, the authors give a weakly fan-planar graph, which is not strongly-fanplanar, i.e. they present a graph that for any fan-planar drawing avoiding Configuration I and II, the third Configuration III is being neglected. Here they use some properties of possible fan-planar drawings of K_7 that have been developed in [12] and that is used as a gadget.

Secondly, the authors claim that the $5n - 10$ density bound which has been proven for strongly fan-planar graphs, i.e. with all three forbidden configurations, also hold for weakly fan-planar graphs, i.e. graphs where Configuration III might be present. The idea is that for any Configuration III that occur in the drawing, several 'independent' parts can be defined, and for these parts, induction on the number of edges can be applied. Reconnecting the two parts lead to the missing edges and the claimed bound.

5   An Outlook to the Future

In this overview we gave an insight on the state of research around the class of fan-planar graphs. Here is a list of tasks for follow-up work in the near future.

1. Concerning the thickness of fan-planar graphs, it is still open what the real bound for the thickness is. In particular, find a graph where all fan-planar drawings have thickness at least 3. Furthermore, find out if the restrictions on the drawings, i.e. Configurations I, II and III, are really necessary to obtain the bound.
2. Characterize the fan-planar graphs that achieve the maximum bound of $5n - 10$ on the number of edges, and consider the recognition question for such graphs as it has been done recently for 2-planar graphs of maximum density [25].
3. Clarify the impact of the single forbidden Configurations I, II and III on the density and other parameters of the respective graphs and their structure.

4. Explore the tasks for straight-line drawings, 1- or 2-bend drawings. Combine the fan-planarity model with the RAC drawing model.
5. Generalize the fan-planar graph model in some way. One possible approach is to define 2-fan-planarity by allowing the intersection of the same edge by even 2 fans. Even the restriction of 2 fans intersecting from opposite/same sides seems interesting.
6. Practical aspects: Define a layout model that takes the characteristics of fan-planarity into account. An idea might be to layout the fans in uniform shapes or in restricted directions, etc.
7. Give a striking application that includes many fans and apply the techniques on fan-planarity.

Acknowledgement. Thanks go to Torsten Ueckerdt and all the (former) members of my group in Tübingen Lena Schlipf, Maximilian Pfister, Axel Kuckuk, Julia Katheder, Henry Förster, Michael Bekos, Patrizio Angelini for the nice collaborations in the past on this fascinating topic.

References

1. Ackerman, E.: On topological graphs with at most four crossings per edge. CoRR 1509.01932 (2015). http://arxiv.org/abs/1509.01932
2. Ackerman, E., Tardos, G.: On the maximum number of edges in quasi-planar graphs. J. Comb. Theory Ser. A **114**(3), 563–571 (2007). https://doi.org/10.1016/j.jcta.2006.08.002
3. Agarwal, P.K., Aronov, B., Pach, J., Pollack, R., Sharir, M.: Quasi-planar graphs have a linear number of edges. Combinatorica **17**(1), 1–9 (1997). http://dx.doi.org/10.1007/BF01196127
4. Alam, M.J., Brandenburg, F.J., Kobourov, S.G.: Straight-line grid drawings of 3-connected 1-planar graphs. In: Wismath, S., Wolff, A. (eds.) GD 2013. LNCS, vol. 8242, pp. 83–94. Springer, Cham (2013). https://doi.org/10.1007/978-3-319-03841-4_8
5. Angelini, P., et al.: Simple k-planar graphs are simple (k+1)-quasiplanar. J. Comb. Theory Ser. B **142**, 1–35 (2020). https://doi.org/10.1016/j.jctb.2019.08.006
6. Angelini, P., Bekos, M.A., Kaufmann, M., Kindermann, P., Schneck, T.: 1-fan-bundle-planar drawings of graphs. Theor. Comput. Sci. **723**, 23–50 (2018). https://doi.org/10.1016/j.tcs.2018.03.005
7. Auer, C., et al.: Outer 1-planar graphs. Algorithmica **74**(4), 1293–1320 (2016). https://doi.org/10.1007/s00453-015-0002-1
8. Bekos, M.A., Cornelsen, S., Grilli, L., Hong, S., Kaufmann, M.: On the recognition of fan-planar and maximal outer-fan-planar graphs. Algorithmica **79**(2), 401–427 (2017). https://doi.org/10.1007/s00453-016-0200-5
9. Bekos, M.A., Gronemann, M., Raftopoulou, C.N.: On the queue number of planar graphs. In: Purchase, H.C., Rutter, I. (eds.) GD 2021. LNCS, vol. 12868, pp. 271–284. Springer, Cham (2021). https://doi.org/10.1007/978-3-030-92931-2_20
10. Bekos, M.A., Kaufmann, M., Raftopoulou, C.N.: On optimal 2- and 3-planar graphs. In: Aronov, B., Katz, M.J. (eds.) 33rd International Symposium on Computational Geometry, SoCG 2017, 4-7 July 2017, Brisbane, Australia. LIPIcs, vol. 77, pp. 16:1–16:16. Schloss Dagstuhl - Leibniz-Zentrum für Informatik (2017). https://doi.org/10.4230/LIPIcs.SoCG.2017.16

11. Biedl, T., Chaplick, S., Kaufmann, M., Montecchiani, F., Nöllenburg, M., Raftopoulou, C.N.: Layered fan-planar graph drawings. In: Esparza, J., Král', D. (eds.) 45th International Symposium on Mathematical Foundations of Computer Science, MFCS 2020, 24-28 August 2020, Prague, Czech Republic. LIPIcs, vol. 170, pp. 14:1–14:13. Schloss Dagstuhl - Leibniz-Zentrum für Informatik (2020). https://doi.org/10.4230/LIPIcs.MFCS.2020.14

12. Binucci, C., et al.: Algorithms and characterizations for 2-layer fan-planarity: from caterpillar to stegosaurus. J. Graph Algorithms Appl. **21**(1), 81–102 (2017). https://doi.org/10.7155/jgaa.00398

13. Binucci, C., et al.: Fan-planarity: properties and complexity. Theor. Comput. Sci. **589**, 76–86 (2015). http://dx.doi.org/10.1016/j.tcs.2015.04.020

14. Bodendiek, R., Schumacher, H., Wagner, K.: Über 1-optimale graphen. Math. Nachr. **117**(1), 323–339 (1984)

15. Borodin, O.V.: A new proof of the 6 color theorem. J. of Graph Theory **19**(4), 507–521 (1995). http://dx.doi.org/10.1002/jgt.3190190406

16. Brandenburg, F.J.: Recognizing optimal 1-planar graphs in linear time. Algorithmica **80**(1), 1–28 (2018). https://doi.org/10.1007/s00453-016-0226-8

17. Brandenburg, F.J.: On fan-crossing graphs. Theor. Comput. Sci. **841**, 39–49 (2020). https://doi.org/10.1016/j.tcs.2020.07.002

18. Brandenburg, F.J.: Fan-crossing free graphs and their relationship to other beyond-planar graphs. Theor. Comput. Sci. **867**, 85–100 (2021). https://doi.org/10.1016/j.tcs.2021.03.031

19. Cheong, O., Förster, H., Katheder, J., Pfister, M., Schlipf, L.: On weakly- and strongly fan-planar graphs. Private Communication (2022)

20. Cheong, O., Har-Peled, S., Kim, H., Kim, H.-S.: On the number of edges of fan-crossing free graphs. In: Cai, L., Cheng, S.-W., Lam, T.-W. (eds.) ISAAC 2013. LNCS, vol. 8283, pp. 163–173. Springer, Heidelberg (2013). https://doi.org/10.1007/978-3-642-45030-3_16

21. Cheong, O., Pfister, M., Schlipf, L.: The thickness of fan-planar graphs is at most three. CoRR abs/2208.12324 (2022). https://doi.org/10.48550/arXiv.2208.12324

22. Didimo, W.: Right angle crossing drawings of graphs. In: Hong, S.-H., Tokuyama, T. (eds.) Beyond Planar Graphs, pp. 149–169. Springer, Singapore (2020). https://doi.org/10.1007/978-981-15-6533-5_9

23. Didimo, W., Eades, P., Liotta, G.: Drawing graphs with right angle crossings. Theor. Comput. Sci. **412**(39), 5156–5166 (2011). http://dx.doi.org/10.1016/j.tcs.2011.05.025

24. Didimo, W., Liotta, G., Montecchiani, F.: A survey on graph drawing beyond planarity. ACM Comput. Surv. **52**(1), 4:1–4:37 (2019). https://doi.org/10.1145/3301281

25. Förster, H., Kaufmann, M., Raftopoulou, C.N.: Recognizing and embedding simple optimal 2-planar graphs. In: Purchase, H.C., Rutter, I. (eds.) GD 2021. LNCS, vol. 12868, pp. 87–100. Springer, Cham (2021). https://doi.org/10.1007/978-3-030-92931-2_6

26. Fox, J., Pach, J., Suk, A.: The number of edges in k-quasi-planar graphs. SIAM J. Discret. Math. **27**(1), 550–561 (2013). https://doi.org/10.1137/110858586

27. Grigoriev, A., Bodlaender, H.L.: Algorithms for graphs embeddable with few crossings per edge. Algorithmica **49**(1), 1–11 (2007). http://dx.doi.org/10.1007/s00453-007-0010-x

28. Hong, S.-H.: Algorithms for 1-planar graphs. In: Hong, S.-H., Tokuyama, T. (eds.) Beyond Planar Graphs, Communications of NII Shonan Meetings, pp. 69–87. Springer, Singapore (2020). https://doi.org/10.1007/978-981-15-6533-5_5

29. Kaufmann, M., Ueckerdt, T.: The density of fan-planar graphs. CoRR 1403.6184 (2014). http://arxiv.org/abs/1403.6184

30. Kaufmann, M., Ueckerdt, T.: The density of fan-planar graphs. Electron. J. Comb. **29**(1) (2022). https://doi.org/10.37236/10521

31. Klemz, B., Knorr, K., Reddy, M.M., Schröder, F.: Simplifying non-simple fan-planar drawings. In: Purchase, H.C., Rutter, I. (eds.) GD 2021. LNCS, vol. 12868, pp. 57–71. Springer, Cham (2021). https://doi.org/10.1007/978-3-030-92931-2_4

32. Kobourov, S.G., Liotta, G., Montecchiani, F.: An annotated bibliography on 1-planarity. Comput. Sci. Rev. **25**, 49–67 (2017). https://doi.org/10.1016/j.cosrev.2017.06.002

33. Pach, J., Tóth, G.: Graphs drawn with few crossings per edge. Combinatorica **17**(3), 427–439 (1997). http://dx.doi.org/10.1007/BF01215922

34. Ringel, G.: Ein Sechsfarbenproblem auf der Kugel. Abhandlungen aus dem Mathematischen Seminar der Universität Hamburg (in German) **29**, 107–117 (1965)

35. Suk, A.: k-quasi-planar graphs. In: van Kreveld, M., Speckmann, B. (eds.) GD 2011. LNCS, vol. 7034, pp. 266–277. Springer, Heidelberg (2012). https://doi.org/10.1007/978-3-642-25878-7_26

36. Suk, A., Walczak, B.: New bounds on the maximum number of edges in k-quasi-planar graphs. Comput. Geom. **50**, 24–33 (2015). http://dx.doi.org/10.1016/j.comgeo.2015.06.001

37. Suzuki, Y.: Re-embeddings of maximum 1-planar graphs. SIAM J. Discrete Math. **24**(4), 1527–1540 (2010). http://dx.doi.org/10.1137/090746835

Computational Geometry

Minimum Ply Covering of Points
with Unit Squares

Stephane Durocher[1], J. Mark Keil[2], and Debajyoti Mondal[2(✉)] (iD)

[1] University of Manitoba, Winnipeg, Canada
stephane.durocher@umanitoba.ca
[2] University of Saskatchewan, Saskatoon, Canada
{keil,dmondal}@cs.usask.ca

Abstract. Given a set P of points and a set U of axis-parallel unit squares in the Euclidean plane, a minimum ply cover of P with U is a subset of U that covers P and minimizes the number of squares that share a common intersection, called the minimum ply cover number of P with U. Biedl et al. [Comput. Geom., 94:101712, 2020] showed that determining the minimum ply cover number for a set of points by a set of axis-parallel unit squares is NP-hard, and gave a polynomial-time 2-approximation algorithm for instances in which the minimum ply cover number is constant. The question of whether there exists a polynomial-time approximation algorithm remained open when the minimum ply cover number is $\omega(1)$. We settle this open question and present a polynomial-time $(8 + \varepsilon)$-approximation algorithm for the general problem, for every fixed $\varepsilon > 0$.

1 Introduction

The *ply* of a set S, denoted $\mathrm{ply}(S)$, is the maximum cardinality of any subset of S that has a non-empty common intersection. The set S *covers* the set P if $P \subseteq \bigcup_{S_i \in S} S_i$. Given sets P and U, a subset $S \subseteq U$ is a *minimum ply cover* of P if S covers P and S minimizes $\mathrm{ply}(S)$ over all subsets of U. Formally:

$$\mathrm{plycover}(P, U) = \operatorname*{arg\,min}_{\substack{S \subseteq U \\ S \text{ covers } P}} \mathrm{ply}(S). \tag{1}$$

The ply of such a set S is called the *minimum ply cover number* of P with U, denoted $\mathrm{ply}^*(P, U)$. Motivated by applications in covering problems, including interference minimization in wireless networks, Biedl et al. [3] introduced the *minimum ply cover problem*: given sets P and U, find a subset $S \subseteq U$ that minimizes (1). They showed that the problem is NP-hard to solve exactly, and remains NP-hard to approximate by a ratio less than two when P is a set of points in $\mathbb{R}^2$ and U is a set of axis-aligned unit squares or a set of unit disks in

This work is supported in part by the Natural Sciences and Engineering Research Council of Canada (NSERC).

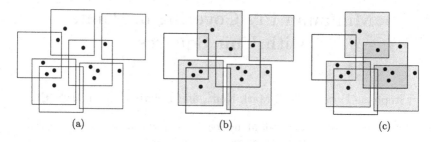

Fig. 1. (a) An input consisting of points and unit squares. (b) A covering of the points with ply 1, which is also the minimum ply cover number for the given input. (c) A covering of the points with ply 2.

$\mathbb{R}^2$. They also provided 2-approximation algorithms parameterized in terms of $\mathrm{ply}^*(P, U)$ for unit disks and unit squares in $\mathbb{R}^2$. Their algorithm for axis-parallel unit squares runs in $O((k + |P|)(2 \cdot |U|)^{3k+1})$ time, where $k = \mathrm{ply}^*(P, U)$, which is polynomial when $\mathrm{ply}^*(P, U) \in O(1)$. Biniaz and Lin [4] generalized this result for any fixed-size convex shape and obtained a 2-approximation algorithm when $\mathrm{ply}^*(P, U) \in O(1)$. The problem of finding a polynomial-time approximation algorithm to the minimum ply cover problem remained open when the minimum ply cover number, $\mathrm{ply}^*(P, U)$, is not bounded by any constant. This open problem is relevant to the motivating application of interference minimization. For example, algorithms for constructing a connected network on a given set of wireless nodes sometimes produce a network with high interference [8]. Selecting a set of network hubs that minimizes interference relates to the ply covering problem in a setting where ply may not be a constant.

Given a set P and a set U of subsets of P, the *minimum membership set cover problem*, introduced by Kuhn et al. [12], seeks to find a subset $S \subseteq U$ that covers P while minimizing the maximum number of elements of S that contain a common point of P. A rich body of research examines the minimum membership set cover problem (e.g., [6,13]). The minimum ply cover problem is a generalization of the minimum membership set cover problem: U is not restricted to subsets of P, and ply is measured at any point covered by U instead of being restricted to points in P. Consequently, the cardinality of a minimum membership set cover is at most the cardinality of a minimum ply cover. Erlebach and van Leeuwen [9] showed that the minimum membership set cover problem remains NP-hard when P is a set of points in $\mathbb{R}^2$ and U are unit squares or unit disks. For unit squares, they gave a 5-approximation algorithm for instances where the optimum objective value is bounded by a constant. Improved approximation algorithms are found in [2] and [10]. We refer the readers to [1,5] for more details on geometric set cover problems.

Our Contribution: In this paper, we consider the minimum ply cover problem for a set P of points in $\mathbb{R}^2$ with a set U of axis-aligned unit squares in $\mathbb{R}^2$ (Fig. 1). We show that for every fixed $\varepsilon > 0$, the minimum ply cover number can be approximated in polynomial time for unit squares within a factor of $(8 + \varepsilon)$.

The algorithm is for the general case, i.e., no assumption on the ply cover of the input instance is needed. Hence, this settles an open question posed in [3] and [4].

Our algorithm overlays a regular grid on the plane and then approximates the ply cover number from the near exact solutions for these grid cells. The most interesting part of the algorithm is to model the idea of bounding the ply cover number with a set of budget points, and to exploit this set's geometric properties to enable dynamic programming to be applied. We show that one can set budget at the corners of a grid cell and check for a solution where the number of squares hit by a corner does not exceed its assigned budget. A major challenge to solve this decision problem is that the squares that hit the four corners may mutually intersect to create a ply that is bigger than any budget set at the corners. We show that an optimal solution can take a few well-behaved forms that can be leveraged to tackle this problem.

2 Minimum Ply Covering with Unit Squares

Let P be a finite set of points in $\mathbb{R}^2$ and let U be a set of axis-parallel unit squares in general position in $\mathbb{R}^2$, i.e., no two squares in U have edges that lie on a common vertical or horizontal line. In this section we give a polynomial-time algorithm to approximate the minimum ply cover number for P with U.

We consider a unit grid $\mathcal{G}$ over the point set P. The rows and columns of the grid are aligned with the x- and y-axes, respectively, and each cell of the grid is a unit square. We choose a grid that is in general position relative to the squares in U. In addition, no grid line intersects the points of P. A grid cell is called *non-empty* if it contains some points of P. We prove that one can first find a near exact ply cover for each non-empty grid cell R and then combine the solutions to obtain an approximate solution for P. We only focus on the ply inside R, because if the ply of a minimum ply cover is realized outside R, then there also exists a point inside R giving the same ply number.

We first show how to find a near exact ply cover when the points are bounded inside a unit square and then show how an approximate ply cover number can be computed for P. We will use the following property of a minimum ply cover. We include the proof in the full version [7] due to space constraint.

Lemma 1. *Let P be a set of points in a unit square R and let U be a set of axis-parallel unit squares such that each square contains either the top left or top right corner of R. Let $W_\ell \subseteq U$ and $W_r \subseteq U$ be the squares that contain the top left and top right corners of R, respectively. Let $S \subseteq U$ be a minimum ply cover of the points in R such that every square in S is necessary. In other words, if a square of S is removed, then the resulting set cannot cover all the points of R. Then S admits the property that one can remove at most one square from S to ensure that squares of $S \cap W_\ell$ do not intersect squares in $S \cap W_r$ (e.g., Fig. 2).*

Ply Cover for Points in a Grid Cell

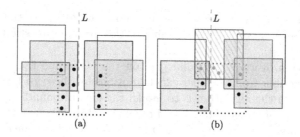

Fig. 2. (a)–(b) Illustration for the configuration of Lemma 1, where $(S \cap W_\ell)$ and $(S \cap W_r)$ are shown in blue and red, respectively. R is shown in dotted line. (Color figure online)

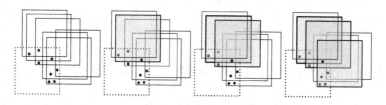

Fig. 3. Illustration for Case 1. The squares taken in the solution are shaded in gray. R is shown in dotted line.

Let R be a 1×1 closed grid cell. Let $Q \subseteq P$ be the set of points in R, and let $W \subseteq U$ be the set of squares that intersect R. Note that by the construction of the grid $\mathcal{G}$, every square in W contains exactly one corner of R. We distinguish some cases depending on the position of the squares in W. In each case we show how to compute either a minimum ply cover or a ply cover with ply at most four more than the minimum ply cover number in polynomial time.

Case 1 (A corner of R intersects all squares in W). In this case we compute a minimum ply cover. Without loss of generality assume that the top right corner of R intersects all the squares in W. We now can construct a minimum ply cover by the following greedy algorithm $\mathcal{G}$.

Step 1: Let z be the leftmost (break ties arbitrarily) uncovered point of Q. Find the square $B \in W$ with the lowest bottom boundary among the squares that contain z.

Step 2: Add B to the solution, remove the points covered by B.

Step 3: Repeat Steps 1 and 2 unless all the points are covered.

Figure 3 illustrates such an example for Case 1. It is straightforward to compute such a solution in $O((|W| + |Q|) \log^2(|W| + |Q|))$ time using standard dynamic data structures, i.e., the point z can be maintained using a range tree and the square B can be maintained by leveraging dynamic segment trees [11].

Lemma 2. *Algorithm $\mathcal{G}$ computes a minimum ply cover.*

Proof. To verify the correctness of the greedy algorithm, first observe that in this case the number of squares in a minimum cardinality cover coincides with a minimum ply cover. We now show that the above greedy algorithm constructs a minimum cardinality cover. We employ an induction on the number of squares in a minimum cardinality cover. Let $W_1, W_2, \ldots, W_k$ be a set of squares in a minimum cardinality cover. First consider the base case where $k = 1$. Since W_1 covers all the points, it also covers z. Since z is the leftmost point and since our choice of square B has the lowest bottom boundary, B must cover all the points. Assume now that if a minimum cardinality cover contains less than k squares, then the greedy algorithm constructs a minimum cardinality cover. Consider now the case when we have k squares in a minimum cardinality cover. For any minimum cardinality cover, if z is covered by a square W_1, then we can replace it with the greedy choice B. The reason is that any point covered by W_1 would also be covered by B. By induction hypothesis, we have a minimum cardinality cover for the points that are not covered by B. Hence the greedy solution must give a minimum cardinality cover. □

Case 2 (Two consecutive corners of R intersect all the squares in W). In this case we compute a minimum ply cover. Without loss of generality assume that the top left and top right corners of R intersect all the squares in W. Let W_ℓ and W_r be the squares of W that intersect the top left corner and top right corner, respectively. We construct a minimum ply cover by considering whether a square of W_ℓ intersects a square of W_r.

If the squares of W_ℓ do not intersect the squares of W_r, then we can reduce it into two subproblems of type Case 1. We solve them independently and it is straightforward to observe that the resulting solution yields a minimum ply cover. Similar to Case 1, here we need $O((|W| + |Q|) \log^2(|W| + |Q|))$ time. Consider now the case when some squares in W_ℓ intersect some squares of W_r. By Lemma 1, there exists a minimum ply cover S such that at least one of the following two properties hold:

C_1 There exists a vertical line L that passes through the left or right side of some square and separates $S \cap W_\ell$ and $S \cap W_r$, as illustrated in Fig. 2(a).

C_2 There exists a square M in S such that after the removal of M from S, one can find a vertical line L that passes through the left or right side of some square and separates $(S \setminus \{M\}) \cap W_\ell$ and $(S \setminus \{M\}) \cap W_r$. This is illustrated in Fig. 2(b), where the square M is shown with the falling pattern.

To find a minimum ply cover, we thus try out all possible L (for C_1), and all possible M and L (for C_2). More specifically, to consider C_1, for each vertical line L passing through the left or right side of some square in W, we independently find a minimum ply cover for the points and squares on the left halfplane of L and right halfplane of L. We then construct a ply cover of Q by taking the union of these two minimum ply covers.

To consider C_2, for each square M, we first delete M and the points it covers. Then for each vertical line L determined by the squares in $(W \setminus \{M\})$, we independently find a minimum ply cover for the points and squares on the left halfplane of L and right halfplane of L. We then construct a ply cover of Q

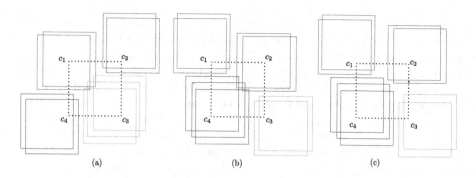

Fig. 4. Illustration for the scenarios that may occur after applying Lemma 1: (a)–(b) DIAGONAL, and (c) DISJOINT. R is shown in dotted line.

by taking the union of these two minimum ply covers and M. Finally, among all the ply covers constructed considering C_1 and C_2, we choose the ply cover with the minimum ply as the minimum ply cover of Q.

Since there are $O(|W|)$ possible choices for L and $O(|W|)$ possible choices for M, the number of ply covers that we construct is $O(|W|^2)$. Each of these ply covers consists of two independent solutions that can be computed in $O((|W| + |Q|) \log^2(|W|+|Q|))$ time using the strategy of Case 1. Hence the overall running time is $O((|W|^3 + |W|^2|Q|) \log^2(|W| + |Q|))$.

Case 3 (Either two opposite corners or at least three corners of R intersect the squares in W). Let S be a minimum ply cover of Q such that all the squares in S are necessary (i.e., removing a square from S will fail to cover Q). Let c_1, c_2, c_3, c_4 be the top-left, top-right, bottom-right and bottom-left corners of R, respectively. Let W_i, where $1 \le i \le 4$, be the squares of W that contain c_i. Similarly, let S_i be the subset of squares in S that contain c_i.

By Lemma 1, one can remove at most four squares from S such that the squares of S_i do not intersect the squares of $S_{(i \bmod 4)+1}$. We assume these squares to be in the solution and hence also remove the points they cover. Consequently, we now have only the following possible scenarios after the deletion.

Diagonal: The squares of S_i do not intersect the squares of $S_{(i \bmod 4)+1}$. The squares of S_1 may intersect the squares of S_3, but the squares of S_2 do not intersect the squares of S_4 (or, vice versa). See Fig. 4(a) and (b).

Disjoint: If two squares intersect, then they belong to the same set, e.g., Fig. 4(c).

We will compute a minimum ply cover in both scenarios. However, considering the squares we deleted, the ply of the final ply cover we compute may be at most four more than the minimum ply cover number.

Case 3.1 (Scenario Diagonal). We now consider the scenario DIAGONAL. Our idea is to perform a search on the objective function to determine the minimum ply cover number. Let k be a guess for the minimum ply cover number. If $k \le 4$, we will show how to leverage Case 1 to verify whether the guess is correct. If

$k > 4$, then one can observe that the ply is determined by a corner of R, as follows. Let H be the common rectangular region of k mutually intersecting squares in the solution. If H does not contain any corner of R, then it lies interior to R. Since H is a rectangular region, we could keep only the squares that determine the boundaries of H to obtain the same point covering with at most 4 squares. Therefore, for $k > 4$, the region determining the ply cover number must include a corner of R. We will use a dynamic program to determine such a ply cover (if exists).

In general, by $T(r, k_1, k_2, k_3, k_4)$ we denote the problem of finding a minimum ply cover for the points in a rectangle r such that the ply is at most $\max\{k_1, k_2, k_3, k_4\}$, and each corner c_i respects its budget k_i, i.e., c_i does not intersect more than k_i squares. We will show that r can always be expressed as a region bounded by at most four squares in W and T returns a feasible ply cover if it exists. To express the original problem, we add four dummy squares in W determined by the four sides of R such that they lie outside of R. Thus $r = R$ is the region bounded by the four dummy squares.

We are now ready to describe the details. Without loss of generality assume that a square $A \in S_4$ intersects a square $B \in S_2$, as shown in Fig. 5(a). We assume A and B to be in a minimum ply cover of R and try out all such pairs. We first consider the case when $k \leq 4$ and the minimum ply cover already contains A and B. We enumerate all $O(|W|^4)$ possible options for $k \leq 4$, S_2, and S_4 with $\mathrm{ply}(S_2 \cup S_4) \leq k$ and for each option, we use Case 1 to determine whether $\mathrm{ply}(W_1)$ and $\mathrm{ply}(W_3)$ are both upper bounded by k. We thus compute the solution to $T(r, k_1, k_2, k_3, k_4)$ and store them in a table $D(r, k_1, k_2, k_3, k_4)$, which takes $O((|W|^5 + |W|^4|Q|) \log^2(|W| + |Q|))$ time.

We now show how to decompose $T(r, k_1, k_2, k_3, k_4)$ into two subproblems assuming that the minimum ply cover already contains A and B. We will use the table D as a subroutine.

The first subproblem consists of the points that lie above A and to the left of B, e.g., Fig. 5(a) and (b). We refer to this set of points by Q_1. The corresponding region r' is bounded by four squares: A, B, and the two (dummy) squares from r. We now describe the squares that need to be considered to cover these points.

- Note that for DIAGONAL, no square in S_1 intersects A or B, hence we can only focus on the squares of W_1 that do not intersect A or B.
- The squares of W_2 that do not intersect Q_1 are removed. The squares of W_2 that contains the bottom left corner of B are removed because including them will make B an unnecessary square in the cover to be constructed.
- Similarly, the squares of W_4 that do not intersect Q_1 or contains the top right corner of A are removed.
- No square in W_3 needs to be considered since to cover a point of Q_1 it must intersect A or B, which is not allowed in DIAGONAL.

The second subproblem consists of the points that lie below B and to the right of A, e.g., Fig. 5(a) and (c). The corresponding region r'' is bounded by four squares: A, B, and the two squares from r. We denote these points by Q_2. The squares to be considered can be described symmetrically.

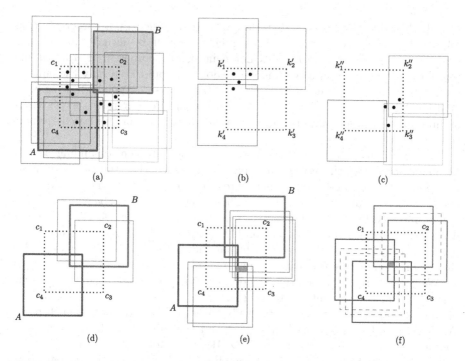

Fig. 5. Illustration for the dynamic program. (a)–(c) Decomposition into subproblems. (d)–(f) Illustration for the $(k + 1)$ mutually intersecting squares. The dashed squares can be safely discarded. R is shown in dotted line.

Let W' and W'' be the set of squares considered to cover Q_1 and Q_2, respectively. By the construction of the two subproblems, we have $Q_1 \cap Q_2 = \varnothing$ and $W' \cap W'' = \varnothing$.

For each corner c_i, we use k'_i and k''_i to denote the budgets allocated for c_i in the first and the second subproblems, respectively. Since we need to ensure that the ply of the problem T is at most $k = \max\{k_1, k_2, k_3, k_4\}$ and each corner c_i respects its budget k_i, we need to carefully distribute the budget among the subproblems when constructing the recurrence formula. Furthermore, let S'_2 and S'_4 be the sets of squares corresponding to c_2 and c_4 that are returned as the solution to the first subproblem. Similarly, define S''_2 and S''_4 for the second subproblem. We now have the following recurrence formula.

$$T(r, k_1, k_2, k_3, k_4) = \min_{\substack{\{A \in W_4, B \in W_2 : A \cap B \neq \varnothing\} \\ k'_1 = k''_1 = k_1, k'_3 = k''_3 = k_3, \\ k'_2 + k''_2 = k_2 - 1, \\ k'_4 + k''_4 = k_4 - 1}} \begin{cases} \begin{aligned} &T(r', k'_1, k'_2, k'_3, k'_4) \cup \\ &T(r'', k''_1, k''_2, k''_3, k''_4) \cup \{A, B\} \end{aligned} & \text{, if } \delta \leq k \\[2ex] \begin{aligned} &T(r', k'_1, k'_2, k'_3, k'_4) \cup \\ &T(r'', k''_1, k''_2, k''_3, k''_4) \cup \beta \end{aligned} & \text{, if } \delta > k \text{ and } k \geq 4 \\[2ex] D(r, k_1, k_2, k_3, k_4) & \text{, if } \delta > k \text{ and } k \leq 3 \end{cases}$$

Here δ is the ply of $(S_2' \cup S_4' \cup S_2'' \cup S_4'' \cup A \cup B)$ and β is the set of squares that remain after discarding unnecessary squares from $(S_2' \cup S_4' \cup S_2'' \cup S_4'' \cup A \cup B)$, i.e., removal of these squares would still ensure that all points are covered by the remaining squares. Since S_1 and S_4 are disjoint, one can also set $k_3' = 0$ in $T(r', k_1', k_2', k_3', k_4')$ and $k_1'' = 0$ in $T(r'', k_1'', k_2'', k_3'', k_4'')$.

If $\delta \leq k$, then the union of $\{A, B\}$ and the squares obtained from the two subproblems must have a ply of at most k for the following two reasons. First, the squares of $S_1 = S_1' \cup S_1''$ (similarly, S_3) cannot intersect the squares of $S_2 \cup S_4 = S_2' \cup S_2'' \cup S_4' \cup S_4''$. Second, by the budget distribution, the ply of S_1 can be at most $k_1 \leq k$ and the ply of S_3 can be at most $k_3 \leq k$.

If $\delta > k$ and $k \leq 4$, then each of S_1, S_2, S_3, S_4 can have at most three rectangles. We can look it up using the table $D(r, k_1, k_2, k_3, k_4)$.

If $\delta > k > 4$, then we can have $k + 1$ mutually intersecting squares in $(S_2' \cup S_4' \cup S_2'' \cup S_4'' \cup A \cup B)$ and in the following we show how to construct a solution with ply cover at most k respecting the budgets, or to determine whether no such solution exists.

If $T(r', k_1', k_2', k_3', k_4')$ and $T(r'', k_1'', k_2'', k_3'', k_4'')$ each returns a feasible solution, then we know that $(k + 1)$ mutually intersecting squares can neither appear in $S_2' \cup S_4'$ nor in $S_2'' \cup S_4''$. Therefore, these $k + 1$ mutually intersecting squares must include either both A and B, or at least one of A and B. We now consider the following options.

Option 1: S_4 and S_2 each contains at least two squares that belong to the set of $k + 1$ mutually intersecting squares. Since the region created by the $k + 1$ mutually intersecting squares is a rectangle, as illustrated in Fig. 5(f), we can keep only the squares that determine the boundaries of this rectangle to obtain the same point covering.

After discarding the unnecessary squares, we only have β squares where $|\beta| = 4 < k$. Thus the ply of the union of $S_1 \cup S_3$ and the remaining β squares is at most k. Hence we can obtain an affirmative solution by taking $T(r', k_1', k_2', k_3', k_4') \cup T(r'', k_1'', k_2'', k_3'', k_4'') \cup \beta$.

Option 2: S_4 only contains A and A intersects all k squares of $S_2' \cup S_2'' \cup B$. Since the $k + 1$ mutually intersecting region is a rectangle, as illustrated in Fig. 5(d), we can keep only the squares that determine the boundaries of this rectangle to obtain the same point covering. After discarding the unnecessary squares, we only have β squares where $|\beta| = 3 < k$. Hence we can handle this case in the same way as in Option 1.

Option 3: S_2 only contains B and B intersects all k squares of $S_4' \cup S_4'' \cup A$. This case is symmetric to Option 2.

In the base case of the recursion, we either covered all the points, or we obtain a set of problems of type Case 1 or of Scenario DISJOINT (Case 3.1.2). The potential base cases corresponding to Case 1 are formed by guessing $O(|W|^2)$ pairs of intersecting squares from opposite corners, as illustrated in Fig. 6(a). The potential $O(|W|^4)$ base cases corresponding to Scenario DISJOINT are formed by two pairs of intersecting squares from opposite corners, as illustrated in Fig. 6(b).

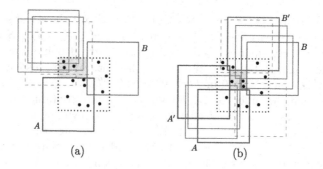

Fig. 6. Illustration for the base cases, where the region corresponding to the base cases are shown in gray. (a) The base case corresponds to Case 1, where we ignore the squares that intersect the chosen squares A and B. (b) An example of the base case that corresponds to scenario DISJOINT, where we need to construct a solution such that no two squares from opposite corners intersect. We ignore all the squares of W_1 or W_3 that intersect the chosen squares A and B, or A' and B', as well as those that makes any of them unnecessary. R is shown in dotted line.

The precomputation of the base cases takes $O(|W|^4 f(|W|, |Q|))$ time, where $f(|W|, |Q|)$ is the time to solve a problem of type Case 1 and of Scenario DISJOINT. We will discuss the details of $f(|W|, |Q|)$ in the proof of Theorem 1.

Since r is determined by at most four squares (e.g., Fig. 6), and since there are four budgets, the solution to the subproblems can be stored in a dynamic programming table of size $O(|W|^4 k^4)$. Computing each entry requires examining $O(|W|^2)$ pairs of squares. Thus the overall running time becomes $O(|W|^6 k^4 + |W|^4 f(|W|, |Q|))$.

Case 3.2 (Scenario Disjoint). In this case, we can find a sequence of empty rectangles $\sigma = (e_1, e_2, \dots)$ from top to bottom such that they do not intersect any square of S, as illustrated in Fig. 7(a)–(b). The idea is again to exploit a dynamic programming with a budget given for each corner of R. A subproblem is expressed by a region determined by at most two squares—one intersecting the left side and the other intersecting the right side of R. In Fig. 7(c), this region is shown in gray. The overall running time for this case is $O(|W|^4 k^4 + |W|^4 \log |Q| + |Q| \log |Q|)$. See the full version [7] for more details.

The following theorem combines all cases and its proof is in full version [7].

Theorem 1. *Given a set Q of points inside a unit square R and a set W of axis-parallel unit squares, a ply cover of size $4 + k^*$ can be computed in $O((|W|^8 (k^*)^4 + |W|^8 \log |Q| + |W|^4 |Q| \log |Q|) \log k^*)$ time, where $k^* = \mathrm{ply}^*(Q, W) \le \min\{|Q|, |W|\}$.*

Covering a General Point Set

Given a set P of points and a set U of axis-parallel unit squares, both in $\mathbb{R}^2$, we now give a polynomial-time algorithm that returns a ply cover of P with U whose ply is at most $(8 + \varepsilon)$ times the minimum ply cover number of P with U. Recall that our algorithm partitions P along a unit grid and applies Theorem 1

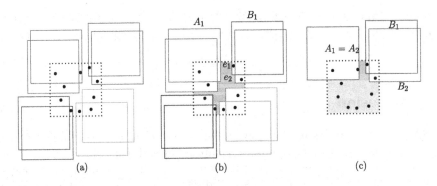

Fig. 7. Illustration for the dynamic program. R is shown in dotted line.

iteratively at each grid cell to select a subset of U that is a near minimum ply cover for the grid cell. Elements of U selected to cover points of P in a given grid cell overlap neighbouring grid cells, which can cause the ply to increase in those neighbouring cells; Lemma 3 allows us to prove Theorem 2 and Corollary 1, showing that the resulting ply is at most $(8 + \varepsilon)$ times the optimal value.

Partition P using a unit grid, i.e., each cell in the partition contains $P \cap [i, i+ i) \times [j, j + 1)$, for some $i, j \in \mathbb{Z}$. Each grid cell has eight grid cells adjacent to it. Let $C_1, \ldots, C_4$ denote the four grid cells in counter-clockwise order that are its diagonal neighbours. We now have the following lemma with the proof in the full version [7].

Lemma 3. *If any point p in a grid cell C is contained in four squares, $\{S_1, \ldots, S_4\} \subseteq U$, such that for each $i \in \{1, \ldots, 4\}$, S_i intersects the cell C_i that is C's diagonal grid neighbour, then $C \subseteq S_1 \cup S_2 \cup S_3 \cup S_4$.*

We now partition P along a unit grid and apply Theorem 1 iteratively to find a near minimum ply cover for each grid cell. For each cell that contains a point p of P, we leverage Lemma 3 to show that at most 8 grid cells can contribute to the ply of p. We thus obtain the following theorem with the proof in the full version [7].

Theorem 2. *Given a set P of points and a set U of axis-parallel unit squares, both in $\mathbb{R}^2$, a ply cover of P using U can be computed in $O((|U|^8(k^*)^4 + |U|^8 \log |P| + |U|^4 |P| \log |P|) \log k^*)$ time whose ply is at most $8k^* + 32$, where $k^* = \text{ply}^*(P, U) \leq \min\{|P|, |U|\}$ denotes the minimum ply cover number of P by U.*

Corollary 1. *Given a set P of points and a set U of axis-parallel unit squares, both in $\mathbb{R}^2$, a ply cover of P using U can be computed in polynomial time whose ply is at most $(8 + \varepsilon)$ times the minimum ply cover number $k^* = \text{ply}^*(P, U)$, for every fixed $\varepsilon > 0$.*

Proof. We use Theorem 2 to find a ply cover with ply at most $8k^* + 32$, and then consider the following two cases. **Case 1.** Suppose $\varepsilon k^* \geq 32$. Then $8k^* + 32 \leq$

$(8+\varepsilon)k^*$. **Case 2.** Suppose $\epsilon k^* < 32$. We apply the 2-approximation algorithm of Biedl et al. [3] in $O(|P| \cdot |U|)^{3k^*+1})$ time, which is polynomial since $k^* \in O(1)$.

$\square$

3 Conclusion

We gave a $(8+\varepsilon)$-approximation polynomial-time algorithm for the minimum ply cover problem with axis-parallel unit squares. Through careful case analysis, it may be possible to further improve the running time of our approximation algorithm presented in Theorem 2. A natural direction for future research would be to reduce the approximation factor or to apply a different algorithmic technique with lower running time. It would also be interesting to examine whether our strategy can be generalized to find polynomial-time approximation algorithms for other covering shapes, such as unit disks or convex shapes of fixed size.

References

1. Agarwal, P.K., Ezra, E., Fox, K.: Geometric optimization revisited. In: Steffen, B., Woeginger, G.J. (eds.) Computing and Software Science - State of the Art and Perspectives, LNCS, vol. 10000, pp. 66–84. Springer, Cham (2019)
2. Basappa, M., Das, G.K.: Discrete unit square cover problem. Discret. Math. Algorithms Appl. 10(6), 1850072:1–1850072:18 (2018)
3. Biedl, T.C., Biniaz, A., Lubiw, A.: Minimum ply covering of points with disks and squares. Comput. Geom. **94**, 101712 (2021)
4. Biniaz, A., Lin, Z.: Minimum ply covering of points with convex shapes. In: Proceedings of the 32nd Canadian Conference on Computational Geometry (CCCG), pp. 2–5 (2020)
5. Chan, T.M., He, Q.: Faster approximation algorithms for geometric set cover. In: Cabello, S., Chen, D.Z. (eds.) Proceedings of the 36th International Symposium on Computational Geometry (SoCG). LIPIcs, vol. 164, pp. 27:1–27:14. Schloss Dagstuhl - Leibniz-Zentrum für Informatik (2020)
6. Demaine, E.D., Feige, U., Hajiaghayi, M., Salavatipour, M.R.: Combination can be hard: approximability of the unique coverage problem. SIAM J. Comput. **38**(4), 1464–1483 (2008)
7. Durocher, S., Keil, J.M., Mondal, D.: Minimum ply covering of points with unit squares. CoRR abs/2208.06122 (2022)
8. Durocher, S., Mehrpour, S.: Interference minimization in k-connected wireless networks. In: Proceedings of the 29th Canadian Conference on Computational Geometry (CCCG), pp. 113–119 (2017)
9. Erlebach, T., van Leeuwen, E.J.: Approximating geometric coverage problems. In: Teng, S. (ed.) Procedings 19th ACM-SIAM Symposium on Discrete Algorithms (SODA), pp. 1267–1276. SIAM (2008)
10. Erlebach, T., van Leeuwen, E.J.: PTAS for weighted set cover on unit squares. In: Serna, M., Shaltiel, R., Jansen, K., Rolim, J. (eds.) APPROX/RANDOM -2010. LNCS, vol. 6302, pp. 166–177. Springer, Heidelberg (2010). https://doi.org/10.1007/978-3-642-15369-3_13

11. van Kreveld, M.J., Overmars, M.H.: Union-copy structures and dynamic segment trees. J. ACM **40**(3), 635–652 (1993)
12. Kuhn, F., von Rickenbach, P., Wattenhofer, R., Welzl, E., Zollinger, A.: Interference in cellular networks: the minimum membership set cover problem. In: Wang, L. (ed.) COCOON 2005. LNCS, vol. 3595, pp. 188–198. Springer, Heidelberg (2005). https://doi.org/10.1007/11533719_21
13. Misra, N., Moser, H., Raman, V., Saurabh, S., Sikdar, S.: The parameterized complexity of unique coverage and its variants. Algorithmica **65**(3), 517–544 (2013)

Overlapping Edge Unfoldings
for Archimedean Solids and (Anti)prisms

Takumi Shiota[✉] and Toshiki Saitoh

School of Computer Science and Systems Engineering, Kyushu Institute
of Technology, Fukuoka, Japan
shiota.takumi779@mail.kyutech.jp, toshikis@ai.kyutech.ac.jp

Abstract. Herein, we discuss the existence of overlapping edge unfoldings for Archimedean solids and (anti)prisms. Horiyama and Shoji showed
that there are no overlapping edge unfoldings for all platonic solids and five
shapes of Archimedean solids. The remaining five Archimedean solids were
also found to have edge unfoldings that overlap. In this study, we propose
a method called rotational unfolding to find an overlapping edge unfolding of a polyhedron. We show that all the edge unfoldings of an icosidodecahedron, a rhombitruncated cuboctahedron, an n-gonal Archimedean
prism, and an m-gonal Archimedean antiprism do not overlap when $3 \leq
n \leq 23$ and $3 \leq m \leq 11$. Our algorithm finds three types of overlapping
edge unfoldings for a snub cube, consisting of two vertices in contact. We
show that an overlapping edge unfolding exists in an n-gonal Archimedean
prism and an m-gonal Archimedean antiprism for $n \geq 24$ and $m \geq 12$. Our
results prove the existence of overlapping edge unfoldings for Archimedean
solids and Archimedean (anti)prisms.

Keywords: Polyhedron · Overlapping edge unfolding · Archimedean
solids · Archimedean (anti)prisms · Enumeration algorithm

1 Introduction

The study of unfoldings of polyhedrons is known to have originated from the
publication "Underweysung der messung mit dem zirckel un richt scheyt" [3]
by Albrecht Dürer in 1525 [4]. Albrecht Dürer drew some edge unfoldings that
cut along the edges of a polyhedron and formed the plane's flat polygon. However, all the edge unfoldings are nonoverlapping polygons, i.e., no two faces in
the polyhedron exhibit overlapping unfoldings. The following open problem is
obtained from this book:

Open Problem 1 ([4], **Open Problem 21.1**). *Does every convex polyhedron
have a nonoverlapping edge unfolding?*

Any convex polyhedron has nonoverlapping unfoldings, i.e., when the polyhedron surface is cut [10,13]. However, Namiki and Fukuda found a convex
polyhedron that has an overlapping edge unfolding [11]. Biedl et al. in 1998 and
Grünbaum in 2003 discovered that there exists a nonconvex polyhedron whose

C.-C. Lin et al. (Eds.): WALCOM 2023, LNCS 13973, pp. 36–48, 2023.
https://doi.org/10.1007/978-3-031-27051-2_4

Table 1. Existence of an overlapping edge unfolding for Archimedean Solids and (Anti)prisms.

Convex polyhedron	Number of edge unfoldings [9]	Is there an overlapping edge unfolding?
Truncated Tetrahedron	261	No [7]
Cuboctahedron	6,912	No [7]
Truncated hexahedron	675,585	No [7]
Truncated octahedron	2,108,512	No [7]
Rhombicuboctahedron	6,272,012,000	No [7]
Icosidodecahedron	1,741,425,868,800	**No [This paper]**
Snub cube	3,746,001,752,064	**Yes [This paper]**
Rhombitruncated cuboctahedron	258,715,122,137,472	**No [This paper]**
Truncated dodecahedron	41,518,828,261,687,500	Yes [8]
Truncated icosahedron	3,127,432,220,939,473,920	Yes [8]
Rhombicosidodecahedron	1,679,590,540,992,923,166,257,971,200	Yes [8]
Snub dodecahedron	7,303,354,923,116,108,380,042,995,304,896,000	Yes [2]
Rhombitruncated icosidodecahedron	181,577,189,197,376,045,928,994,520,239,942,164,480	Yes [8]
n-gonal Archimedean prism	$\begin{cases} \frac{1}{8\sqrt{3}}\{2\sqrt{3}n + \sqrt{3}(2+\sqrt{3})^n \\ \quad +(2+\sqrt{3})^{\lfloor\frac{n}{2}\rfloor}(4+2\sqrt{3}) \\ \quad +(2-\sqrt{3})^{\lfloor\frac{n}{2}\rfloor}(2\sqrt{3}-4) \qquad (\text{if } n \text{ is odd}) \\ \quad +\sqrt{3}((2-\sqrt{3})^n-2)\} \\ 11 \qquad\qquad\qquad\qquad (\text{if } n=4) \\ \frac{1}{24}\{3(2+\sqrt{3})^n + 4\sqrt{3}(2+\sqrt{3})^{\frac{n}{2}} \\ \quad +3(2-\sqrt{3})^n - 4\sqrt{3}(2-\sqrt{3})^{\frac{n}{2}} \quad (\text{otherwise}) \\ \quad +6n-6\} \end{cases}$	No ($3 \le n \le 23$) [This paper] Yes ($n \ge 24$) [This paper]
n-gonal Archimedean antiprism	$\begin{cases} 11 \qquad\qquad\qquad (\text{if } n=3) \\ \frac{1}{10}\left\{\left(\frac{1+\sqrt{5}}{2}\right)^{4n} + \left(\frac{1+\sqrt{5}}{2}\right)^{-4n} - 2\right\} \quad (\text{otherwise}) \\ \quad +\frac{(3+\sqrt{5})^n-(3-\sqrt{5})^n}{2^{n+1}\sqrt{5}} \end{cases}$	No ($3 \le n \le 11$) [This paper] Yes ($n \ge 12$) [This paper]

(a) Truncated dodecahedron   (b) Truncated icosahedron

Fig. 1. Examples of overlapping edge unfoldings [8]. The right edge unfolding can be obtained by cutting along the thick line of the left convex polyhedron.

every edge unfolding overlaps [1,6]. Some studies have reported on the existence and/or the number of overlapping edge unfoldings for convex regular-faced polyhedrons. A snub dodecahedron has an overlapping edge unfolding [2]. Horiyama and Shoji presented an algorithm that enumerates overlapping edge unfoldings for a polyhedron. Their algorithm first enumerates edge unfoldings using binary decision diagrams and then checks the overlapping by numerical calculations for each unfolding. They found overlapping edge unfoldings for a truncated dodecahedron, truncated icosahedron, rhombicosidodecahedron, and rhombitruncated icosidodecahedron (Some are shown in Fig. 1). In addition, they confirmed that platonic solids and five shapes of Archimedean solids do not have overlapping edge unfoldings [7,8] (see Table 1). The edge unfoldings are represented as spanning trees of a polyhedral graph. The algorithm by Horiyama and Shoji first

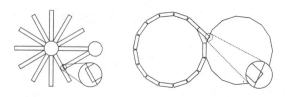

Fig. 2. Overlapping edge unfoldings for an n-gonal prisms [12].

enumerates the spanning trees to find overlapping edge unfoldings; however, if a polyhedron has an excessive number of spanning trees, it is difficult to enumerate overlapping edge unfoldings even if only a small number of them exist. Thus, they considered isomorphism of unfoldings and enumeration of paths instead of enumeration of the spanning trees to reduce the search space for finding overlapping edge unfolding [7]. However, it remains to be clarified a snub cube, an icosidodecahedron, or a rhombitruncated cuboctahedron has overlapping edge unfoldings. Schlickenrieder showed that n-gonal prisms have overlapping edge unfoldings, as shown in Fig. 2 [12]. However, the side faces of n-gonal prisms are not regular; therefore, the overlapping edge unfoldings for n-gonal Archimedean prisms or n-gonal Archimedean antiprisms have not been studied. DeSplinter et al. recently studied the edge unfoldings for high-dimensional cubes and showed that a spanning tree of a Roberts graph can represent an edge unfolding [5]. They proposed a rolling and unfolding method, in which the cubes are rotated on a spanning tree and the edges are cut to ensure that they do not overlap.

Our Contributions. Herein, we propose a method for determining an overlapping edge unfolding called *rotational unfolding* for a polyhedron. The basic principle of our method is the same as that of the rolling and unfolding method. First, a polyhedron is put on a plane, and the following three steps are performed repeatedly: the bottom edges are cut, the polyhedron is rotated in the plane, and overlapping edge unfoldings are searched. The rolling and unfolding method is suitable for determining edge unfoldings for high-dimensional cubes but not for general shapes. Therefore, we extend the method to n-gon by proposing pruning methods on the rotational unfolding using the distance property and symmetry of a polyhedron to determine overlapping unfoldings efficiently. As a result, we obtain the following:

- We show that all the edge unfoldings of an icosidodecahedron and a rhombitruncated cuboctahedron do not overlap and that a snub cube has only three types of overlapping edge unfoldings, as shown in Fig. 3, with two vertices of faces in contact with each other. These are indicated in bold in Table 1. These results are used to the determine the existence of overlapping edge unfoldings for Archimedean solids.
- We find a new type of overlapping edge unfoldings for a truncated icosahedron, as shown in Fig. 4, and show that only one and two types of edge unfoldings exist in a truncated dodecahedron and truncated icosahedron, respectively.

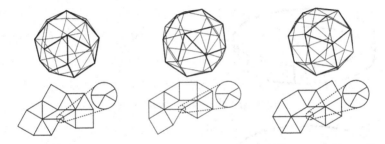

Fig. 3. Three types of edge unfoldings have two faces in contact with the snub cube. The edge unfolding can be obtained by cutting each snub cube along the thick line.

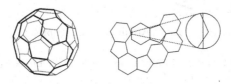

Fig. 4. A new overlapping edge unfolding in a truncated icosahedron. The right edge unfolding is obtained by cutting along the thick line of the left convex polyhedron.

– Through rotational unfolding, we show that overlapping edge unfoldings do not exist for n-gonal Archimedean prisms and m-gonal Archimedean antiprisms for $3 \leq n \leq 23$ and $3 \leq m \leq 11$ by rotational unfolding. We also demonstrate that overlapping edge unfoldings exist in n-gonal Archimedean prisms and m-gonal Archimedean antiprisms for $n \geq 24$ and $m \geq 12$.

2   Preliminaries

Let $G = (V, E)$ be a simple graph where V is a set of vertices and $E \subseteq V \times V$ is a set of edges. A sequence of vertices $(v_1, \ldots, v_k)$ is a *path* if all vertices in the sequence are distinct and every consecutive two vertices are adjacent. A graph is *connected* if there exists a path between any two vertices of the graph. If a graph $T = (V_T, E_T)$ is connected and $|E_T| = |V_T| - 1$, the graph is called a *tree*. A tree $T = (V_T, E_T)$ is a *spanning tree* of $G = (V, E)$ if $V_T = V$ and $E_T \subseteq E$.

A *polyhedron* is a three-dimensional object consisting of at least four polygons, called *faces*, joined at their edges. A *convex polyhedron* is a polyhedron with the interior angles of all two faces less than π. An *n prism* is a polyhedron composed of two identical n-sided polygons, called *bases*, facing each other, and n parallelograms, called *side faces*, connecting the corresponding edges of the two bases. An *n antiprism* is a polyhedron composed of two bases of congruent n-sided polygons and $2n$-sided alternating triangles. An *n-gonal (anti)prism* is an n (anti)prism if the bases are n-sided regular polygons and an *n-gonal Archimedean (anti)prism* is an n-gonal (anti)prism if the side faces are also regular.

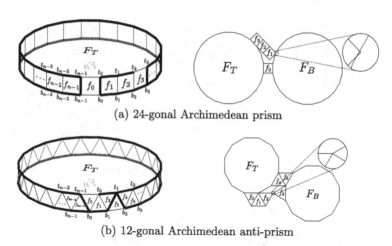

(a) 24-gonal Archimedean prism

(b) 12-gonal Archimedean anti-prism

Fig. 5. An overlapping edge unfolding in an n-gonal Archimedean (anti-)prism. The right edge unfolding is obtained by cutting along the thick line of the left convex polyhedron.

Let P be a polyhedron. P can be viewed as a graph $G_P = (V_P, E_P)$, where V_P is a set of vertices and E_P is a set of edges of P. An *unfolding* (also called a net, a development, or a general unfolding) of the polyhedron P is a flat polygon formed by cutting P's edges or faces and unfolding it into a plane. An edge unfolding of P is an unfolding formed by cutting only edges. We have the following lemma for an edge unfolding of P.

Lemma 1 (see e.g., [4], Lemma 22.1.1). *The cut edges of an edge unfolding of P form a spanning tree of G_P.*

This lemma implies that a spanning tree of G_P corresponds to an edge unfolding of P. Two faces in P are *neighbors* if they contain a common edge. A *dual graph* of P is a graph where each vertex of the dual graph corresponds to a face in the polyhedron, and two vertices are adjacent if and only if the corresponding two faces are neighbors. A spanning tree of the dual graph of P can also be considered an edge unfolding [12].

The following proposition is used to determine whether an edge unfolding of a polyhedron P is overlapping.

Proposition 1 ([8]). *If for any two faces in an edge unfolding, the circumscribed circles of the two faces do not overlap, then the edge unfolding is not overlapping.*

This proposition is useful for efficiently checking the overlapping of an edge unfolding, and it is a necessary condition for overlapping edge unfoldings. If the circumscribed circles of two faces of P intersect, we use numerical calculations to check the overlapping.

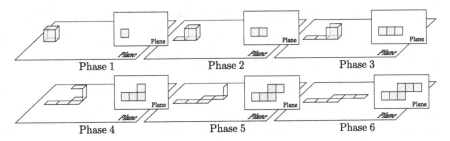

Fig. 6. Illustration of rotational unfolding.

3   Rotational Unfolding

In this section, we propose an algorithm for detecting overlapping edge unfoldings for a polyhedron P. A spanning tree $T(U)$ of a dual graph $D(P)$ of P represents an edge unfolding U. We can determine all overlapping edge unfoldings by enumerating all spanning trees of $D(P)$ and then check the overlapping of the corresponding unfoldings. However, a polyhedron generally contains a large number of spanning trees. Our algorithm employs Lemma 2 to enumerate the paths rather than the spanning trees to efficiently search for overlapping edge unfoldings.

Lemma 2 ([5,7]). *Let U be an overlapping edge unfolding of a polyhedron P, and $T(U)$ be a spanning tree corresponding to U of the dual graph $D(P)$. There exist two vertices $v, v' \in T(U)$ such that a path from v to v' in $T(U)$ represents a consecutive sequence of faces in U with overlapping the two faces corresponding to v and v'.*

For a polyhedron P, we present a simple and recursive procedure called rotational unfolding to find paths and check their overlap. In this procedure, we first place P in the plane. The *start face* f_s of P is the bottom face. We rotate P and unfold the current bottom in the rotational unfolding. Let f_ℓ be the current bottom face, called the *last face*. In the first step of the procedure, f_ℓ is the start face f_s. The rotational unfolding first checks whether there exists a neighbor face of f_ℓ in P. Then, for each neighbor face f, we run the following three steps: we cut the edges of f_ℓ except for the edge sharing f, roll the polyhedron P to be the bottom f, and check the overlap between f_s and f. To check the overlapping of edge unfoldings, we compute the coordinate of the outer center of f from that of f_ℓ and the angle of the shared edge. Then, we check the overlap between f_s and f using Proposition 1 or numerical calculations. Let v_{f_s} and v_f be the vertices corresponding to the face f_s and f of $D(P)$, respectively. If f_s and f overlap, we output a part of edge unfolding corresponding to a path from v_{f_s} to v_f. Otherwise, we run the procedure recursively. Figure 6 illustrates the rotational unfolding procedure.

Although the number of paths is smaller than that of spanning trees, it is still large. To reduce the search space, we implement three methods for speeding

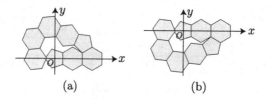

Fig. 7. (a) and (b) are symmetric with respect to the x−axis.

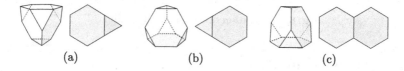

Fig. 8. Cases of the first two faces in the rotational unfolding.

up the search. The first method uses the simple distance property. Let D be the Euclidean distance between the outer center of f_s and that of f, r_s and r be the circumscribed circle radii of f_s and f, respectively, and W be the sum of circumscribed circle diameters of the remaining faces in P. For f_s and a face in P to overlap, the distance between f_s and f have to be smaller than W; that is, if $W + r_s + r < D$, f_s does not overlap any other faces in P for any unfolding because f_s is too far from the other faces in P. Thus, if $W + r_s + r < D$, we prune the search.

The second method uses the symmetry of the polyhedron. Figure 7 shows a symmetric edge unfolding. If a polyhedron has such symmetric unfoldings, we only compute one of them to check if a self-overlapping edge unfolding exists. To implement this pruning, we maintain the y-coordinate of the outer center of the last face before it becomes non-zero. We prune the search if the y-coordinate becomes negative for the first time. Note that, this pruning does not work for a snub cube and a snub dodecahedron because they do not have a mirror symmetry.

In the third method, we run the rotational unfolding by fixing a few steps of the search. In the algorithm, we first choose a start face. We only need to consider restricted patterns of the first few faces of polyhedrons. For example, in the case of a truncated tetrahedron, which consists of regular triangles and regular hexagons, as shown in Fig. 8, we only consider three patterns of the start and next face pairs: (a) a triangle and a hexagon, (b) a hexagon and a triangle, and (c) a hexagon and a hexagon. We find paths from these patterns as start faces. Note that we cannot consider only the first face shape to find the patterns. For example, both of the start faces of Fig. 9(a) and (b) are squares; however, we need to consider both cases because the remaining polyhedrons are not isomorphic. Thus, for each polyhedron shape, we have to find start patterns based on symmetry.

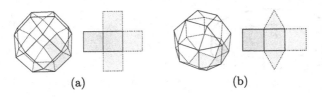

Fig. 9. Cases of the first three faces in the rotational unfolding for a rhombitruncated cuboctahedron.

4   Archimedean Solids

We implemented rotational unfolding in C++ and adapted it to Archimedean solids to find their overlapping edge unfoldings. We obtain the following theorem from our experiments.

Theorem 1.

(a) An icosidodecahedron and a rhombitruncated cuboctahedron have no overlapping edge unfoldings.
(b) A snub cube has three types of overlapping unfoldings with two vertices of faces in contact, as shown in Fig. 3.

An overlapping edge unfolding exists for a truncated dodecahedron and a truncated icosahedron [8]. Our algorithm finds the other overlapping unfoldings for truncated icosahedron, as shown in Fig. 4, and we verify that it has no other types of overlapping edge unfoldings.

5   Archimedean Prisms

In this section, we prove Theorem 2.

Theorem 2. *Let n be a natural number and $P_R(n)$ be an n-gonal Archimedean prism.*

(a) If $3 \leq n \leq 23$, $P_R(n)$ has no overlapping edge unfoldings.
(b) For $n \geq 24$, there exists an overlapping edge unfolding in $P_R(n)$.

We demonstrate the case of no overlapping edge unfolding of Theorem 2(a) for every $n \in \{3, \ldots, 23\}$ of $P_R(n)$ using rotational unfolding.

Theorem 2(b) can be proven by constructing an overlapping edge unfolding for $P_R(n)$. Let F_T and F_B be the top and bottom faces of $P_R(n)$, respectively, and $f_0, \cdots, f_{n-1}$ be the sides, which are numbered counterclockwise viewing from the top face F_T. For $i \in \{0, \ldots, n-1\}$, let t_i and b_i be vertices on F_T and F_B such that they share two faces f_i and f_{i+1}, where $f_n = f_0$. For $n = 24$, $P_R(n)$ has an overlapping edge unfolding, as shown in Fig. 5(a) (right), consisting of faces $\{F_B, f_0, F_T, f_3, f_2, f_1\}$ obtained by cutting along the thick line of $P_R(n)$, as shown in Fig. 5(a) (left). For $25 \leq n \leq 28$, $P_R(n)$ has an overlapping edge unfolding similar to $P_R(24)$.

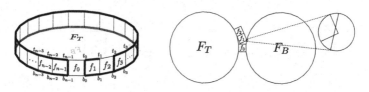

Fig. 10. An overlapping edge unfolding in the 29-gonal Archimedean prism. The right edge unfolding is obtained by cutting along the thick line of the left convex polyhedron.

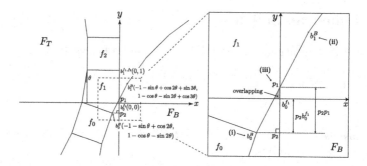

Fig. 11. Magnified image of overlapping areas in the edge unfolding of $P_R(n)$

It remains to be shown that an overlapping edge unfolding of $P_R(n)$ exists for $n \geq 29$. We prove that the edge unfolding consisting of faces $\{F_B, f_0, F_T, f_2, f_1\}$ overlap, as shown in Fig. 10(right), by cutting along the thick line of $P_R(n)$ shown in Fig. 10(left).

Lemma 3. *For $n \geq 29$, if we cut the edges $(t_0, t_1), (t_0, b_0), (b_0, b_1),$ and (b_1, b_2) and do not cut $(t_{n-1}, t_0), (b_{n-1}, b_0),$ and (t_1, t_2) of $P_R(n)$, any edge unfolding is overlapping.*

Proof. Figure 11 shows a part of edge unfolding consisting of $\{F_B, f_0, F_T, f_2, f_1\}$. We define t_i^T and b_i^B for $i \in \{0, \dots, n-1\}$ as vertices on F_T and F_B in the edge unfolding such that they are t_i and b_i in $P_R(n)$, respectively. Let S be a subset of faces $\{f_0, \dots, f_{n-1}\}$. The vertices t_i and b_i in $P_R(n)$ that are shared by S in the edge unfolding are denoted as t_i^S and b_i^S, respectively. Here, we set $b_0^{f_1}$ and $b_1^{f_1, f_2}$ as $(0, 0)$ and $(0, 1)$ in the plane, respectively. We can obtain the following three conditions.

(i) Point b_0^B exists in the third quadrant.
(ii) Point b_1^B exists in the first quadrant.
(iii) Let p_1 be an intersection point of the segment $b_0^B b_1^B$ and the y-axis. The y-coordinate of p_1 is positive.

The y-coordinate of p_1 is within $(0, -1)$ to $(0, 1)$ because the length of the line segment $b_0^B b_1^B$ is one if the conditions (i) and (ii) are satisfied. And if the y-coordinate of p_1 is positive, the line segment $b_0^{f_1} b_1^{f_1, f_2}$ intersects the line segment $b_0^B b_1^B$. Therefore, the faces f_1 and F_B overlap if the three conditions are satisfied.

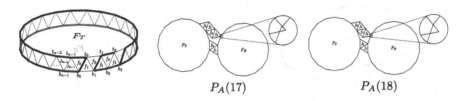

$P_A(17)$ $P_A(18)$

Fig. 12. Overlapping edge unfoldings of $P_A(17)$ and $P_A(18)$, consisting of the set of faces $\{F_T, f_0, f_1, F_B, f_5, f_4, f_3, f_2\}$. The center and right edge unfolding are obtained by cutting along the thick line of the left convex polyhedron.

We will show that the three conditions are satisfied. We define the angle $\theta = \frac{2\pi}{n}$ as the exterior angle of the regular n-sided polygon. The range of θ is $0 < \theta \le \frac{2\pi}{29}$ because $n \ge 29$. We make the following claim.

Claim 1. *The coordinates of b_0^B and b_1^B are $(-1 - \sin\theta + \cos 2\theta, 1 - \cos\theta - \sin 2\theta)$ and $(-1 - \sin\theta + \cos 2\theta + \sin 3\theta, 1 - \cos\theta - \sin 2\theta + \cos 3\theta)$, respectively.*

From the claim and differential analysis, we can show conditions **(i)** and **(ii)**.

Let p_2 be an intersection point of the perpendicular line from point b_0^B to the y-axis and y-axis; that is, the coordinates of p_2 are $(0, 1 - \cos\theta - \sin 2\theta)$. To show condition **(iii)**, we give the following claim.

Claim 2. *The length of the line segment $p_2 p_1$ is greater than that of $p_2 b_0^{f_1}$.*

Thus, conditions **(i)**–**(iii)** hold; that is, an overlapping edge unfolding exists for $P_R(n)$, where $n \ge 29$.   □

6 Archimedean Anti-prisms

In this section, we prove Theorem 3.

Theorem 3. *Let n be a natural number and $P_A(n)$ be an n-gonal Archimedean antiprism.*

(a) If $3 \le n \le 11$, $P_A(n)$ has no overlapping edge unfoldings.
(b) For $n \ge 12$, there exists an overlapping edge unfolding in $P_A(n)$.

We demonstrate the no overlapping edge unfolding of Theorem 3(a) for every $n \in \{3, \ldots, 11\}$ of $P_A(n)$ using rotational unfolding.

Theorem 3(b) can be proven by constructing an overlapping edge unfolding for $P_A(n)$. Let F_T and F_B be the top and bottom faces of $P_A(n)$, respectively, and $f_0, \cdots, f_{2n-1}$ be the sides, which are numbered counterclockwise viewing from the top face F_T. For $i \in \{0, \ldots, n-1\}$, let t_i and b_i be vertices on F_T and F_B such that they share three faces f_{2i}, f_{2i+1}, and f_{2i+2} and f_{2i-1}, f_{2i}, and f_{2i+1}, where $f_{-1} = f_{2n-1}$ and $f_{2n} = f_0$. For $n = 12$, $P_A(n)$ has an overlapping edge unfolding, as shown in Fig. 5(b) (right), consisting of faces $\{f_3, F_B, f_5, f_4, F_T, f_0, f_1, f_2\}$ obtained by cutting along the thick line of $P_A(n)$, as shown in Fig. 5(b) (left). For $13 \le n \le 16$, $P_A(n)$ has an overlapping edge unfolding similar to $P_A(12)$.

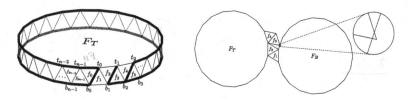

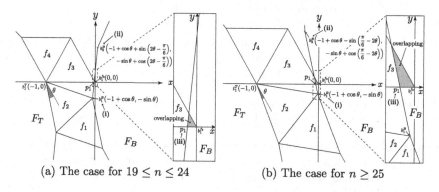

Fig. 13. An overlapping edge unfolding in the 19-gonal Archimedean prism. We obtain the right edge unfolding by cutting along the thick line of the left convex polyhedron.

(a) The case for $19 \leq n \leq 24$ (b) The case for $n \geq 25$

Fig. 14. Magnified image of overlapping areas in the edge unfolding of $P_A(n)$

For $n \in \{17, 18\}$, $P_A(n)$ has overlapping edge unfoldings consisting of faces $\{F_T, f_0, f_1, F_B, f_5, f_4, f_3, f_2\}$, as shown in Fig. 12(left), obtained by cutting along the thick line $P_A(n)$, as shown in Fig. 12(center and right).

It remains to be shown that an overlapping edge unfolding of $P_A(n)$ exists for $n \geq 19$. We can prove that the edge unfolding consisting of faces $\{F_B, f_1, f_2, F_T, f_4, f_3\}$ contains overlapping, as shown in Fig. 13(right), by cutting along the thick line of $P_A(n)$, as shown in Fig. 13(left).

Lemma 4. *For $n \geq 19$, if we cut the edges (t_1, b_1) and (b_1, b_2) and do not cut the edges $(b_0, b_1), (t_0, t_1), (t_0, b_1), (t_1, t_2)$, and (t_1, b_2) of $P_A(n)$, any edge unfolding is overlapping.*

Proof. Figure 14 shows a part of edge unfolding consisting of a set of faces $\{F_B, f_1, f_2, F_T, f_4, f_3\}$. We define t_i^T and b_i^B for $i \in \{0 \cdots n-1\}$ as vertices on F_T and F_B in the edge unfolding such that they are t_i and b_i in $P_A(n)$, respectively. Let S be a subset of faces $\{f_0, \ldots, f_{2n-1}\}$. The vertices t_i and b_i are vertices in $P_A(n)$ that are shared by S in the edge unfolding are denoted as t_i^S and b_i^S, respectively. Here, we set $b_1^{f_3}$ and t_1^T as $(0, 0)$ and $(-1, 0)$ in the plane, respectively. We will show that the following three conditions are satisfied.

(i) Point b_1^B exists in the third quadrant.
(ii) The y-coordinate of point b_2^B is positive.
(iii) Let p_1 be an intersection point of the segment $b_1^B b_2^B$ and the x-axis. The x-coordinate of point p_1 is in $-1 < p_1 < 0$.

Face f_3 is a triangle such that the bottom is $(-1, 0)$ to $(0, 0)$. From conditions **(i)** and **(ii)**, there exists an intersection point p_1 of the segment the segment $b_1^B b_2^B$ and the x-axis. Moreover, if p_1 is within $(-1, 0)$ to $(0, 0)$, the line segment $b_1^B b_2^B$ intersects f_3; that is, f_3 and F_B overlap.

We define the angle $\theta = \frac{2\pi}{n}$ as the exterior angle of the regular n-sided polygon. The range of θ is $0 < \theta \leq \frac{2\pi}{19}$ because $n \geq 19$. We obtain the following claim.

Claim 3. *The coordinate of b_1^B is $(-1 + \cos\theta, -\sin\theta)$, the coordinate of b_2^B is*

$$\begin{cases} \left(-1 + \cos\theta + \sin\left(2\theta - \frac{\pi}{6}\right), -\sin\theta + \cos\left(2\theta - \frac{\pi}{6}\right)\right) & \text{if } 19 \leq n \leq 24 \\ \left(-1 + \cos\theta - \sin\left(\frac{\pi}{6} - 2\theta\right), -\sin\theta + \cos\left(\frac{\pi}{6} - 2\theta\right)\right) & \text{if } n \geq 25 \end{cases}$$

and the x-coordinate of p_1 is

$$\begin{cases} \cos\left(\frac{\pi}{6} - \theta\right) / \cos\left(2\theta - \frac{\pi}{6}\right) - 1 & \text{if } 19 \leq n \leq 24 \\ \cos\left(\frac{\pi}{6} - \theta\right) / \cos\left(\frac{\pi}{6} - 2\theta\right) - 1 & \text{if } n \geq 25. \end{cases}$$

From the claim and differential analysis, we can show the conditions **(i)**–**(iii)**. Thus, the conditions **(i)**–**(iii)** hold; that is, an overlapping edge unfolding exists for $P_A(n)$, where $n \geq 19$.   □

Acknowledgments. This work was supported in part by JSPS KAKENHI Grant Numbers JP18H04091, JP19K12098, and 21H05857.

References

1. Biedl, T.C., et al.: Unfolding some classes of orthogonal polyhedra. In: 10th Canadian Conference on Computational Geometry (1998)
2. Croft, H.T., Falconer, K.J., Guy, R.K.: Unsolved Problems in Geometry, reissue edn. Springer, Heidelberg (1991)
3. Dürer, A.: Underweysung der messung, mit dem zirckel und richtscheyt in linien ebenen unnd gantzen corporen (1525)
4. Demaine, E.D., O'Rourke, J.: Geometric Folding Algorithms: Linkages, Origami, Polyhedra. Cambridge University Press, Cambridge (2007)
5. DeSplinter, K., Devadoss, S.L., Readyhough, J., Wimberly, B.: Nets of higher-dimensional cubes. In: 32nd Canadian Conference on Computational Geometry (2020)
6. Grünbaum, B.: Are your polyhedra the same as my polyhedra? In: Aronov, B., Basu, S., Pach, J., Sharir, M. (eds.) Discrete and Computational Geometry, vol. 25, pp. 461–488. Springer, Heidelberg (2003). https://doi.org/10.1007/978-3-642-55566-4_21
7. Hirose, K.: Hanseitamentai no tenkaizu no kasanari ni tsuite (On the overlap of Archimedean solids). Saitama Univ. graduation thesis. Supervisor: Takashi Horiyama (2015). (in Japanese)
8. Horiyama, T., Shoji, W.: Edge unfoldings of platonic solids never overlap. In: 23rd Canadian Conference on Computational Geometry (2011)

9. Horiyama, T., Shoji, W.: The number of different unfoldings of polyhedra. In: Cai, L., Cheng, S.-W., Lam, T.-W. (eds.) ISAAC 2013. LNCS, vol. 8283, pp. 623–633. Springer, Heidelberg (2013). https://doi.org/10.1007/978-3-642-45030-3_58
10. Mount, D.M.: On finding shortest paths on convex polyhedra. Technical report, Center for Automation Research, University of Maryland College Park (1985)
11. Namiki, M., Fukuda, K.: Unfolding 3-dimensional convex polytopes. A package for Mathematica 1.2 or 2.0. Mathematica Notebook (1993)
12. Schlickenrieder, W.: Nets of polyhedra. Ph.D. thesis, Technische Universität Berlin, Berlin (1997)
13. Sharir, M., Schorr, A.: On shortest paths in polyhedral spaces. SIAM J. Comput. 15(1), 193–215 (1986)

Flipping Plane Spanning Paths

Oswin Aichholzer[1] , Kristin Knorr[2] , Wolfgang Mulzer[2] ,
Johannes Obenaus[2(✉)] , Rosna Paul[1(✉)] , and Birgit Vogtenhuber[1]

[1] Institute of Software Technology, Graz University of Technology, Graz, Austria
{oaich,ropaul,bvogt}@ist.tugraz.at
[2] Institut für Informatik, Freie Universität Berlin, Berlin, Germany
{kristin.knorr,wolfgang.mulzer,johannes.obenaus}@fu-berlin.de

Abstract. Let S be a planar point set in general position, and let $\mathcal{P}(S)$ be the set of all plane straight-line paths with vertex set S. A flip on a path $P \in \mathcal{P}(S)$ is the operation of replacing an edge e of P with another edge f on S to obtain a new valid path from $\mathcal{P}(S)$. It is a long-standing open question whether for every given point set S, every path from $\mathcal{P}(S)$ can be transformed into any other path from $\mathcal{P}(S)$ by a sequence of flips. To achieve a better understanding of this question, we show that it is sufficient to prove the statement for plane spanning paths whose first edge is fixed. Furthermore, we provide positive answers for special classes of point sets, namely, for wheel sets and generalized double circles (which include, e.g., double chains and double circles).

Keywords: Flips · Plane spanning paths · Generalized double circles

1 Introduction

Reconfiguration is a classical and widely studied topic with various applications in multiple areas. A natural way to provide structure for a reconfiguration problem is by studying the so-called *flip graph*. For a class of objects, the flip graph has a vertex for each element and adjacencies are determined by a local flip operation (we will give the precise definition shortly). In this paper we are concerned with transforming plane spanning paths via edge flips.

Let S be a set of n points in the plane in general position (i.e., no three points are collinear), and let $\mathcal{P}(S)$ be the set of all plane straight-line spanning paths for S, i.e., the set of all paths with vertex set S whose straight-line embedding on S is crossing-free. A *flip* on a path $P \in \mathcal{P}(S)$ is the operation of removing

This work was initiated at the 2nd Austrian Computational Geometry Reunion Workshop in Strobl, June 2021. We thank all participants for fruitful discussions. J.O. is supported by ERC StG 757609. O.A. and R.P. are supported by FWF grant W1230. B.V. is supported by FWF Project I 3340-N35. K.K. is supported by the German Science Foundation (DFG) within the research training group 'Facets of Complexity' (GRK 2434). W.M. is partially supported by the German Research Foundation within the collaborative DACH project *Arrangements and Drawings* as DFG Project MU 3501/3-1, and by ERC StG 757609.

C.-C. Lin et al. (Eds.): WALCOM 2023, LNCS 13973, pp. 49–60, 2023.
https://doi.org/10.1007/978-3-031-27051-2_5

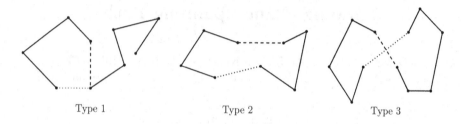

Type 1   Type 2   Type 3

Fig. 1. The three types of flips in plane spanning paths.

an edge e from P and replacing it by another edge f on S such that the graph $(P \setminus e) \cup f$ is again a path from $\mathcal{P}(S)$. Note that the edges e and f might cross. The *flip graph* on $\mathcal{P}(S)$ has vertex set $\mathcal{P}(S)$ and two vertices are adjacent if and only if the corresponding paths differ by a single flip. The following conjecture will be the focus of this paper:

Conjecture 1 (Akl et al. [3]). For every point set S in general position, the flip graph on $\mathcal{P}(S)$ is connected.

Related Work. For further details on reconfiguration problems in general we refer the reader to the surveys of Nishimura [10] and Bose and Hurtado [4]. Connectivity properties of flip graphs have been studied extensively in a huge variety of settings, see, e.g., [6–9,11] for results on triangulations, matchings and trees.

In our setting of plane spanning paths, flips are much more restricted, making it more difficult to prove a positive answer. Prior to our work only results for point sets in convex position and very small point sets were known. Akl et al. [3], who initiated the study of flip connectivity on plane spanning paths, showed connectedness of the flip graph on $\mathcal{P}(S)$ if S is in convex position or $|S| \leq 8$. In the convex setting, Chang and Wu [5] derived tight bounds concerning the diameter of the flip graph, namely, $2n - 5$ for $n = 3, 4$, and $2n - 6$ for $n \geq 5$.

For the remainder of this paper, we consider the flip graph on $\mathcal{P}(S)$ (or a subset of $\mathcal{P}(S)$). Moreover, unless stated otherwise, the word *path* always refers to a path from $\mathcal{P}(S)$ for an underlying point set S that is clear from the context.

Flips in Plane Spanning Paths. Let us have a closer look at the different types of possible flips for a path $P = v_1, \ldots, v_n \in \mathcal{P}(S)$ (see also Fig. 1). When removing an edge $v_{i-1}v_i$ from P with $2 \leq i \leq n$, there are three possible new edges that can be added in order to obtain a path (where, of course, not all three choices will necessarily lead to a plane path in $\mathcal{P}(S)$): v_1v_i, $v_{i-1}v_n$, and v_1v_n. A flip of *Type 1* is a valid flip that adds the edge v_1v_i (if $i > 2$) or the edge $v_{i-1}v_n$ (if $i < n$). It results in the path $v_{i-1}, \ldots, v_1, v_i, \ldots, v_n$, or the path $v_1, \ldots, v_{i-1}, v_n, \ldots, v_i$. That is, a Type 1 flip inverts a contiguous chunk from one of the two ends of the path. A flip of *Type 2* adds the edge v_1v_n and has the additional property that the edges $v_{i-1}v_i$ and v_1v_n do not cross. In this case, the path P together with the edge v_1v_n forms a plane cycle. If a Type 2 flip is

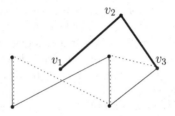

Fig. 2. Example where the flip graph is disconnected if the first three vertices of the paths are fixed. No edge of the solid path can be flipped, but there is at least one other path (dotted) with the same three starting vertices.

possible for one edge $v_{i-1}v_i$ of P, then it is possible for all edges of P. A Type 2 flip can be simulated by a sequence of Type 1 flips, e.g., flip v_1v_2 to v_1v_n, then flip v_2v_3 to v_1v_2, then v_3v_4 to v_2v_3, etc., until flipping $v_{i-1}v_i$ to $v_{i-2}v_{i-1}$. A flip of *Type 3* also adds the edge v_1v_n, but now the edges v_1v_n and $v_{i-1}v_i$ cross. Note that a Type 3 flip is only possible if the edge v_1v_n crosses exactly one edge of P, and then the flip is possible only for the edge $v_{i-1}v_i$ that is crossed.

Contribution. We approach Conjecture 1 from two directions. First, we show that it is sufficient to prove flip connectivity for paths with a fixed starting edge. Second, we verify Conjecture 1 for several classes of point sets, namely wheel sets and generalized double circles (which include, e.g., double chains and double circles).

Towards the first part, we define, for two distinct points $p, q \in S$, the following subsets of $\mathcal{P}(S)$: let $\mathcal{P}(S, p)$ be the set of all plane spanning paths for S that start at p, and let $\mathcal{P}(S, p, q)$ be the set of all plane spanning paths for S that start at p and continue with q. Then for any S, the flip graph on $\mathcal{P}(S, p, q)$ is a subgraph of the flip graph on $\mathcal{P}(S, p)$, which in turn is a subgraph of the flip graph on $\mathcal{P}(S)$. We conjecture that all these flip graphs are connected:

Conjecture 2. For every point set S in general position and every $p \in S$, the flip graph on $\mathcal{P}(S, p)$ is connected.

Conjecture 3. For every point set S in general position and every $p, q \in S$, the flip graph on $\mathcal{P}(S, p, q)$ is connected.

Towards Conjecture 1, we show that it suffices to prove Conjecture 3:

Theorem 1. *Conjecture 2 implies Conjecture 1.*

Theorem 2. *Conjecture 3 implies Conjecture 2.*

Note that the analogue of Conjecture 3 for paths where the first $k \geq 3$ vertices are fixed, does not hold: Fig. 2 shows a counterexample with 7 points and $k = 3$.

Towards the flip connectivity for special classes of point sets, we consider wheel sets and generalized double circles. A point set is in *wheel configuration* if it has exactly one point inside the convex hull. For generalized double circles we defer the precise definition to Sect. 4, however, intuitively speaking a generalized double circle is obtained by replacing each edge of the convex hull by a flat enough concave chain of arbitrary size (as depicted on the right). We show that the flip graph is connected in both cases:

Theorem 3. *(⋆) Let S be a set of n points in wheel configuration. Then the flip graph (on $\mathcal{P}(S)$) is connected with diameter at most $2n - 4$.*

Theorem 4. *(⋆) Let S be a set of n points in generalized double circle configuration. Then the flip graph (on $\mathcal{P}(S)$) is connected with diameter $O(n^2)$.*

Finally, we remark that using the order type database [1], we are able to computationally verify Conjecture 1 for every set of $n \leq 10$ points in general position (even when using only Type 1 flips).[1]

Notation. We denote the convex hull of a point set S by $\mathrm{CH}(S)$. All points $p \in S$ on the boundary of $\mathrm{CH}(S)$ are called *extreme points* and the remaining points are called *interior* points.

The proofs of results marked by a (⋆) are omitted or only sketched in this version. All full proofs can be found in the arxiv version [2].

2   A Sufficient Condition

In this section we prove Theorem 1 and Theorem 2.

Lemma 1. *(⋆) Let S be a point set in general position and $p, q \in S$. Then there exists a path $P \in \mathcal{P}(S)$ which has p and q as its end vertices.*

Theorem 1. *Conjecture 2 implies Conjecture 1.*

Proof. Let S be a point set and $P_s, P_t \in \mathcal{P}(S)$. If P_s and P_t have a common endpoint, we can directly apply Conjecture 2 and the statement follows. So assume that P_s has the endpoints v_a and v_b, and P_t has the endpoints v_c and v_d, which are all distinct. By Lemma 1 there exists a path P_m having the two endpoints v_a and v_c. By Conjecture 2 there is a flip sequence from P_s to P_m with the common endpoint v_a, and again by Conjecture 2 there is a further flip sequence from P_m to P_t with the common endpoint v_c. This concludes the proof. ☐

---

[1] The source code is available at https://github.com/jogo23/flipping_plane_spanning_paths.

Towards Theorem 2, we first have a closer look at what edges form *viable* starting edges. For a given point set S and points $p, q \in S$, we say that pq forms a *viable* starting edge if there exists a path $P \in \mathcal{P}(S)$ that starts with pq. For instance, an edge connecting two extreme points that are not consecutive along $\mathrm{CH}(S)$ is not a viable starting edge. The following lemma shows that these are the only non-viable starting edges.

Lemma 2. *(⋆) Let S be a point set in general position and $u, v \in S$. The edge uv is a viable starting edge if and only if one of the following is fulfilled: (i) u or v lie in the interior of $\mathrm{CH}(S)$, or (ii) u and v are consecutive along $\mathrm{CH}(S)$.*

The following lemma is the analogue of Lemma 1:

Lemma 3. *(⋆) Let S be a point set in general position and $v_1 \in S$. Further let $S' \subset S$ be the set of all points $p \in S$ such that $v_1 p$ forms a viable starting edge. Then for two points $q, r \in S'$ that are consecutive in the circular order around v_1, there exists a plane spanning cycle containing the edges $v_1 q$ and $v_1 r$.*

Theorem 2. *Conjecture 3 implies Conjecture 2.*

Proof. Let S be a point set and $v_1 \in S$. Further let $P, P' \in \mathcal{P}(S, v_1)$. If P and P' have the starting edge in common, then we directly apply Conjecture 3 and are done. So let us assume that the starting edge of P is $v_1 v_2$ and the starting edge of P' is $v_1 v_2'$. Clearly $v_2, v_2' \in S'$ holds. Sort the points in S' in radial order around v_1. Further let $v_x \in S'$ be the next vertex after v_2 in this radial order and C be the plane spanning cycle with edges $v_1 v_2$ and $v_1 v_x$, as guaranteed by Lemma 3.

By Conjecture 3, we can flip P to $C \setminus v_1 v_x$. Then, flipping $v_1 v_2$ to $v_1 v_x$ we get to the path $C \setminus v_1 v_2$, which now has $v_1 v_x$ as starting edge. We iteratively continue this process of "rotating" the starting edge until reaching $v_1 v_2'$.   □

Theorems 1 and 2 imply that it suffices to show connectedness of certain subgraphs of the flip graph. A priori it is not clear whether this is an easier or a more difficult task – on the one hand we have smaller graphs, making it easier to handle. On the other hand, we may be more restricted concerning which flips we can perform, or exclude certain "nice" paths.

3   Flip Connectivity for Wheel Sets

Akl et al. [3] proved connectedness of the flip graph if the underlying point set S is in convex position. They showed that every path in $\mathcal{P}(S)$ can be flipped to a *canonical path* that uses only edges on the convex hull of S. To generalize this approach to other classes of point sets, we need two ingredients: (i) a set of *canonical paths* that serve as the target of the flip operations and that have the property that any canonical path can be transformed into any other canonical path by a simple sequence of flips, usually of constant length; and (ii) a strategy to flip any given path to some canonical path.

Recall that a set S of $n \geq 4$ points in the plane is a *wheel set* if there is exactly one interior point $c_0 \in S$. We call c_0 the *center* of S and classify the edges on S as follows: an edge incident to the center c_0 is called a *radial* edge, and an edge along $CH(S)$ is called *spine* edge (the set of spine edges forms the *spine*, which is just the boundary of the convex hull here). All other edges are called *inner* edges. The *canonical paths* are those that consist only of spine edges and one or two radial edges.

We need one observation that will also be useful later. Let S be a point set and $P = v_1, \ldots, v_n \in \mathcal{P}(S)$. Further, let v_i $(i \geq 3)$ be a vertex such that no edge on S crosses $v_1 v_i$. We denote the face bounded by $v_1, \ldots, v_i, v_1$ by $\Phi(v_i)$.

Observation 5. *Let S be a point set, $P = v_1, \ldots, v_n \in \mathcal{P}(S)$, and v_i $(i \geq 3)$ be a vertex such that no edge on S crosses $v_1 v_i$. Then all vertices after v_i (i.e., $\{v_{i+1}, \ldots, v_n\}$) must entirely be contained in either the interior or the exterior of $\Phi(v_i)$.*

Theorem 3. *($\star$) Let S be a set of n points in wheel configuration. Then the flip graph (on $\mathcal{P}(S)$) is connected with diameter at most $2n - 4$.*

Proof (Sketch). Let $P = v_1, \ldots, v_n \in \mathcal{P}(S)$ be a non-canonical path and w.l.o.g., let $v_1 \neq c_0$. We show how to apply suitable flips to increase the number of spine edges of P. By Lemma 2, $v_1 v_2$ can only be radial or a spine edge. In the former case we can flip the necessarily radial edge $v_2 v_3$ to the spine edge $v_1 v_3$. In the latter case, let v_a with $a \neq 2$ be a neighbor of v_1 along the convex hull. Then, either $v_{a-1} v_a$ is not a spine edge and hence, we can flip it to $v_1 v_a$, or otherwise we show, using Observation 5, that P actually already is a canonical path.   □

4   Flip Connectivity for Generalized Double Circles

The proof for generalized double circles is in principle similar to the one for wheel sets but much more involved. For a point set S and two extreme points $p, q \in S$, we call a subset $CC(p, q) \subset S$ *concave chain* (chain for short) for S, if (i) $p, q \in CC(p, q)$; (ii) $CC(p, q)$ is in convex position; (iii) $CC(p, q)$ contains no other extreme points of S; and (iv) every line ℓ_{xy} through any two points $x, y \in CC(p, q)$ has the property that all points of $S \setminus CC(p, q)$ are contained in the open halfplane bounded by ℓ_{xy} that contains neither p nor q. Note that the extreme points p and q must necessarily be consecutive along $CH(S)$. If there is no danger of confusion, we also refer to the spanning path from p to q along the convex hull of $CC(p, q)$ as the concave chain.

A point set S is in *generalized double circle* position if there exists a family of concave chains such that every inner point of S is contained in exactly one chain and every extreme point of S is contained in exactly two chains. We denote the class of generalized double circles by GDC. For $S \in$ GDC, it is not hard to see that the union of the concave chains forms an uncrossed spanning cycle (cf. the full version [2]). Figure 3 gives an illustration of generalized double circles.

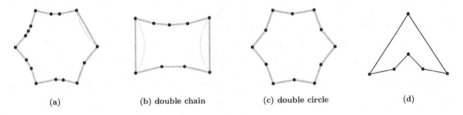

Fig. 3. (a–c) Examples of generalized double circles (the uncrossed spanning cycle is depicted in orange). (d) A point set that is *not* a generalized double circle, but admits an uncrossed spanning cycle. (Color figure online)

Before diving into the details of the proof of Theorem 4, we start by collecting preliminary results in a slightly more general setting, namely for point sets S fulfilling the following property:

(P1) there is an *uncrossed* spanning cycle C on S, i.e., no edge joining two points of S crosses any edge of C.

A point set fulfilling (P1) is called *spinal* point set. When considering a spinal point set S, we first fix an uncrossed spanning cycle C, which we call *spine* and all edges in C *spine edges*. For instance, generalized double circles are spinal point sets and the spine is precisely the uncrossed spanning cycle formed by the concave chains as described above. Whenever speaking of the spine or spine edges for some point set without further specification, the underlying uncrossed cycle is either clear from the context, or the statement holds for any choice of such a cycle. Furthermore, we call all edges in the exterior/interior of the spine *outer/inner edges*.

We define the *canonical paths* to be those that consist only of spine edges. Note that this definition also captures the canonical paths used by Akl et al. [3], and that any canonical path can be transformed into any other by a single flip (of Type 2). Two vertices incident to a common spine edge are called *neighbors*.

Valid Flips. We collect a few observations which will be useful to confirm the validity of a flip. Whenever we apply more than one flip, the notation in subsequent flips refers to the original path and not the current (usually we apply one or two flips in a certain step). Figure 4 gives an illustration of Observation 6.

Observation 6. *Let S be a spinal point set, $P = v_1, \ldots, v_n \in \mathcal{P}(S)$, and v_1, v_a $(a \neq 2)$ be neighbors. Then the following flips are valid (under the specified additional assumptions):*

(a) *flip $v_{a-1}v_a$ to v_1v_a*

(b) *flip v_av_{a+1} to $v_{a-1}v_{a+1}$*   *(if the triangle $\triangle v_{a-1}v_av_{a+1}$ is empty and (b) is performed subsequently after the flip in (a))*

(c) *flip v_av_{a+1} to v_1v_{a+1}*   *(if the triangle $\triangle v_1v_av_{a+1}$ is empty and $v_{a-1}v_a$ is a spine edge)*

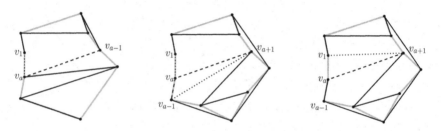

Fig. 4. *Left to right:* Illustration of the three flips in Observation 6. The spine is depicted in orange and edge flips are indicated by replacing dashed edges for dotted (in the middle, the two flips must of course be executed one after the other). (Color figure online)

Strictly speaking, in Observation 6(c) we do not require $v_{a-1}v_a$ to be a spine edge, but merely to be an edge not crossing v_1v_{a+1}. The following lemma provides structural properties for generalized double circles, if the triangles in Observation 6(b, c) are non-empty, i.e., contain points from S (see also Fig. 5 (left)):

Lemma 4. *(⋆) Let $S \in \mathsf{GDC}$ and $p, q, x \in S$ such that p and q are neighbors. Further, let the triangle $\triangle pqx$ be non-empty. Then the following holds:*

(i) At least one of the two points p, q is an extreme point (say p),
(ii) x does not lie on a common chain with p and q, but shares a common chain with either p or q (the latter may only happen if q is also an extreme point).

Combinatorial Distance Measure. In contrast to the proof for wheel sets, it may now not be possible anymore to directly increase the number of spine edges and hence, we need a more sophisticated measure. Let C be the spine of a spinal point set S and $p, q \in S$. Further let $o \in \{\mathrm{cw}, \mathrm{ccw}\}$ be an orientation. We define the *distance* between p, q *in direction* o, denoted by $d^o(p, q)$, as the number of spine edges along C that lie between p and q in direction o. Furthermore, we define the *distance* between p and q to be

$$d(p, q) = \min\{d^{\mathrm{cw}}(p, q), d^{\mathrm{ccw}}(p, q)\}.$$

Note that neighboring points along the spine have distance 1. Using this notion, we define the *weight* of an edge to be the distance between its endpoints and the (overall) weight of a path on S to be the sum of its edge weights.

Our goal is to perform weight-decreasing flips. To this end, we state two more preliminary results (see also Fig. 5 (middle) and (right)):

Observation 7. *Let S be a spinal point set, p, q, r be three neighboring points in this order (i.e., q lies between p and r), and $s \in S \setminus \{p, q, r\}$ be another point. Then $d(p, s) < d(q, s)$ or $d(r, s) < d(q, s)$ holds.*

Combining Observation 6 and Observation 7, it is apparent that we can perform weight-decreasing flips whenever $\triangle v_{a-1}v_av_{a+1}$ and $\triangle v_1v_av_{a+1}$ are empty.

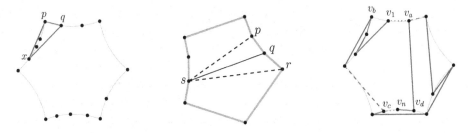

Fig. 5. *Left:* Illustration of Lemma 4. If p and q are neighbors, x has to lie on the depicted chain in order to obtain a non-empty triangle $\triangle pqx$. *Middle:* Illustration of Observation 7. One of the dashed edges has smaller weight than the solid: $d(s, q) = 4$; $d(s, p) = 4$; $d(s, r) = 3$. *Right:* Illustration of Lemma 5. The initial path is depicted by solid and dashed edges. Flipping the dashed edges to the dotted edges increases the number of spine edges. (Color figure online)

Lemma 5. *($\star$) Let S be a spinal point set, $P = v_1, \ldots, v_n \in \mathcal{P}(S)$, and v_a, v_b ($a, b \neq 2$) be neighbors of v_1 as well as v_c, v_d ($c, d \neq n - 1$) be neighbors of v_n. If $\max(a, b) > \min(c, d)$, then the number of spine edges in P can be increased by performing at most two flips, which also decrease the overall weight of P.*

Note that v_b or v_d in Lemma 5 may not exist, if the first or last edge of P is a spine edge. Lemma 5 essentially enables us to perform weight decreasing flips whenever the path traverses a neighbor of v_n before it reached both neighbors of v_1. We are now ready to prove Theorem 4, but briefly summarize the proof strategy from a high-level perspective beforehand:

High Level Proof Strategy. To flip an arbitrary path $P \in \mathcal{P}(S)$ to a canonical path, we perform iterations of suitable flips such that in each iteration we either

(i) increase the number of spine edges along P, while not increasing the overall weight of P, or
(ii) decrease the overall weight of P, while not decreasing the number of spine edges along P.

Note that for the connectivity of the flip graph it is not necessary to guarantee the non increasing overall weight in the first part. However, this will provide us with a better bound on the diameter of the flip graph.

Theorem 4. *($\star$) Let S be a set of n points in generalized double circle configuration. Then the flip graph (on $\mathcal{P}(S)$) is connected with diameter $O(n^2)$.*

Proof (Sketch). Let $P = v_1, \ldots, v_n \in \mathcal{P}(S)$ be a non-canonical path. We show how to iteratively transform P to a canonical path by increasing the number of spine edges or decreasing its overall weight. Let v_a ($a \neq 2$) be a neighbor of v_1.

We can assume, w.l.o.g, that v_1 and v_n are not neighbors (i.e., $a < n$), since otherwise we can flip an arbitrary (non-spine) edge of P to the spine edge $v_1 v_n$ (performing a Type 2 flip). Furthermore, we can also assume w.l.o.g., that

Fig. 6. Illustration of Case 1. If $v_1 v_2$ is not a spine edge and $\triangle v_1 v_a v_{a+1}$ is empty, we make progress by flipping the dashed edges to the dotted. (Color figure online)

$v_{a-1} v_a$ is a spine edge, since otherwise we can flip $v_{a-1} v_a$ to the spine edge $v_1 v_a$ (Observation 6(a)). This also implies that the edge $v_a v_{a+1}$, which exists because $a < n$, is not a spine edge, since v_a already has the two neighbors v_{a-1} and v_1.

We distinguish two cases – $v_1 v_2$ being a spine edge or not:

Case 1: $v_1 v_2$ is not a spine edge.

This case is easier to handle, since we are guaranteed that both neighbors of v_1 are potential candidates to flip to. In order to apply Observation 6, we require $\triangle v_1 v_a v_{a+1}$ to be empty. If that is the case we apply the following flips (see also Fig. 6):

$$\text{flip } v_a v_{a+1} \text{ to } v_1 v_{a+1} \qquad \text{and} \qquad \text{flip } v_1 v_2 \text{ to } v_1 v_a,$$

where the first flip results in the path $v_a, \dots, v_1, v_{a+1}, \dots, v_n$ (and is valid by Observation 6(c)) and the second flip results in the path $v_2, \dots, v_a, v_1, v_{a+1}, \dots, v_n$ (valid due to Observation 6(a)). Together, the number of spine edges increases, while the overall weight does not increase.

If $\triangle v_1 v_a v_{a+1}$ is not empty we need to be more careful, using Lemma 4 (details can be found in the full version [2]).

Case 2: $v_1 v_2$ is a spine edge.

In this case we will consider P from both ends v_1 and v_n. Our general strategy here is to first rule out some easier cases and collect all those cases where we cannot immediately make progress. For these remaining "bad" cases we consider the setting from both ends of the path.

Again, we skip the analysis of the easier cases and just summarize the six "bad" cases. These "bad" cases always involve v_1, v_a, or v_{a-1} being an extreme point. Instead of spelling all these cases out, we give an illustration in Fig. 7.

In the remainder of the proof we settle these "bad" cases by arguing about both ends of the path, i.e., we consider all $\binom{6}{2} + 6 = 21$ combinations of "bad" cases.

We exclude several combinations as follows. By Lemma 5, we can assume that $a < c$ holds (otherwise there are weight decreasing flips) and hence, no "bad" case where v_{a+1} is in the interior of $\Phi(v_a)$ can be combined with a "bad"

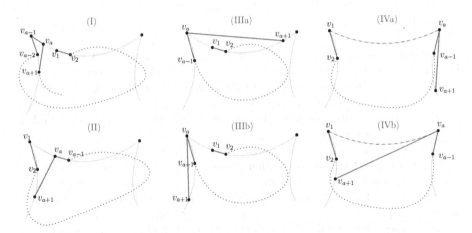

Fig. 7. The six "bad" cases. The solid edges depict the fixed edges of the corresponding "bad" case and the red arcs (here and in the following) indicate that there is no vertex other than the two extreme points lying on this chain. (Color figure online)

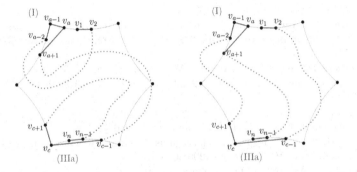

Fig. 8. (I) and (IIIa) cannot be combined in a plane manner (left), except if the path traverses a neighbor of v_n before those of v_1, i.e., $c < a$ holds (right). (Color figure online)

case having v_n or v_c as extreme point (Observation 5). This excludes (almost) all combinations involving (I), (II), or (IVb); see Fig. 8 for an example.

For the remaining cases, we try to decrease the weight of P by flipping $v_a v_{a+1}$ either to $v_1 v_{a+1}$ or $v_{a-1} v_{a+1}$ (see Observation 7). If these flips are valid they are either weight-decreasing or we can identify disjoint regions that must each contain at least $n/2$ vertices, which will result in a contradiction. Again, we skip the details of this analysis.

Iteratively applying the above process transforms P to a canonical path and the $O(n^2)$ bound for the required number of flips also follows straightforwardly. □

5   Conclusion

In this paper, we made progress towards a positive answer of Conjecture 1, though it still remains open in general. We approached Conjecture 1 from two directions and believe that Conjecture 3 might be easier to tackle, e.g. for an inductive approach. For all our results we used only Type 1 and Type 2 flips (which can be simulated by Type 1 flips). It is an intriguing question whether Type 3 flips are necessary at all.

Concerning the approach of special classes of point sets, of course one can try to further adapt the ideas to other classes. Most of our results hold for the setting of spinal point sets; the main obstacle that remains in order to show flip connectivity for the point sets satisfying condition (P1) would be to adapt Lemma 4. A proof for general point sets, however, seems elusive at the moment.

Lastly, there are several other directions for further research conceivable, e.g. non-straight-line drawings.

References

1. Aichholzer, O., Aurenhammer, F., Krasser, H.: Enumerating order types for small point sets with applications. Order **19**, 265–281 (2002). https://doi.org/10.1023/A:1021231927255
2. Aichholzer, O., Knorr, K., Mulzer, W., Obenaus, J., Paul, R., Vogtenhuber, B.: Flipping plane spanning paths (2022). https://doi.org/10.48550/ARXIV.2202.10831
3. Akl, S.G., Islam, M.K., Meijer, H.: On planar path transformation. Inf. Process. Lett. **104**(2), 59–64 (2007). https://doi.org/10.1016/j.ipl.2007.05.009
4. Bose, P., Hurtado, F.: Flips in planar graphs. Comput. Geom. **42**(1), 60–80 (2009). https://doi.org/10.1016/j.comgeo.2008.04.001
5. Chang, J.M., Wu, R.Y.: On the diameter of geometric path graphs of points in convex position. Inf. Process. Lett. **109**(8), 409–413 (2009). https://doi.org/10.1016/j.ipl.2008.12.017
6. Hernando, C., Hurtado, F., Noy, M.: Graphs of non-crossing perfect matchings. Graphs Comb. **18**(3), 517–532 (2002)
7. Houle, M., Hurtado, F., Noy, M., Rivera-Campo, E.: Graphs of triangulations and perfect matchings. Graphs Comb. **21**, 325–331 (2005). https://doi.org/10.1007/s00373-005-0615-2
8. Lawson, C.L.: Transforming triangulations. Discret. Math. **3**(4), 365–372 (1972)
9. Nichols, T.L., Pilz, A., Tóth, C.D., Zehmakan, A.N.: Transition operations over plane trees. Discret. Math. **343**(8), 111929 (2020). https://doi.org/10.1016/j.disc.2020.111929
10. Nishimura, N.: Introduction to reconfiguration. Algorithms **11**(4) (2018). https://doi.org/10.3390/a11040052
11. Wagner, K.: Bemerkungen zum Vierfarbenproblem. Jahresbericht der Deutschen Mathematiker-Vereinigung **46**, 26–32 (1936). http://eudml.org/doc/146109

Away from Each Other

Tetsuya Araki and Shin-ichi Nakano[✉]

Gunma University, Kiryu 376-8515, Japan
nakano@gunma-u.ac.jp

Abstract. We consider the following three problems where k is a constant. For those problems there are cases where k is typically a small constant.

Given a polygon with n edges on a plane we want to find k points in the polygon so that the minimum pairwise Euclidean distance of the k points is maximized. Intuitively, for an island, we want to locate k drone bases far away from each other in flying distance to avoid congestion in the sky. In this paper we give an $O(((1/\epsilon)^2 + n/\epsilon)^k)$ time $1/(1 + \epsilon)$ approximation algorithm to solve the problem, where $\epsilon < 1$ is a positive number. This is the first PTAS for the problem.

Given a set of n straight line segments on a plane we want to find k points on the straight line segments so that the minimum pairwise Euclidean distance of the k points is maximized. Intuitively, for some road network, we want to locate k drone bases far away from each other to avoid congestion in the sky and also each base face a road. In this paper we design the first PTAS for the problem.

Given a polygon with n edges on a plane we want to find k points in the polygon so that the minimum length of paths inside the polygon connecting two points among the k points is maximized. Intuitively, for an island, we want to locate k coffee shops far away from each other to avoid self competition for walking customers. In this paper we design the first PTAS for the problem.

1 Introduction

The facility location problem and many of its variants have been studied [15,16]. Typically, given a set of points on which facilities can be placed and an integer k, we want to place k facilities on some points so that a designated objective function is minimized. By contrast in the *dispersion problem*, we want to place facilities so that a designated objective function is maximized.

Our Results. In this paper we consider three dispersion problems on a plane. Fix a constant integer k.

Given a polygon P with n edges on a plane, we want to find k points in P so that the minimum pairwise Euclidean distance of the k points is maximized. See an example in Fig. 1. We call the problem *the k-dispersion problem in a polygon*. Note that the k points may contain a point in a polygon which is not on the

C.-C. Lin et al. (Eds.): WALCOM 2023, LNCS 13973, pp. 61–70, 2023.
https://doi.org/10.1007/978-3-031-27051-2_6

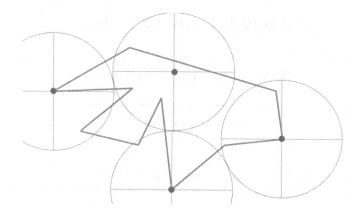

Fig. 1. A solution of a k-dispersion problem in a polygon with $k = 4$.

boundary, for instance if we choose 5 points in a square the 5 points consist of the four corner points and the center of the square.

Intuitively, for an island, we want to locate k drone bases far away from each other to avoid congestion in the sky.

In this paper we give an $O(((1/\epsilon)^2 + n/\epsilon)^k)$ time $1/(1 + \epsilon)$ approximation algorithm to solve the problem, where $\epsilon < 1$ is a positive number. Thus the problem has a PTAS.

The algorithm first computes a set of (grid) points in P and computes an optimal k points among the set of points by a brute force algorithm. By choosing the gap size of the grid suitably we ensure the approximation ratio.

If k is a part of the inputs, not a constant, then a similar problem has no PTAS [6].

We consider the following two more problems.

Given a set of n (connected) straight line segments on a plane we want to find k points on the straight line segments so that the minimum pairwise Euclidean distance of the k points is maximized. See an example in Fig. 2. We call the problem *the k-dispersion problem on straight line segments*. In this paper we design a PTAS for the problem.

Given a polygon with n edges on a plane we want to find k points in the polygon so that the minimum length of paths inside the polygon (where a path is a sequence of straight line segments in the polygon) connecting two points among the k points is maximized. See an example in Fig. 3. We call the problem *the k-dispersion problem in a polygon with the geodesic distance*. In this paper we design the first PTAS for the problem.

Related Results. Given a set C of n points, a distance function d and an integer k with $k < n$, the *max-min k-dispersion problem* computes a subset $S \subset C$ with $|S| = k$ such that the cost $cost(S) = \min_{\{u,v\} \subset S}\{d(u,v)\}$ is maximized. Several results are known for the max-min k-dispersion problem. The problem is NP-hard [18]. If d is a metric (the distance satisfies the triangle inequality)

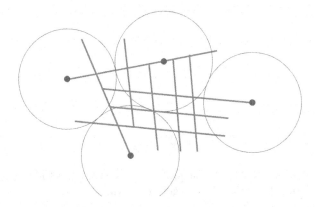

Fig. 2. A solution of a k–dispersion problem in a set of straight line segments with $k = 4$.

a polynomial-time approximation algorithm with approximation ratio two is known [25], and it is NP-hard to compute a solution with approximation ratio less than two [25]. An exponential time exact algorithm is known [3]. If P is a set of n points on a line one can solve the problem in $O(kn)$ time by dynamic programming [27], in $O(n \log \log n)$ time by sorted matrix search method [2], and in $O((2k^2)^k n)$ time by the pigeonhole principle [4]. For the max-sum version several results are also known [8,10–12,19,23,25]. For a variety of related problems, see [6,12]. See more applications, including *result diversification*, in [11,25,26].

Given a set of n disjoint intervals on a line the *max-min dispersion problem on intervals* computes one point from each interval so that the minimum pairwise distance of the n points is maximized. If the disjoint intervals are given in the sorted order on the line, two $O(n)$ time algorithms to solve the problem are known [7,24]. Given a set of n intervals on a line and a constant integer k with $k < n$, even if the disjoint intervals are given in any (unsorted) order, one can compute k points from the intervals in $O((2k^2)^k n)$ time so that the minimum pairwise distance of the k points is maximized [5].

Given a set of disjoint disks with arbitrary radii, the *dispersion problem on disks* is the problem to compute one point in each disk so that the minimum distance among the points is maximized. The problem is NP-hard, and some approximation algorithms are known [9,17,20], also an $O((n/\epsilon^2)^k)$ time $1/(1+\epsilon)$ approximation algorithm is known [5].

The dispersion problem in a polygon is similar to the following packing problem. Given an integer k and a disk, the circle packing in a circle problem computes the maximum radius of k identical disks which can be packed without overlapping into the given disk. Given an integer k, a disk with radius r and a number $dist$, if we have an algorithm to decide whether one can locate k points in the disk so that the minimum distance among them is $dist$ or more, then, by using the algorithm, we can decide whether one can pack k identical disks with

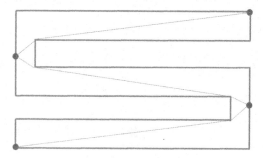

Fig. 3. A solution of a k-dispersion problem in a polygon with $k = 4$ where the minimum length of paths inside the polygon connecting two points among the k points is maximized.

radii $dist/2$ or more into a disk with radius $r + dist/2$. Only the following result is known for the time complexity of the circle packing in a circle problem. It is in EXPTIME, but whether it is in PSPACE or not is open [1]. For similar problems the following are known. It is NP-hard to decide whether a given set of (possibly not identical) circles can be packed into a given square [14, Corollary 7.2]. It is NP-hard to decide whether a given set of (possibly not identical) circles can be packed into a given circle [21, Corollary 6.2].

The remainder of this paper is organized as follows. In Sect. 2 we give an $O(((1/\epsilon)^2 + n/\epsilon)^k)$ time $1/(1 + \epsilon)$ approximation algorithm for the k-dispersion problem in a polygon with n edges. In Sect. 3 we design algorithms to solve two more similar problems. Finally Sect. 4 is a conclusion.

2 k-Dispersion in a Polygon

Fix a constant integer k. Given a polygon P with n edges on a plane we want to find k points in P so that the minimum pairwise Euclidean distance of the k points is maximized. Let $\epsilon < 1$ be a positive number. In this section we give an $O(((1/\epsilon)^2 + n/\epsilon)^k)$ time $1/(1 + \epsilon)$ approximation algorithm to solve the problem.

We need some notations. For a set S of points let $cost(S)$ be the minimum pairwise Euclidean distance among the points in S. Let P^* be a set of k points in P with the maximum $cost(P^*)$.

Let W be the difference between the x-coordinates of a leftmost point and a rightmost point in P. Similarly let H be the difference between the y-coordinates of a topmost point and a bottommost point in P. Without loss of generality we can assume that $W \geq H$.

We have the following lemma.

Lemma 1. $cost(P^*) > W/k$

Proof. Fix a directed path starting at a leftmost point p_ℓ and ending at a rightmost point p_r in P. Let $p_0 = p_\ell, p_1, \cdots, p_{k-1} = p_r$ be the points on the directed

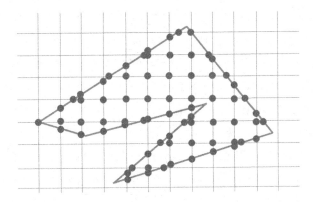

Fig. 4. An example of the set G of points in P.

path such that p_i is the first point having its x-coordinate $x(p_\ell) + Wi/(k-1)$ for $i = 0, 1, \cdots, k-1$. Now the distance between any two points of the k points is at least $W/(k-1)$, so more than W/k. □

By a standard plane sweep algorithm (see page 1022 of textbook [13]) we can sweep P by a horizontal sweep line from top to bottom in $O(n \log n)$ time, and during the sweep we can maintain the edges of P intersecting the current horizontal sweep line with the left-to-right order of the intersection points with the current horizontal sweep line. During the sweep we construct a set G of points in P, as follows. We consider a grid of gap size $W\epsilon/ck$ on P, where a constant c is explained later, and stop the horizontal sweep line at each horizontal line on the grid, that is, when the y-coordinate of the horizontal sweep line is $W\epsilon i/ck$ for each integer i, and append to G the points on the current horizontal sweep line which are (1) the intersection points with the vertical grid lines (the number of such points is at most $(1 + ck/\epsilon)^2$ in total), and (2) the intersection points with edges of P (the number of such points is at most $(1 + ck/\epsilon)n$ in total). Similarly we sweep P by a vertical sweep line from left to right, and append to G the points on the current vertical sweep line which are (3) the intersection points with edges of P. Now $|G| \leq (1 + ck/\epsilon)^2 + 2(1 + ck/\epsilon)n$ holds. An example of G is shown in Fig. 4.

Let $G(P^*)$ be the set of points derived from P^* by choosing a nearest point in G for each point in P^*. We choose c large enough so that (c1) $cost(G(P^*)) > cost(P^*)/(1 + \epsilon)$ holds and (c2) no two points in P^* have the common nearest point in G. (If two points in P^* have the common nearest point in G then $cost(G(P^*)) = 0$. We wish to prohibit this case.)

If we set c as $c \geq 2\sqrt{2}(1 + \epsilon)$ then the following holds by Lemma 1.

$$\frac{1}{c} \leq \frac{1}{2\sqrt{2}(1+\epsilon)}$$

$$\frac{2\sqrt{2}}{c}(1+\epsilon) \leq 1$$

$$\frac{2\sqrt{2}\,W(1+\epsilon)}{c}\frac{}{k} \leq \frac{W}{k} < cost(P^*)$$

$$\frac{2\sqrt{2}W\epsilon}{ck} < cost(P^*)\frac{\epsilon}{1+\epsilon}$$

$$= cost(P^*)\frac{(1+\epsilon)-1}{1+\epsilon}$$

$$= cost(P^*) - \frac{1}{1+\epsilon}cost(P^*)$$

$$\frac{1}{1+\epsilon}cost(P^*) < cost(P^*) - \frac{2\sqrt{2}W\epsilon}{ck} \leq cost(G(P^*))$$

Thus (c1) holds. Note that $cost(P^*) - \frac{2\sqrt{2}W\epsilon}{ck} \leq cost(G(P^*))$ holds, since the distance between a point in P^* and its nearest point in G is at most $\sqrt{2}\frac{W\epsilon}{ck}$.

If we set c as $c \geq (1+2\sqrt{2})\epsilon$ then the following holds.

$$\frac{1}{(1+2\sqrt{2})\epsilon} \geq \frac{1}{c}$$

$$1 \geq \frac{(1+2\sqrt{2})\epsilon}{c}$$

$$\frac{W}{k} \geq \frac{(1+2\sqrt{2})W\epsilon}{ck}$$

$$\frac{W}{k} - \frac{2\sqrt{2}W\epsilon}{ck} \geq \frac{W\epsilon}{ck}$$

Now the following holds by Lemma 1.

$$cost(G(P^*)) \geq cost(P^*) - \frac{2\sqrt{2}W\epsilon}{ck} > \frac{W}{k} - \frac{2\sqrt{2}W\epsilon}{ck} \geq \frac{W\epsilon}{ck}$$

Note that again $cost(P^*) - \frac{2\sqrt{2}W\epsilon}{ck} \leq cost(G(P^*))$ holds.

Thus, for any two points p and q in P^*, let p' and q' be the nearest points in G, respectively, then the distance between p' and q' is at least $W\epsilon/ck$, so (c2) holds.

We set c large enough to satisfy the above two conditions.

Algorithm. Let G_A be the set G' of k points in G maximizing $cost(G')$. If we find a set G_A in $O(|G|^k)$ time by a brute force algorithm we have $cost(G_A) \geq cost(G(P^*)) \geq cost(P^*)/(1+\epsilon)$.

Now we have the following theorem. Note that k is a constant.

Theorem 1. *One can compute a set G_A of k points in a given polygon P with n edges with $cost(G_A) \geq cost(P^*)/(1 + \epsilon)$ in $O((1/\epsilon^2 + n/\epsilon)^k)$ time, where P^* is a set of k points in P maximizing $cost(P^*)$.*

3   More Problems

In this section we design two algorithms to solve two more problems.

Given a set L of n straight line segments on a plane, we want to find k points in L so that the minimum pairwise Euclidean distance of the k points is maximized. We assume that L are connected, that means, for any pair of points in L, there is a path in L from a point to the other point. We set W' as the difference between the x-coordinate of a leftmost end point and the x-coordinate of a rightmost end point in L. Similarly let H' be the difference between the y-coordinate of a topmost end point and the y-coordinate of a bottommost end point in L. Without loss of generality we can assume that $W' \geq H'$.

As before we define that for a set S of points let $cost(S)$ be the minimum pairwise Euclidean distance among the points in S. Let P^* be a set of k points in L with the maximum $cost(P^*)$.

We have the following lemma.

Lemma 2. $cost(P^*) > W'/k$

Proof. Similar to Lemma 1.

Now we explain the algorithm.

By a standard plane sweep algorithm we sweep L by a horizontal sweep line from top to bottom in $O(n \log n)$ time. During the sweep we construct a set G of points in L, as follows. We consider a grid of gap size $W'\epsilon/ck$ on L, where c is a constant, and stop the horizontal sweep line at each horizontal line on the grid, and append to G the points on the current horizontal sweep line which are (2') the intersection points with straight line segments in L (the number of such points is at most $(1 + ck/\epsilon)n$ in total). Similarly we sweep L by a vertical sweep line from left to right, and append to G the points on the current vertical sweep line which are (3') the intersection points with straight line segments in L. Now $|G| \leq 2(1 + ck/\epsilon)n$ holds. Then again by a brute force algorithm compute G_A which is the set G' of k points in G maximizing $cost(G')$.

Now we have the following theorem.

Theorem 2. *Given a set L of n straight line segments on a plane, one can compute a set P_A of k points on L with $cost(P_A) \geq cost(P^*)/(1+\epsilon)$ in $O((n/\epsilon)^k)$ time, where P^* is a set of k points on L maximizing $cost(P^*)$.*

Proof. Similar to Theorem 1.

By replacing the Euclidean distance by the length of a shortest path inside a polygon, we can define the following problem.

Given a polygon P with n edges on a plane, we want to find k points in P so that the minimum length of the shortest paths inside P (where a path is a sequence of straight line segments in the polygon) connecting two points among the k points is maximized.

By preprocessing the polygon in $O(n \log n)$ time, one can compute, for any pair of query points in P, the length of the shortest path inside P in $O(\log n)$ time [22]. Thus given a set of k points in P we can compute in $O(k^2 \log n)$ time the minimum length of the shortest paths inside P connecting two points among the k points.

For a set S of points in P let $cost'(S)$ be the minimum length of the shortest paths inside P connecting two points among S. Let P^* be a set of k points in P with the maximum $cost'(P^*)$.

We can solve the problem as follows. First by a standard plane sweep algorithm we sweep P by a horizontal sweep line from top to bottom in $O(n \log n)$ time and by a vertical sweep line from left to right in $O(n \log n)$ time, and construct a set of points G in P as the algorithm in Sect. 2. For this problem we additionally append to G the end points of each segment of P. So $|G|$ increased by n. Now $|G| \le (1 + ck/\epsilon)^2 + (3 + 2ck/\epsilon)n$ holds. We need to append these points to G to ensure $cost'(P^*) - \frac{2\sqrt{2}W\epsilon}{ck} \le cost'(G(P^*))$ for this problem. Then by a brute force algorithm compute G_A which is the set G' of k points in G maximizing $cost'(G')$. We can find G_A in $O(|G|^k)$ time.

We again choose c large enough so that (c1") $cost'(G(P^*)) > cost'(P^*)/(1+\epsilon)$ holds and (c2") no two points in P^* have the common nearest point in G.

We have the following theorem.

Theorem 3. *One can compute a set G_A of k points in a given polygon P with n edges with $cost'(G_A) \ge cost'(P^*)/(1 + \epsilon)$ in $O((1/\epsilon^2 + n/\epsilon)^k k^2 \log n)$ time, where $d'(u, v)$ is the length of the shortest path inside P connecting two points u and v, and $cost'(S) = \min_{\{u,v\} \subset S}\{d'(u, v)\}$ and P^* is a set of k points in P maximizing $cost'(P^*)$.*

Proof. Similar to Theorem 1.

4   Conclusion

In this paper we have designed an algorithm to solve the k-dispersion problem in a polygon. For a fixed constant integer k, given a polygon with n edges, our algorithm computes a set G_A of k points in the polygon with $cost(G_A) \ge cost(P^*)/(1 + \epsilon)$ in $O(((1/\epsilon)^2 + n/\epsilon)^k)$ time, where P^* is an optimal solution. Thus the problem has a PTAS.

Then we have defined two natural dispersion problems. For a constant integer k we can design a PTAS to compute k points on given straight line segments on a plane so that the minimum pairwise Euclidean distance of the k points is maximized, and a PTAS to compute k points in a given polygon so that the minimum length of the shortest paths inside P connecting two points among the k points is maximized.

Each algorithm is simple but the first PTAS to solve a natural problem. We hope further improvements will continue.

References

1. Alt, H.: Computational aspects of packing problems. Bull. EATCS (118) (2016)
2. Akagi, T., Nakano, S.: Dispersion on the line, IPSJ SIG Technical Reports, 2016-AL-158-3 (2016)
3. Akagi, T., et al.: Exact algorithms for the max-min dispersion problem. In: Chen, J., Lu, P. (eds.) FAW 2018. LNCS, vol. 10823, pp. 263–272. Springer, Cham (2018). https://doi.org/10.1007/978-3-319-78455-7_20
4. Araki, T., Nakano, S.: Max–min dispersion on a line. J. Comb. Optim. **44**, 1824–1830 (2020). https://doi.org/10.1007/s10878-020-00549-5
5. Araki, T., Miyata, H., Nakano, S.: Dispersion on intervals. In: Proceedings of CCCG2021 (2021)
6. Baur, C., Fekete, S.P.: Approximation of geometric dispersion problems. In: Jansen, K., Rolim, J. (eds.) APPROX 1998. LNCS, vol. 1444, pp. 63–75. Springer, Heidelberg (1998). https://doi.org/10.1007/BFb0053964 Also in Algorithmica 30, 451–470 (2001)
7. Biedl, T., Lubiw, A., Naredla, A.M., Ralbovsky, P.D., Stroud, G.: Dispersion for intervals: a geometric approach. In: Proceedings of SOSA 2021 (2021)
8. Birnbaum, B., Goldman, K.J.: An improved analysis for a greedy remote-clique algorithm using factor-revealing LPs. Algorithmica **50**, 42–59 (2009)
9. Cabello, S.: Approximation algorithms for spreading points. J. Algorithms **62**, 49–73 (2007)
10. Cevallos, A., Eisenbrand, F., Zenklusen, R.: Max-sum diversity via convex programming. In: Proceedings of SoCG 2016, pp. 26:1–26:14 (2016)
11. Cevallos, A., Eisenbrand, F., Zenklusen, R.: Local search for max-sum diversification. In: Proceedings of SODA 2017, pp. 130–142 (2017)
12. Chandra, B., Halldorsson, M.M.: Approximation algorithms for dispersion problems. J. Algorithms **38**, 438–465 (2001)
13. Cormen, T.H., Leiserson, C.E., Rivest, R.L., Stein, C.: Introduction to Algorithms. MIT Press, Cambridge (2009)
14. Demaine, E.D., Fekete, S.P., Lang, R.J.: Circle packing for origami design is hard. In: Proceedings of the 5th International Conference on Origami in Science, Mathematics and Education (OSME 2010), pp. 609–626 (2010). Also, CoRR abs/1008.1224
15. Drezner, Z.: Facility Location: A Survey of Applications and Methods. Springer, Heidelberg (1995)
16. Drezner, Z., Hamacher, H.W.: Facility Location: Applications and Theory. Springer, Heidelberg (2004)
17. Dumitrescu, A., Jiang, M.: Dispersion in disks. Theory Comput. Syst. **51**, 125–142 (2012)
18. Erkut, E.: The discrete p-dispersion problem. Eur. J. Oper. Res. **46**, 48–60 (1990)
19. Fekete, S.P., Meijer, H.: Maximum dispersion and geometric maximum weight cliques. Algorithmica **38**, 501–511 (2004)
20. Fiala, J., Kratochvil, J., Proskurowski, A.: Systems of distant representatives. Discret. Appl. Math. **145**, 306–36 (2005)

21. Fekete, S.P., Keldenich, P., Scheffer, C.: Packing disks into disks with optimal worst-case density. In: Proceedings 35th International Symposium on Computational Geometry, SoCG 2019. LIPIcs, vol. 129, pp. 35:1–35:19 (2019)
22. Guibas, L.J., Hershberger, J.: Optimal shortest path queries in a simple polygon. J. Comput. Syst. Sci. **39**, 126–152 (1989)
23. Hassin, R., Rubinstein, S., Tamir, A.: Approximation algorithms for maximum dispersion. Oper. Res. Lett. **21**, 133–137 (1997)
24. Li, S., Wang, H.: Dispersing points on intervals. In: Proceedings of ISAAC 2016, Article no. 52 (2016). https://doi.org/10.4230/LIPIcs.ISAAC.2016.52
25. Ravi, S.S., Rosenkrantz, D.J., Tayi, G.K.: Heuristic and special case algorithms for dispersion problems. Oper. Res. **42**, 299–310 (1994)
26. Sydow, M.: Approximation guarantees for max sum and max min facility dispersion with parameterised triangle inequality and applications in result diversification. Mathematica Applicanda **42**, 241–257 (2014)
27. Wang, D.W., Kuo, Y.-S.: A study on two geometric location problems. Inf. Process. Lett. **28**, 281–286 (1988)

Piercing Diametral Disks Induced by Edges of Maximum Spanning Trees

A. Karim Abu-Affash[1(⊠)], Paz Carmi[2], and Meytal Maman[2]

[1] Department of Software Engineering, Shamoon College of Engineering,
Be'er Sheva, Israel
abuaa1@sce.ac.il
[2] Department of Computer Science, Ben-Gurion University, Be'er Sheva, Israel
carmip@cs.bgu.ac.il

Abstract. Let P be a set of points in the plane and let T be a maximum-weight spanning tree of P. For an edge (p, q), let D_{pq} be the diametral disk induced by (p, q), i.e., the disk having the segment $\overline{pq}$ as its diameter. Let $\mathcal{D}_T$ be the set of the diametral disks induced by the edges of T. In this paper, we show that one point is sufficient to pierce all the disks in $\mathcal{D}_T$, thus, the set $\mathcal{D}_T$ is Helly. Actually, we show that the center of the smallest enclosing circle of P is contained in all the disks of $\mathcal{D}_T$, and thus the piercing point can be computed in linear time.

Keywords: Maximum spanning tree · Piercing set · Helly's theorem · Fingerhut's conjecture

1 Introduction

Let P be a set of points in the plane and let $G = (P, E)$ be the complete graph over P. A *maximum-weight spanning tree* T of P is a spanning tree of G with maximum edge weight, where the weight of an edge $(p, q) \in E$ is the Euclidean distance between p and q, and denoted by $|pq|$. For an edge (p, q), let D_{pq} denote the *diametral* disk induced by (p, q), i.e., the disk having the segment $\overline{pq}$ as its diameter. Let $\mathcal{D}_T$ be the set of the diametral disks obtained by the edges of T, i.e., $\mathcal{D}_T = \{D_{pq} : (p, q) \in E_T\}$, where E_T is the set of the edges of T. In this paper, we prove that the disks in $\mathcal{D}_T$ have a non-empty intersection.

1.1 Related Works

Let $\mathcal{F}$ be a set of geometric objects in the plane. A set S of points in the plane *pierces* $\mathcal{F}$ if every object in $\mathcal{F}$ contains a point of S, in this case, we say that S is a *piercing set* of $\mathcal{F}$. The piercing problem, i.e., finding a minimum cardinality set S that pierces a set of geometric objects, has attracted researchers for the past century.

This work was partially supported by Grant 2016116 from the United States – Israel Binational Science Foundation.

A famous result is Helly's theorem [7,8], which states that for a set $\mathcal{F}$ of convex objects in the plane, if every three objects have a non-empty intersection, then there is one point that pierces all objects in $\mathcal{F}$. The problem of piercing pairwise intersecting objects has been also studied, particularly when the objects are disks in the plane. It has been proven by Danzer [4] and by Stacho [11,12] that a set of pairwise intersecting disks in the plane can be pierced by four points. However, these proofs are involved and it seems that they can not lead to an efficient algorithm. Recently, Har-Peled et al. [6] showed that every set of pairwise intersecting disks in the plane can be pierced by five points and gave a linear time algorithm for finding these points. Carmi et al. [2] improved this result by showing that four points are always sufficient to pierce any set of pairwise intersecting disks in the plane, and also gave a linear time algorithm for finding these points.

In 1995, Fingerhut [5] conjectured that for any maximum-weight perfect matching $M = \{(a_1, b_1), (a_2, b_2), \ldots, (a_n, b_n)\}$ of $2n$ points in the plane, there exists a point c, such that $|ca_i| + |cb_i| \leq \alpha \cdot |a_i b_i|$, for every $1 \leq i \leq n$, where $\alpha = \frac{2}{\sqrt{3}}$. That is, the set of the ellipses E_i with foci at a_i and b_i, and contain all the points x, such that $|a_i x| + |xb_i| \leq \frac{2}{\sqrt{3}} \cdot |a_i b_i|$, have a non-empty intersection. Recently, Bereg et al. [1] considered a variant of this conjecture. They proved that there exists a point that pierces all the disks whose diameters are the edges of M. The proof is accomplished by showing that the set of disks is Helly, i.e., for every three disks there is a point in common.

1.2   Our Contribution

A common and natural approach to prove that all the disks in $\mathcal{D}_T$ have a non-empty intersection is using Helly's Theorem, i.e., to show that every three disks have a non-empty intersection. However, we use a different approach and show that all the disks in $\mathcal{D}_T$ have a non-empty intersection by characterizing a specific point that pierces all the disks in $\mathcal{D}_T$. More precisely, we prove the following theorem.

Theorem 1. *Let C^* be the smallest enclosing circle of the points of P and let c^* be its center. Then, c^* pierces all the disks in $\mathcal{D}_T$.*

This approach is even stronger since it implies a linear-time algorithm for finding the piercing point, using Megiddo's linear-time algorithm [9] for computing the smallest enclosing circle of P.

The result in this paper can be considered as a variant of Fingerhut's Conjecture. That is, for a maximum-weight spanning tree (instead of a maximum-weight perfect matching) and $\alpha = \sqrt{2}$ (instead of $\alpha = \frac{2}{\sqrt{3}}$) the conjecture holds.

2   Preliminaries

Let P be a set of points in the plane, let T be a maximum-weight spanning tree of P, and let $\mathcal{D}_T$ be the set of the diametral disks induced by the edges of T.

Let C^* be the smallest enclosing circle of the points of P, and let r^* and c^* be its radius and its center, respectively. We assume, w.l.o.g., that $r^* = 1$ and c^* is located at the origin $(0,0)$. Let $\mathcal{D}^*$ be the disk having C^* as its boundary. Let A_1, A_2, A_3, and A_4 (resp., Q_1, Q_2, Q_3, and Q_4) be the four arcs (resp., the four quarters) obtained by dividing C^* (resp., $\mathcal{D}^*$) by the x and the y-axis; see Fig. 1 for an illustration.

Lemma 1. *Each one of the arcs A_1 and A_3 contains at least one point of P or each one of the arcs A_2 and A_4 contains at least one point of P.*

Proof. By definition, there are at least two points of P on C^*. If there are exactly two points p and q on C^*, then the segment $\overline{pq}$ is a diameter of C^*, and clearly, p and q are on non-adjacent arcs of C^*; see Fig. 1(a). Otherwise, there are at least three points of P on C^*; see Fig. 1(b). In this case, there are three points p, q, and t on C^*, such that the triangle $\triangle pqt$ contains c^*. Thus, every angle in this triangle is acute, and therefore two points from p, q, t are on non-adjacent arcs of C^*. $\qquad\qquad\qquad\qquad\qquad\qquad\qquad\qquad\qquad\qquad\qquad\qquad\blacksquare$

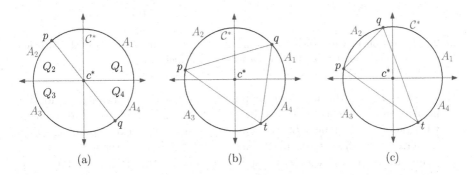

Fig. 1. The smallest enclosing circle C^* of P. (a) Two points on C^*. (b) and (c) Three points on C^*.

Lemma 2. *Let p and q be two points in Q_3, such that p is on the negative x-axis, the angle $\angle pc^*q < \frac{\pi}{2}$, and $|c^*p| \geq |c^*q|$; see Fig. 2. Then,*

(i) for every point t on $A_1 \cup A_2$, we have $|qt| > |pq|$,
(ii) for every point t on $A_1 \cup A_4$, we have $|pt| > |pq|$, and
(iii) for every two points t and t' on A_2 and A_4, respectively, we have $|tt'| > |pq|$.

Proof. (i) Let $a = (-1,0)$ and $b = (1,0)$ be the intersection points of C^* with the negative and the positive x-axis, respectively; see Fig. 2(a). Let D_q be the disk with center q and radius $|qa|$. Since $|c^*q| \leq |c^*p|$, we have $\angle c^*pq \leq \angle c^*qp$, and thus $\angle c^*pq \leq \frac{\pi}{2}$. Hence, $\angle qpa > \frac{\pi}{2}$, and thus $|qa| > |qp|$. Let q' be the intersection point of the line passing through a and q with the y-axis, and let $D_{q'}$ be the disk with center q' and radius $|q'a|$; see Fig. 2(a). Since $D_{q'}$ intersects C^* at the points a and b, the arc $A_1 \cup A_2$ is outside $D_{q'}$ (this is correct for every disk centered at a point x on the negative y-axis and has a radius $|xa|$). Thus, for every point t on $A_1 \cup A_2$, we have $|q't| \geq |qa|$. Since D_q is contained in $D_{q'}$, this is also correct for D_q. Therefore, for every point t on $A_1 \cup A_2$, we have $|qt| \geq |qa| > |qp|$.

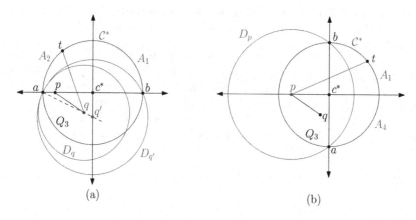

(a)   (b)

Fig. 2. Illustration of the proof of Lemma 2. (a) Any point t on $A_1 \cup A_2$ satisfies $|qt| > |pq|$. (b) Any point t on $A_1 \cup A_4$ satisfies $|pt| > |pq|$.

(ii) Let a and b be the intersection points of C^* with the negative and the positive y-axis, respectively; see Fig. 2(b). Let D_p be the disk centered at p with radius $|pa|$. Hence, D_p contains Q_3, and thus for every point $z \in Q_3$, we have $|pz| < |pa|$, particularly $|pq| < |pa|$. Since D_p intersects C^* at the points a and b, the arc $A_1 \cup A_4$ is outside D_p (this is correct for every disk centered at a point x on the negative x-axis and has a radius $|xa|$). Therefore, for every point t on $A_1 \cup A_4$, we have $|pt| > |pa| > |pq|$.

(iii) Since $\angle pc^*q < \frac{\pi}{2}$, we have $|pq| < \sqrt{2}$. Moreover, by the location of t and t', we have $|tt'| \geq \sqrt{2}$. Therefore, $|tt'| > |pq|$.

Notice that Lemma 2 holds for every two points p and q inside C^*, such that $\angle pc^*q < \frac{\pi}{2}$. This is true since we can always rotate the points of P around c^* (and reflect them with respect to the x-axis if needed) until the farthest point from c^* among p and q lays on the negative x-axis and the other point lays inside Q_3.

Corollary 1. *Lemma 2 holds for every two points p and q inside C^*, such that $\angle pc^*q < \frac{\pi}{2}$.*

3   Proof of Theorem 1

Let $G = (P, E)$ be the complete graph over P and let $T = (P, E_T)$ be the maximum-weight spanning tree of P (i.e., of G). A maximum-weight spanning tree can be computed by Kruskal's algorithm [3] (or by the algorithm provided by Monma et al. [10]) which uses the fact that for any cycle C in G, if the weight of an edge $e \in C$ is less than the weight of each other edge in C, then e cannot be an edge in any maximum-weight spanning tree of P. Kruskal's algorithm works as follows. It sorts the edges in E in non-increasing order of their weight, and then goes over these edges in this order and adds an edge (p, q) to E_T if it does not produce a cycle in T. Based on this fact, we prove that for every edge $(p, q) \in E_T$, the disk D_{pq} contains c^*. More precisely, we prove that for each edge $(p, q) \in E_T$ the angle $\angle pc^*q$ is at least $\frac{\pi}{2}$.

Lemma 3. *For every edge* $(p, q) \in E_T$, *we have* $\angle pc^*q \geq \frac{\pi}{2}$.

Proof. Let (p, q) be an edge in E_T. We show that if $\angle pc^*q < \frac{\pi}{2}$, then there is a cycle in G in which the edge (p, q) has the minimum weight among the edges of this cycle, and thus (p, q) can not be in a maximum-weight spanning tree of P. Assume towards a contradiction that $\angle pc^*q < \frac{\pi}{2}$, and assume, w.l.o.g., that p and q are in Q_3, p is on the x-axis, and $|c^*p| > |c^*q|$. We distinguish between two cases:

(i) If there is a point t on A_1, then, by Lemma 2, we have $|tp| > |pq|$ and $|tq| > |pq|$. Thus, the edges (t, p), (p, q), and (q, t) form a cycle and the edge (p, q) has a weight less than the weight of each other edge in this cycle; see Fig. 3(a). This contradicts that $(p, q) \in E_T$.

(ii) Otherwise, by Lemma 1, there exist two points t and t' on A_2 and A_4, respectively. By Lemma 2, we have $|tq| > |pq|$, $|t'p| > |pq|$ and $|tt'| > |pq|$. Thus, the edges (t, t'), (t', p), (p, q), and (q, t) form a cycle and the edge (p, q) has a weight less than the weight of each other edge in this cycle; see Fig. 3(b). This contradicts that $(p, q) \in E_T$.

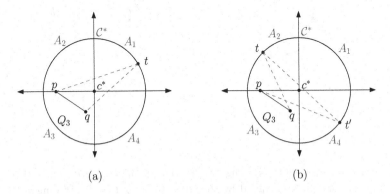

(a)   (b)

Fig. 3. Illustration of the proof of Lemma 3. (a) (p, q) is of minimum weight in the cycle $<t, p, q>$. (b) (p, q) is of minimum weight in the cycle $<t, p, q, t'>$.

4 Conclusion

In this paper, we have shown that the diametral disks obtained by the edges of a maximum-weight spanning tree of a set of points P have a non-empty intersection. We showed that the disks can be pierced by the center of the smallest enclosing circle of P, which can be computed in linear time [9].

Fingerhut [5] conjectured that for any maximum-weight perfect matching $M = \{(a_1, b_1), (a_2, b_2), \ldots, (a_n, b_n)\}$ of $2n$ points in the plane, the set of the ellipses E_i with foci at a_i and b_i, and contains all the points x, such that $|a_i x| + |x b_i| \leq \alpha \cdot |a_i b_i|$, for every $1 \leq i \leq n$, where $\alpha = \frac{2}{\sqrt{3}}$, have a non-empty intersection. The smallest known value for α is $\alpha = \sqrt{2}$, which was provided by Bereg et al. [1].

In this paper, we considered a variant of Fingerhut's Conjecture for maximum-weight spanning tree instead of maximum-weight perfect matching. We showed that for any maximum-weight spanning tree T and $\alpha = \sqrt{2}$, there exists a point c^*, such that for every edge (a, b) in T, $|c^* a| + |c^* b| \leq \alpha \cdot |ab|$. In Fig. 4(a), we show an example of a maximum-weight spanning tree, such that for any $\alpha < \frac{1+\sqrt{3}}{2}$, the conjecture does not hold. This provides a lower bound on α. Moreover, in Fig. 4(b), we show an example of a maximum-weight spanning tree for which the center c^* of the smallest enclosing circle does not satisfy the inequality for $\alpha = \frac{1+\sqrt{3}}{2}$. This means that our approach does not work for $\alpha = \frac{1+\sqrt{3}}{2}$, but does not mean that the conjecture does not hold for $\alpha = \frac{1+\sqrt{3}}{2}$. Even though the gap between $\sqrt{2} \approx 1.414$ and $\frac{1+\sqrt{3}}{2} \approx 1.366$ is very small, it is an interesting open question to find the exact value for α for which the conjecture holds.

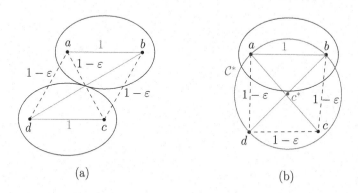

(a)   (b)

Fig. 4. A maximum-weight spanning tree of the points $\{a, b, c, d\}$ (red edges) and $\alpha = \frac{1+\sqrt{3}}{2}$. (a) The ellipses defined by the edges (a, b) and (c, d) are tangent to each other. (b) The ellipse defined by the edge (a, b) does not contain the point c^*. (Color figure online)

References

1. Bereg, S., Chacón-Rivera, O., Flores-Peñaloza, D., Huemer, C., Pérez-Lantero, P., Seara, C.: On maximum-sum matchings of points. J. Glob. Optim. **85**, 111–128 (2022)
2. Carmi, P., Katz, M.J., Morin, P.: Stabbing pairwise intersecting disks by four points. CoRR, abs/1812.06907 (2018)
3. Cormen, T.H., Leiserson, C.E., Rivest, R.L., Stein, C.: Introduction to Algorithms, 3rd edn. The MIT Press, Cambridge (2009)
4. Danzer, L.: Zur Lösung des Gallaischen Problems über Kreisscheiben in der Euklidischen Ebene. Studia Sci. Math. Hungar **21**(1–2), 111–134 (1986)
5. Eppstein, D.: Geometry junkyard. https://www.ics.uci.edu/~eppstein/junkyard/maxmatch.html
6. Har-Peled, S., et al.: Stabbing pairwise intersecting disks by five points. Discret. Math. **344**(7), 112403 (2021)
7. Helly, E.: Über Mengen konvexer Körper mit gemeinschaftlichen Punkten. Jahresber. Dtsch. Math.-Ver. **32**, 175–176 (1923)
8. Helly, E.: Über Systeme von abgeschlossenen Mengen mit gemeinschaftlichen Punkten. Monatshefte Math. **37**(1), 281–302 (1930)
9. Megiddo, N.: Linear-time algorithms for linear programming in $\mathbb{R}^3$ and related problems. SIAM J. Comput. **12**(4), 759–776 (1983)
10. Monma, C., Paterson, M., Suri, S., Yao, F.: Computing Euclidean maximum spanning trees. Algorithmica **5**(1–4), 407–419 (1990)
11. Stacho, L.: Über ein Problem für Kreisscheiben Familien. Acta Sci. Math. (Szeged) **26**, 273–282 (1965)
12. Stacho, L.: A solution of Gallai's problem on pinning down circles. Mat. Lapok **32**(1–3), 19–47 (1981/84)

Reflective Guarding a Gallery

Arash Vaezi[1(✉)], Bodhayan Roy[2], and Mohammad Ghodsi[1,3]

[1] Department of Computer Engineering, Sharif University of Technology,
Tehran, Iran
avaezi@ce.sharif.edu, ghodsi@sharif.edu
[2] Indian Institute of Technology Kharagpur, Kharagpur, India
broy@maths.iitkgp.ac.in
[3] Institute for Research in Fundamental Sciences (IPM), Tehran, Iran

Abstract. This paper studies a variant of the Art Gallery problem in which the "walls" can be replaced by *reflecting edges*, which allows the guards to see further and thereby see a larger portion of the gallery. Given a simple polygon P, first, we consider one guard as a point viewer, and we intend to use reflection to add a certain amount of area to the visibility polygon of the guard. We study visibility with specular and diffuse reflections where the specular type of reflection is the mirror-like reflection, and in the diffuse type of reflection, the angle between the incident and reflected ray may assume all possible values between 0 and π. Lee and Aggarwal already proved that several versions of the general Art Gallery problem are NP-hard. We show that several cases of adding an area to the visible area of a given point guard are NP-hard, too.

Second (A primary version of the second result presented here is accepted in EuroCG 2022 [1] whose proceeding is not formal), we assume that all edges are reflectors, and we intend to decrease the minimum number of guards required to cover the whole gallery.

Chao Xu proved that even considering r specular reflections, one may need $\lfloor \frac{n}{3} \rfloor$ guards to cover the polygon. Let r be the maximum number of reflections of a guard's visibility ray.

In this work, we prove that considering r *diffuse* reflections, the minimum number of *vertex or boundary* guards required to cover a given simple polygon $\mathcal{P}$ decreases to $\lceil \frac{\alpha}{1+\lfloor \frac{r}{8} \rfloor} \rceil$, where α indicates the minimum number of guards required to cover the polygon without reflection. We also generalize the $\mathcal{O}(\log n)$-approximation ratio algorithm of the vertex guarding problem to work in the presence of reflection.

1 Introduction

Consider a simple polygon $\mathcal{P}$ with n vertices and a point viewer q inside $\mathcal{P}$. Suppose $C(\mathcal{P})$ denotes $\mathcal{P}$'s topological closure (the union of the interior and the

B. Roy—The author is supported by an ISIRD Grant from Sponsored Research and Industrial Consultancy, IIT Kharagpur, and a MATRICS grant from Science and Engineering Research Board.

boundary of $\mathcal{P}$). Two points x and y are visible to each other, if and only if the line segment $\overline{xy}$ lies completely in $C(\mathcal{P})$. The visibility polygon of q, denoted as $VP(q)$, consists of all points of $\mathcal{P}$ visible to q. Many problems concerning visibility polygons have been studied so far. There are linear-time algorithms to compute $VP(q)$ [2,3]. Edges of $VP(q)$ that are not edges of $\mathcal{P}$ are called *windows*.

If some of the edges of $\mathcal{P}$ are made into mirrors, then $VP(q)$ may enlarge. Klee first introduced visibility in the presence of mirrors in 1969 [4]. He asked whether every polygon whose edges are all mirrors is illuminable from every interior point. In 1995 Tokarsky constructed an all-mirror polygon inside which there exists a dark point [5]. Visibility with reflecting edges subject to different types of reflections has been studied earlier [6]: (1) *Specular-reflection*: in which the direction light is reflected is defined by the law-of-reflection. Since we are working in the plane, this law states that the angle of incidence and the angle of reflection of the visibility rays with the normal through the polygonal edge are the same. (2) *Diffuse-reflection*: that is to reflect light with all possible angles from a given surface. The diffuse case is where the angle between the incident and reflected ray may assume all possible values between 0 and π.

Some papers have specified the maximum number of allowed reflections via mirrors in between [7]. In multiple reflections, we restrict the path of a ray coming from the viewer to turn at polygon boundaries at most r times. Each time this ray will reflect based on the type of reflection specified in a problem (specular or diffuse).

Every edge of $\mathcal{P}$ can potentially become a reflector. However, the viewer may only see some edges of $\mathcal{P}$. When we talk about an edge, and we want to consider it as a reflector, we call it a *reflecting edge* (or a *mirror-edge* considering specular reflections). Each edge has the potential of getting converted into a reflecting edge in a final solution of a visibility extension problem (we use the words "reflecting edge" and "reflected" in general, but the word "mirror" is used only when we deal with specular reflections).

Two points x and y inside $\mathcal{P}$ can see each other through a reflecting edge e, if and only if they are reflected visible with a specified type of reflection. We call these points *reflected visible* (or *mirror-visible*).

The Art Gallery problem is to determine the minimum number of guards that are sufficient to see every point in the interior of an art gallery room. The art gallery can be viewed as a polygon $\mathcal{P}$ of n vertices, and the guards are stationary points in $\mathcal{P}$. If guards are placed at vertices of $\mathcal{P}$, they are called *vertex guards*. If guards are placed at any point of $\mathcal{P}$, they are called *point guards*. If guards are allowed to be placed along the boundary of $\mathcal{P}$, they are called *boundary-guards* (on the perimeter). To know more details on the history of this problem see [8].

The Art Gallery problem was proved to be *NP*-hard first for polygons with holes by [9]. For guarding simple polygons, it was proved to be *NP*-complete for vertex guards by [10]. This proof was generalized to work for point guards by [11]. The class $\exists\mathbb{R}$ consists of problems that can be reduced in polynomial time to the problem of deciding whether a system of polynomial equations with integer coefficients and any number of real variables has a solution. It can be easily seen that $NP \subseteq \exists\mathbb{R}$. The article [12] proved that the Art Gallery problem is $\exists\mathbb{R}$-complete. Sometimes irrational coordinates are required to describe an optimal solution [13].

Ghosh [14] provided an $\mathcal{O}(\log n)$-approximation algorithm for guarding polygons with or without holes with *vertex* guards. King and Kirkpatrick obtained an approximation factor of $\mathcal{O}(\log\log(OPT))$ for vertex guarding or perimeter guarding simple polygons [15]. To see more information on approximating various versions of the Art Gallery problem see [16], or [17].

Result 1. *Given a simple polygon $\mathcal{P}$ and a query point q as the position of a single viewer (guard), consider extending the area of the visibility polygon of q ($VP(q)$) by choosing an appropriate subset of edges and make them reflecting edges so that q can see the whole $\mathcal{P}$.*

A) To extend the surface area of $VP(q)$ by exactly a given amount, the problem is NP-complete.

B) To extend the surface area of $VP(q)$ using the minimum number of diffuse reflecting edges and by at least a given amount, the problem is NP-hard.

Result 2. *Suppose that in the Art Gallery problem a given polygon, possibly with holes, can be guarded by α vertex guards without reflections, then the gallery can be guarded by at most $\lceil \frac{\alpha}{1+\lfloor \frac{r}{8} \rfloor} \rceil$ guards when r diffuse reflections are permitted.*

For both the diffuse and specular reflection the Art Gallery problem considering r diffuse reflection is solvable in $\mathcal{O}(n^{8^{r+1}+10})$ time with an approximation ratio of $\mathcal{O}(\log n)$.

1.1   Our Settings

Every guard can see a point if the point is directly visible to the guard or if it is reflected visible. This is a natural and non-trivial extension of the classical Art Gallery setting. The problem of visibility via reflection has many applications in wireless networks, and Computer Graphics, in which the signal and the view ray can reflect on walls several times, and it loses its energy after each reflection. There is a large literature on geometric optics (such as [18–20]), and on the chaotic behavior of a reflecting ray of light or a bouncing billiard ball (see, e.g., [21–24]). Particularly, regarding the Art Gallery problem, reflection helps in decreasing the number of guards (see Fig. 1).

Sections 2 and 3 study the problem of extending the surface area of the visibility polygon $VP(q)$ of a point guard q inside a polygon $\mathcal{P}$ by means of reflecting edges. Section 3 considers the scenario in which the visibility polygon of the source needs to be extended at least k units of area where k is a given value. However, to make the problem more straightforward, one may consider adding a specific area with an exact given surface area to the visibility of the source. Section 2 considers extending the visibility polygon of q exactly k units of area.

A special reflective case of the general Art Gallery problem is described by Chao Xu in 2011 [25]. Since we want to generalize the notion of guarding a

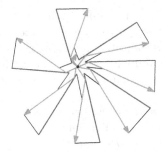

Fig. 1. This figure illustrates a situation where a single guard is required if we use reflection-edges; $\Theta(n)$ guards are required if we do not consider reflection. Red segments illustrate the reflected-edges. (Color figure online)

simple polygon, if the edges become mirrors instead of walls, the light loses intensity every time it gets reflected on the mirror. Therefore after r reflections, it becomes undetectable to a guard. Chao Xu proved that regarding multiple specular reflections, for any n, there exist polygons with n vertices that need $\lfloor \frac{n}{3} \rfloor$ guards. Section 4 deals with the same problem but regarding diffuse reflection. G. Barequet et al. [26] proved the minimum number of diffuse reflections sufficient to illuminate the interior of any simple polygon with n walls from any interior-point light source is $\lfloor \frac{n}{2} \rfloor - 1$. E. Fox-Epstein et al. [27] proved that to make a simple polygon in a general position visible for a single point light source, we need at most $\lfloor \frac{n-2}{4} \rfloor$ diffuse reflections on the edges of the polygon, and this is the best possible bound. These two papers consider a *single point viewer*; however, in Sect. 4 we considered helping the art gallery problem with diffuse reflection on the edges of the polygon. So, we want to decrease the minimum number of guards required to cover a given simple polygon using r diffuse reflections. We will prove that we can reduce the optimal number with the help of diffuse reflection. Note that we do not assume general positions for the given polygon. For more information on combining reflection with the art gallery problem see [7,28–30], and [6].

2   Expanding *VP(q)* by *Exactly k* Units of Area

We begin this section with the following theorem, and the rest of the section covers the proof of this theorem.

Theorem 1. *Given a simple polygon $\mathcal{P}$, a point $q \in \mathcal{P}$, and an integer $k > 0$, the problem of choosing any number of, say l, reflecting edges of $\mathcal{P}$ in order to expand $VP(q)$ by exactly k units of area is NP-complete in the following cases:*

1. *Specular-reflection where a ray can be reflected only once.*
2. *Diffuse-reflection where a ray can be reflected any number of times.*

Clearly, it can be verified in polynomial time if a given solution adds precisely k units to $VP(q)$. Therefore, the problem is in *NP*.

Consider an instance of the Subset-Sum problem (*InSS*), which has $val(1), val(2) ..., val(m)$ non-negative integer values, and a target number $\mathcal{T}$. Suppose $m \in \Theta(n)$, where n indicates the number of vertices of $\mathcal{P}$. The Subset-Sum Problem involves determining whether a subset from a list of integers can sum to a target value $\mathcal{T}$. Note that the variant in which all inputs are positive is NP-complete as well [31].

In the following subsections, we will show that the Subset-Sum problem is reducible to this problem in polynomial time. Thus, we deduce that our problem in the cases mentioned above is *NP*-complete.

2.1 *NP*-Hardness for Specular Reflections

The reduction polygon $\mathcal{P}$ consists of two rectangular chambers attached side by side. The chamber to the right is taller, while the chamber to the left is shorter but quite broad. The query point q is located in the right chamber (see Fig. 2). The left chamber has left-leaning triangles attached to its top and bottom edges. In the reduction from Subset-Sum, the areas of the bottom spikes correspond to the weights of the sets (the values of *InSS*). The top triangles are narrow and have negligible areas. Their main purpose is to house the edges which may be turned into reflecting edges so that q can see the bottom spikes.

To describe the construction formally, consider *InSS*. Denote the ith value by $val(i)$ and the sum of the values till the ith value, $\sum_{k=1}^{i} val(k)$, by $sum(i)$. We construct the reduction polygon in the following steps:

(1) Place the query point q at the origin $(0,0)$.
(2) Consider the x-axis as the bottom edge of the left rectangle.
(3) Denote the left, right and bottom points of the ith bottom spike by $llt(i)$, $rlt(i)$ and $blt(i)$ respectively. Set the coordinates for $llt(i)$ at $(i+2(sum(i-1)), 0)$, $rlt(i)$ at $(i+2(sum(i)), 0)$, and those for $blt(i)$ at $(i+2(sum(i)), -1)$.
(4) The horizontal polygonal edges between the top triangles are good choices for mirrors, so we call them *mirror-edges*. The ith mirror-edge lies between the $(i-1)^{th}$ and i^{th} top spikes. Denote the left and right endpoints of the ith mirror-edge by $lm(i)$ and $rm(i)$ respectively. Set the coordinates of $lm(i)$ at $(\frac{(i+2(sum(i-1)))}{2}, 2(sum(m) + m))$, and those of $rm(i)$ at $(\frac{(i+2(sum(i)))}{2}, 2(sum(m) + m))$.

Denote the topmost point of the ith top spike by $ut(i)$ and set its coordinates at $(\frac{(i+2(sum(i)))}{2}, 4(sum(m) + 2m))$.

2.2 Properties of the Reduction Polygon

In this subsection, we discuss properties that follow from the above construction of the reduction polygon. We have the following lemmas.

Lemma 1. *The query point q can see the region enclosed by the ith bottom spike only through a specular reflection through the ith mirror-edge.*

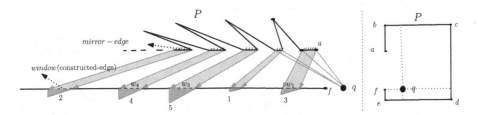

Fig. 2. Two main components of the reduction polygon is illustrated.

Proof. See the full version of the paper [32] for the proof.

Lemma 2. *All coordinates of the reduction polygon are rational and take polynomial time to compute.*

Proof. This too follows from the construction, as the number of sums used is linear, and each coordinate is derived by at most one division from such a sum.

Lemma 3. *The problem of extending the visibility polygon of a query point inside a simple polygon via single specular reflection is NP-complete.*

Proof. From Lemma 1 it follows that a solution for the problem exists if an only if a solution exists for the corresponding Subset-Sum problem. From Lemma 2 it follows that the reduction can be carried out in polynomial time, thus proving the claim.

Observation 1. *The multiple reflection case of the first case of the problem mentioned in Theorem 1 is still open. That is the above-mentioned reduction (presented in Subsect. 2.1) does not work if more than one reflection is allowed.*

Proof. See the proof in the complete version of the paper [32].

2.3   *NP*-Hardness for Diffuse Reflections

This subsection deals with the second part of Theorem 1. Considering diffuse reflection, since rays can be reflected into wrong spikes (a spike which should get reflected visible via another reflecting edge) the previous reduction does not work. Considering multiple plausible reflections, the problem becomes even harder. These rays have to be excluded by an appropriate structure of the polygonal boundary.

The construction presented in this subsection works in the case of multiple reflections, too. Again we reduce the Subset-Sum problem to our problem (Result 1(A) considering diffuse reflections).

As before, we place the query point q at the origin, $(0,0)$. The main polygon $\mathcal{P}$ used for the reduction is primarily a big rectangle, with around two-thirds of it being to the right of q (see Fig. 3). On top of this rectangle are m *"double triangle"* structures. Each double triangle structure consists of triangles sharing

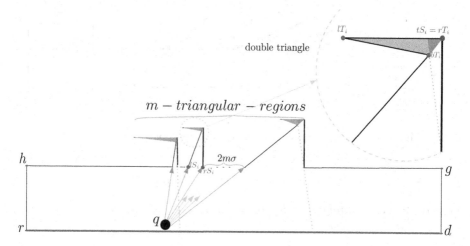

Fig. 3. A schema of the reduction polygon. Note that $m \in \Theta(n)$. The main polygon is a rectangle, with gadgets on its top edge. The green region plus the grey region are in one triangle which its surface area equals to a value of $InSS$. (Color figure online)

some of their interiors. The lower among these two triangles, referred to as the *second triangles*, is right-angled and has its base on the main rectangle, with its altitude to the right and hypotenuse to the left. The upper triangle, referred to as the *"top triangle"*, is inverted, i.e., its base or horizontal edge is at the top. One of its vertices is the top vertex of the second triangle, and another of its vertices merges into the hypotenuse of the second triangle. Its third vertex juts out far to the left, at the same vertical level as the top vertex of the second triangle, making the top triangle a very narrow triangle. The area of the top triangle equals the value of the ith set in the Subset-Sum problem (val_i).

We make each top triangle to get diffusely reflected visible by only some specific reflecting edges. However, as mentioned previously, there can be a troublemaker shared reflected visible area in each top triangle. This area is illustrated in green in Fig. 3. We know that every value of the Subset-Sum problem is an integer. We manage to set the coordination of the polygon so that the sum of the surface area of all the green regions gets equal to a value less than 1, and all of these regions are entirely reflected visible to the lower edge of the main rectangle. As a result, seeing the green areas through a reflection via the bottom red edge cannot contribute towards seeing exactly an extra region of k units of area. Remember that k is an integer.

Formally, denote the ith top and second triangles by T_i and S_i respectively. Denote the top, left and right vertices of S_i by tS_i, lS_i and rS_i respectively. Denote the sum of values of all subsets of the Subset-Sum problem by σ. In fact, $\sigma = \sum_{i=1}^{m} val_i$. In general, the triangle S_i has a base length of i units, and its base is $2m^2\sigma$ units distance away from those of S_{i-1} and S_{i+1}. Therefore the coordinates of lS_i and rS_i are given by $((2m^2\sigma)\frac{i(i+1)}{2} - i, m^2(m+1)\sigma)$

and $((2m^2\sigma)\frac{i(i+1)}{2}, m^2(m+1)\sigma)$ respectively. For any vertex v of the reduction polygon, let us denote the x and y coordinates of v by $x(v)$ and $y(v)$ respectively. The vertex tS_i is obtained by drawing a ray originating at q and passing through lS_i, and having it intersect with the vertical line passing through rS_i. This point of intersection is tS_i with coordinates $(x(rS_i), m^2(m+1)\sigma + \frac{m^2(m+1)\sigma}{x(lS_i)})$.

Denote the leftmost and bottom-most vertices of T_i by lT_i and bT_i respectively. Recall that bT_i lies on the hypotenuse of S_i, and lT_i has the same y-coordinate as tS_i. Moreover, we place bT_i in such a way, that the sum of the total regions of all top triangles seen from the base of the main rectangle (the rd edge in Fig. 3) is less than 1. Intuitively, bT_i divides the hypotenuse of S_i in the $m^2 - 1 : 1$ ratio. Accordingly, the coordinates of bT_i are: $(1 + \frac{i(i-1)}{2} + (m^2 - 1)\frac{(x(tS_i)-x(lS_i))}{(m^2)}, m + (m^2 - 1)\frac{(y(tS_i)-y(lS_i))}{(m^2)})$.

Next, we set coordinates of lT_i in a way that the total surface area of T_i gets equal to the value of the ith subset. Denote the value of the ith subset by val_i. Then, the coordinates of lT_i are given by $(x(tS_i) - 2\frac{val_i}{y(tS_i)-y(bT_i)}, y(tS_i))$. Finally, the coordinates of the four vertices of the main rectangle holding all the double triangle gadgets, are given by $(-x(rS_m), m^2(m+1)\sigma)$, $(-x(rS_m), -1)$, $(2(x(rS_m)), -1)$ and $(2(x(rS_m)), m^2(m+1)\sigma)$.

Lemma 4. *The reduction stated in Subsect. 2.3 proves that the problem of adding exactly k units of area to a visibility polygon via only a single diffuse-reflection per ray, is NP-complete. The reduction polygon has rational coordinates with size polynomial with respect to n.*

Proof. See the full version of the paper [32] for the proof.

We can use the above-mentioned reduction in case multiple reflection is allowed. See the following corollary.

Corollary 2. *The reduction stated in Subsect. 2.3 works if multiple reflections is allowed.*

Proof. See the full version of the paper [32] for the proof.

3   Expanding at Least k Units of Area

In this section, we modify the reduction of Lee and Lin [10] and use it to infer that the problem of extending the visibility polygon of a given point by a region of area at least k units of area with the *minimum* number of reflecting edges is NP-hard, where k is a given amount (Result 1(B)). The idea is that the potential vertex guards are replaced with edges that can reflect the viewer (q). We need an extra reflecting edge, though. Only a specific number of edges can make an invisible region (a spike) entirely reflected visible to the viewer. Converting the correct minimum subset of these edges to the reflecting edges determines the optimal solution for the problem.

The specular-reflection cases of the problem are still open. Nonetheless, it was shown by Aronov [6] in 1998 that in such a polygon where all of its edges are mirrors, the visibility polygon of a point can contain holes. And also, when we consider at most r specular reflections for every ray, we can compute the visibility polygon of a point inside that within $\mathcal{O}(n^{2r} \log n)$ of time complexity and $\mathcal{O}(n^{2r})$ of space complexity [7].

Conjecture 1. Given a simple polygon $\mathcal{P}$, and a query point q inside the polygon, and a positive value k, the problem whether l of the edges of the polygon can be turned to reflecting edges so that the area added to $VP(q)$ through (single/multiple) diffuse-reflections increases at least k units of area is NP-hard.

Proof Idea. See the full version of the paper for the proof [32].

4   Regular Visibility vs Reflection

This section deals with Result 2. Under some settings, visibility with reflections can be seen as a general case of regular visibility. For example, consider guarding a polygon $\mathcal{P}$ with vertex guards, where all the edges of the polygons are diffuse reflecting edges, and r reflections are allowed for each ray. Let S be the set of guards in an optimal solution if we do not consider reflection. Since we allow multiple vertices of the polygon to be collinear, we slightly change the notion of a visibility polygon for convenience. Consider the visibility polygon of a vertex (see Fig. 4). It may include lines containing points that do not have any interior point of the visibility polygon within a given radius. So, given a vertex x of $\mathcal{P}$, we consider the union of the interior of the original $VP(x)$ (denoted by $int\,VP(x)$), and the limit points of $int\,VP(x)$, as our new kind of visibility polygon of x. Clearly, a guard set of $\mathcal{P}$ gives a set of the new kind of visibility polygons whose union is $\mathcal{P}$.

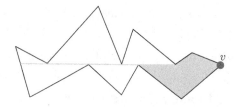

Fig. 4. The visibility polygon of a point v.

Consider any guard $v \in S$. The visibility polygon $VP(v)$ of v must have at least one window[1]. Otherwise, v is the only guard of $\mathcal{P}$. Consider such a window, say, w. Let x be a point of intersection of w with the polygonal boundary. Then there must be at least another guard $u \in S$ such that x lies in both $VP(u)$ and $VP(v)$. The following lemmas discuss how $VP(u)$ and $VP(v)$ can be united

---

[1] A window is an edge of a viewer's visibility polygon, which is not a part of an edge of the main polygon.

using a few diffuse reflections, and how the whole polygon can be seen by a just a fraction of the optimal guards, depending on the number of reflections allowed per ray.

Theorem 3. *If $\mathcal{P}$ can be guarded by α vertex guards without reflections, then $\mathcal{P}$ can be guarded by at most $\lceil \frac{\alpha}{1+\lfloor \frac{r}{8} \rfloor} \rceil$ guards when r diffuse reflections are permitted.*

To prove this theorem see the following lemmas first:

Lemma 5. *If $\mathcal{S}$ is an optimal vertex guard set of polygon $\mathcal{P}$ and $|\mathcal{S}| > 1$ then for every guard $u \in \mathcal{S}$ there exists a different guard $v \in \mathcal{S}$" such that u and v can see each other through five diffuse reflections. Furthermore, u and v can fully see each other's visibility polygons with eight diffuse reflections.*

Proof. See the full version of the paper [32] for the proof.

Now we build a graph G as follows. We consider the vertex guards in $\mathcal{S}$ as the vertices of G, and add an edge between two vertices of G if and only if the two corresponding vertex guards in $\mathcal{S}$ can see each other directly or through at most five reflections. We have the following Lemma.

Lemma 6. *The graph G is connected.*

Proof. See the full version of the paper [32] for the proof.

Consider any optimal vertex guard set $\mathcal{S}$ for the Art Gallery problem on the polygon $\mathcal{P}$, where $|\mathcal{S}| = \alpha$. Build a graph G on $\mathcal{S}$ as it was mentioned before Lemma 6. Due to Lemma 6, G is connected. Find a spanning tree T of G and root it at any vertex. Denote the ith level of vertices of T by L_i. Given a value of r, divide the levels of T into $1 + \lfloor \frac{r}{8} \rfloor$ classes, such that the class $\mathcal{C}_i$ contains all the vertices of all levels of T of the form $L_{i+x(1+\lfloor \frac{r}{8} \rfloor)}$, where $x \in \mathbb{Z}_0^+$. By the pigeonhole principle, one of these classes will have at most $\lceil \frac{\alpha}{1+\lfloor \frac{r}{8} \rfloor} \rceil$ vertices. Again, by Lemma 5, given any vertex class $\mathcal{C}$ of T, all of $\mathcal{P}$ can be seen by the vertices of $\mathcal{C}$ when r diffuse reflections are allowed. The theorem follows.

Corollary 4. *The above bound (mentioned in Theorem 3) holds even if the guards are allowed to be placed anywhere on the boundary of the polygon.*

Proof. The proof follows directly from the proof of Theorem 3 since Lemmas 5 and 6 are valid for boundary guards as well.

Observation 2. *The above bound (mentioned in Theorem 3) does not hold in the case of arbitrary point guards.*

Proof. See the full version of the paper [32] for the proof.

Finding an approximate solution to the vertex guard problem with r diffuse reflections is harder than approximating the standard problem. Reflection may change the position of guards remarkably. Here, we have a straight-forward generalization of Ghosh's discretization algorithm presented in [33].

Theorem 5. *For vertex guards, the art gallery problem considering r reflections, for both the diffuse and specular reflection are solvable in $\mathcal{O}(n^{8^{r+1}+10})$ time giving an approximation ratio of $\mathcal{O}(\log n)$.*

Proof. See the full version of the paper [32] for the proof.

5   Conclusion

In this paper, we deal with a variant of the Art Gallery problem in which the guards are empowered with reflecting edges. Many applications consider one source and they want to make that source visible via various access points. Consider a WiFi network in an organization where due to some policies all the personnel should be connected to one specific network. The access points should receive the signal from one source and deliver it to places where the source cannot access. The problem is to minimizing the access points.

The gallery is denoted by a given simple polygon $\mathcal{P}$. This article mentioned a few versions of the problem of adding an area to the visibility polygon of a given point guard inside $\mathcal{P}$ as a viewer. Although we know that reflection could be helpful, we proved that several versions of the problem are *NP*-hard or *NP*-complete.

Nonetheless, we proved that although specular reflection might not help decrease the *minimum* number of guards required for guarding a gallery, diffuse reflection can decrease the optimal number of guards.

References

1. Vaezi, A., Roy, B., Ghodsi, M.: Reflection helps guarding an art gallery. In: The 38th European Workshop on Computational Geometry, Perugia, Italy, March 2022, hal-03674221, pp. 3:1–3:7 (2022)
2. Guibas, L.J., Hershberger, J., Leven, D., Sharir, M., Tarjan, R.E.: Linear-time algorithms for visibility and shortest path problems inside triangulated simple polygons. Algorithmica **2**, 209–233 (1987)
3. Lee, D.T.: Visibility of a simple polygon. Comput. Vis. Graph. Image Process. **22**, 207–221 (1983)
4. Klee, V.: Is every polygonal region illuminable from some point? Comput. Geom. Am. Math. Monthly **76**, 180 (1969)
5. Tokarsky, G.T.: Polygonal rooms not illuminable from every point. Am. Math. Monthly **102**, 867–879 (1995)
6. Aronov, B., Davis, A.R., Dey, T.K., Pal, S.P., Prasad, D.: Visibility with one reflection. Discret. Comput. Geom. **19**, 553–574 (1998)
7. Davis, B.A.A.R., Dey, T.K., Pal, S.P., Prasad, D.: Visibility with multiple specular reflections. Discret. Comput. Geom. **20**, 62–78 (1998)
8. Urrutia, J.: Handbook of Computational Geometry (2000)
9. O'Rourke, J., Supowit, K.: Some NP-hard polygon decomposition problems. IEEE Trans. Inf. Theory **29**(2), 181–189 (1983)
10. Lee, D.T., Lin, A.: Computational complexity of art gallery problems. IEEE Trans. Inf. Theory **32**, 276–282 (1986)

11. Aggarwal, A.: The art gallery theorem: its variations, applications and algorithmic aspects. Ph.D. thesis, The Johns Hopkins University, Baltimore, MD (1984)
12. Abrahamsen, M., Adamaszek, A., Miltzow, T.: The art gallery problem is ∃ℝ-complete. In: Proceedings of the 50th Annual ACM SIGACT Symposium on Theory of Computing, STOC (2018)
13. Abrahamsen, M., Adamaszek, A., Miltzow, T.: Irrational guards are sometimes needed (2017)
14. Ghosh, S.K.: Approximation algorithms for art gallery problems. In: Proceedings of the Canadian Information Processing Society Congress, pp. 429–436 (1987)
15. King, J., Kirkpatrick, D.: Improved approximation for guarding simple galleries from the perimeter. Discret. Comput. Geom. **46**, 252–269 (2011)
16. Ashur, S., Filtser, O., Katz, M.J.: A constant-factor approximation algorithm for vertex guarding a WV-polygon. J. Comput. Geom. **12**(1), 128–144 (2021)
17. Vaezi, A.: A constant-factor approximation algorithm for point guarding an art gallery. arXiv:2112.01104 (2021)
18. Guenther, R.: Modern Optics. Wiley, New York (1990)
19. Born, M., Wolf, E.: Principles of Optics, 6th edn. Pergamon Press, Oxford (1980)
20. Newton, I.: Opticks, or a Treatise of the Reflections, Refractions, Inflections and Colours of Light, 4th edn. London (1730)
21. Boldrighini, C., Keane, M., Marchetti, F.: Billiards in polygons. Ann. Probab. **6**, 532–540 (1978)
22. Gutkin, E.: Billiards in polygons. Phys. D **19**, 311–333 (1986)
23. Kerckhoff, S., Masur, H., Smillie, J.: Ergodicity of billiard flows and quadratic differentials. Ann. Math. **124**, 293–311 (1986)
24. Kozlov, V.V., Treshchev, D.V.: Billiards: A Genetic Introduction to the Dynamics of Systems with Impacts. Translations of Mathematical Monographs, vol. 89, pp. 62–78. American Mathematical Society, Providence (1991)
25. Xu, C.: A generalization of the art gallery theorem with reflection and a cool problem (2011). https://chaoxuprime.com/posts/2011-06-06-a-generalization-of-the-art-gallery-theorem-with-reflection-and-a-cool-problem.html
26. Barequet, G., et al.: Diffuse reflection diameter in simple polygons. Discret. Appl. Math. **210**, 123–132 (2016). Seventh Latin-American Algorithms, Graphs, and Optimization Symposium, LAGOS 2013, Playa del Carmen, México, p. 2013
27. Fox-Epstein, E., Tóth, C.D., Winslow, A.: Diffuse reflection radius in a simple polygon. Algorithmica **76**, 910–931 (2016)
28. Vaezi, A., Ghodsi, M.: Visibility extension via reflection-edges to cover invisible segments. Theor. Comput. Sci. **789**, 22–33 (2019)
29. Vaezi, A., Ghodsi, M.: How to extend visibility polygons by mirrors to cover invisible segments. In: Poon, S.-H., Rahman, M.S., Yen, H.-C. (eds.) WALCOM 2017. LNCS, vol. 10167, pp. 42–53. Springer, Cham (2017). https://doi.org/10.1007/978-3-319-53925-6_4
30. Vaezi, A., Ghodsi, M.: Extending visibility polygons by mirrors to cover specific targets. In: EuroCG2013, pp. 13–16 (2013)
31. Kleinberg, J., Tardos, E.: Algorithm Design. Pearson Education India (2006)
32. Vaezi, A., Roy, B., Ghodsi, M.: Visibility extension via reflection (2020)
33. Ghosh, S.K.: Approximation algorithms for art gallery problems in polygons. Discret. Appl. Math. **158**(6), 718–722 (2010)

Improved and Generalized Algorithms for Burning a Planar Point Set

Prashant Gokhale[1], J. Mark Keil[2], and Debajyoti Mondal[2]($\boxtimes$) (iD)

[1] Indian Institute of Science, Bangalore, India
prashantag@iisc.ac.in
[2] University of Saskatchewan, Saskatoon, Canada
{keil,dmondal}@cs.usask.ca

Abstract. Given a set P of points in $\mathbb{R}^2$, a point burning process is a discrete time process to burn all the points of P where fires must be initiated at the points of P. Specifically, the point burning process starts with a single burnt point from P, and at each subsequent step, burns all the points in $\mathbb{R}^2$ that are within one unit distance from the currently burnt points, as well as one other unburnt point of P (if exists). The point burning number of P is the smallest number of steps required to burn all the points of P. If we allow the fire to be initiated anywhere in $\mathbb{R}^2$, then the burning process is called an anywhere burning process. One can think of the anywhere burning problem as finding the minimum integer r such that P can be covered with disks of radii $0, 1, 2, \ldots, r$. A burning process provides a simple model for distributing commodities to the locations in P by sending a daily bulk shipment to a distribution center, i.e., the place where we initiate a fire. The burning number corresponds to the minimum number of days to reach all locations. In this paper we show that both point and anywhere burning problems admit PTAS in one dimension. We then show that in two dimensions, point burning and anywhere burning are $(1.96+\varepsilon)$ and $(1.92+\varepsilon)$ approximable, respectively, for every $\varepsilon > 0$, which improves the previously known $(2 + \varepsilon)$ approximation for these problems. We then generalize the results by allowing the points to have different fire spreading rates, and prove that even if the burning sources are given as input, finding a point burning sequence itself is NP-hard. Finally, we obtain a 2-approximation for burning the maximum number of points in a given number of steps.

Keywords: Computational geometry · Burning number · Approximation algorithm · NP-hard

1 Introduction

Graph burning was introduced by Bonato et al. [4] as a simplified model to investigate the spread of influence in a network. Given a finite, simple, undirected

This work is supported in part by the Natural Sciences and Engineering Research Council of Canada (NSERC). The work of P. Gokhale was supported by a MITACS Globalink Internship at the University of Saskatchewan.

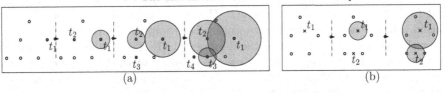

Fig. 1. Illustration for (a) point burning and (b) anywhere burning. The burning sources are illustrated in labelled dots and cross marks, respectively.

graph G, the burning process on G is defined as a discrete time process as follows. Initially, at time $t = 0$, all vertices in the graph are unburnt. Once a node is burnt, it remains so until the end of the process. At time t, where $t > 0$ is a positive integer, the process burns all the neighbors of the currently burnt vertices, as well as one more unburnt vertex (if exists). This process stops when all vertices are burnt. The graph burning problem seeks to minimize the number of steps required to burn the whole graph. We refer the reader to [3] for a survey on graph burning.

Keil et al. [10] introduced two geometric variants of this problem — point burning and anywhere burning, where the goal is to burn a given set of points P in $\mathbb{R}^2$. The *point burning model* allows for initiating fires only at points of P. The burning process starts by burning one point of P, and then at each subsequent step, the fire burns all unburnt points of $\mathbb{R}^2$ that are within one unit of any currently burnt point and a new unburnt point in P is chosen to initiate the fire. The *point burning number* of P is the smallest number of steps required to burn all the points of P. Figure 1(a) illustrates this model, where the burnt points are shaded in red. In the *anywhere burning model*, the burning process is the same but the fires can be started at any point in $\mathbb{R}^2$, and the corresponding burning number is called *anywhere burning number*. Figure 1(b) illustrates this model. Note that we may not have an unburnt point to initiate fire at the last step. Note that one can think of the anywhere burning problem as finding the minimum integer r such that P can be covered with disks of radii $0, 1, 2, \ldots, r$. The point burning problem is the same with the additional constraint that the centers of the disks must be chosen from P.

The geometric version of the burning process can provide a simple model of supply management where products need to be shipped in bulk to distribution centers. Consider a business that needs to maintain a continuous supply of perishable goods to a set of P locations. Each day it can manage to send one large shipment to a hub location that distributes the goods further to the nearby locations over time. The point burning considers only the points of P as potential hubs, whereas anywhere burning allows to create a hub at any point in $\mathbb{R}^2$. The burning number indicates the minimum number of days needed to distribute the goods to all locations. For example, in Fig. 1(a), the hubs are t_1, t_2, t_3 and t_4, and the business can keep sending the shipments to the hubs after every three days in the same order.

The graph burning problem is known to be NP-hard for forests of paths [2] and APX-hard for subcubic graphs [11]. However, the problem is approximable within a factor of 3 [5], which has recently been improved further to $(3 - 2/b)$ where b is the burning number of the input graph [7]. The introduction of point and anywhere burning naturally raises the question of whether one can prove analogous results for these problems. Keil et al. [10] showed that both problems are NP-hard, but approximable within a factor of $(2 + \varepsilon)$, for every $\varepsilon > 0$. However, a number of interesting problems are yet to be explored. For example, can we find better approximation algorithms? Does there exist a PTAS for these problems? Can we maximize the number of burnt points within a given time limit? What happens if the points have different rates for spreading the fire? This is relevant in practice when the distribution capabilities vary across different distribution centers. Can we find a burning sequence in polynomial time if the burning sources are given? This last question is known to be NP-complete for graph burning [11].

Contribution: In this paper, we obtain the following results.

- We show that in one dimension one can find a PTAS for both point and anywhere burning. In two dimensions, we improve the approximation ratio for point and anywhere burning to $(1.96296+\varepsilon)$ and $(1.92188+\varepsilon)$, respectively.
- We consider a generalization where the fire spreading rates vary across the given points. We show how to adapt the existing approximation algorithms to obtain constant-factor approximation for point burning if the ratio of the largest and the smallest rate is a constant.
- We prove that even if the burning sources are given as input, finding a point burning sequence itself is NP-hard. This problem was known to be NP-hard for graph burning, but this result does not hold in geometric setting.
- Our NP-hardness result implies that given a set of q burning sources, it is NP-hard to find a point burning sequence that maximizes the number of burnt points within q steps. In contrast, we show how to adapt a set cover technique to obtain a 2-approximation for burning the maximum number of points in a given number of steps. This result holds even when a set of burning sources are specified at the input.

2   Burning Number in One Dimension

In this section we consider the case when the points of P are on a line. Assume that the points are ordered in increasing x-coordinate and let $A[i]$ be the x-coordinate of the ith point from the left. Let δ^* be the burning number.

2.1   PTAS for Anywhere Burning

We now provide a polynomial-time approximation scheme (PTAS), i.e., a $(1+\varepsilon)$-approximation algorithm for every $\varepsilon > 0$, for the anywhere burning problem. Intuitively, we can visualize this problem as a covering problem with intervals in one dimension.

Our strategy is to make a guess δ for the burning number starting from 1. We keep increasing the guess by 1 as long as we can prove the current δ to be a lower bound on the burning number. At some point when we are unable to establish δ as a lower bound, we show how to find an approximate solution.

Note that for a δ, we have δ intervals of length 0, 2, 4, ... , $2(\delta - 1)$ to cover all the points. We group these intervals into t different groups as follows. The first group will have $\frac{\delta}{t}$ intervals with length at most $2\left(\frac{\delta}{t}\right)$. Generally, the jth group will have $\frac{\delta}{t}$ intervals with length larger than $2(j - 1)\left(\frac{\delta}{t}\right)$ and at most $2j\left(\frac{\delta}{t}\right)$. For each group, we now relax all its intervals such that their length is equal to the largest interval in the group, i.e., $2j\left(\frac{\delta}{t}\right)$. We use the notation $S(\delta)$ to denote this new set of intervals. We now use dynamic programming to check if there is a placement of the intervals in $S(\delta)$ so that every point is covered. If not, then we are sure that δ is not the burning number, as we had relaxed every interval. Otherwise, we will use these intervals to obtain an approximate solution. Later, we will show how to choose t to obtain a PTAS.

Let $V = (v_1, \ldots, v_{t+1})$ be a $(t+1)$-tuple of integers. Let $D(V)$ be the problem of covering v_{t+1} points of P from the left with v_j intervals of group j, where $1 \le j \le t$. We use $P(v_{t+1})$ to denote these points that are to be covered. Assume that the rest of the points, i.e., $P \setminus P(v_{t+1})$ are already covered by the rest of the intervals of $S(\delta)$. We can then express $D(V)$ using the following recursion with trivial values for the base cases: $D(V) = \bigvee_{j=1}^{t} D(W^j)$.

Here $W^j = (w_1, \ldots, w_{t+1})$ is similar to V except at two places: w_j and w_{t+1}. The jth element w_j is set to $(v_j - 1)$, because we have now used one more interval I of group j to cover some points of $P(v_{t+1})$. Note that the best position for this interval is when its right end coincides with the rightmost point z of $P(v_{t+1})$. It is straightforward to prove this formally with the observation that for every covering we can shift the rightmost interval to the left unless its right end point coincides with z. Therefore, we set w_{t+1} to be the number of remaining points that remains to be covered after placing I.

We store the solution to the subproblems using a multidimensional table L where $L[V]$ stores the solution to $D(V)$. Using table lookup, the time taken to compute a cell of the table is $O(t)$. Since each of the t groups has $\frac{\delta}{t}$ intervals, and since P has n points, the size of L is $O\left(n\left(\frac{\delta}{t}\right)^t\right)$. Thus the overall running time of the dynamic programming algorithm is $O\left(nt\left(\frac{\delta}{t}\right)^t\right)$.

Note that the question of covering P using $S(\delta)$ is obtained from $L[U^\delta]$, where $U^\delta = (u_1, \ldots, u_{t+1})$ be a $(t+1)$-tuple with $u_j = \frac{\delta}{t}$ for $1 \le j \le t$ and $u_{t+1} = n$. We now describe the process of guessing δ and arriving at an approximate answer.

- Start guessing from $\delta = 1$
- Use the dynamic programming to compute $L[U^\delta]$, i.e., the solution to the relaxed covering question for $S(\delta)$ in $O\left(nt\left(\frac{\delta}{t}\right)^t\right)$ time.
- If $L[U^\delta]$ does not contain an affirmative answer, then we know that $\delta^* > \delta$. We thus iterate again by increasing value of δ by 1.

– If $L[U^\delta]$ contains an affirmative answer, stop and return the approximate burning number to be $\delta\left(1+\frac{2}{t}\right)$. At this point we know that $\delta \leq \delta^*$. We construct the burning sequence as follows: First, burn the midpoints of all intervals of length 2δ in the covering solution (i.e., the largest intervals), then burn the midpoints of all intervals of length $2(t-1)\left(\frac{\delta}{t}\right)$ in the solution and so on. However, since we used relaxed intervals in the dynamic programming, we keep burning for an extra $\frac{2\delta}{t}$ steps to ensure that each interval reaches its relaxed size (i.e., burns all points of P).

Observe that we took $\delta + \frac{2\delta}{t} = \delta\left(1+\frac{2}{t}\right)$ steps to burn all the points. Since $\delta \leq \delta^*$, we have $\delta\left(1+\frac{2}{t}\right) \leq \delta^*\left(1+\frac{2}{t}\right)$. Therefore, our algorithm achieves an approximation factor of $1+\frac{2}{t}$ for anywhere burning. Since $\delta^* \leq n$, the total running time of our algorithm is bounded by $O\left(n^2\left(\frac{n}{t}\right)^t\right)$. Given an $\varepsilon > 0$, we choose t such that $\varepsilon = \frac{2}{t}$. Thus we get an approximation factor of $(1+\varepsilon)$ and a running time of $O\left(n^{2+\frac{2}{\varepsilon}}\left(\frac{\varepsilon}{2}\right)^{\frac{2}{\varepsilon}}\right)$. We thus obtain the following theorem.

Theorem 1. *Given a set P of n points on a line and a positive constant $\varepsilon > 0$. One can approximate the anywhere burning number of P within a factor of $(1+\varepsilon)$ and compute the corresponding burning sequence in polynomial time.*

2.2   PTAS for Point Burning

We can slightly modify the algorithm for anywhere burning problem to obtain a PTAS for the point burning problem. We refer the reader to the full version [9].

Theorem 2. *Given a set P of n points on a line and a positive constant $\varepsilon > 0$. One can approximate the point burning number of P within a factor of $(1+\varepsilon)$ and compute the corresponding burning sequence in polynomial time.*

3   Burning Number in Two Dimensions

In this section we assume that the points of P are in $\mathbb{R}^2$. We first give a $(1.92188+\varepsilon)$-approximation algorithm for anywhere burning (Sect. 3.1) and then a $(1.96296+\varepsilon)$-approximation algorithm for point burning (Sect. 3.2). Note that this improves the previously known $(2+\varepsilon)$-approximation factor for these problems [10].

3.1   Anywhere Burning

Our algorithm for anywhere burning is inspired by the $(2+\varepsilon)$-approximation algorithm of Keil et al. [10], where we improve the approximation factor by using a geometric covering argument.

Keil et al. [10] leverage the discrete unit disk cover problem to obtain an approximation algorithm for anywhere burning. The input to the *discrete unit disk cover problem* is a set of points P and a set of unit disks $\mathcal{U}$ in $\mathbb{R}^2$. The goal is to choose the smallest set $U \subset \mathcal{U}$ that covers all the points of P. There exists a PTAS for the discrete unit disk cover problem [12].

Let δ^* be the actual burning number. Keil et al. [10] iteratively guess the anywhere burning number δ from 1 to n. For each δ, they construct a set of n disks, each of radius δ, that are centered at the points of P, and $\binom{n}{3} + \binom{n}{2}$ additional disks, where each disk is of radius δ and is centered at the center of a circle determined by either two or three points of P. The reason is that any solution to the anywhere burning can be perturbed to obtain a subset of the discretized disks. Then they compute a $(1 + \varepsilon)$ approximation U'_δ for the discrete unit disk cover U_δ. If $\frac{|U'_\delta|}{(1+\varepsilon)} > \delta$, then δ cannot be the burning number as otherwise, one could construct a smaller discrete unit disk cover by choosing disks that are centered at the burning sources of an optimal burning sequence. At this point, the guess is increased by one. The iteration stops when $\frac{|U'_\delta|}{(1+\varepsilon)} \leq \delta$, where we know that $\delta \leq \delta^*$. At this point, Keil et al. [10] show how to construct a burning sequence of length $(2 + \varepsilon)$.

We now describe a new technique for constructing the burning sequence. To burn all points in P, we use $1.92188\delta(1 + \varepsilon)$ steps. We first choose the centers of a 0.92188 fraction of U'_δ disks and burn them in arbitrary order. Later, we will see that the number 0.92188 relates to a geometric covering result [13]. This requires $0.92188|U'_\delta| = 0.92188\delta(1 + \varepsilon)$ steps. We then burn for another $\delta(1 + \varepsilon)$ steps. This will ensure the previously chosen $0.92188|U'_\delta|$ fires to have a radius of at least δ. Therefore, these fires will burn all the points that are covered by the corresponding disks of the discrete unit disk cover solution. We are now left with $(1 - 0.92188)|U'_\delta| = 0.07812|U'_\delta|$ disks in U'_δ that need to be covered using the next $\delta(1 + \varepsilon)$ steps. Observe that $(1 - 0.6094) = 0.3906$ fraction of these $\delta(1 + \varepsilon)$ fires have radius at least 0.6094. Since a unit disk can be covered by 5 equal disks[1] of radius at most 0.6094 [13], we can cover the remaining $\frac{0.3906}{5}|U'_\delta| = 0.07812|U'_\delta|$ disks of U'_δ.

Since $\delta \leq \delta^*$, We have the following theorem.

Theorem 3. *Given a set P of points in $\mathbb{R}^2$ and an $\varepsilon > 0$, one can compute an anywhere burning sequence in polynomial time where the length of the sequence is at most $1.92188(1 + \varepsilon)$ times the anywhere burning number of P.*

3.2   Point Burning

The algorithm for anywhere burning can be easily adapted to provide a $(1.96296 + \varepsilon)$ approximation ratio for point burning. We refer the reader to the full version [9] for further details.

Theorem 4. *Given a set P of points in $\mathbb{R}^2$ and an $\varepsilon > 0$, one can compute a point burning sequence in polynomial time where the length of the sequence is at most $\frac{53(1+\varepsilon)}{27} \approx 1.96(1 + \varepsilon)$ times the point burning number of P.*

---

[1] http://oeis.org/A133077.

4   Generalizations for Point Burning

In this section we consider two generalizations of the problem.

The first one is *point burning with non-uniform rates* (Sect. 4.1). Specifically, for each i from 1 to n, the ith point in P is assigned a positive integer (*rate*) r_i. If a fire starts at the ith point q, then the fire will spread with a rate of r_i per step. Specifically, at the kth step, where k is a positive integer, the fire burns all points of $\mathbb{R}^2$ that are within kr_1 units of q. Note that point burning with uniform rates, i.e., $r_1 = \ldots = r_n$ reduces to point burning. The second one is k-*burning number*, i.e., when k points can be burned at each step (Sect. 4.2). This version has previously been considered for graph burning and graph k-burning number is known to be 3 approximable [11].

4.1   Point Burning with Non-uniform Rates

Let h be the ratio of the fastest rate to the slowest rate, i.e., $h = \max_{1 \leq i,j \leq n} \frac{r_i}{r_j}$ (intuitively, it is the maximum ratio over all pairs of rates). In this section we show that for every fixed h, point burning number with non-uniform rates is approximable within a constant factor. We will use the concept of dominating set in a disk graph. A *disk graph* is a geometric intersection graph where the vertices correspond to a set of disks in the plane and there is an edge if and only if the corresponding pair of disks intersect. A *dominating set* in a disk graph is a subset S of vertices such that every vertex is either in S or has a neighbour in S. Gibson and Parwani [8] provides a PTAS for finding a minimum dominating set in disk graphs, where the disks can have different radii.

We are now ready to present the algorithm. For a positive integer m, let G_m be the disk graph obtained by constructing for each point $t \in P$, a disk centered at t with radius $\frac{m}{2}r_t$. Let D_m be a minimum dominating set of G_m.

We start guessing the burning number δ from 1 to n, and for each guess, we compute a $(1 + \varepsilon)$-approximate dominating set $E_{\delta-1}$ of $G_{\delta-1}$. We now have $|E_{\delta-1}| \leq (1+\varepsilon)|D_{\delta-1}|$. If $\delta < \frac{|E_{\delta-1}|}{(1+\varepsilon)} \leq |D_{\delta-1}|$, then we can claim that δ burning sources are not enough to burn all the points and can increase the guess by 1. Suppose for a contradiction that all the points can be burned in δ steps. We can then choose the disks corresponding to the burning sources to obtain a dominating set with less than $|D_{\delta-1}|$ disks, a contradiction.

Once we get $\delta \geq \frac{|E_{\delta-1}|}{(1+\varepsilon)}$, we stop. At this point, we know that $\delta \leq \delta^*$. We now construct a burning sequence by first burning all the points in $E_{\delta-1}$ (in an arbitrary order) and then continuing the burning for $h(\delta - 1)$ more steps. We need to show all the points of P are burned. Take some point $p \in P$, if $p \in E_{\delta-1}$, then it is clearly burned. Otherwise, p is dominated by a point $q \in E_{\delta-1}$. Here the Euclidean distance between p and q is at most $\frac{\delta-1}{2}(r_p + r_q)$.

If $r_q \geq r_p$, then the radius for the fire initiated at q is $r_q(\delta-1)$ which is larger than $\frac{\delta-1}{2}(r_p + r_q)$. Therefore, p must be burned.

Otherwise, assume that $r_q < r_p$. Here the distance between p and q is at most $\frac{\delta-1}{2}(r_p + r_q) \leq (\delta - 1)r_p$. Since $q \in E_{\delta-1}$, by our burning strategy, the fire at q will continue to burn for at least $h(\delta - 1)$ steps. Therefore, its radius is at least $r_q h(\delta - 1)$ steps. Since h is the maximum ratio of the burning rates, $r_q h(\delta - 1) \geq r_q(\frac{r_p}{r_q})(\delta - 1) = (\delta - 1)r_p$. Hence the point p must be burned.

The number of rounds taken by our algorithm is $|E_{\delta-1}| + h(\delta - 1) \leq (1 + \varepsilon)\delta^* + h(\delta^* - 1) = (1 + h + \varepsilon)\delta^*$. We thus obtain the following theorem.

Theorem 5. *Let P be a set of points in $\mathbb{R}^2$, where each point is assigned a burning rate. Let h be the ratio of the fastest rate to the slowest rate. Given an $\varepsilon > 0$, one can compute a point burning sequence in polynomial time where the sequence length is at most $(1 + h + \varepsilon)$ times the point burning number of P.*

4.2 k-Burning with Non-uniform Rates

We now consider the point burning model when k points are allowed to burn at each step and the goal is to compute the k-burning number, i.e., minimum number of rounds to burn all points of P. Our algorithm for this model is the same as in the previous section except that we stop iterating the guess when $k\delta \geq \frac{|E_{\delta-1}|}{(1+\varepsilon)}$. The reason we keep iterating in the case when $k\delta < \frac{|E_{\delta-1}|}{(1+\varepsilon)} < |D_{\delta-1}|$ is that burning all points in $k\delta$ steps would imply the existence of a dominating set of size smaller than $|D_{\delta-1}|$. We thus obtain the following theorem.

Theorem 6. *Let P be a set of points in $\mathbb{R}^2$, where each point is assigned a burning rate, and let h be the ratio of the fastest rate to the slowest rate. Given an $\varepsilon > 0$ and a positive integer $k > 0$, one can compute a point k-burning sequence in polynomial time where the length of the sequence is at most $(1 + h + \varepsilon)$ times the point burning number of P.*

5   NP-Hardness

In this section we show that computing a point burning sequence is NP-hard even if we are given the burning sources.

We will reduce the NP-Hard problem LSAT [1], which is a 3-SAT formula where each clause (viewed as a set of literals) intersects at most one other clause, and, moreover, if two clauses intersect, then they have exactly one literal in common. Given an LSAT instance, one can sort the literals such that each clause corresponds to at most three consecutive literals, and each clause may share at most one of its literals with another clause, in which case this literal is extreme in both clauses [1].

Let I be an instance of LSAT with m clauses and n variables. Without loss of generality we may assume for every variable, both its positive and negative literals appear in I. Otherwise, we could set a truth value to the variable to satisfy some clauses and remove them to obtain an LSAT instance I' which is satisfiable if and only if I is satisfiable. We will construct a point set P with $4n + m$ points and identify $2n$ points to be used as the burning sources. We will show that I is satisfiable if and only if P has a point burning sequence that uses only the given $2n$ points as the burning sources.

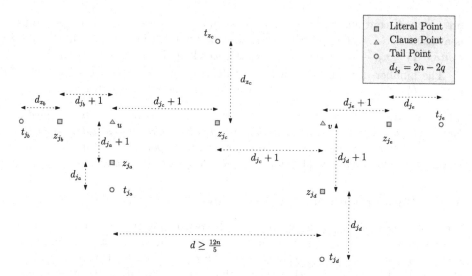

Fig. 2. Illustration for the construction of the point set for a pair of intersecting clauses: $(j_a \vee j_b \vee j_c)$ and $(j_c \vee j_d \vee j_e)$.

Construction of the Point Set: The point set includes one point per clause, which is called a *clause point* and one point per literal, which is called a *literal point*. For each literal, we add a point, which we call a *tail point* of that literal point. For example, consider a pair of clauses with k distinct literals and assume that these clauses have a literal in common. The corresponding point set consists of two clause points, k literal points and k tail points. Figure 2 gives an illustration with $k = 5$.

We now describe the construction in detail. We call a clause (or a pair of clauses that share a common literal) *independent* if it does not intersect any other clause. We place these independent elements (clauses or pairs of clauses) far from each other such that points in one element are at least n^2 units apart from the points in any other element.

If a literal is shared among a pair of clauses, then we call it an *intersection literal*. If each of the two clauses corresponding to an intersection literal is of size three (i.e., contains three literals), then we call intersection literal *heavy* and otherwise, we call it *light*. Note that there can be at most $\frac{2n}{5}$ heavy literals (as each such literal corresponds to a pair of clauses with 5 distinct literals, which must be independent by the property of LSAT). We now relabel the variables (with labels $1, \ldots, n$) such that variables that correspond to heavy literals get lower labels. For a variable with label k, we label its literals as j_k and $\overline{j_k}$. Thus for any heavy literal j_k, we have $k \leq \frac{2n}{5}$.

We first describe the construction of the points corresponding to a heavy literal. Let $(j_a \vee j_b \vee j_c)$ and $(j_c \vee j_d \vee j_e)$ be the corresponding clauses. For an integer q let d_{j_q} to be the distance of $(2n-2q)$ units. In general, when placing the literal point z_{j_q}, we ensure that it is at distance $(d_{j_q}+1)$ from its corresponding clause point. When placing a tail point t_{j_q}, we ensure that it is at a distance d_{j_q} from its corresponding literal point z_{j_q}. Therefore, in the following we only describe the position of the literal, clause and tail points relative to each other.

We place a clause point u for $(j_a \vee j_b \vee j_c)$. We place the literal points z_{j_b} and z_{j_c} to the left and right of the u (on the horizontal line through u), respectively. We place the literal point z_{j_a} vertically below u. We then place the clause point v and its literal points to the right of the intersection literal z_{j_c} symmetrically, as shown in Fig. 2. We place the tail points t_{j_b} and t_{j_e} to the left and right of z_{j_b} and z_{j_e}, respectively. The tail points t_{j_a} and t_{j_d} are placed vertically below z_{j_a} and z_{j_d}, respectively. Finally, the tail point t_{j_c} is placed vertically above z_{j_c}.

For the light literals, we have at most 4 distinct literal points to place. The construction is the same as above and we intentionally put three literal points on the line passing through the clause points and the remaining one (if exists) below its corresponding clause point.

For each independent clause, the construction is again the same as for $(j_a \vee j_b \vee j_c)$. If it contains two literals, then we place both literal points on the horizontal line passing through the clause point.

Remark 1. For a literal z_q, the construction ensures the following properties.

P_1: The distance of t_{z_q} is strictly greater than $2n - 2q + 1$ from all literal points except for its own literal point.

P_2: The distance of z_{j_q} is strictly greater than $2n - 2q + 1$ from all tail points except for its own tail point.

Remark 1 is straightforward to verify from the construction except for the case when a heavy literal correspond to two literal points below the line through the clause points. It may initially appear that if the construction places them in close proximity, then the nearest tail point of one literal point may be the tail point of the other literal point. However, such a scenario does not appear due to our initial relabelling of the literals, i.e., the smallest distance between the vertical lines through these literal points is at least $d = 2(2n - 2(\frac{2n}{5})) = 2(\frac{6n}{5}) = \frac{12n}{5} > 2n$ (Fig. 2).

Reduction: First we show that if I is satisfiable, then P can be burnt in $2n$ steps using only the literal points as sources. For every r from 1 to n, if j_r is true then burn z_{j_r} at the $(2r - 1)$th step and burn $z_{\overline{j_r}}$ at the $2r$th step. If j_r is false, then swap the steps. Any tail point t_{j_a} is at a distance of $(2n - 2a)$ from z_{j_a}. Even if j_a is false, it still has $(2n - 2a)$ steps left to burn, which will be enough to burn the tail point. It only remains to show that all clause points are burned. Let u be any clause point, as I is satisfiable, at least one of its variable j_r must be true, which will burn for at least $2n - (2r - 1) = 2n - 2r + 1$ steps and thus burn u.

We now assume that there is a burning sequence that burns all the points of P using the literal points as sources. To construct a satisfying truth assignment for the LSAT, we use the following lemma and the proof is included in the full version [9].

Lemma 1. *For a variable with label r, its literal points z_{j_r} and $z_{\overline{j_r}}$ are burnt in the $(2r-1)$th and $2r$th step, respectively (or in the reverse order).*

We now construct a satisfying truth assignment as follows. If z_{j_r} or $z_{\overline{j_r}}$ is burnt in odd step, we set it to be true. Otherwise, we set it to be false. We now show that I is satisfied. Assume for a contradiction that there is a clause c which is not satisfied, i.e., all of its associated literal points are burnt in even steps. By Lemma 1, each literal point z_{j_a} corresponding to c burns for $(2n-2a)$ steps, which is not enough to burn the clause point. This contradicts our initial assumption of a valid burning sequence. This completes the NP-hardness reduction.

Theorem 7. *Given a set of points P in the plane and a subset S of P, it is NP-hard to construct a point burning sequence using only the points of S that burns all the points of P within $|S|$ rounds.*

5.1   Burning Maximum Number of Points

Our NP-hardness result implies that given a subset $S = \{s^1, \ldots, s^q\}$ of q points from P, it is NP-hard to burn the maximum number of points by only burning the points of S within q rounds. We now show how to obtain a 2-approximation for this problem. For every point s^j, where $1 \leq j \leq q$, we consider q sets. The ith set Δ_i^j, where $1 \leq i \leq q$, contains the points that are covered by the disk of radius i centered at p, i.e., these points are within a distance of i from p. We thus have a collection of sets $\{\Delta_1^1, \ldots, \Delta_q^1, \ldots, \Delta_1^q, \ldots, \Delta_q^q\}$, which can be partitioned into q groups based on radius, i.e., the ith group contains the sets $\{\Delta_i^1, \ldots, \Delta_i^q\}$. To burn the maximum number of points by burning S, we need to select one subset from each radius group so that the cardinality of the union of these sets is maximized. This is exactly the maximum set cover problem with group budget constraints, which is known to be 2-approximable [6].

Theorem 8. *Given a set P of n points in the plane and a subset S of q points from P, one can compute a point burning sequence using S within q rounds in polynomial time that burns at least half of the maximum number of points that can be burned using S within q rounds.*

If we want to burn maximum number of points within q rounds, then we can set S to be equal to P to have a 2-approximate solution. We thus have the following corollary.

Corollary 1. *Given a set of n points in the plane and a positive integer $q < n$, one can compute a point burning sequence in polynomial time that burns at least half of the maximum number of points that can be burned within q rounds.*

6   Conclusion

In this paper we have shown that point burning and anywhere burning problems admit PTAS in one dimension and improved the known approximation factors in two dimensions. To improve the previously known approximation factor in two dimensions we used a geometric covering argument. We believe our covering strategy can be refined further by using a tedious case analysis. However, this would not provide a PTAS. Therefore, the most intriguing question in this context is whether these problems admit PTAS in two dimensions.

We have also proven that the problem of burning the maximum number of points within a given number of rounds is NP-hard, but 2-approximable by a known result on set cover with group budget constraints. It would be interesting to design a better approximation algorithm leveraging the geometric structure.

References

1. Arkin, E.M., et al.: Selecting and covering colored points. Discret. Appl. Math. **250**, 75–86 (2018)
2. Bessy, S., Bonato, A., Janssen, J.C.M., Rautenbach, D., Roshanbin, E.: Burning a graph is hard. Discret. Appl. Math. **232**, 73–87 (2017)
3. Bonato, A.: A survey of graph burning. Contrib. Discret. Math. **16**(1), 185–197 (2021)
4. Bonato, A., Janssen, J., Roshanbin, E.: Burning a graph as a model of social contagion. In: Bonato, A., Graham, F.C., Prałat, P. (eds.) WAW 2014. LNCS, vol. 8882, pp. 13–22. Springer, Cham (2014). https://doi.org/10.1007/978-3-319-13123-8_2
5. Bonato, A., Kamali, S.: Approximation algorithms for graph burning. In: Gopal, T.V., Watada, J. (eds.) TAMC 2019. LNCS, vol. 11436, pp. 74–92. Springer, Cham (2019). https://doi.org/10.1007/978-3-030-14812-6_6
6. Chekuri, C., Kumar, A.: Maximum coverage problem with group budget constraints and applications. In: Jansen, K., Khanna, S., Rolim, J.D.P., Ron, D. (eds.) APPROX/RANDOM -2004. LNCS, vol. 3122, pp. 72–83. Springer, Heidelberg (2004). https://doi.org/10.1007/978-3-540-27821-4_7
7. Garcia-Diaz, J., Sansalvador, J.C.P., Rodríguez-Henríquez, L.M., Cornejo-Acosta, J.A.: Burning graphs through farthest-first traversal. IEEE Access **10**, 30395–30404 (2022)
8. Gibson, M., Pirwani, I.A.: Algorithms for dominating set in disk graphs: breaking the $\log n$ barrier. In: de Berg, M., Meyer, U. (eds.) ESA 2010. LNCS, vol. 6346, pp. 243–254. Springer, Heidelberg (2010). https://doi.org/10.1007/978-3-642-15775-2_21
9. Gokhale, P., Keil, J.M., Mondal, D.: Improved and generalized algorithms for burning a planar point set. CoRR abs/2209.13024 (2022)
10. Keil, J.M., Mondal, D., Moradi, E.: Burning number for the points in the plane. In: Proceedings of the 34th Canadian Conference on Computational Geometry (2022)
11. Mondal, D., Rajasingh, A.J., Parthiban, N., Rajasingh, I.: APX-hardness and approximation for the k-burning number problem. Theor. Comput. Sci. **932**, 21–30 (2022)
12. Mustafa, N.H., Ray, S.: Improved results on geometric hitting set problems. Discret. Comput. Geom. **44**(4), 883–895 (2010). https://doi.org/10.1007/s00454-010-9285-9
13. Neville, E.H.: On the solution of numerical functional equations illustrated by an account of a popular puzzle and of its solution. Proc. Lond. Math. Soc. **2**(1), 308–326 (1915)

On the Longest Flip Sequence to Untangle Segments in the Plane

Guilherme D. da Fonseca[1] (ID), Yan Gerard[2] (ID), and Bastien Rivier[2(✉)] (ID)

[1] Aix-Marseille Université and LIS, Marseille, France
guilherme.fonseca@lis-lab.fr
[2] Université Clermont Auvergne and LIMOS, Clermont-Ferrand, France
{yan.gerard,bastien.rivier}@uca.fr

Abstract. A set of segments in the plane may form a Euclidean TSP tour or a matching, among others. Optimal TSP tours as well as minimum weight perfect matchings have no crossing segments, but several heuristics and approximation algorithms may produce solutions with crossings. To improve such solutions, we can successively apply a flip operation that replaces a pair of crossing segments by non-crossing ones. This paper considers the maximum number $\mathbf{D}(n)$ of flips performed on n segments. First, we present reductions relating $\mathbf{D}(n)$ for different sets of segments (TSP tours, monochromatic matchings, red-blue matchings, and multigraphs). Second, we show that if all except t points are in convex position, then $\mathbf{D}(n) = \mathcal{O}(tn^2)$, providing a smooth transition between the convex $\mathcal{O}(n^2)$ bound and the general $\mathcal{O}(n^3)$ bound. Last, we show that if instead of counting the total number of flips, we only count the number of distinct flips, then the cubic upper bound improves to $\mathcal{O}(n^{8/3})$.

Keywords: Planar geometry · Matching · Reconfiguration · Euclidean TSP

1 Introduction

In the Euclidean Travelling Salesman Problem (TSP), we are given a set P of n points in the plane and the goal is to produce a closed tour connecting all points of minimum total Euclidean length. The TSP problem, both in the Euclidean and in the more general graph versions, is one of the most studied NP-hard optimization problems, with several approximation algorithms, as well as powerful heuristics (see for example [2,13,17]). Multiple PTAS are known for the Euclidean version [3,26,30], in contrast to the general graph version that unlikely admits a PTAS [11]. It is well known that the optimal solution for the Euclidean TSP is a simple polygon, i.e., has no crossing segments, and in some situation a crossing-free solution is necessary [10]. However, most approximation algorithms (including Christofides and the PTAS), as well as a variety of

This work is supported by the French ANR PRC grant ADDS (ANR-19-CE48-0005).

C.-C. Lin et al. (Eds.): WALCOM 2023, LNCS 13973, pp. 102–112, 2023.
https://doi.org/10.1007/978-3-031-27051-2_10

simple heuristics (nearest neighbor, greedy, and insertion, among others) may produce solutions with pairs of crossing segments. In practice, these algorithms may be supplemented with a local search phase, in which crossings are removed by iteratively modifying the solution.

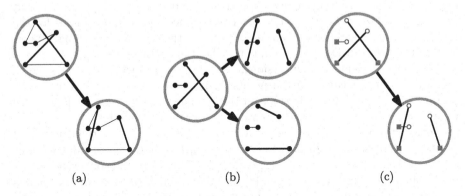

Fig. 1. Examples of flips in a (a) TSP tour, (b) monochromatic matching, and (c) red-blue matching. (Color figure online)

Given a Euclidean TSP tour, a *flip* is an operation that removes a pair of crossing segments and adds a new pair of segments preserving a tour (Fig. 1(a)). If we want to find a tour without crossing segments starting from an arbitrary tour, it suffices to find a crossing, perform a flip, and repeat until there are no crossings. It is easy to see that the process will eventually finish, as the length of the tour can only decrease when we perform a flip. Since a flip may create several new crossings, it is not obvious how to bound the number of flips performed until a crossing-free solution is obtained. Let $\mathbf{D}_{\mathrm{TSP}}(n)$ denote the maximum number of flips successively performed on a TSP tour with n segments. An upper bound of $\mathbf{D}_{\mathrm{TSP}}(n) = \mathcal{O}(n^3)$ is proved in [24], while the best lower bound known is $\mathbf{D}_{\mathrm{TSP}}(n) = \Omega(n^2)$. In contrast, if the points P are in convex position, then tight bounds of $\Theta(n^2)$ are easy to prove.

In this paper, we show that we can consider a conceptually simpler problem of flips in matchings (instead of Hamiltonian cycles), in order to bound the number of flips to both problems. Next, we describe this *monochromatic matching version*.

Consider a set of n line segments in the plane defining a matching M on a set P of $2n$ points. In this case, a *flip* replaces a pair of crossing segments by a pair of non-crossing ones using the same four endpoints (Fig. 1(b)). Notice that, in contrast to the TSP version, one of two possible pairs of non-crossing segments is added. As previously, let $\mathbf{D}_{\mathrm{MM}}(n)$ denote the maximum number of flips successively performed on a monochromatic matching with n segments. In Sect. 2, we show that $\mathbf{D}_{\mathrm{MM}}(n) \leq \mathbf{D}_{\mathrm{TSP}}(3n) \leq \mathbf{D}_{\mathrm{MM}}(3n)$, hence it suffices to prove asymptotic bounds for $\mathbf{D}_{\mathrm{MM}}(n)$ in order to bound $\mathbf{D}_{\mathrm{TSP}}(n)$.

A third and last version of the problem that we consider is the *red-blue matching version*, in which the set P is partitioned into two sets of n points each called *red* and *blue*, with segments only connecting points of different colors (Fig. 1(c)). Let $\mathbf{D}_{\text{RB}}(n)$ denote the analogous maximum number of flips successively performed on a red-blue matching with n segments. The red-blue matching version has been thoroughly studied [6,12]. In Sect. 2, we also show that $\mathbf{D}_{\text{MM}}(n) \leq \mathbf{D}_{\text{RB}}(2n) \leq \mathbf{D}_{\text{MM}}(2n)$ and, as a consequence, asymptotic bounds for the monochromatic matching version also extend to the red-blue matching version. We use the notation $\mathbf{D}(n)$ for bounds that hold in all three versions.

For all the aforementioned versions, special cases arise when we impose some constraint on the location of the points P. In the *convex case*, P is in *convex position*. Then it is known that, for all three versions, $\mathbf{D}(n) = \Theta(n^2)$ [6,8]. This tight bound contrasts with the gap for the general case bounds.

For all three versions in the convex case, $\mathbf{D}(n) \leq \binom{n}{2}$ as the number of crossings decreases at each flip. The authors have recently shown that without convexity $\mathbf{D}_{\text{RB}}(n) \geq 1.5\binom{n}{2} - \frac{n}{4}$ [12], which is higher than the convex bound. A major open problem conjectured in [8] is to determine if the non-convex bounds are $\Theta(n^2)$ as the convex bounds. Unfortunately, the best upper bound known for the non-convex case remains $\mathbf{D}(n) = \mathcal{O}(n^3)$ [24] since 1981[1], despite recent work on this specific problem [6,8,12]. The best lower bound known is $\mathbf{D}(n) = \Omega(n^2)$ [8,12].

The argument for the convex case bound of $\mathbf{D}(n) \leq \binom{n}{2}$ breaks down even if all but one point are in convex position, as the number of crossings may not decrease. In Sect. 3, we present a smooth transition between the convex and the non-convex cases. We show that, in all versions, if there are t points anywhere and the remaining points are in convex position with a total of n segments, then the maximum number of flips is $\mathcal{O}(tn^2)$.

Finally, in Sect. 4, we use a balancing argument similar to the one of Erdős et al. [15] to show that if, instead of counting the number of flips, we count the number of distinct flips (two flips are the same if they change the same set of four segments), then we get a bound of $\mathcal{O}(n^{8/3})$.

1.1   Related Reconfiguration Problems

Combinatorial reconfiguration studies the step-by-step transition from one solution to another, for a given combinatorial problem. Many reconfiguration problems are presented in [20].

It may be tempting to use an alternative definition for a flip in order to remove crossings and reduce the length of a TSP tour. The *2OPT flip* is not restricted to crossing segments, as long as it decreases the Euclidean length of the tour. However, the number of 2OPT flips performed may be exponential [14].

Another important parameter $\mathbf{d}(n)$ is the *minimum* number of flips needed to remove crossings from any set of n segments. When the points are in convex

---

[1] While the paper considers only the TSP version, the proof of the upper bound also works for the matching versions, as shown in [8].

position, then it is known that $\mathbf{d}(n) = \Theta(n)$ in all three versions [6,12,28,32]. If the red points are on a line, then $\mathbf{d}_{RB}(n) = \mathcal{O}(n^2)$ [6,12]. In the monochromatic matching version, if we can choose which pair of segments to add in a flip, then $\mathbf{d}_{MM}(n) = \mathcal{O}(n^2)$ [8]. For all remaining cases, the best bounds known are $\mathbf{d}(n) = \Omega(n)$ and $\mathbf{d}(n) = \mathcal{O}(n^3)$.

It is also possible to relax the flip definition to all operations that replace two segments by two others with the same four endpoints, whether they cross or not [4,5,7,9,16,31]. This definition has also been generalized to multigraphs with the same degree sequence [18,19,22].

In the context of triangulations, a flip consists of removing one segment and adding another one while preserving a triangulation. Reconfiguration problems for triangulations are also widely studied [1,21,23,25,27,29].

1.2   Definitions

Consider a set of points P. We say that two segments $s_1, s_2 \in \binom{P}{2}$ *cross* if they intersect in exactly one point which is not an endpoint of either s_1 or s_2. Furthermore, a line ℓ and a segment s *cross* if they intersect in exactly one point that is not an endpoint of s.

Let s_1, s_1', s_2, s_2' be four segments of $\binom{P}{2}$ forming a cycle with s_1, s_2 crossing. We define a *flip* $f = s_1, s_2 \rightarrow s_1', s_2'$ as the function that maps any set (or multiset) of segments M containing the two crossing segments s_1 and s_2 to $f(M) = M \cup \{s_1', s_2'\} \setminus \{s_1, s_2\}$ provided that $f(M)$ satisfies the property required by the version of the problem in question (being a monochromatic matching, a red-blue matching, a TSP tour...). This leads to the most general version of the problem, called the *multigraph* version. We note that a flip preserves the degree of every point. However, a flip may not preserve the multiplicity of a segment, which is why, in certain versions, we must consider multisets and multigraphs and not just sets and graphs.

A *flip sequence* of *length* m is a sequence of flips $f_1, \ldots, f_m$ with a corresponding sequence of (multi-)sets of segments $M_0, \ldots, M_m$ such that $M_i = f_i(M_{i-1})$ for $i = 1, \ldots, m$. Unless mentioned otherwise, we assume general position for the points in P (no three collinear points).

Given a property Π over a multiset of n line segments, we define $\mathbf{D}_\Pi(n)$ as the maximum length of a flip sequence such that every multiset of n segments in the sequence satisfies property Π. We consider the following properties Π: TSP for Hamiltonian cycle, RB for red-blue matching, MM for monochromatic matching, and G for multigraph. Notice that if a property Π is stronger than a property Π', then $\mathbf{D}_\Pi(n) \leq \mathbf{D}_{\Pi'}(n)$.

2   Reductions

In this section, we provide a series of inequalities relating the different versions of $\mathbf{D}(n)$. We show that all different versions of $\mathbf{D}(n)$ have the same asymptotic behavior.

Theorem 1. *For all positive integer n, we have the following relations*

$$\mathbf{D}_{\mathrm{MM}}(n) = \mathbf{D}_{\mathrm{G}}(n), \tag{1}$$

$$2\mathbf{D}_{\mathrm{MM}}(n) \leq \mathbf{D}_{\mathrm{RB}}(2n) \leq \mathbf{D}_{\mathrm{MM}}(2n), \tag{2}$$

$$2\mathbf{D}_{\mathrm{MM}}(n) \leq \mathbf{D}_{\mathrm{TSP}}(3n) \leq \mathbf{D}_{\mathrm{MM}}(3n). \tag{3}$$

Proof. Equality 1 can be rewritten $\mathbf{D}_{\mathrm{G}}(n) \leq \mathbf{D}_{\mathrm{MM}}(n) \leq \mathbf{D}_{\mathrm{G}}(n)$. Hence, we have to prove six inequalities. The right-side inequalities are immediate, since the left-side property is stronger than the right-side property (using G instead of MM for inequality 3).

The proofs of the remaining inequalities follow the same structure: given a flip sequence of the left-side version, we build a flip sequence of the right-side version, having similar length and number of points.

We first prove the inequality $\mathbf{D}_{\mathrm{G}}(n) \leq \mathbf{D}_{\mathrm{MM}}(n)$. A point of degree δ larger than 1 can be replicated as δ points that are arbitrarily close to each other in order to produce a matching of $2n$ points. This replication preserves the crossing pairs of segments (possibly creating new crossings). Thus, for any flip sequence in the multigraph version, there exists a flip sequence in the monochromatic matching version of equal length, yielding $\mathbf{D}_{\mathrm{G}}(n) \leq \mathbf{D}_{\mathrm{MM}}(n)$.

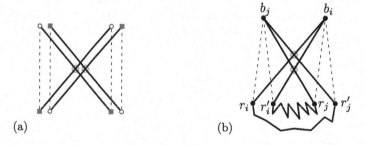

Fig. 2. (a) Two red-blue flips to simulate a monochromatic flip. (b) Two TSP flips to simulate a monochromatic flip. (Color figure online)

The left inequality of (2) is obtained by duplicating the monochromatic points of the matching M into two arbitrarily close points, one red and the other blue. Then each segment of M is also duplicated into two red-blue segments. We obtain a bichromatic matching M' with $2n$ segments. A crossing in M corresponds to four crossings in M'. Flipping this crossing in M amounts to choose which of the two possible pairs of segments replaces the crossing pair. It is simulated by flipping the two crossings in M' such that the resulting pair of double segments corresponds to the resulting pair of segments of the initial flip. These two crossings always exist and it is always possible to flip them one after the other as they involve disjoint pairs of segments. Figure 2(a) shows this construction. A sequence of m flips on M provides a sequence of $2m$ flips on M'. Hence, $2\mathbf{D}_{\mathrm{MM}}(n) \leq \mathbf{D}_{\mathrm{RB}}(2n)$.

To prove the left inequality of (3), we start from a red-blue matching M with $2n$ points and n segments and build a tour T with $3n$ points and $3n$ segments. We then show that the flip sequence of length m on M provides a flip sequence of length $2m$ on T. We build T in the following way. Given a red-blue segment $rb \in M$, the red point r is duplicated in two arbitrarily close points r and r' which are adjacent to b in T. We still need to connect the points r and r' in order to obtain a tour T. We define T as the tour $r_1, b_1, r'_1, \ldots, r_i, b_i, r'_i, \ldots, r_n, b_n, r'_n \cdots$ where r_i is matched to b_i in M (Fig. 2(b)).

We now show that a flip sequence of M with length m provides a flip sequence of T with length $2m$. For a flip $r_i b_i, r_j b_j \twoheadrightarrow r_i b_j, r_j b_i$ on M, we perform two successive flips $r_i b_i, r'_j b_j \twoheadrightarrow r_i b_j, r'_j b_i$ and $r'_i b_i, r_j b_j \twoheadrightarrow r'_i b_j, r_j b_i$ on T.

The tour then becomes $r_1, b_1, r'_1, \ldots, r_i, b_j, r'_i, \ldots, r_j, b_i, r'_j, \ldots, r_n, b_n, r'_n, \cdots$ on which we can apply the next flips in the same way. Hence, $2\mathbf{D}_{\mathsf{MM}}(n) \leq \mathbf{D}_{\mathsf{TSP}}(3n)$, concluding the proof. $\qquad\square$

3   Near Convex Sets

In this section, we bridge the gap between the $\mathcal{O}(n^2)$ bound on the length of flip sequences for a set P of points in convex position and the $\mathcal{O}(n^3)$ bound for P in general position. We prove the following theorem in the monochromatic matching version; the translations to the other versions follows from the reductions from Sect. 2, noticing that all reductions preserve the number of points in non-convex position up to constant factors.

Theorem 2. *In the monochromatic matching version with n segments, if all except t points of P are in convex position, then the length of a flip sequence is $\mathcal{O}(tn^2)$.*

Proof. The proof strategy is to combine the potential Φ_X used in the convex case with the potential Φ_L used in the general case. Given a matching M, the potential $\Phi_X(M)$ is defined as the number of crossing pairs of segments in M. Since there are n segments in M, $\Phi_X(M) \leq \binom{n}{2} = \mathcal{O}(n^2)$. Unfortunately, with points in non-convex position, a flip f might *increase* Φ_X, i.e. $\Phi_X(f(M)) \geq \Phi_X(M)$ (as shown in Fig. 1).

The potential Φ_L is derived from the line potential introduced in [24] but instead of using the set of all the $\mathcal{O}(n^2)$ lines through two points of P, we use a subset of $\mathcal{O}(tn)$ lines in order to take into account that only t points are in non-convex position. More precisely, let the potential $\Phi_\ell(M)$ of a line ℓ be the number of segments of M crossing ℓ. Note that $\Phi_\ell(M) \leq n$. The potential $\Phi_L(M)$ is then defined as follows: $\Phi_L(M) = \sum_{\ell \in L} \Phi_\ell(M)$.

We now define the set of lines L as the union of L_1 and L_2, defined hereafter. Let C be the subset containing the $2n - t$ points of P which are in convex position. Let L_1 be the set of the $\mathcal{O}(tn)$ lines through two points of P, at least one of which is not in C. Let L_2 be the set of the $\mathcal{O}(n)$ lines through two points of C which are consecutive on the convex hull boundary of C.

Let the potential $\Phi(M) = \Phi_X(M) + \Phi_L(M)$. We have the following bounds: $0 \le \Phi(M) \le \mathcal{O}(tn^2)$. To complete the proof of Theorem 2, we show that any flip decreases Φ by at least 1 unit.

We consider an arbitrary flip $f = p_1p_3, p_2p_4 \twoheadrightarrow p_1p_4, p_2p_3$. Let p_x be the point of intersection of p_1p_3 and p_2p_4. It is shown in [12,24] that f never increases the potential Φ_ℓ of a line ℓ. More precisely, we have the following three cases:

- The potential Φ_ℓ decreases by 1 unit if the line ℓ separates the final segments p_1p_4 and p_2p_3 and exactly one of the four flipped points belongs to ℓ. We call these lines f-critical (Fig. 3(a)).
- The potential Φ_ℓ decreases by 2 units if the line ℓ strictly separates the final segments p_1p_4 and p_2p_3. We call these lines f-dropping (Fig. 3(b)).
- The potential Φ_ℓ remains stable in the remaining cases.

Notice that, if a point q lies in the triangle $p_1p_xp_4$, then the two lines qp_1 and qp_4 are f-critical (Fig. 3(a)).

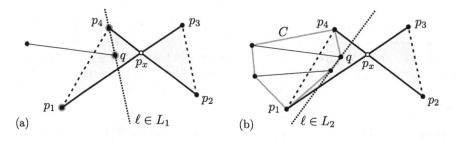

Fig. 3. (a) An f-critical line ℓ for a flip $f = p_1p_3, p_2p_4 \twoheadrightarrow p_1p_4, p_2p_3$. This situation corresponds to case (2a) with $\ell \in L_1$. (b) An f-dropping line ℓ. This situation corresponds to case (2b) with $\ell \in L_2$.

To prove that Φ decreases, we have the following two cases.

Case 1. If Φ_X decreases, as the other term Φ_L does not increase, then their sum Φ decreases as desired.

Case 2. If not, then Φ_X increases by an integer k with $0 \le k \le n-1$, and we know that there are $k+1$ new crossings after the flip f. Each new crossing involves a distinct segment with one endpoint, say q_i ($0 \le i \le k$), inside the non-simple polygon p_1, p_4, p_2, p_3 (Fig. 3). Next, we show that each point $q \in \{q_0, \ldots, q_k\}$ maps to a distinct line in L which is either f-dropping or f-critical, thus proving that the potential Φ_L decreases by at least $k + 1$ units.

We assume without loss of generality that q lies in the triangle $p_1p_xp_4$. We consider the two following cases.

Case 2a. If at least one among the points q, p_1, p_4 is not in C, then either qp_1 or qp_4 is an f-critical line $\ell \in L_1$ (Fig. 3(a)).

Case 2b. If not, then q, p_1, p_4 are all in C, and the two lines through q in L_2 are both either f-dropping (the line ℓ in Fig. 3(b)) or f-critical (the line qp_4 in

Fig. 3(b)). Consequently, there are more lines $\ell \in L_2$ that are either f-dropping or f-critical than there are such points $q \in C$ in the triangle $p_1 p_x p_4$, and the theorem follows.   □

4   Distinct Flips

In this section, we prove the following theorem in the monochromatic matching version, yet, the proof can easily be adapted to the other versions. We remark that two flips are considered *distinct* if the sets of four segments in the flips are different.

Theorem 3. *In all versions with n segments, the number of distinct flips in any flip sequence is $\mathcal{O}(n^{8/3})$.*

The proof of Theorem 3 is based on a balancing argument from [15] and is decomposed into two lemmas that consider a flip f and two matchings M and $M' = f(M)$. Similarly to [24], let L be the set of lines defined by all pairs of points in $\binom{P}{2}$. For a line $\ell \in L$, let $\Phi_\ell(M)$ be the number of segments of M crossed by ℓ and $\Phi_L(M) = \sum_{\ell \in L} \Phi_\ell(M)$. Notice that $\Phi(M) - \Phi(M')$ depends only on the flip f and not on M or M'. The following lemma follows immediately from the fact that $\Phi_L(M)$ takes integer values between 0 and $\mathcal{O}(n^3)$ [24].

Lemma 1. *For any integer k, the number of flips f in a flip sequence with $\Phi(M) - \Phi(M') \geq k$ is $\mathcal{O}(n^3/k)$.*

Lemma 1 bounds the number of flips (distinct or not) that produce a large potential drop in a flip sequence. Next, we bound the number of distinct flips that produce a small potential drop. The bound considers all possible flips on a fixed set of points and does not depend on a particular flip sequence.

Lemma 2. *For any integer k, the number of distinct flips f with $\Phi(M) - \Phi(M') < k$ is $\mathcal{O}(n^2 k^2)$.*

Proof. Let F be the set of flips with $\Phi(M) - \Phi(M') < k$ where $M' = f(M)$. We need to show that $|F| = \mathcal{O}(n^2 k^2)$. Consider a flip $f = p_1 p_3, p_2 p_4 \twoheadrightarrow p_1 p_4, p_2 p_3$ in F. Next, we show that there are at most $4k^2$ such flips with a fixed final segment $p_1 p_4$. Since there are $\mathcal{O}(n^2)$ possible values for $p_1 p_4$, the lemma follows. We show only that there are at most $2k$ possible values for p_3. The proof that there are at most $2k$ possible values for p_2 is analogous.

We sweep the points in $P \setminus \{p_4\}$ by angle from the ray $p_1 p_4$. As shown in Fig. 4, let $q_1, \ldots, q_k$ be the first k points produced by this sweep in one direction, $q_{-1} \ldots, q_{-k}$ in the other direction and $Q = \{q_{-k} \ldots, q_{-1}, q_1, \ldots, q_k\}$. To conclude the proof, we show that p_3 must be in Q. Suppose $p_3 \notin Q$ for the sake of a contradiction and assume without loss of generality that p_3 is on the side of q_i with positive i. Then, consider the lines $L' = \{p_1 q_1, \ldots, p_1 q_k\}$. Notice that $L' \subseteq L, |L'| = k$, and for each $\ell \in L'$ we have $\Phi_\ell(M) > \Phi_\ell(M')$, which contradicts the hypothesis that $\Phi(M) - \Phi(M') < k$.   □

Theorem 3 is a consequence of Lemmas 1 and 2 with $k = n^{1/3}$.

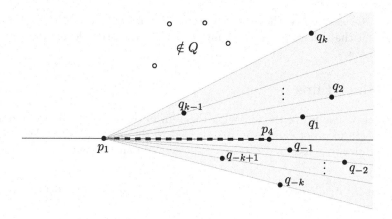

Fig. 4. Illustration for the proof of Lemma 2.

5   Conclusion and Open Problems

In Sect. 2, we showed a relationship among several different bounds on the maximum number of flips for a set of n line segments. This result shows how upper bounds to the monochromatic matching version can be easily transferred to different versions. But they can also be applied to transfer lower bounds among different versions. For example, the lower bound of $\frac{3}{2}\binom{n}{2} - \frac{n}{4}$ for the red-blue matching case [12] implies a lower bound of $\frac{1}{3}\binom{n}{2} - \frac{n}{2}$ for TSP. It is not clear if the constants in the TSP lower bound may be improved, perhaps by a more direct approach (or perhaps the lower bounds are not even asymptotically tight). However, we showed that all these versions are related by constant factors.

We can also use the results from Sect. 2 to convert the bounds from Sect. 3 to the general multigraph version, and hence to spanning trees and other types of graphs. In this case, the number t of points in non-convex position needs to be replaced by the sum of the degrees of the points in non-convex position. If the graphs are dense, or have non-constant degree, we do not know of any non-trivial lower bounds or better upper bounds.

Another key property that we did not consider in this paper is the length $\mathbf{d}(n)$ of the shortest flip sequence to untangle any set of n segments. In general, we only know that $\mathbf{d}(n) = \Omega(n)$ and $\mathbf{d}(n) \leq \mathbf{D}(n) = \mathcal{O}(n^3)$, for all versions. Whether similar reductions are possible is an elusive question. Furthermore, if the n points are in convex position, then $\mathbf{d}_{TSP}(n) = \Theta(n)$ for all versions. It is unclear if a transition in the case of t points in non-convex position is possible.

The result of Theorem 3 is based on the $\mathcal{O}(n^2k^2)$ bound from Lemma 2. A better analysis of the dual arrangement could potentially improve this bound, perhaps to $\mathcal{O}(n^2k)$.

The main open question, though, is whether the $\mathcal{O}(n^3)$ bound to both $\mathbf{D}(n)$ and $\mathbf{d}(n)$ can be improved for points in general position. The $\mathcal{O}(n^{8/3})$ bound on the number of distinct flips presented in Sect. 4 is a hopeful step in this direction, at least for $\mathbf{d}(n)$. We were not able to find a set of line segments that requires the same pair of segments to be flipped twice in order to be untangled.

References

1. Aichholzer, O., Mulzer, W., Pilz, A.: Flip distance between triangulations of a simple polygon is NP-complete. Discrete Comput. Geom. **54**(2), 368–389 (2015). https://doi.org/10.1007/s00454-015-9709-7
2. Applegate, D.L., Bixby, R.E., Chvátal, V., Cook, W.J.: The Traveling Salesman Problem. Princeton University Press, Princeton (2011)
3. Arora, S.: Polynomial time approximation schemes for Euclidean TSP and other geometric problems. In: 37th Conference on Foundations of Computer Science, pp. 2–11 (1996)
4. Bereg, S., Ito, H.: Transforming graphs with the same degree sequence. In: Ito, H., Kano, M., Katoh, N., Uno, Y. (eds.) KyotoCGGT 2007. LNCS, vol. 4535, pp. 25–32. Springer, Heidelberg (2008). https://doi.org/10.1007/978-3-540-89550-3_3
5. Bereg, S., Ito, H.: Transforming graphs with the same graphic sequence. J. Inf. Process. **25**, 627–633 (2017)
6. Biniaz, A., Maheshwari, A., Smid, M.: Flip distance to some plane configurations. Comput. Geom. **81**, 12–21 (2019). https://arxiv.org/abs/1905.00791
7. Bonamy, M., et al.: The perfect matching reconfiguration problem. In: 44th International Symposium on Mathematical Foundations of Computer Science. LIPIcs, vol. 138, pp. 80:1–80:14 (2019)
8. Bonnet, É., Miltzow, T.: Flip distance to a non-crossing perfect matching (2016). https://arxiv.org/abs/1601.05989
9. Bousquet, N., Joffard, A.: Approximating shortest connected graph transformation for trees. In: Chatzigeorgiou, A., et al. (eds.) SOFSEM 2020. LNCS, vol. 12011, pp. 76–87. Springer, Cham (2020). https://doi.org/10.1007/978-3-030-38919-2_7
10. Buchin, M., Kilgus, B.: Fréchet distance between two point sets. Comput. Geom. **102**, 101842 (2022)
11. Chlebík, M., Chlebíková, J.: Approximation hardness of travelling salesman via weighted amplifiers. In: Du, D.-Z., Duan, Z., Tian, C. (eds.) COCOON 2019. LNCS, vol. 11653, pp. 115–127. Springer, Cham (2019). https://doi.org/10.1007/978-3-030-26176-4_10
12. Das, A.K., Das, S., da Fonseca, G.D., Gerard, Y., Rivier, B.: Complexity results on untangling red-blue matchings. In: 15th Latin American Theoretical Informatics Symposium (LATIN 2022). LNCS, vol. 13568. Springer, Cham (2022). https://doi.org/10.1007/978-3-031-20624-5_44, https://arxiv.org/abs/2202.11857
13. Davendra, D.: Traveling Salesman Problem: Theory and Applications. BoD-Books on Demand, Norderstedt (2010)
14. Englert, M., Röglin, H., Vöcking, B.: Worst case and probabilistic analysis of the 2-Opt algorithm for the TSP. Algorithmica **68**(1), 190–264 (2013). https://doi.org/10.1007/s00453-013-9801-4
15. Erdös, P., Lovász, L., Simmons, A., Straus, E.G.: Dissection graphs of planar point sets. In: A Survey of Combinatorial Theory, pp. 139–149. Elsevier (1973)
16. Erdős, P.L., Király, Z., Miklós, I.: On the swap-distances of different realizations of a graphical degree sequence. Comb. Probab. Comput. **22**(3), 366–383 (2013)
17. Gutin, G., Punnen, A.P.: The Traveling Salesman Problem and Its Variations, vol. 12. Springer Science & Business Media, Berlin (2006)

18. Hakimi, S.L.: On realizability of a set of integers as degrees of the vertices of a linear graph. I. J. Soc. Ind. Appl. Math. **10**(3), 496–506 (1962)
19. Hakimi, S.L.: On realizability of a set of integers as degrees of the vertices of a linear graph II. uniqueness. J. Soc. Ind. Appl. Math. **11**(1), 135–147 (1963)
20. van den Heuvel, J.: The complexity of change. Surv. Comb. **409**, 127–160 (2013)
21. Hurtado, F., Noy, M., Urrutia, J.: Flipping edges in triangulations. Discrete Comput. Geom. **22**(3), 333–346 (1999)
22. Joffard, A.: Graph domination and reconfiguration problems. Ph.D. thesis, Université Claude Bernard Lyon 1 (2020)
23. Lawson, C.L.: Transforming triangulations. Discret. Math. **3**(4), 365–372 (1972)
24. van Leeuwen, J.: Untangling a traveling salesman tour in the plane. In: 7th Workshop on Graph-Theoretic Concepts in Computer Science (1981)
25. Lubiw, A., Pathak, V.: Flip distance between two triangulations of a point set is NP-complete. Comput. Geom. **49**, 17–23 (2015)
26. Mitchell, J.S.: Guillotine subdivisions approximate polygonal subdivisions: a simple polynomial-time approximation scheme for geometric TSP, k-mst, and related problems. SIAM J. Comput. **28**(4), 1298–1309 (1999)
27. Nishimura, N.: Introduction to reconfiguration. Algorithms **11**(4), 52 (2018)
28. Oda, Y., Watanabe, M.: The number of flips required to obtain non-crossing convex cycles. In: Ito, H., Kano, M., Katoh, N., Uno, Y. (eds.) KyotoCGGT 2007. LNCS, vol. 4535, pp. 155–165. Springer, Heidelberg (2008). https://doi.org/10.1007/978-3-540-89550-3_17
29. Pilz, A.: Flip distance between triangulations of a planar point set is APX-hard. Comput. Geom. **47**(5), 589–604 (2014)
30. Rao, S.B., Smith, W.D.: Approximating geometrical graphs via "spanners" and "banyans". In: Proceedings of the 30th Annual ACM Symposium on Theory of Computing, pp. 540–550 (1998)
31. Will, T.G.: Switching distance between graphs with the same degrees. SIAM J. Discret. Math. **12**(3), 298–306 (1999)
32. Wu, R., Chang, J., Lin, J.: On the maximum switching number to obtain non-crossing convex cycles. In: 26th Workshop on Combinatorial Mathematics and Computation Theory, pp. 266–273 (2009)

String Algorithm

Inferring Strings from Position Heaps
in Linear Time

Koshiro Kumagai[✉], Diptarama Hendrian, Ryo Yoshinaka,
and Ayumi Shinohara

Graduate School of Information Sciences, Tohoku University, Sendai, Japan
koshiro_kumagai@shino.ecei.tohoku.ac.jp,
{diptarama,ryoshinaka,ayumis}@tohoku.ac.jp

Abstract. Position heaps are index structures of text strings used for
the string matching problem. They are rooted trees whose edges and
nodes are labeled and numbered, respectively. This paper is concerned
with variants of the inverse problem of position heap construction and
gives linear-time algorithms for those problems. The basic problem is to
restore a text string from a rooted tree with labeled edges and numbered
nodes. In the variant problems, the input trees may miss edge labels or
node numbers which we must restore as well.

Keywords: Position heaps · Reverse engineering · Enumeration

1 Introduction

The string matching problem searches for occurrences of a pattern P in a text
T. It has been widely studied for many years and many efficient algorithms have
been proposed. Those techniques can be classified into mainly two approaches.
The first one is to construct data structures from P by preprocessing P. For
example, the Knuth-Morris-Pratt algorithm [19] constructs border arrays, the
Boyer-Moore method [5] constructs suffix tables, and the Z-algorithm [16] constructs prefix tables which is the dual notion of suffix tables. The other approach
is preprocessing T to create indexing structures, such as suffix trees [25], suffix
arrays [21], LCP arrays [21], suffix graphs [3], compact suffix graphs [4], and
position heaps [12]. Indexing structures are advantageous when searching for
many different patterns in a text.

The reverse engineering of those data structures has also been widely studied.
Studying reverse engineering deepens our insight into those data structures. For
example, it may enable us to design an algorithm generating indexing structures
with specific structural characteristics, which should be useful for verifying other
software processing them. The early studies targeted border arrays [9,10,14].
Later, Clément et al. [8] proposed a linear time algorithm for inferring strings
from prefix tables. Those data structures are produced by preprocessing patterns. The reverse engineering for indexing structures has been studied for suffix
arrays [2,11], LCP arrays [18], suffix graphs [2], and suffix trees [6,17,24]. The
techniques used in [17] and [24] involve finding an Eulerian cycles on a graph
modifying an input tree.

© The Author(s), under exclusive license to Springer Nature Switzerland AG 2023
C.-C. Lin et al. (Eds.): WALCOM 2023, LNCS 13973, pp. 115–126, 2023.
https://doi.org/10.1007/978-3-031-27051-2_11

In this paper, we discuss the reverse engineering of another type of indexing structures, called *position heaps* [12,20]. The position heap of a string T is a rooted tree with labeled edges and numbered nodes. Actually, Ehrenfeucht et al. [12] and Kucherov [20] gave different definitions of position heaps. By either definition, position heaps can be constructed in linear time online assuming the alphabet size to be constant. In addition, we can find all occurrence positions of a pattern P in $O(|P|^2 + k)$ time, where k is the output size. Moreover, by augmenting position heaps with additional data structures, we can improve the searching time to $O(|P| + k)$.

We consider the following four types of reverse engineering of Kucherov's position heaps [20]. The first problem is to restore a source text T from an input edge-labeled and node-numbered rooted tree so that the input should be the position heap $\mathrm{PH}(T)$ of T. While this problem allows at most one solution, the other problems may have many possible solutions. In the second problem, input trees miss edge labels. In the third problem, input trees miss node numberings. Instance trees of the fourth problem miss both edge labels and node numberings but have potential *suffix links* among nodes, which play an important role in the construction of position heaps. We show that all the problems above can be solved in linear time in the input size. Among those, we devote the most pages to the third problem. We reduce the problem to finding a special type of Eulerian cycle over the input tree augmented with suffix links. By showing the problem of finding an Eulerian cycle of this special type is linear-time solvable, we conclude that restoring a text from a position heap without node numbers is linear-time solvable. This can be seen analogous to the techniques used in [17] and [24] for the suffix tree reverse engineering. In addition, we present formulas for counting the number of possible text strings, which can be computed in polynomial time. Moreover, we show efficient algorithms for enumerating all possible text strings in output linear time.

2   Preliminaries

Let Σ be a finite alphabet and let the size of Σ be constant. For a string w over Σ, the length of w is denoted by $|w|$. The *empty string* ε is the string of length 0. Throughout this paper, strings are 1-indexed. For $1 \leq i \leq j \leq |w|$, we let $w[i]$ be the i-th letter of w, and $w[i : j]$ be the substring of w which starts at position i and ends at position j. In particular, we denote $w[i : |w|]$ by $w[i :]$ and $w[1 : j]$ by $w[: j]$. The concatenation of two strings s and t is denoted by st.

Let $\mathbb{N}_0$ and $\mathbb{N}_1$ be the set of natural numbers including and excluding 0, respectively. We denote the cardinality of a set X by $|X|$.

2.1   Graphs

A *directed multigraph* G is a tuple (V, E, Γ) where V is the node set, $E \subseteq V \times V$ is the edge set, and $\Gamma \colon E \to \mathbb{N}_1$ gives each edge its multiplicity. The *head* and the *tail* of an edge $(u, v) \in E$ are v and u, respectively. This paper disallows self-loops: $(v, v) \notin E$ for any $v \in V$. When $\Gamma(e) = 1$ for all $e \in E$, G is called a *directed graph* and is simply denoted by (V, E). An *edge-labeled multigraph* is a tuple (V, E, Γ, Ψ) where $\Psi \colon E \to \Sigma$ for an alphabet Σ. A sequence $p = \langle e_1, \ldots, e_\ell \rangle$ of edges is called a v_0-v_ℓ *path* if there are $v_0, \ldots, v_\ell \in V$ such that $e_i = (v_{i-1}, v_i)$ for all $i \in \{1, \ldots, \ell\}$. Note that, the same node may occur more than once in a path in this paper. We call p a v_0-*cycle* when $v_0 = v_\ell$. For a t-u path p_1 and a u-v path p_2, we denote by $p_1 \cdot p_2$ the concatenation of p_1 and p_2, which will be a t-v path. By extending the domain of Ψ to sequences of edges, we define the *path label* $\Psi(p)$ of p to be the string $\Psi(e_1) \cdots \Psi(e_\ell)$. When there exists just one v_0-v_ℓ path, we call its label the v_0-v_ℓ path label and denote it by $\Psi((v_0, v_\ell)) \in \Sigma^*$.

A directed graph G is a t-*rooted tree* $(t \in V)$ if there exists exactly one t-v path for all $v \in V$. We call t the *root* of G. Similarly, G is a t-*oriented tree* if there exists exactly one v-t path for all $v \in V$. We call t the *sink* of G. For a t-rooted tree $G = (V, E)$, if $(u, v) \in E$, then u is the *parent* of v and v is a *child* of u. For two nodes $u, v \in V$ such that a u-v path exists, v is a *descendant* of u, and u is an *ancestor* of v. The *depth* of v is the length of the unique path from the root to v. We denote the set of all descendants of v as $\mathcal{D}_G(v)$.

Two directed multigraphs $G = (V, E, \Gamma)$ and $G' = (V', E', \Gamma')$ are *isomorphic*, denoted by $G \equiv G'$, if there is a bijection ϕ over V such that $V' = \phi(V)$, $E' = \{ (\phi(u), \phi(v)) \mid (u, v) \in E \}$, and $\Gamma'((\phi(u), \phi(v))) = \Gamma((u, v))$. The definition of isomorphism is naturally extended and applied for edge-labeled directed multigraphs. When G is a rooted tree, we can verify $G \equiv G'$ in linear time. If $V' = V$ and G' is a t-oriented tree, then G' is a t-*oriented spanning tree* of G.

Let $G = (V, E, \Gamma)$ be a directed multigraph. For a node $v \in V$, $\delta_G^-(v)$ and $\delta_G^+(v)$ are the sets of edges whose heads and tails are v, respectively. We denote the sum of the multiplicities of edges contained in $\delta_G^-(v)$ and $\delta_G^+(v)$ by $\Delta_G^-(v) = \sum_{e \in \delta_G^-(v)} \Gamma(e)$ and $\Delta_G^+(v) = \sum_{e \in \delta_G^+(v)} \Gamma(e)$, respectively. A cycle p is *Eulerian* when p contains e just $\Gamma(e)$ times for all $e \in E$. We also call a directed multigraph *Eulerian* if it has an Eulerian cycle. It is well-known that G is Eulerian if and only if G is connected and $\Delta_G^-(v) = \Delta_G^+(v)$ for all $v \in V$ [13]. Therefore, we can check whether G is Eulerian in $O(|V| + |E|)$ time. We often drop the subscript G from $\mathcal{D}_G$, δ_G^+, Δ_G^- etc. when G is clear from the context.

2.2 Position Heaps

A position heap is an index structure with which one can efficiently solve the pattern matching problem. In this paper, we follow Kucherov's definition [20]. Let T be a string of length n ending with a unique letter, i.e., $T[i] \neq T[n]$ for all $i \in \{1, \ldots, n - 1\}$. The position heap $\mathrm{PH}(T)$ of T is an edge-labeled rooted tree (V, E, Ψ) defined as follows. Let h_0 be ε, and h_i be the shortest prefix of $T[i :]$ not contained in $\{h_0, \ldots, h_{i-1}\}$ for all $i \in \{1, \ldots, n\}$. Since T ends with a unique letter, $T[i :] \neq h_j$ for any $j < i$, and thus h_i is always defined. Then, define $V = \{0, \ldots, n\}$, $E = \{(i, j) \mid h_i c = h_j$ for some $c \in \Sigma\}$, and $\Psi((i, j)) = c$ if $h_i c = h_j$. Clearly, a position heap is 0-rooted and h_i is the 0-i path label for

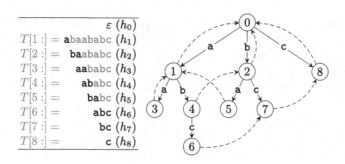

Fig. 1. PHS(abaababc) (dashed arrows are suffix links)

all $i \in \{0, \ldots, n\}$. Moreover, we have $i \leq j$ if node i is an ancestor of node j. We call T the *source text* of PH(T). Kucherov showed that one can determine whether a pattern P occurs in T in $O(|P|^2)$ time using PH(T). Moreover, we can determine it in $O(|P|)$ time with auxiliary data structures.

In Kucherov's algorithm for constructing position heaps, the mapping $\mathcal{S} \colon V \setminus \{0\} \to V$ called *suffix links* plays an important role. It is defined by $\mathcal{S}(i) = j$ such that $h_i = ch_j$ for some $c \in \Sigma$ for $i > 0$. The suffix links are well-defined. It is clear that the depth of node i is the depth of node $\mathcal{S}(i)$ plus 1. We often treat $\mathcal{S}$ as a subset of $V \times V$. We denote the position heap augmented with its suffix links by PHS(T) = $(V, E, \Psi, \mathcal{S})$. Figure 1 shows PHS($T$) for T = abaababc.

2.3 Problem Definitions

In this paper, we consider the following inverse problems of position heap construction. The first problem is inferring the source text T from a position heap.

Problem 1 (Inferring source texts from node-numbered edge-labeled trees).
Input: An edge-labeled rooted tree (V, E, Ψ) with $V = \{0, \ldots, |V| - 1\}$.
Output: A string T such that PH(T) = (V, E, Ψ) if such T exists. Otherwise, "invalid".

We will also consider the problem where edge labels are missing.

Problem 2 (Inferring source texts from node-numbered trees).
Input: A rooted tree (V, E) with $V = \{0, \ldots, |V| - 1\}$.
Output: A string T such that PH(T) = (V, E, Ψ) for some Ψ if such T exists. Otherwise, "invalid".

The third problem is inferring source texts T from trees whose nodes are not numbered but edges are labeled.

Problem 3 (Inferring source texts from edge-labeled trees).
Input: An edge-labeled rooted tree (V, E, Ψ).
Output: A string T such that PH(T) $\equiv$ (V, E, Ψ) if such T exists. Otherwise, "invalid".

In the end, we will address the problem where the input trees miss both node numbers and edge labels but have potential suffix links.

Problem 4 (Inferring source texts from trees with links).
Input: A pair $(G, \mathcal{S})$ of a rooted tree $G = (V, E)$ and a partial map $\mathcal{S} \colon V \rightarrowtail V$.
Output: A string T such that $\mathrm{PHS}(T) \equiv (V, E, \Psi, \mathcal{S})$ for some Ψ if such T exists. Otherwise, "invalid".

Figure 2 shows examples of instances of Problem 2 and 3 and Fig. 3 shows all possible answers for the instance of Fig. 2(b).

3   Proposed Algorithms

3.1   Inferring Source Texts from Node-Numbered Edge-Labeled Trees

Solving Problem 1 is easy. Given an edge-labeled tree (V, E, Ψ) where $V = \{0, \ldots, n\}$, let h_i be the 0–i path label on G for every $i \in V$. If the input is the position heap of some string T, it must hold $T[i] = h_i[1]$. Therefore, by DFS on G remembering the initial letter of each path label, we can construct the candidate string T in linear time. Then, we can verify whether $\mathrm{PH}(T) = (V, E, \Psi)$ in linear time, since the position heap of T can be constructed in linear time [20].

Theorem 1. *Problem 1 is solvable in linear time.*

3.2   Inferring Source Texts from Node-Numbered Trees

Figure 2(a) shows an input to an instance of Problem 2. The following procedure solves Problem 2. We label the outgoing edges of the root with arbitrary but distinct letters of Σ. Then, we construct an output candidate T following the method for Problem 1 in the previous subsection.

Theorem 2. *Problem 2 is solvable in linear time.*

There can be many correct outputs for input unless it is invalid. The number of possible source texts to output equals the number of how to attach the labels to edges from the root r. Since the number of letters that appear in T equals $\Delta^+(r)$, the number of possible texts is $|\Sigma|! \, / \left(|\Sigma| - \Delta^+(r)\right)!$. One can enumerate such T in output linear time because one can enumerate all $\Delta^+(r)$-permutations of Σ in output linear time [23].

3.3   Inferring Source Texts from Edge-Labeled Trees

Compared to the previous two problems, solving Problem 3 in linear time requires more elaborate arguments. In this subsection, we assume that two distinct outgoing edges of a node have different labels, since otherwise obviously the input cannot be extended to a position heap. We will investigate the structural properties of position heaps augmented with the suffix links, and see that the text T will appear as the label of a path with a specific property over $\mathrm{PHS}(T)$.

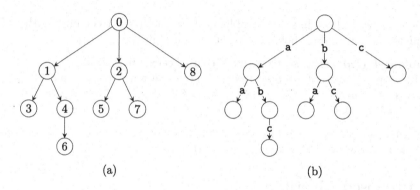

Fig. 2. Examples of inputs to instances of (a) Problem 2 and (b) Problem 3.

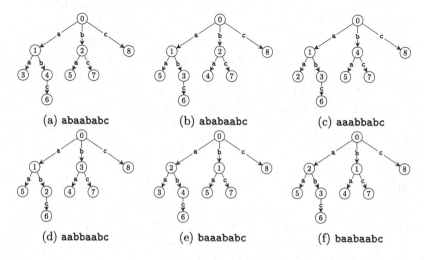

Fig. 3. All possible answers to Problem 3 when the graph in Fig. 2(b) is given.

Lemma 1. *Let* $\mathrm{PHS}(T) = (V, E, \Psi, \mathcal{S})$ *with* $V = \{0, \ldots, n\}$. *We have* $\mathcal{S}(n) = 0$ *and* $i + 1 \in \mathcal{D}(\mathcal{S}(v))$ *for all* $i \in V \setminus \{0, n\}$.

Proof. We show the lemma by induction on the depth of node i. When the depth of node i is 1, $\mathcal{S}(i)$ is the root 0, which is an ancestor of every node including $i + 1$. Note that the depth of node n is 1 since T ends with a unique letter. Suppose the depth of node $i < n$ is two or more. In this case, let the 0–i path label h_i be awb for some $a, b \in \Sigma$ and $w \in \Sigma^*$. Let j be the parent of i, for which $h_j = aw$. Let $i_s = \mathcal{S}(i)$ and $j_s = \mathcal{S}(j)$, i.e., $h_{i_s} = wb$ and $h_{j_s} = w$. By the induction hypothesis, we have $j + 1 \in \mathcal{D}(j_s)$, i.e., $h_{j+1} = h_{j_s} w'$ for some $w' \in \Sigma^*$, which implies that $j + 1 \geq j_s$. Together with the fact that $i > j$, we have $i + 1 > j_s$. Since h_i and h_{i+1} are prefixes of $T[i :]$ and $T[i+1 :]$, respectively, either $h_i[2 :] = wb$ is a prefix of h_{i+1} or the other way around. The fact $h_{j_s} = w$ and $i + 1 > j_s$ implies that wb is a prefix of h_{i+1}. That is, h_{i_s} is a prefix of h_{i+1}, which means $i + 1 \in \mathcal{D}(i_s)$. □

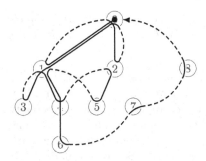
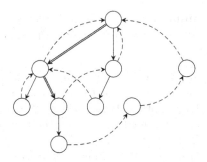

Fig. 4. The T-trace cycle of PHS(T) with $T =$ **abaababc**, which is an answer to the input graph in Fig. 2(b). Dashed lines represent suffix links.

Fig. 5. The trace graph of the input graph in Fig. 2(b). The multiplicities of doubled edges are 2 and the others are 1. Dashed arrows show suffix links.

Hereafter, by a *path/cycle* of PHS(T) $= (V, E, \Psi, \mathcal{S})$, we mean a path/cycle of $(V, E \cup \mathcal{S})$. We call elements of $E \cup \mathcal{S}$ *arcs* while reserving the term edges for elements of E. From Lemma 1, for all $i \in \{1, \ldots, n-1\}$, PHS(T) has a special i–$(i+1)$ path which starts with the suffix link followed by zero or some number of edges. We define a cycle by concatenating all special i–$(i+1)$ paths.

Definition 1. *For* PHS(T) $= (V, E, \Psi, \mathcal{S})$, *let* $f_i = (i, \mathcal{S}(i))$, p_0 *the path from* 0 *to* 1, *and* p_i *the path from* $\mathcal{S}(i)$ *to* $i+1$ *for* $i > 0$. *The* T-trace cycle *of* PHS(T) *is the sequence* $p_0 \cdot f_1 \cdot p_1 \cdots f_{n-1} \cdot p_{n-1} \cdot f_n$.

Figure 4 shows the T-trace cycle of PHS(T) for $T =$ **abaababc**. Note that the T-trace cycle is a cycle in the graph $(V, E \cup \mathcal{S})$, where each element of $\mathcal{S}$ appears exactly once. Since following an edge from E and a suffix link from $\mathcal{S}$ increases and decreases the depth by one, respectively, the total numbers of occurrences of edges and suffix links in the T-trace cycle should be balanced. That is, the T-trace cycle contains exactly n occurrences of edges from E. The following lemma explains why we call the cycle T-trace cycle.

Lemma 2. *Let* $e \in E$ *be the* i-th *occurrence of an edge in the* T-trace cycle of PHS(T). *Then* $\Psi(e) = T[i]$.

Proof. Suppose the i-th edge $e = (u, v)$ in the T-trace cycle $p = p_0 \cdot f_1 \cdots p_{n-1} \cdot f_n$ occurs in the p_j segment. In other words, p can be written as $p' \cdot (u, v) \cdot p''$, where p' contains j suffix links and $i-1$ edges. Then, the depth of v is $i - j$. Moreover, the edge e is on the path from the root to the node $j + 1$, whose label is a prefix of $T[j + 1 :]$. That is, $\Psi(e)$ is the $(i - j)$-th letter of $T[j + 1 :]$. Hence, $\Psi(e) = T[(j + 1) + (i - j) - 1] = T[i]$. $\square$

Lemma 2 allows us to spell T by following the T-trace cycle without referring to node numbers. To solve Problem 3, we will construct the T-trace cycle of PHS(T) $\equiv (V, E, \Psi, \mathcal{S}_G)$ for some T from the input graph $G = (V, E, \Psi)$. For this end, we first reconstruct the suffix links $\mathcal{S}$.

Lemma 3. *From an edge-labeled rooted tree $G = (V, E, \Psi)$, one can uniquely construct S in linear time such that $\mathrm{PHS}(T) \equiv (V, E, \Psi, S)$ for some T if any exist.*

Proof. We recover the suffix links of nodes from shallower to deeper. Let r be the root of G. From the definition of suffix links, we have $S(v) = r$ for every node of depth 1. For $e = (u, v) \in E$ with $\Psi(e) = c$, we assume $S(u)$ has already been determined. Let aw be the r–u path label where $a \in \Sigma$ and $w \in \Sigma^*$. The r–$S(u)$ path label is w and the r–v path label is awc. Therefore, the r–$S(v)$ path label is wc. Hence, an edge $(S(u), S(v))$ labeled c exists. So, for the node $t \in V$ such that $(S(u), t) \in E$ and $\Psi((S(u), t)) = c$, we determine $S(v) = t$.   □

If we fail to give a suffix link to any of the nodes by the procedure described in the proof of Lemma 3, the answer to Problem 3 is "invalid".

While the T-trace cycle contains just one occurrence of each suffix link, the numbers of occurrences of respective edges vary. Actually, one can uniquely determine the multiplicity of each edge in the T-trace cycle from G.

Lemma 4. *Let $\sigma(e)$ be the number of occurrences of e in the T-trace cycle for all $e \in E$. Then, it holds that*

$$\sigma(e) = 1 - \left| \{u \in V \mid S_G(u) = v\} \right| + \sum_{e' \in \delta_G^+(v)} \sigma(e') \qquad (1)$$

where v is the head of e.

Note that $\delta_G^+(v)$ contains no suffix links of $\mathrm{PHS}(T)$.

Proof. The T-trace cycle must include the same number of occurrences of arcs coming into and going out from node v. Since each suffix link occurs just once in the T-trace cycle, we obtain the lemma.   □

Lemma 5. *The system of Eqs. (1) in σ has a unique solution. Moreover, it can be computed in linear time.*

Proof. One can uniquely determine the value of $\sigma(e)$ inductively on the height of $e \in E$. Then, the linear-time computation is obvious.   □

Let us call a cycle p of (V, E, Ψ, S) a *legitimate cycle* if it is the T-trace cycle for some T. Based on Lemmas 4 and 5, we define the directed multigraph for which every legitimate cycle is Eulerian.

Definition 2 (Trace graph). *The trace graph $\mathcal{G}(G)$ of an edge-labeled tree $G = (V, E, \Psi)$ is a tuple (V, E', S, Γ) where $E' = \{e \in E \mid \sigma(e) > 0\}$ and $\Gamma \colon E' \cup S \to \mathbb{N}_1$ is defined by*

$$\Gamma(e) = \begin{cases} 1 & \text{if } e \in S, \\ \sigma(e) & \text{if } e \in E', \end{cases}$$

where S and σ are given in Lemmas 3 and 5, respectively.

Figure 5 shows the trace graph of Fig. 2(b). The doubled arrows have multiplicity 2 and the others have 1. The dashed arrows are suffix links.

From the definition, it is obvious that the T-trace cycle is an r-Eulerian cycle of $\mathcal{G}(G)$ where r is the root of G. However, not every Eulerian cycle of $\mathcal{G}(G)$ can be a legitimate cycle. Recall that in the definition of the T-trace cycle, the suffix link of every node u proceeds all outgoing edges of u. We say that an Eulerian cycle p of $\mathcal{G}(G)$ *respects* $\mathcal{S}$ if no edges of $\delta_G^+(u)$ occur before $(u, \mathcal{S}(u))$ in p.

Lemma 6. *A cycle p is an r-Eulerian cycle respecting $\mathcal{S}$ if and only if p is the T-trace cycle of some T.*

Proof. ($\Longleftarrow$) By definition.

($\Longrightarrow$) Let $n = |V|$ and r be the root of G. Let p_i and f_{i+1} be the sequences of edges and the suffix links for $i = 0, \ldots, n-1$ so that $p = p_0 \cdot f_1 \cdot p_1 \ldots p_{n-1} \cdot f_n$. Since p ends at r and only suffix links point to r, p always ends with a suffix link. We define the bijection $\Lambda : V \rightarrow \{0, \ldots, n\}$ such that $\Lambda(r) = 0$ and $\Lambda(s) = i$ if $f_i = (s, \mathcal{S}(s))$ for all $s \in V \setminus \{r\}$. Let s_i be the node such that $\Lambda(s_i) = i$.

We first show $\Lambda(u) < \Lambda(v)$ for all $(u, v) \in E$ by induction on $\Lambda(v)$. Suppose the claim holds true for all v such that $\Lambda(v) < i$. Then, we will show the claim holds for the edge whose head is s_i. If $|p_i| \geq 1$, the edge (s_k, s_i) occurs just before $f_i = (s_i, \mathcal{S}(s_i))$ in p. Since p respects $\mathcal{S}$, $f_k = (s_k, \mathcal{S}(s_k))$ occurs before (s_k, s_i). Thus, we have $k < i$. If $|p_i| = 0$, $f_{i-1} = (s_{i-1}, s_i)$. Let the parents of s_{i-1} and s_i be s_j and s_k, respectively. By the induction hypothesis, $j < i - 1$. By the definition of $\mathcal{S}$, $f_j = (s_j, s_k) \in \mathcal{S}$. Since p respects $\mathcal{S}$, $f_k = (s_k, \mathcal{S}(s_k))$ appears either before f_j or right after f_j. That is, $k \leq j + 1$ holds. Therefore, $k < i$.

Now, we define a string T by $T[i] = \Psi(e_i)$ where e_i is the i-th edge in p for $i = 1, \ldots, n$, and define h_i inductively to be the shortest prefix of $T[i :]$ which is not in $\{h_0, \ldots, h_{i-1}\}$ where $h_0 = \varepsilon$. We will show by induction on i that for all $j \leq i$, the s_0-s_j path label $\Psi((s_0, s_j))$ is $h_j = T[j : x_j]$ where $x_j = |p_0 \ldots p_{j-1}|$. This implies $(V, E, \Psi) \equiv \mathrm{PH}(T)$ when $i = n$. Then the constructed $\mathcal{S}$ is the correct suffix links of $\mathrm{PH}(T)$ by Lemma 3 and thus p is the T-trace cycle.

Let $g_i = \Psi((s_0, s_i))$. The claim clearly holds for $i = 0$ by $g_0 = h_0 = \varepsilon$. Suppose the claim holds true for i. That is, $g_i = h_i = T[i : x_i]$ where $x_i = |p_0 \ldots p_{i-1}|$. Let $u = \mathcal{S}(s_i)$. By the definition of $\mathcal{S}$, we have $\Psi((s_0, u)) = g_i[2 :] = T[i+1 : x_i]$. By the definition of T, $\Psi((u, s_{i+1})) = p_i = T[x_i + 1 : x_i + |p_i|] = T[x_i + 1 : x_{i+1}]$, where $x_{i+1} = |p_0 \ldots p_i|$. By concatenating these two paths, we obtain $g_{i+1} = \Psi((s_0, s_{i+1})) = T[i+1 : x_{i+1}]$. Since the labels of all proper ancestors of s_{i+1} are at most i, all prefixes of g_{i+1} appears in $\{g_0, \ldots, g_i\} = \{h_0, \ldots, h_i\}$. That is, g_{i+1} is the least prefix of $T[i+1 :]$ not in $\{h_0, \ldots, h_i\}$, i.e., $g_{i+1} = h_{i+1}$. $\square$

Therefore, to find a source text T, it is enough to find an r-Eulerian cycle over $(V, E, \mathcal{S}, \Gamma)$ that respects $\mathcal{S}$ where r is the root. We show that this problem can be solved in linear time on general graphs.

Problem 5 (The ECP (Eulerian cycle with priority edges) problem).
Input: A tuple (G, F, r) of a directed multigraph $G = (V, E, \Gamma)$, an edge subset $F \subseteq E$, and a start node $r \in V$ such that $|F \cap \delta^+(v)| \leq 1$ for all $v \in V$ and

$\Gamma(e) = 1$ for all $e \in F$.

Output: An r-Eulerian cycle that respects F if any. Otherwise, "invalid".

We call edges of F *priority edges*. Without loss of generality, we may assume a node has a priority outgoing edge only if it has another outgoing edge. If a node has only one outgoing edge and it has priority, then one can remove it from F and make it a non-priority edge. This does not affect possible solutions. In what follows, we show how to solve the ECP problem in linear time.

First, let us review a linear-time algorithm for constructing an r-Eulerian cycle. The following procedure gives a justification for the so-called BEST theorem [1,7], which counts the number of Eulerian cycles in a directed multigraph.

1. Construct an arbitrary r-oriented spanning tree H of G,
2. Starting from r, choose an arbitrary unused edge to follow next, except that an edge in H can be chosen only when it is the only remaining choice, until we follow all the edges of G.

This process guarantees to find an Eulerian cycle without getting stuck. We modify this procedure so that the output shall respect F.

1. Construct an arbitrary t-oriented spanning tree H of $(V, E \setminus F)$,
2. Starting from r, choose an arbitrary unused edge to follow next, except that
   - choose an unused priority edge if the current node has any,
   - an edge in H can be chosen only when it is the only remaining choice,
   until we follow all the edges of G.

Theorem 3. *We can compute an answer to the ECP problem in linear time.*

One can count the number of r-ECPs by modifying the BEST theorem formula. Letting $G' = (V, E \setminus F, \Gamma')$ with the restriction Γ' of Γ to $E \setminus F$, the number of r-ECPs is given as

$$\Delta_{G'}^+(r) \cdot \prod_{v \in V} \frac{(\Delta_{G'}^+(v) - 1)!}{\prod_{e \in \delta_{G'}^+(v)} \Gamma'(e)!} \cdot \sum_{(V,E') \in \mathcal{T}_{G'}(r)} \prod_{e \in E'} \Gamma'(e) \tag{2}$$

where $\mathcal{T}_{G'}(r)$ is the set of r-oriented spanning trees of G'. One can compute (2) in polynomial time by the matrix-tree theorem [22].

Theorem 4. *We can calculate the number of r-ECPs in polynomial time.*

One can also enumerate r-ECPs. We have already described a linear-time nondeterministic algorithm to find an r-ECP. Gabow and Myers proposed an algorithm [15] to enumerate spanning trees in output linear time. By searching all the possible choices of the procedure, we enumerate all the r-ECPs.

Theorem 5. *We can enumerate r-ECPs in linear time per solution.*

Corollary 1. *Problem 3 is solvable in linear time. Moreover, one can count and enumerate all possible answers in polynomial time and output linear time, respectively.*

Proof. The first claim follows from Theorem 3. By Theorems 4 and 5, it suffices to show that two distinct legitimate cycles p and p' over a trace graph give different source texts. Suppose e and e' are the first mismatch of p and p'. Since choosing a suffix link is obligatory, $e \neq e'$ implies $e, e' \in E$. Since distinct edges with the same tail have distinct labels, $\Psi(e) \neq \Psi(e')$, and thus those two cycles spell different source texts. $\square$

3.4   Inferring Source Texts from Trees with Links

Instance trees of Problem 4 miss both node numbers and edge labels but have possible suffix links. This problem can be solved by combining ideas for solving Problems 2 and 3. We first label the outgoing edges of the root node with arbitrary distinct letters. Then, the other edge labels are uniquely determined by the definition of suffix links, as long as the input is valid. Now, the algorithm for Problem 3 can be applied. Similarly one can solve the counting and enumerating variants of Problem 4.

Theorem 6. *We can solve Problem 4 in linear time. Moreover, one can count the number of output strings in polynomial time, and enumerate all output strings in linear time per each.*

4   Conclusion

We studied four types of reverse engineering problems on Kucherov's position heaps [20] and showed that all problems can be solved in linear time. One can think of an even more restrictive variant, where the input tree has no edge labels, no node numbers, and no suffix links. In this setting, we need to find "valid" suffix links, which seems a challenging task.

One can also study the reverse engineering problems of position heaps based on the definition by Ehrenfeucht et al. [12]. We conjecture that those problems can be solved by quite similar techniques presented in this paper.

Another interesting direction of future work is to study the reverse engineering of augmented position heaps [12].

Acknowledgements. The authors deeply appreciate the anonymous reviewers helpful comments. This work was supported by JSPS KAKENHI Grant Numbers JP19K20208 (DH), JP18H04091 (RY), JP18K11150 (RY), JP19K12098 (RY), JP20H05703 (RY), and JP21K11745 (AS).

References

1. van Aardenne-Ehrenfest, T., de Bruijn, N.G.: Circuits and trees in oriented linear graphs. Simon Stevin: Wis- en Natuurkundig Tijdschrift **28**, 203–217 (1951)
2. Bannai, H., Inenaga, S., Shinohara, A., Takeda, M.: Inferring strings from graphs and arrays. In: Proceedings of the MFCS 2003, pp. 208–217 (2003)

3. Blumer, A., Blumer, J., Haussler, D., Ehrenfeucht, A., Chen, M.T., Seiferas, J.: The smallest automaton recognizing the subwords of a text. Theoret. Comput. Sci. **40**, 31–55 (1985)

4. Blumer, A., Blumer, J., Haussler, D., McConnell, R., Ehrenfeucht, A.: Complete inverted files for efficient text retrieval and analysis. J. ACM **34**(3), 578–595 (1987)

5. Boyer, R.S., Moore, J.S.: A fast string searching algorithm. Commun. ACM **20**, 762–772 (1977)

6. Cazaux, B., Rivals, E.: Reverse engineering of compact suffix trees and links: a novel algorithm. J. Discret. Algorithms **28**, 9–22 (2014)

7. Charalambides, C.A.: Enumerative Combinatorics, vol. 2. Chapman and Hall/CRC, Boca Raton (2018)

8. Clément, J., Crochemore, M., Rindone, G.: Reverse engineering prefix tables. In: Proceedings of the STACS 2009, pp. 289–300 (2009)

9. Duval, J.P., Lecroq, T., Lefebvre, A.: Border array on bounded alphabet. J. Autom. Lang. Comb. **10**(1), 51–60 (2005)

10. Duval, J.P., Lecroq, T., Lefebvre, A.: Efficient validation and construction of border arrays and validation of string matching automata. RAIRO Theor. Inform. Appl. **43**(2), 281–297 (2009)

11. Duval, J.P., Lefebvre, A.: Words over an ordered alphabet and suffix permutations. RAIRO Theor. Inform. Appl. **36**(3), 249–259 (2002)

12. Ehrenfeucht, A., McConnell, R.M., Osheim, N., Woo, S.W.: Position heaps: a simple and dynamic text indexing data structure. J. Discret. Algorithms **9**(1), 100–121 (2011)

13. Fleischner, H.: Eulerian Graphs and Related Topics, vol. 1. Elsevier, Amsterdam (1990)

14. Franek, F., Lu, W., Ryan, P.J., Smyth, W.F., Sun, Y., Yang, L.: Verifying a border array in linear time. J. Comb. Math. Comb. Comput. **42**, 223–236 (2002)

15. Gabow, H.N., Myers, E.W.: Finding all spanning trees of directed and undirected graphs. SIAM J. Comput. **7**(3), 280–287 (1978)

16. Gusfield, D.: Algorithms on Strings, Trees, and Sequences: Computer Science and Computational Biology. Cambridge University Press, Cambridge (1997)

17. I, T., Inenaga, S., Bannai, H., Takeda, M.: Inferring strings from suffix trees and links on a binary alphabet. Discret. Appl. Math. **163**, 316–325 (2014)

18. Kärkkäinen, J., Piatkowski, M., Puglisi, S.J.: String inference from longest-common-prefix array. In: Proceedings of the ICALP 2017, pp. 62:1–62:14 (2017)

19. Knuth, D.E., Morris, J.J.H., Pratt, V.R.: Fast string searching in strings. SIAM J. Comput. **6**(2), 323–350 (1977)

20. Kucherov, G.: On-line construction of position heaps. J. Discret. Algorithms **20**, 3–11 (2013)

21. Manber, U., Myers, G.: Suffix arrays: a new method for on-line string searches. SIAM J. Comput. **22**(5), 935–948 (1993)

22. Moore, C., Mertens, S.: The Nature of Computation. OUP Oxford, Oxford (2011)

23. Sedgewick, R.: Permutation generation methods. ACM Comput. Surv. (CSUR) **9**(2), 137–164 (1977)

24. Starikovskaya, T., Vildhøj, H.W.: A suffix tree or not a suffix tree? J. Discret. Algorithms **32**, 14–23 (2015)

25. Weiner, P.: Linear pattern matching algorithm. In: Proceedings of the 14th IEEE Symposium on Switching and Automata Theory, pp. 1–11 (1973)

Internal Longest Palindrome Queries in Optimal Time

Kazuki Mitani[1], Takuya Mieno[2(✉)], Kazuhisa Seto[3], and Takashi Horiyama[3]

[1] Graduate School of Information Science and Technology,
Hokkaido University, Sapporo, Japan
`kazukida199911204649@eis.hokudai.ac.jp`
[2] Department of Computer and Network Engineering,
University of Electro-Communications, Chofu, Japan
`tmieno@uec.ac.jp`
[3] Faculty of Information Science and Technology,
Hokkaido University, Sapporo, Japan
`{seto,horiyama}@ist.hokudai.ac.jp`

Abstract. Palindromes are strings that read the same forward and backward. Problems of computing palindromic structures in strings have been studied for many years with a motivation of their application to biology. The longest palindrome problem is one of the most important and classical problems regarding palindromic structures, that is, to compute the longest palindrome appearing in a string T of length n. The problem can be solved in $\mathcal{O}(n)$ time by the famous algorithm of Manacher [Journal of the ACM, 1975]. In this paper, we consider the problem in the internal model. The internal longest palindrome query is, given a substring $T[i..j]$ of T as a query, to compute the longest palindrome appearing in $T[i..j]$. The best known data structure for this problem is the one proposed by Amir et al. [Algorithmica, 2020], which can answer any query in $\mathcal{O}(\log n)$ time. In this paper, we propose a linear-size data structure that can answer any internal longest palindrome query in constant time. Also, given the input string T, our data structure can be constructed in $\mathcal{O}(n)$ time.

Keywords: String algorithms · Palindromes · Internal queries

1 Introduction

Palindromes are strings that read the same backward as forward. Palindromes have been widely studied with the motivation of their application to biology [22]. Computing and counting palindromes in a string are fundamental tasks. Manacher [26] proposed an $\mathcal{O}(n)$-time algorithm that computes all maximal palindromes in the string of length n. Droubay et al. [16] showed that any string of length n contains at most $n + 1$ distinct palindromes (including the empty string). Then, Groult et al. [21] proposed an $\mathcal{O}(n)$-time algorithm to enumerate

Partially supported by JSPS KAKENHI Grant Numbers JP20H05964.

the number of distinct palindromes in a string. The above $\mathcal{O}(n)$-time algorithms are time-optimal since reading the input string of length n takes $\Omega(n)$ time.

Regarding the longest palindrome computation, Funakoshi et al. [18] considered the problem of computing the longest palindromic substring of the string T' after a single character insertion, deletion, or substitution is applied to the input string T of length n. Of course, using $\mathcal{O}(n)$ time, we can obtain the longest palindromic substring of T' from scratch. However, this idea is naïve and appears to be inefficient. To avoid such inefficiency, Funakoshi et al. [18] proposed an $\mathcal{O}(n)$-space data structure that can compute the solution for any editing operation given as a query in $\mathcal{O}(\log(\min\{\sigma, \log n\}))$ time where σ is the alphabet size. Amir et al. [7] considered the dynamic longest palindromic substring problem, which is an extension of Funakoshi et al.'s problem where up to $\mathcal{O}(n)$ sequential editing operations are allowed. They proposed an algorithm that solves this problem in $\mathcal{O}(\sqrt{n}\log^2 n)$ time per a single character edit w.h.p. with a data structure of size $\mathcal{O}(n \log n)$, which can be constructed in $\mathcal{O}(n \log^2 n)$ time. Furthermore, Amir and Boneh [6] proposed an algorithm running in poly-logarithmic time per a single character substitution.

Internal queries are queries about substrings of the input string T. Let us consider a situation where we solve a certain problem for each of k different substrings of T. If we run an $\mathcal{O}(|w|)$-time algorithm from scratch for each substring w, the total time complexity can be as large as $\mathcal{O}(kn)$. To be more efficient, by performing an appropriate preprocessing on T, we construct some data structure for the query to output each solution efficiently. Such efficient data structures for palindromic problems are known. Rubinchik and Shur [30] proposed an algorithm that computes the number of distinct palindromes in a given substring of an input string of length n. Their algorithm runs in $\mathcal{O}(\log n)$ time with a data structure of size $\mathcal{O}(n \log n)$, which can be constructed in $\mathcal{O}(n \log n)$ time. Amir et al. [7] considered a problem of computing the longest palindromic substring in a given substring of the input string of length n; it is called the internal longest palindrome query. Their algorithm runs in $\mathcal{O}(\log n)$ time with a data structure of size $\mathcal{O}(n \log n)$, which can be constructed in $\mathcal{O}(n \log^2 n)$ time.

This paper proposes a new algorithm for the internal longest palindrome query. The algorithm of Amir et al. [7] uses 2-dimensional orthogonal range maximum queries [3, 4, 12]; furthermore, time and space complexities of their algorithm are dominated by this query. Instead of 2-dimensional orthogonal range maximum queries, by using palindromic trees [31], weighted ancestor queries [19], and range maximum queries [17], we obtain a time-optimal algorithm.

Theorem 1. *Given a string T of length n over a linearly sortable alphabet, we can construct a data structure of size $\mathcal{O}(n)$ in $\mathcal{O}(n)$ time that can answer any internal longest palindrome query in $\mathcal{O}(1)$ time.*

Here, an alphabet is said to be *linearly sortable* if any sequence of n characters from Σ can be sorted in $\mathcal{O}(n)$ time. For example, the integer alphabet $\{1, 2, \ldots, n^c\}$ for some constant c is linearly sortable because we can sort a sequence from the alphabet in linear time by using a radix sort with base n. We also assume the word-RAM model with word size $\omega \geq \log n$ bits for input size n.

Related Work. Internal queries have been studied on many problems not only those related to palindromic structures. For instance, Kociumaka et al. [25] considered the internal pattern matching queries that are ones for computing the occurrences of a substring U of the input string T in another substring V of T. Besides, internal queries for string alignment [13,32–34], longest common prefix [1,5,20,28], and longest common substring [7] have been studied in the last two decades. See [24] for an overview of internal queries. We also refer to [2,10,11,14,15,23] and references therein.

2 Preliminaries

2.1 Strings and Palindromes

Let Σ be an alphabet. An element of Σ is called a character, and an element of Σ^* is called a string. The empty string ε is the string of length 0. The length of a string T is denoted by $|T|$. For each i with $1 \leq i \leq |T|$, the i-th character of T is denoted by $T[i]$. For each i and j with $1 \leq i, j \leq |T|$, the string $T[i]T[i+1]\cdots T[j]$ is denoted by $T[i..j]$. For convenience, let $T[i'..j'] = \varepsilon$ if $i' > j'$. If $T = xyz$, then x, y, and z are called a prefix, substring, and suffix of T, respectively. They are called a proper prefix, a proper substring, and a proper suffix of T if $x \neq T$, $y \neq T$, and $z \neq T$, respectively. The string y is called an infix of T if $x \neq \varepsilon$ and $z \neq \varepsilon$. The reversal of string T is denoted by T^R, i.e., $T^R = T[|T|]\cdots T[2]T[1]$. A string T is called a palindrome if $T = T^R$. Note that ε is also a palindrome. For a palindromic substring $T[i..j]$ of T, the center of $T[i..j]$ is $\frac{i+j}{2}$. A palindromic substring $T[i..j]$ is called a maximal palindrome in T if $i = 1$, $j = |T|$, or $T[i-1] \neq T[j+1]$. In what follows, we consider an arbitrary fixed string T of length $n > 0$. In this paper, we assume that the alphabet Σ is linearly sortable. We also assume the word-RAM model with word size $\omega \geq \log n$ bits.

Let z be the number of palindromic suffixes of T. Let $suf(T) = (s_1, s_2, \ldots, s_z)$ be the sequence of the lengths of palindromic suffixes of T sorted in increasing order. Further let $dif_i = s_i - s_{i-1}$ for each i with $2 \leq i \leq z$. For convenience, let $dif_1 = 0$. Then, the sequence $(dif_1, \ldots, dif_z)$ is monotonically non-decreasing (Lemma 7 in [27]). Let $(suf_1, suf_2, \ldots, suf_p)$ be the partition of $suf(T)$ such that for any two elements s_i, s_j in $suf(T)$, $s_i, s_j \in suf_k$ for some k iff $dif_i = dif_j$. By definition, each suf_k forms an arithmetic progression. It is known that the number p of arithmetic progressions satisfies $p \in \mathcal{O}(\log n)$ [9,27]. For $1 \leq k \leq p$ and $1 \leq \ell \leq |suf_k|$, $suf_{k,\ell}$ denote the ℓ-th term of suf_k. Figure 1 shows an example of the above definitions.

2.2 Tools

In this section, we list some data structures used in our algorithm in Sect. 3.

	s	dif	
ε	0	0	$suf_{1,1}$ ⊣ suf_1
a	1	1	$suf_{2,1}$ ⊣ suf_2
aba	3	2	$suf_{3,1}$
ababa	5	2	$suf_{3,2}$ ⊢ suf_3
abababa	7	2	$suf_{3,3}$
ababababaababababa	14	7	$suf_{4,1}$
ababababaababababaababababa	21	7	$suf_{4,2}$ ⊢ suf_4
ababababaababababaababababababababaababababaababababa	43	22	$suf_{5,1}$ ⊣ suf_5

Fig. 1. Palindromic suffixes of $T =$ ababababaababababaababababababababaababababaababababa and the partition $(suf_1, \ldots, suf_5)$ of their lengths. Three integers $s_3 = 3, s_4 = 5$, and $s_5 = 7$ are represented by a single arithmetic progression suf_3 since $dif_3 = dif_4 = dif_5 = 2$. Since s_4 is the second smallest term in suf_3, $suf_{3,2} = s_4$.

Palindromic Trees and Series Trees. The palindromic tree of T is a data structure that represents all distinct palindromes in T [31]. The palindromic tree of T, denoted by paltree(T), has d ordinary nodes and one auxiliary node $\bot$ where $d \leq n + 1$ is the number of all distinct palindromes in T. Each ordinary node v corresponds to a palindromic substring of T (including the empty string ε) and stores its length as weight(v). For the auxiliary node $\bot$, we define weight($\bot$) = -1. For convenience, we identify each node with its corresponding palindrome. For an ordinary node v in paltree(T) and a character c, if nodes v and cvc exist, then an edge labeled c connects these nodes. The auxiliary node $\bot$ has edges to all nodes corresponding to length-1 palindromes. Each node v in paltree(T) has a suffix link that points to the longest palindromic proper suffix of v. Let link(v) be the string pointed to by the suffix link of v. We define link(ε) = link($\bot$) = $\bot$. See the Fig. 2(a) for example. For each node v corresponding to a non-empty palindrome in paltree(T), let $\delta_v = |v| - |$link(v)$|$ be the difference between the lengths of v and its longest palindromic proper suffix. For convenience, let $\delta_\varepsilon = 0$. Each node v corresponding to a non-empty palindrome has a series link that points to the longest palindromic proper suffix u of v such that $\delta_u \neq \delta_v$. Let serieslink(v) be the string pointed to by the series link of v.

Let LSufPal be an array of length n such that LSufPal[j] stores a pointer to the node in paltree(T) corresponding to the longest palindromic suffix of $T[1..j]$ for each $1 \leq j \leq n$. The definition of LSufPal is identical to the array node[1] defined in [31], and it was shown that node[1] can be computed in $\mathcal{O}(n)$ time. Hence, LSufPal can be computed in $\mathcal{O}(n)$ time. Let LPrePal be an array of length n such that LPrePal[i] stores a pointer to the node in paltree(T) corresponding to the longest palindromic prefix of $T[i..n]$ for each $1 \leq i \leq n$. LPrePal can be computed in $\mathcal{O}(n)$ time as well as LSufPal.

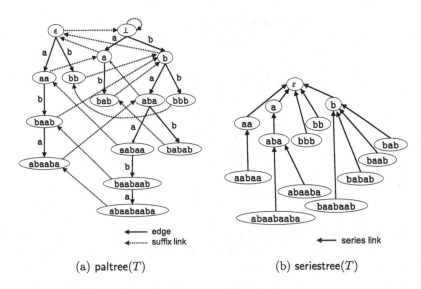

(a) paltree(T)   (b) seriestree(T)

Fig. 2. Illustration for the palindromic tree and the series tree of string $T =$ abaabaababbb Since $\delta_{\text{abaabaaba}} = |\text{abaabaaba}| - |\text{abaaba}| = 3$, $\delta_{\text{abaaba}} = |\text{abaaba}| - |\text{aba}| = 3$, and $\delta_{\text{aba}} = |\text{aba}| - |\text{a}| = 2$, then serieslink(abaabaaba) = aba. abaabaaba stores the arithmetic progression representing $\{6, 9\}$, abaaba stores the arithmetic progression representing $\{6\}$ and aba stores the arithmetic progression representing $\{3\}$.

Theorem 2 (Proposition 4.10 in [31]). *Given a string T over a linearly sortable alphabet, the palindromic tree of T, including its suffix links and series links, can be constructed in $\mathcal{O}(n)$ time. Also, LSufPal and LPrePal can be computed in $\mathcal{O}(n)$ time.*

Let us consider the subgraph $\mathcal{S}$ of paltree(T) that consists of all ordinary nodes and reversals of all series links. By the definition, $\mathcal{S}$ has no cycle and $\mathcal{S}$ is connected (any node is reachable from the node ε), i.e., it forms a tree. We call the tree $\mathcal{S}$ the series tree of T, and denote it by seriestree(T). By definition of series links, the set of lengths of palindromic suffixes of v that are longer than $|\text{serieslink}(v)|$ can be represented by an arithmetic progression. Each node v stores the arithmetic progression, represented by a triple consisting of its first term, its common difference, and the number of terms. Arithmetic progressions for all nodes can be computed in linear time by traversing the palindromic tree. It is known that the length of a path consisting of series links is $\mathcal{O}(\log n)$ [31]. Hence, the height of seriestree(T) is $\mathcal{O}(\log n)$. See the Fig. 2(b) for illustration.

Weighted Ancestor Query. A rooted tree whose nodes are associated with integer weights is called a monotone weighted tree if the weight of every non-root node is not smaller than the parent's weight. Given a monotone weighted tree $\mathcal{T}$ for preprocess and a node v and an integer k for query, a weighted ancestor query (WAQ) returns the ancestor u closest to the root of v such that the weight of u

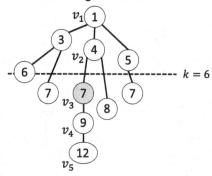

Fig. 3. Illustration for weighted ancestor query. Integers in nodes denote the weights. Given a node v_5 in a monotone weighted tree T and an integer $k = 6$ for query, WAQ returns the node v_3 since v_3 is an ancestor of v_5, weight$(v_3) > k = 6$, and the weight of the parent v_2 of v_3 is not greater than $k = 6$.

is greater than k. Let $\mathsf{WAQ}_T(v, k)$ be the output of the weighted ancestor query for tree T, node v, and integer k. See Fig. 3 for a concrete example.

It is known that there is an $\mathcal{O}(N)$-space data structure that can answer any weighted ancestor query in $\mathcal{O}(\log \log N)$ time where N is the number of nodes in the tree [8]. In general, the query time $\mathcal{O}(\log \log N)$ is known to be optimal within $\mathcal{O}(N)$ space [29]. On the other hand, if the height of the input tree is low enough, the query time can be improved:

Theorem 3 (Proposition 15 in [19]). *Let ω be the word size. Given a monotone weighted tree with N nodes and height $\mathcal{O}(\omega)$, one can construct an $\mathcal{O}(N)$ space data structure in $\mathcal{O}(N)$ time that can answer any weighted ancestor query in constant time.*

In this paper, we use weighted ancestor queries only on the series tree of T whose height is $\mathcal{O}(\log n) \subseteq \mathcal{O}(\omega)$, where ω is the word size, thus we will apply Theorem 3. Note that we assume the word-RAM model with word size $\omega \geq \log n$ bits.

Range Maximum Query. Given an integer array A of length m for preprocess and two indices i and j with $1 \leq i \leq j \leq m$ for query, range maximum query returns the index of a maximum element in the sub-array $A[i..j]$. Let $\mathrm{RMQ}_A(i, j)$ be the output of the range maximum query for array A and indices i, j. In other words, $\mathrm{RMQ}_A(i, j) = \arg\max_k \{A[k] \mid i \leq k \leq j\}$. The following result is known:

Theorem 4 (Theorem 5.8 in [17]). *Let m be the size of the input array A. There is a data structure of size $2m + o(m)$ bits that can answer any range maximum query on A in constant time. The data structure can be constructed in $\mathcal{O}(m)$ time.*

3   Internal Longest Palindrome Queries

In this section, we propose an efficient data structure for the internal longest palindrome query defined as follows:

Internal longest palindrome query

Preprocess: A string T of length n.
Query input: Two indices i and j with $1 \leq i \leq j \leq n$.
Query output: The longest palindromic substring in $T[i..j]$.

Our data structure requires only $\mathcal{O}(n)$ words of space, and can answer any internal longest palindrome query in constant time. To answer queries efficiently, we classify all palindromic substrings of T into palindromic prefixes, palindromic infixes, and palindromic suffixes. First, we compute the longest palindromic prefix and the longest palindromic suffix of $T[i..j]$. Second, we compute a palindromic infix that is a candidate for the answer. As we will discuss in a later subsection, this candidate may not be the longest palindromic infix of $T[i..j]$. Finally, we compare the three above palindromes and output the longest one.

3.1   Palindromic Suffixes and Prefixes

First, we compute the longest palindromic suffix of $T[i..j]$. In the preprocessing, we build seriestree(T) and a data structure for the weighted ancestor queries on seriestree(T), and compute LSufPal as well. The query algorithm consists of three steps:

Step 1: Obtain the longest palindromic suffix of $T[1..j]$.
   We obtain the longest palindromic suffix v of $T[1..j]$ from LSufPal$[j]$. If $|v| \leq |T[i..j]|$, then v is the longest palindromic suffix of $T[i..j]$. Then we return $T[j - |v| + 1..j]$ and the algorithm terminates. Otherwise, we continue to Step 2.

Step 2: Determine the group to which the desired length belongs.
   Let ℓ be the length of the longest palindromic suffix of $T[i..j]$ we want to know. We use the longest palindromic suffix v of $T[1..j]$ obtained in Step 1. First, we find the shortest palindrome u that is an ancestor of v in seriestree(T) and has length at least $|T[i..j]|$. Such a palindrome (equivalently the node) u can be found by weighted ancestor query on the series tree, i.e., $u = \mathsf{WAQ}_{\mathsf{seriestree}(T)}(v, j - i)$. Then $|u|$ is an upper bound of ℓ. Let suf_α be the group such that $|u| \in suf_\alpha$. If the smallest element $suf_{\alpha,1}$ in suf_α is at most $|T[i..j]|$, the length ℓ belongs to the same group suf_α as $|u|$. Otherwise, the length ℓ belongs to the previous group $suf_{\alpha-1}$.

Step 3: Calculate the desired length.
   Let suf_β be the group to which the length ℓ belongs, which is determined in Step 2. Since suf_β is an arithmetic progression, i.e. $suf_{\beta,\gamma} = suf_{\beta,1} + (\gamma - 1)dif_\beta$ for $1 \leq \gamma \leq |suf_\beta|$, the desired length ℓ can be computed by using a constant number of arithmetic operations. Then we return $T[j - \ell + 1..j]$.

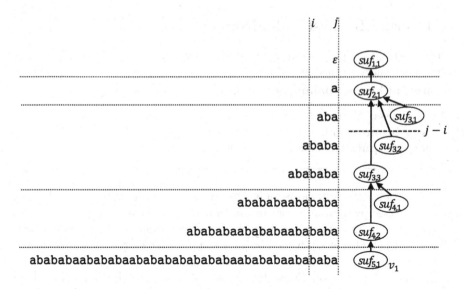

Fig. 4. Illustration for how to compute the longest palindromic suffix of $T[i..j]$, when $T[1..j] = $ ababababaababababaababababababaababababaababababa. The graph on the right hand depicts a part of the series tree of a string T, and the lengths of palindromes are written inside the nodes. In Step 1, we obtain the length $suf_{5,1}$ of the longest palindromic suffix v_1 of $T[1..j]$. In Step 2, we find $suf_{3,3}$ by $\mathsf{WAQ}_{\mathsf{seriestree}(T)}(v_1, j - i)$. Since $suf_{3,1} > j - i + 1$, the desired length belongs to suf_3. In Step 3, since suf_3 is an arithmetic progression, we can find that $suf_{3,1}$ is the longest palindromic suffix of $T[i..j]$ in constant time.

See Fig. 4 for illustration. Now, we show the correctness of the algorithm and analyze time and space complexities.

Lemma 1. *We can compute the longest palindromic suffix and prefix of $T[i..j]$ in $\mathcal{O}(1)$ time with a data structure of size $\mathcal{O}(n)$ that can be constructed in $\mathcal{O}(n)$ time.*

Proof. In the preprocessing, we build $\mathsf{seriestree}(T)$, LSufPal, LPrePal and a data structure of weighted ancestor query on $\mathsf{seriestree}(T)$ in $\mathcal{O}(n)$ time (Theorem 2 and 3). Recall that since the height of $\mathsf{seriestree}(T)$ is $\mathcal{O}(\log n) \subseteq \mathcal{O}(\omega)$, we can apply Theorem 3 to the series tree. Again, by Theorem 2 and 3, the space complexity is $\mathcal{O}(n)$ words of space.

In what follows, let ℓ be the length of the longest palindromic suffix of $T[i..j]$. In Step 1, we can obtain the longest palindromic suffix v of $T[1..j]$ by just referring to LSufPal$[j]$. If $|v| \leq |T[i..j]|$, v is also the longest palindromic suffix of $T[i..j]$, i.e., $\ell = |v|$. Otherwise, v is not a substring of $T[i..j]$. In Step 2, we first query $\mathsf{WAQ}_{\mathsf{seriestree}(T)}(v, j-i)$. The resulting node u corresponds to a palindromic suffix of $T[1..j]$, which is longer than $|T[i..j]|$. Let suf_α and suf_β be the groups to which $|u|$ and ℓ belong to, respectively. If the smallest element $suf_{\alpha,1}$ in suf_α is at most $j - i + 1$, then the desired length ℓ satisfies $suf_{\alpha,1} \leq \ell \leq |u|$. Namely,

$\beta = \alpha$. Otherwise, if s is greater than $j - i + 1$, ℓ is not in suf_α but is in $suf_{\alpha-x}$ for some $x > 1$. If we assume that ℓ belong to $suf_{\alpha-y}$ for some $y \geq 2$, the length of $\mathsf{serieslink}(u)$ belonging to $suf_{\alpha-1}$ is longer than $T[i..j]$. However, it contradicts that u is the answer of $\mathsf{WAQ}_{\mathsf{seriestree}(T)}(v, j-i)$. Hence, if s is greater than $j-i+1$, then the length ℓ is in $suf_{\alpha-1}$. Namely, $\beta = \alpha - 1$. In Step 3, we can compute ℓ in constant time since we know the arithmetic progression suf_β to which ℓ belongs. More specifically, ℓ is the largest element that is in suf_β and is at most $j - i + 1$.

Throughout the query algorithm, all operations, including WAQ and operations on arithmetic progressions, can be done in constant time. Thus the query algorithm runs in constant time. □

We can compute the longest palindromic prefix of $T[i..j]$ in a symmetric way using LPrePal instead of LSufPal.

3.2   Palindromic Infixes

Next, we compute the longest palindromic infix except for ones that are obviously shorter than the longest palindromic prefix or the longest palindromic suffix of the query substring. We show that to find the desired palindromic infix, it suffices to consider maximal palindromes whose centers are between the centers of the longest palindromic prefix and the longest palindromic suffix of $T[i..j]$. Let t be the ending position of the longest palindromic prefix and s be the starting position of the longest palindromic suffix. Namely, $T[i..t]$ is the longest palindromic prefix and $T[s..j]$ is the longest palindromic suffix of $T[i..j]$.

Lemma 2. *Let w be a palindromic infix of $T[i..j]$ and c be the center of w. If $c < \frac{i+t}{2}$ or $c > \frac{s+j}{2}$, w cannot be the longest palindromic substring of $T[i..j]$.*

Proof. Palindrome w is a proper substring of $T[i..t]$ (resp. $T[s..j]$) if $c < \frac{i+t}{2}$ (resp. $c > \frac{s+j}{2}$). Then, w is shorter than $T[i..t]$ or $T[s..j]$ (see also Fig. 5). □

Then, we consider palindromes whose centers are between the centers of the longest palindromic prefix and the longest palindromic suffix of $T[i..j]$.

Lemma 3. *Let w be a palindromic substring of T and c be the center of w. If $\frac{i+t}{2} < c < \frac{s+j}{2}$, then w is a palindromic infix of $T[i..j]$.*

Proof. Let $w = T[p..q]$. Then, $c = \frac{p+q}{2}$. To prove that w is a palindromic infix, we show that $p > i$ and $q < j$. For the sake of contradiction, we assume $p \leq i$. If $\frac{i+t}{2} < c \leq \frac{i+j}{2}$, there exists a palindromic prefix w_1 whose center is c. This contradicts that $T[i..t]$ is the longest palindromic prefix of $T[i..j]$ since $T[i..t]$ is a substring of w_1 (see also Fig. 6). Otherwise, if $\frac{i+j}{2} < c < \frac{s+j}{2}$, there exists a palindromic suffix whose w_2 center is c. This contradicts that $T[s..j]$ is the longest palindromic suffix of $T[i..j]$ since $T[i..t]$ is a substring of w_2 (see also Fig. 6). Therefore, $p > i$. We can show $q < j$ in a symmetric way. □

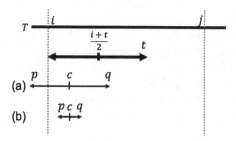

Fig. 5. Illustration for Lemma 2. Two-way arrows denote palindromic substrings of T. $T[i..t]$ is the longest palindromic prefix of $T[i..j]$. A palindrome whose center c is less than $\frac{i+t}{2}$ is either (a) not a substring of $T[i..j]$ or (b) shorter than the longest palindromic prefix of $T[i..j]$ as shown in this figure.

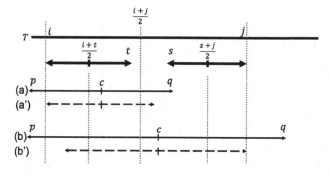

Fig. 6. Illustration for contradictions in the proof of Lemma 3. $T[i..t]$ is the longest palindromic prefix and $T[s..j]$ is the longest palindromic suffix of $T[i..j]$. If a palindrome as (a) exists, there exists a palindromic prefix (a') of $T[i..j]$ that is longer than $T[i..t]$, a contradiction. Similarly, the existence of a palindrome as (b) leads to a contradiction.

By Lemmas 2 and 3, when a palindromic infix w of $T[i..j]$ is the longest palindromic substring of $T[i..j]$, the center of w must be located between $\frac{i+t}{2}$ and $\frac{s+j}{2}$. Furthermore, w is a maximal palindrome in T. In other words, w is the longest maximal palindrome in T whose center c satisfies $\frac{i+t}{2} < c < \frac{s+j}{2}$. To find such a (maximal) palindrome, we build a succinct RMQ data structure on the length-$(2n-1)$ array MP that stores the lengths of maximal palindromes in T. For each integer and half-integer $c \in \{1, 1.5, \ldots, n - 0.5, n\}$, $\mathsf{MP}[2c-1]$ stores the length of the maximal palindrome whose center is c. By doing so, when the indices t and s are given, we can find a candidate for the longest palindromic substring which is an infix of $T[i..j]$ in constant time. More precisely, the length of the candidate is $\mathsf{MP}[\mathsf{RMQ_{MP}}(i + t, s + j - 2)]$ since the center c of the candidate satisfies $\frac{i+t}{2} < c < \frac{s+j}{2}$ ($i + t - 1 < 2c - 1 < s + j - 1$). By Manacher's algorithm [26], MP can be constructed in $\mathcal{O}(n)$ time. Then, we obtain the following lemma.

Lemma 4. *Given the longest palindromic prefix $T[i..t]$ and the longest palindromic suffix $T[s..j]$ of $T[i..j]$, we can compute the longest palindromic infix of $T[i..j]$ whose centers are between the centers of $T[i..t]$ and $T[s..j]$ in $\mathcal{O}(1)$ time with a data structure of size $\mathcal{O}(n)$ that can be constructed in $\mathcal{O}(n)$ time.*

By Lemmas 1 and 4, we have shown our main theorem:

Theorem 1. *Given a string T of length n over a linearly sortable alphabet, we can construct a data structure of size $\mathcal{O}(n)$ in $\mathcal{O}(n)$ time that can answer any internal longest palindrome query in $\mathcal{O}(1)$ time.*

References

1. Abedin, P., et al.: A linear-space data structure for range-LCP queries in polylogarithmic time. Theor. Comput. Sci. **822**, 15–22 (2020)
2. Abedin, P., Ganguly, A., Pissis, S.P., Thankachan, S.V.: Efficient data structures for range shortest unique substring queries. Algorithms **13**(11), 1–9 (2020)
3. Agarwal, P.K.: Range Searching. In: Handbook of Discrete and Computational Geometry, pp. 1057–1092. Chapman and Hall/CRC, Boca Raton (2017)
4. Alstrup, S., Brodal, G.S., Rauhe, T.: New data structures for orthogonal range searching. In: 2013 IEEE 54th Annual Symposium on Foundations of Computer Science, p. 198. IEEE Computer Society (2000)
5. Amir, A., Apostolico, A., Landau, G.M., Levy, A., Lewenstein, M., Porat, E.: Range LCP. J. Comput. Syst. Sci. **80**(7), 1245–1253 (2014)
6. Amir, A., Boneh, I.: Dynamic palindrome detection. arXiv preprint arXiv:1906.09732 (2019)
7. Amir, A., Charalampopoulos, P., Pissis, S.P., Radoszewski, J.: Dynamic and internal longest common substring. Algorithmica **82**(12), 3707–3743 (2020)
8. Amir, A., Landau, G.M., Lewenstein, M., Sokol, D.: Dynamic text and static pattern matching. ACM Trans. Algorithms **3**(2), 19 (2007)
9. Apostolico, A., Breslauer, D., Galil, Z.: Parallel detection of all palindromes in a string. Theor. Comput. Sci. **141**(1), 163–173 (1995)
10. Babenko, M., Gawrychowski, P., Kociumaka, T., Kolesnichenko, I., Starikovskaya, T.: Computing minimal and maximal suffixes of a substring. Theor. Comput. Sci. **638**, 112–121 (2016)
11. Badkobeh, G., Charalampopoulos, P., Kosolobov, D., Pissis, S.P.: Internal shortest absent word queries in constant time and linear space. Theor. Comput. Sci. **922**, 271–282 (2022)
12. Bentley, J.L.: Multidimensional divide-and-conquer. Commun. ACM **23**(4), 214–229 (1980)
13. Charalampopoulos, P., Gawrychowski, P., Mozes, S., Weimann, O.: An almost optimal edit distance oracle. In: 48th International Colloquium on Automata, Languages, and Programming (ICALP 2021). Schloss Dagstuhl-Leibniz-Zentrum für Informatik (2021)
14. Charalampopoulos, P., Kociumaka, T., Mohamed, M., Radoszewski, J., Rytter, W., Waleń, T.: Internal dictionary matching. Algorithmica **83**(7), 2142–2169 (2021)
15. Charalampopoulos, P., Kociumaka, T., Wellnitz, P.: Faster approximate pattern matching: a unified approach. In: 2020 IEEE 61st Annual Symposium on Foundations of Computer Science (FOCS), pp. 978–989. IEEE (2020)

16. Droubay, X., Justin, J., Pirillo, G.: Episturmian words and some constructions of de Luca and Rauzy. Theor. Comput. Sci. **255**(1–2), 539–553 (2001)
17. Fischer, J., Heun, V.: Space-efficient preprocessing schemes for range minimum queries on static arrays. SIAM J. Comput. **40**(2), 465–492 (2011)
18. Funakoshi, M., Nakashima, Y., Inenaga, S., Bannai, H., Takeda, M.: Computing longest palindromic substring after single-character or block-wise edits. Theor. Comput. Sci. **859**, 116–133 (2021)
19. Ganardi, M.: Compression by contracting straight-line programs. In: Mutzel, P., Pagh, R., Herman, G. (eds.) 29th Annual European Symposium on Algorithms (ESA 2021). Leibniz International Proceedings in Informatics (LIPIcs), vol. 204, pp. 45:1–45:16. Schloss Dagstuhl - Leibniz-Zentrum für Informatik (2021)
20. Ganguly, A., Patil, M., Shah, R., Thankachan, S.V.: A linear space data structure for range LCP queries. Fund. Inform. **163**(3), 245–251 (2018)
21. Groult, R., Prieur, É., Richomme, G.: Counting distinct palindromes in a word in linear time. Inf. Process. Lett. **110**(20), 908–912 (2010)
22. Gusfield, D.: Algorithms on stings, trees, and sequences: computer science and computational biology. ACM SIGACT News **28**(4), 41–60 (1997)
23. Kociumaka, T.: Minimal suffix and rotation of a substring in optimal time. In: 27th Annual Symposium on Combinatorial Pattern Matching (CPM 2016), vol. 54, pp. 28:1–28:12. Schloss Dagstuhl-Leibniz-Zentrum fuer Informatik (2016)
24. Kociumaka, T.: Efficient data structures for internal queries in texts. Ph.D. thesis, University of Warsaw (2018)
25. Kociumaka, T., Radoszewski, J., Rytter, W., Waleń, T.: Internal pattern matching queries in a text and applications. In: Proceedings of the Twenty-Sixth Annual ACM-SIAM Symposium on Discrete Algorithms, pp. 532–551. SIAM (2014)
26. Manacher, G.: A new linear-time "on-line" algorithm for finding the smallest initial palindrome of a string. J. ACM (JACM) **22**(3), 346–351 (1975)
27. Matsubara, W., Inenaga, S., Ishino, A., Shinohara, A., Nakamura, T., Hashimoto, K.: Efficient algorithms to compute compressed longest common substrings and compressed palindromes. Theor. Comput. Sci. **410**(8–10), 900–913 (2009)
28. Matsuda, K., Sadakane, K., Starikovskaya, T., Tateshita, M.: Compressed orthogonal search on suffix arrays with applications to range LCP. In: 31st Annual Symposium on Combinatorial Pattern Matching (CPM 2020). Schloss Dagstuhl-Leibniz-Zentrum für Informatik (2020)
29. Pătrașcu, M., Thorup, M.: Time-space trade-offs for predecessor search. In: Proceedings of the Thirty-Eighth Annual ACM Symposium on Theory of Computing, pp. 232–240 (2006)
30. Rubinchik, M., Shur, A.M.: Counting palindromes in substrings. In: Fici, G., Sciortino, M., Venturini, R. (eds.) SPIRE 2017. LNCS, vol. 10508, pp. 290–303. Springer, Cham (2017). https://doi.org/10.1007/978-3-319-67428-5_25
31. Rubinchik, M., Shur, A.M.: Eertree: an efficient data structure for processing palindromes in strings. Eur. J. Comb. **68**, 249–265 (2018)
32. Sakai, Y.: A substring-substring LCS data structure. Theor. Comput. Sci. **753**, 16–34 (2019)
33. Sakai, Y.: A data structure for substring-substring LCS length queries. Theor. Comput. Sci. **911**, 41–54 (2022)
34. Tiskin, A.: Semi-local string comparison: algorithmic techniques and applications. Math. Comput. Sci. **1**(4), 571–603 (2008)

Finding the Cyclic Covers of a String

Roberto Grossi[1] , Costas S. Iliopoulos[2] , Jesper Jansson[3] , Zara Lim[2(✉)] ,
Wing-Kin Sung[4] , and Wiktor Zuba[5]

[1] Dipartimento di Informatica, Università di Pisa, Pisa, Italy
roberto.grossi@unipi.it
[2] Department of Informatics, King's College London, London, UK
{costas.iliopoulos,zara.lim}@kcl.ac.uk
[3] Graduate School of Informatics, Kyoto University, Kyoto, Japan
jj@i.kyoto-u.ac.jp
[4] Department of Computer Science, National University of Singapore,
Singapore, Singapore
ksung@comp.nus.edu.sg
[5] Centrum Wiskunde & Informatica, Amsterdam, The Netherlands
wiktor.zuba@cwi.nl

Abstract. We introduce the concept of cyclic covers, which generalizes
the classical notion of covers in strings. Given any nonempty string X of
length n, a factor W of X is called a cyclic cover if every position of X
belongs to an occurrence of a *cyclic shift* of W. Two cyclic covers are
distinct if one is not a cyclic shift of the other. The *cyclic cover problem*
requires finding all distinct cyclic covers of X. We present an algorithm
that solves the cyclic cover problem in $\mathcal{O}(n \log n)$ time. This is based
on finding a well-structured set of standard occurrences of a constant
number of factors of a cyclic cover candidate W, computing the regions
of X covered by cyclic shifts of W, extending those factors, and taking
the union of the results.

Keywords: String · Cyclic string · Cover · Periodicity · Regularities

1 Introduction

String periodicities and repetitions have been thoroughly studied in many fields
such as string combinatorics, pattern matching and automata theory [25,26]
which can be linked to its importance across various applications, in addition
to its theoretical aspects. Detection algorithms and data structures for repeated
patterns and regularities span across several fields of computer science [13,18],
for example computational biology, pattern matching, data compression, and
randomness testing.

Covers of strings have also been extensively studied in similar fields of combinatorics. The concept originates from quasiperiodicity, a generalization of periodicity which also allows those identical strings to overlap [5]. A factor W of a
nonempty string X is called a cover if every position of X belongs to some occurrence of W in X. Furthermore, a cover W must also be a border (i.e. appearing

C.-C. Lin et al. (Eds.): WALCOM 2023, LNCS 13973, pp. 139–150, 2023.
https://doi.org/10.1007/978-3-031-27051-2_13

as both a prefix and a suffix) of the string X. Moore and Smyth [30] developed a linear-time algorithm which computes all covers of a string. Apostolico et al. [6] developed a linear-time algorithm for finding the shortest cover, which Breslauer developed into an on-line algorithm [8]. Li and Smyth [24] produced an on-line algorithm for the all-covers problem. Related string factorization problems include antiperiods [2] and anticovers [1], in addition to approximate [3] and partial [22] covers and seeds [21]. Other combinatorial covering problems consider applications to graphs [11,31].

Cyclic strings have been commonly studied throughout various computer science and mathematical fields, mostly occurring in the field of combinatorics. A cyclic string is a string that does not have an initial or terminal position; instead, the two ends of the string are joined together, and the string can be viewed as a necklace of letters. A cyclic string of length n can be also viewed as a traditional linear string, which has the left- and right-most letters wrapped around and stuck together. Under this notion, the same cyclic string can be seen as n linear strings, which would all be considered equivalent. One of the earliest studies of cyclic strings occurs in Booth's linear time algorithm [7] for computing the lexicographically smallest cyclic factor of a string. Other closely related works reference terms such as 'Lyndon factorization' and 'canonization' [4,10,14,16,27, 28,33]. Some recent advances on cyclic strings can be found in [12]. Aside from combinatorics, cyclic strings have applications within Computational Biology, such as detecting DNA viruses with circular structures [34,35].

We introduce the concept of a *cyclic cover* of a nonempty string X of length $n = |X|$, which generalizes the notion of a cover under cyclic shifts. A factor W of X is called a cyclic cover if every position of X belongs to an occurrence of a *cyclic shift* of W. Figure 1 displays an example where X has a cyclic cover where factors have length $\ell = 3, 4, 7, 10, 13, 16$ (cyclic occurrences of the shortest two are shown on the figure).

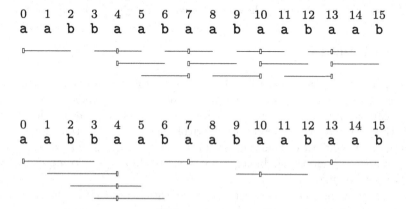

Fig. 1. String $X = $ aabbaabaabaabaab has a cyclic cover of length 3, as $X[0..2]$, $X[3..5]$, $X[4..6]$, $X[5..7]$, $X[6..8]$, $X[7..9]$, $X[8..10]$, $X[9..11]$, $X[10..12]$, $X[11..13]$, $X[12..14]$, $X[13..15]$ are all cyclic shifts of the same factor, and cover all positions of X. Similarly X has a cyclic cover of length 4.

Looking at the example, we observe that if two distinct factors W and Z, of the same length ℓ, are cyclic covers of X, then W must be a cyclic shift of Z, and vice versa. For this, we say that two cyclic covers are distinct if one is *not* the cyclic shift of the other (in other words, they must have different length as factors of X). Moreover, if a cyclic cover of length ℓ exists, then the prefix of X of length ℓ is also a cyclic cover and, consequently, is the representative of all factors of length ℓ that are cyclic covers of X (as the latter ones are all cyclic shifts of the prefix). Because of this, it is enough to give ℓ as output.

Our Contribution. We introduce the following *cyclic cover problem*: given an input string X of length n, find all the distinct cyclic covers of X, namely, the prefixes of X that are cyclic covers (actually, their lengths ℓ). Under this definition, we have at most n distinct cyclic covers whereas there might be $\Theta(n^2)$ distinct factors that are cyclic covers.[1] In the example of Fig. 1, the output of the cyclic cover problem is $\ell = 3, 4, 7, 10, 13, 16$.

We show that for a string of length n, the cyclic cover problem can be solved in $\mathcal{O}(n \log n)$ time. We assume, that the input string is over a polynomially bounded integer alphabet, and the word RAM model of computation with word size $\mathcal{O}(\log n)$ (both restrictions follow from the restrictions of cited and used data structures).

The rest of the paper is organized as follows. In Sect. 2, we present the preliminary concepts. Section 3 shows our findings for cyclic covers. Finally, we present concluding remarks in Sect. 4.

2 Preliminaries

2.1 Basic Definitions

A *string* X of length $n = |X|$ is a sequence of n letters over an integer alphabet $\Sigma = \{0, \ldots, n^{\mathcal{O}(1)}\}$. The letter at position i, for $0 \leq i < n$, is denoted as $X[i]$. A positive integer $p < n$ is called a period of X if $X[i] = X[i + p]$ for all $i = 0, \ldots, n - p - 1$. By $X[i \mathinner{.\,.} j]$, we denote a *factor* of X equal to $X[i] \cdots X[j]$, whereby if $i > j$, then it is the empty string. The factor $X[i \mathinner{.\,.} j]$ is a *prefix of* X if $i = 0$, and a *suffix of* X if $j = n - 1$. If $X[0 \mathinner{.\,.} b - 1] = X[n - b \mathinner{.\,.} n - 1]$, the factor $X[0 \mathinner{.\,.} b - 1]$ is called a *border* of X. A factor W is *periodic* if its smallest period is at most $|W|/2$, and W is *highly-periodic* if its smallest period is at most $|W|/4$. An important property used throughout the paper is Fine and Wilf's periodicity lemma.

Lemma 1 ([15]). *If p, q are periods of a string X of length $|X| \geq p + q - gcd(p, q)$, then $gcd(p, q)$ is also a period of X.*

A factor U is a *cyclic shift* of a factor W if $W = AB$ and $U = BA$ for some strings A and B. In that case, we also say that U is a *d-cyclic shift* of W where

---

[1] For example, for $X = a^k b a^k$ and $k > 1$, all factors $a^i b a^j$ are cyclic covers of X, for $i, j \geq 0$ such that $i + j \geq k$. They are represented by the prefixes of length $i + j + 1$.

$d = |A|$. (Clearly, $d = 0$ implies that $U = W$ and so there is no cyclic shift.) A factor W is called a *cyclic cover* of X if, for every position i ($0 \le i < n$), there exists a factor $X[l .. l+|W|-1]$ that is a cyclic shift of W and contains position i (i.e. $0 \le l \le i \le l+|W|-1 < n$). Two cyclic covers are distinct if they are not cyclic shifts of one another. As observed in the introduction, the distinct cyclic covers are represented by (the lengths of) the prefixes of X.

We denote by $lcp(X[i .. j], X[k .. l])$ the length of the longest common prefix of factors $X[i .. j]$ and $X[k .. l]$. Also, we denote by $lcp^r(X[i .. j], X[k .. l])$ the length of the longest common suffix of $X[i .. j]$ and $X[k .. l]$. Both lcp and lcp^r can be computed in $\mathcal{O}(1)$ time after an $\mathcal{O}(n)$-time preprocessing of X [19].

2.2 The IPM Data Structure

A useful data structure called the *Internal Pattern Matching* (IPM) data structure was introduced in [23]. The following three lemmas summarize some of its properties. Let us denote by $occ(W, Z)$ the (possibly empty) list of positions j such that $W = Z[j .. j + |W| - 1]$.

Lemma 2 ([20,23]). *Given a string X of length n, the IPM data structure of X after $\mathcal{O}(n)$ time and space construction computes $occ(A, B)$ for any factors A and B of X where $|A| \le |B| \le 2|A|$, in $\mathcal{O}(1)$ time. Furthermore, the list of positions is presented as an arithmetic progression.*

Lemma 3 ([20,23]). *Given a string X of length n, the IPM data structure of X after $\mathcal{O}(n)$ time and space construction determines if A is a cyclic shift of B in $\mathcal{O}(1)$ time, for any two factors A and B of X.*

Lemma 4 ([23]). *Given a string X of length n, the 2-Period data structure of X after $\mathcal{O}(n)$ time and space construction determines if A is periodic and if that is the case computes its shortest period in $\mathcal{O}(1)$ time for any factor A of X.*

In [23] the structures of Lemmas 2 and 3 are constructed in $\mathcal{O}(n)$ expected time. These constructions were made worst-case in [20]. The structure of Lemma 4 was constructed in $\mathcal{O}(n)$ worst-case time already in [23].

3 Cyclic Covers

Consider a string $X[0 .. n-1]$ and its length-ℓ factor $W[0 .. \ell - 1]$. A straightforward approach to verify if X is cyclically covered by W leads to a quadratic algorithm, as follows: we apply Lemma 3 to test whether $X[i .. i+\ell-1]$ is a cyclic shift of $X[0 .. \ell - 1]$, for all $i = 0, 1, \ldots, n - \ell$. If the cyclic shifts of $X[0 .. \ell - 1]$ cover all positions of X, we report $X[0 .. \ell - 1]$ as a cyclic cover of $X[0 .. n-1]$. Such verification takes $\mathcal{O}(n - \ell)$ time. The cyclic cover problem can be solved by verifying all $\ell \in \{1, \ldots, n-1\}$, which takes $\mathcal{O}(n^2)$ time in total.

Below, we show that this problem can be solved in $\mathcal{O}(n \log n)$ time. Before we detail the algorithm, we first outline 3 techniques:

1. Section 3.1 gives a function $FindFixedCover(W, X, i, j)$ that verifies if X is cyclically covered by W with the constraint that $W[i]$ aligns to $X[j]$.
2. Based on the function $FindFixedCover(W, X, i, j)$, Sect. 3.2 gives an $\mathcal{O}(n/\ell)$-time algorithm that finds regions in X covered by W when W is highly-periodic.
3. Based on the function $FindFixedCover(W, X, i, j)$, Sect. 3.3 gives an $\mathcal{O}(n/\ell)$-time algorithm that finds regions in X covered by W when W is not highly-periodic.

3.1 Find Regions in X Covered by Cyclic Shifts of W with the Constraint that $W[i]$ Aligns to $X[j]$

Consider a string $X[0 .. n - 1]$ and its length-ℓ factor $W[0 .. \ell - 1]$. For any $j' \in [j - \ell + 1 .. j]$, a length-$\ell$ factor $X[j' .. j' + \ell - 1]$ of X is called a cyclic shift of $W[0 .. \ell - 1]$ with $W[i]$ aligned to $X[j]$ if the i-cyclic shift of W equals the $(j - j')$-cyclic shift of $X[j' .. j' + \ell - 1]$. The lemma below computes the region $X[\alpha .. \beta]$ in X that is cyclically covered by W with the constraint that $W[i]$ aligns to $X[j]$.

Lemma 5. *Consider a string $X[0 .. n - 1]$ and its length-ℓ factor $W[0 .. \ell - 1]$. Let $\ell_1 = lcp(W[i .. \ell - 1]W[0 .. i - 1], X[j .. n - 1])$ and $\ell_2 = lcp^r(W[i + 1 .. \ell - 1] W[0 .. i], X[0 .. j]) - 1$. If $\ell_1 + \ell_2 \geq \ell$, then $X[j - \ell_2 .. j + \ell_1 - 1]$ is cyclically covered by W with the constraint that $W[i]$ aligns to $X[j]$; otherwise, such a cyclic cover does not exist.*

Proof. Let $U = W[i .. \ell - 1]W[0 .. i - 1]$. Observe that $X[j - \ell_2 .. j + \ell_1 - 1] = U^2[\ell - \ell_2 .. \ell + \ell_1 - 1]$ from the definitions of ℓ_1 and ℓ_2. Every factor of length ℓ of this string is a cyclic shift of U, hence also of W, thus it is cyclically covered by W, with the constraint that $U[0]$ aligns to $X[j]$ (hence $W[i]$ aligns to $X[j]$).

If $\ell_1 + \ell_2 < \ell$, then X does not contain any cyclic shift of U, with $U[0]$ aligned to $X[j]$ since $U[\ell - \ell_2 - 1] \neq X[j - \ell_2 - 1]$ and $U[\ell_1] \neq X[j + \ell_1]$, and any such cyclic shift (as a factor of X) would contain one of those two positions (those positions are less then ℓ positions apart, and position j is in between them). □

For example, given the word $X = babbbababb$, the factor $W = bbab$ and the constraint that $X[3]$ aligns with $W[1]$, $X[1 .. 6]$ is is cyclically covered by W (see Fig. 2).

$W = bbab$ and $X = abbbab$ where $W[1]$ aligns with $X[3]$. Thus W cyclically covers the region $X[1 .. 6]$.

Based on the above lemma, we denote $FindFixedCover(W, X, i, j)$ as the function that returns the region $X[\alpha .. \beta]$ that is cyclically covered by W with the constraint that $W[i]$ aligns to $X[j]$. If no such cyclic cover exists, the function returns an empty region.

Lemma 6. *After linear time preprocessing, we can compute $FindFixedCover(W, X, i, j)$ for any factor W of X in $\mathcal{O}(1)$ time.*

$$\ell_1$$

$$\overbrace{\qquad\qquad\qquad}$$

	1	2	3	0
	b	a	b	b

0	1	2	3	4	5	6	7	8	9
b	a	b	b	b	a	b	a	b	b

	2	3	0	1
	a	b	b	b

$$\underbrace{\qquad\qquad\qquad}_{\ell_2}$$

Fig. 2. Given $X = babbbababb$, $W = bbab$ and the constraint that $X[3]$ aligns with $W[1]$, the factor $X[1..6]$ is cyclically covered by W. The lengths ℓ_1 and ℓ_2 denote $lcp(W[1..3]W[0], X[4..9])$ and $lcp^r(W[2..3]W[0..1], X[0..4]) - 1$, respectively.

Proof. We first build the lcp data structures for X and for its reverse in linear time and then use them to compute ℓ_1 and ℓ_2. Even though $W[i..\ell-1]W[0..i-1]$ does not need to occur in X in such a case, we simply compute ℓ_1 in two steps. If $lcp(W[i..\ell-1], X[j..n-1]) < \ell-i$, then it represents the sought value. Otherwise $\ell_1 = (\ell-i) + lcp(W[0..i-1], X[j+\ell-i..n-1])$. Note that ℓ_2 is computed analogously.   □

3.2   Finding Regions in X that are Cyclically Covered by a Highly-Periodic Factor W

Lemma 8 describes how to find regions that are cyclically covered by $W[0..\ell-1]$ if W is of period q where $q \le \ell/4$. To show it, we make use of Lemma 7 from [29] (see also [9,17]) to represent occurrences in a convenient way. Below, we let (j_1, q, m) denote the arithmetic progression $j_1, j_2, \ldots, j_m$ with $j_{s+1} = j_s + q$, where $1 \le s < m$.

Lemma 7 ([29], **Lemma 3.1**). *Suppose the minimum period of $W[0..\ell-1]$ is q. For a length-2ℓ factor Y, $occ(W, Y)$ equals a single arithmetic progression (j_1, q', m'). If $m' \ge 3$, then $q' = q$.*

Lemma 8. *Suppose the smallest period of $W[0..\ell-1]$ is $q \le \ell/4$. We can find which parts of $X[i..i+\ell-1]$ are cyclically covered by W in $\mathcal{O}(1)$ time.*

Proof. Any cyclic shift of W that covers any position of $X[i..i+\ell-1]$ must be fully contained inside $X[i-\ell..i+2\ell-1]$, hence we are going to restrict our search to that region.

Let $Y = W[0..\lfloor \ell/2q \rfloor q - 1]$, which is $W[0..q-1]^{\lfloor \ell/2q \rfloor}$. Note that $\ell/3 < |Y| \le \ell/2$, and also $|Y| \ge 2q$, hence q is its smallest period (a smaller period would imply a smaller period of X by Lemma 1). Any cyclic shift of W must contain Y as a factor.

We first find the occurrences of Y in $X[i-\ell..i+2\ell-1]$. By Lemma 2, these occurrences can be found in $\mathcal{O}(1)$ time by computing $occ(Y, X[i'..i'+2|Y|])$ for

$i' \in \{i - \ell + h|Y| \mid h = 0, 1, 2, \dots \lfloor 3\ell/|Y| \rfloor\}$. Since $3\ell/|Y| < 9$ we have at most 9 arithmetic progressions with period q (by Lemma 7) plus up to 18 standalone occurrences.

For each standalone occurrence starting at position j we can simply run $FindFixedCover(W, X, 0, j)$ separately. Processing of the arithmetic progressions is a little more complex however.

To see how to do it efficiently, we first prove a crucial claim. For an arithmetic progression (j_1, q, m) and $1 \leq s \leq m$, let $X[\alpha_s .. \beta_s] = FindFixedCover(W, X, 0, j_s)$. We claim that the following inequalities hold: $\beta_s \leq \beta_{s+1}$ and $\alpha_s \leq \alpha_{s+1}$. The former inequality $\beta_s = j_s + lcp(W, X[j_s .. n-1]) - 1 \leq j_s + lcp(W[0 .. q-1]W, X[j_s .. n-1]) - 1 = j_s + q + lcp(W, X[j_{s+1} .. n-1]) - 1 = \beta_{s+1}$ is simple to see as W is a prefix of $W[0 .. q-1]W$. For the latter inequality $\alpha_s \leq \alpha_{s+1}$, notice that if $|W|$ is a multiple of q, then we can simply apply a proof symmetric to the one for β's. Otherwise $\alpha_{s+1} < \alpha_s \leq j_s$ for $s \geq 1$ would imply a non-trivial border of W of length q, which in turn would imply that $W - |q|$ is a period of W. By Lemma 1, we have $gcd(q, |W| - q) < q$, which is a contradiction. This completes the proof of the claim.

Due to this claim, the region obtained for this sequence is $X[\alpha_1 .. \beta_m]$, and only two calls of $FindFixedCover$ are needed.   □

In conclusion, as factors $X[k\ell .. (k+1)\ell - 1]$ for $k \in [0, \lfloor \frac{n}{\ell} \rfloor - 1]$ and $X[n - \ell .. n - 1]$ contain all positions of X, we have the following corollary.

Corollary 1. *After an $\mathcal{O}(n)$ time preprocessing of the string $X[0 .. n - 1]$, for any highly-periodic factor $W[0 .. \ell - 1]$, we can compute the regions in X which are cyclically covered by W in $\mathcal{O}(n/\ell)$ time.*

3.3   Finding Regions in X that Are Cyclically Covered by a Non-highly-periodic Factor W

The lemma below states that factors that are not highly-periodic do not occur frequently in X, and follows directly from the definition of a period.

Lemma 9. *Consider a string $X[0 .. n - 1]$ and a non-highly-periodic factor $W[0 .. \ell - 1]$. Any two occurrences of W in X are at distance at least $\ell/4$.*

Let W' be some factor of W. If a cyclic shift of W contains W', we call it a W'-containing cyclic shift of W.

Consider $W[0 .. \ell - 1] = W_l W_r$ where $|W_l| = \lfloor \ell/2 \rfloor$. The following lemma gives a way to find all regions in X covered by cyclic shifts of W.

Lemma 10. *Consider $W[0 .. \ell - 1] = W_l W_r$ where $|W_l| = \lfloor \ell/2 \rfloor$. For a string X, let A be the set of all regions in X covered by W_l-containing cyclic shifts of $W_l W_r$, and let B be the set of all regions in X covered by W_r-containing cyclic shifts of $W_r W_l$. Then $A \cup B$ forms the set of all regions in X that are cyclically covered by W.*

Proof. Observe that every cyclic shift of W must contain either W_l or W_r. Hence, the lemma follows. $\square$

Below we focus on describing an algorithm that finds all regions in X covered by W_l-containing cyclic shifts of W_lW_r. All regions in X covered by W_r-containing cyclic shifts of W_rW_l can be found by an analogous algorithm.

To find all regions in X covered by W_l-containing cyclic shifts of W_lW_r, we consider two cases: W_l is highly-periodic or not.

If W_l is not highly-periodic, then it has $\mathcal{O}(n/\ell)$ occurrences in $X[0 .. n-1]$ (Lemma 9). Thus we can find all these occurrences in $\mathcal{O}(n/\ell)$ time given the IPM data structure (Lemma 3). Then, using $FindFixedCover()$, the regions in X covered by these W_l-containing cyclic shifts of W can be found using $\mathcal{O}(n/\ell)$ time.

For a highly periodic W_l, let $q_l \leq \ell/8$ denote its shortest period, and let d_l denote the longest prefix of W which is q_l-periodic. Let us also denote $W_{l'} = W[0 .. d_l - 1]$ and $W_{r'} = W[d_l .. \ell - 1]$. Notice that if $W_{r'}W_{l'}$ is highly-periodic we can simply reduce our problem to the case with a highly-periodic W as any cyclic shift of W is also a cyclic shift of $W_{r'}W_{l'}$. Notice, also, that a W_l-containing cyclic shift of W (d-cyclic shift of W for $d = 0$ or $d \geq |W_l|$) is always a $W_{l'}W[d_l]$-containing factor of W (for $d = 0$ or $d > d_l$) or a $W_{r'}W_l$-containing factor of W (for $|W_l| \leq d \leq d_l$).

Now it is enough to show that, for a highly-periodic W_l when W and $W_{r'}W_{l'}$ are not highly-periodic, $W_{l'}W[d_l]$ and $W_{r'}W_l$ are not highly-periodic as well.

Lemma 11. $W_{l'}W[d_l]$ *is non-periodic (hence also non-highly-periodic).*

Proof. By contradiction, suppose that $W[0 .. d_l] = W_{l'}W[d_l]$ has period $q' \leq (d_l + 1)/2$. This means that $W_{l'}$ has both periods q_l and q'. Since $q_l + q' \leq \ell/8 + (d_l + 1)/2 \leq d_l$, we have that $gcd(q_l, q')$ is also a period of $W_{l'}$ by Lemma 1.

We observe that q' cannot be a multiple of q_l as in this case $W[d_l] = W[d_l - q'] = W[d_l - q]$, which contradicts the definition of d_l. Hence we get $gcd(q_l, q') < q_l$, which in turn contradicts the fact that q_l is the shortest period of $W_{l'}$. $\square$

Lemma 12. $W_{r'}W_l$ *is not highly-periodic.*

Proof. Suppose, on the contrary, that $W_{r'}W_l$ has period $q' \leq |W_{r'}W_l|/4 \leq \ell/4$. This means, that W_l has both periods q_l and q'. Since $q_l + q' \leq \ell/2$ by Lemma 1 $gcd(q_l, q')$ is also a period of W_l.

If q' is a multiple of q_l, then $W_{r'}W_{l'}$ is also $q' \leq \ell/4$ periodic contrary to the assumptions, otherwise $gcd(q_l, q') < q_l$ which contradicts that q_l is the shortest period of W_l. $\square$

Now, we are ready to describe a function $FindCyclicCover(W_lW_r, X)$ that returns all regions in X that are covered by W_l-containing cyclic shifts of W. This function is described in Algorithm 1.

Algorithm 1. $FindCyclicCover(W_l, W_r, X)$

Output: Regions in X covered by W_l-containing cyclic shifts of W

1: If $W = W_l W_r$ or $W_{r'} W_{l'}$ is of period $\leq \ell/4$, we apply Corollary 1 to find the regions of X covered by W using $\mathcal{O}(n/\ell)$ time and return the answer.
2: $Ans = \emptyset$
3: **if** W_l is not highly-periodic **then**
4:      Find $j_1, \ldots, j_m$ such that $X[j_s \ldots j_s + |W_l| - 1] = W_l$ using $\mathcal{O}(n/\ell)$ time.
5:      For each j_s, $Ans = Ans \cup FindFixedCover(W, X, 0, j_s)$
6: **else**
7:      Find $j_1, \ldots, j_m$ such that $X[j_s \ldots j_s + d_l] = W_{l'} W[d_l]$ using $\mathcal{O}(n/\ell)$ time.
8:      For each j_s, $Ans = Ans \cup FindFixedCover(W, X, 0, j_s)$
9:      Find $j_1, \ldots, j_m$ such that $X[j_s \ldots j_s + |W_{r'} W_l| - 1] = W_{r'} W_l$ using $\mathcal{O}(n/\ell)$ time.
10:     For each j_s, $Ans = Ans \cup FindFixedCover(W, X, d_l, j_s)$
11: **end if**
12: Return Ans

Lemma 13 summarizes the time complexity of $FindCyclicCover(W_l, W_r, X)$.

Lemma 13. *Given the lcp, IPM and 2-Period data structures of X, we can compute $FindCyclicCover(W_l, W_r, X)$ (and $FindCyclicCover(W_r, W_l, X)$) in $\mathcal{O}(n/\ell)$ time.*

Proof. Let us first assume that we know an occurrence in X of any given string. To check whether W and W_l are (highly-)periodic, it is enough to perform the 2-Period queries (Lemma 4). Later, with the use of a single lcp query ($lcp(X, X[q_l \ldots n-1])$ in this case), one can compute d_l. $W_{r'} W_{l'}$ can only be highly periodic if W_l is periodic with the same period, hence a check of whether it is highly periodic only requires a comparison between parts of W_l and W_r which takes $\mathcal{O}(1)$ time in total. After determining which method to use, the algorithm performs $\mathcal{O}(n/l)$ $FindFixedCover()$ queries, which results in a $\mathcal{O}(n/l)$ total time complexity.

In general, we do not know the occurrences of some of the strings (for example $W_r W_l$), or even if they occur in X at all. To address this issue and be able to use the internal data structures we make some adjustments.

For the cyclic shifts of W, namely, $W_r W_l$, $W_{r'} W_{l'}$ and its counterpart used by $FindCyclicCover(W_r, W_l, X)$, we only need to check whether they are highly-periodic and employ the lcp (or lcp^r) with another string. To address the first point, it is sufficient to check whether their longest factor which appears in W is periodic, and whether the period can be extended to the whole string (with lcp queries). This factor must be of length at least $\ell/2$; hence, it must be periodic if the whole string is highly-periodic. Its shortest period is the only candidate for the shortest period (of length at most $\ell/4$) of the whole string. As for the second point, lcp, this is only used by Lemma 6, where this problem has already been solved.

Another string which does not need to appear in X is $W_{r'} W_l$ (symmetrically $(W_r W_l)[0 \ldots d_r]$ used by $FindCyclicCover(W_r, W_l, X)$). We make use of this string only if W_l is highly periodic. Using the lcp^r query, we can find how far this period extends to the left in $W_{r'}W_l$. Now, instead of looking for the whole $W_{r'}W_l$ in the parts of X, we simply look for W_l. If a whole arithmetic sequence (j_1, q_l, m) of occurrences is found, then we know that only one of those occurrences can be extended to the whole $W_{r'}W_l$ (with j_{k+1}, where k is equal to the number of periods of W_l at the end of $W_{r'}$). This way we can process the whole X in $\mathcal{O}(n/\ell)$ time.   □

Theorem 1 (Cyclic cover problem). *Given a string X of length n, over an integer alphabet, we can find all integers $\ell > 0$ such that the prefix $W = X[0 .. \ell - 1]$ is a cyclic cover of X, in $\mathcal{O}(n \log n)$ total time.*

Proof. In the preprocessing step we construct the Internal Data Structure answering lcp, IPM and 2-Period queries in $\mathcal{O}(n)$ time (Lemmas 2 and 4). For any fixed ℓ, let $W_l = X[0 .. \lfloor \ell/2 \rfloor - 1]$ and $W_r = X[\lfloor \ell/2 \rfloor .. \ell - 1]$. We can check if $W = X[0 .. \ell - 1]$ is a cyclic cover of $X[0 .. n - 1]$ by applying $FindCyclicCover(W_l, W_r, X)$ and $FindCyclicCover(W_r, W_l, X)$. Lemma 13 shows that these two functions run in $\mathcal{O}(\frac{n}{\ell})$ time. The total time to test $\ell = 1, \ldots, n$ is upper bounded by $\mathcal{O}(\sum_{\ell=1}^{n} \frac{n}{\ell}) = \mathcal{O}(n \log n)$.   □

4   Concluding Remarks

In this paper we showed that all distinct cyclic covers can be found in $\mathcal{O}(n \log n)$ time. The techniques introduced in our solution can also give (much simpler) algorithms for two other related problems.

The first one is to find all cyclic borders of X, namely, all values of ℓ such that prefix $X[0 .. \ell - 1]$ is a cyclic shift of suffix $X[n - \ell .. n - 1]$. It can be solved in $\mathcal{O}(n)$ time by simply using Lemma 3 n times.

The second problem is to find all the cyclic factorizations, which are a special case of the cyclic covers: X is partitioned into factors of length ℓ, for all feasible ℓ, so that each resulting factor is a cyclic shift of the others. We obtain an $\mathcal{O}(n \log \log n)$ time algorithm by using Lemma 3 $\mathcal{O}(\frac{n}{\ell})$ times for every length ℓ that divides n (denoted as $\ell | n$). The complexity follows from the bound $\sum_{\ell | n} \frac{n}{\ell} = \mathcal{O}(n \log \log n)$ given in [32, Thm.2].

References

1. Alzamel, M., et al.: Finding the anticover of a string. In: 31st Annual Symposium on Combinatorial Pattern Matching (CPM 2020), vol. 161 (2020)
2. Alzamel, M., et al.: Online algorithms on antipowers and antiperiods. In: Brisaboa, N.R., Puglisi, S.J. (eds.) SPIRE 2019. LNCS, vol. 11811, pp. 175–188. Springer, Cham (2019). https://doi.org/10.1007/978-3-030-32686-9_13
3. Amir, A., Levy, A., Lubin, R., Porat, E.: Approximate cover of strings. Theor. Comput. Sci. **793**, 59–69 (2019). https://doi.org/10.1016/j.tcs.2019.05.020
4. Apostolico, A., Crochemore, M.: Fast parallel Lyndon factorization with applications. Math. Syst. Theor. **28**(2), 89–108 (1995). https://doi.org/10.1007/BF01191471

5. Apostolico, A., Ehrenfeucht, A.: Efficient detection of quasiperiodicities in strings. Theor. Comput. Sci. **119**(2), 247–265 (1993)
6. Apostolico, A., Farach, M., Iliopoulos, C.S.: Optimal superprimitivity testing for strings. Inf. Process. Lett. **39**(1), 17–20 (1991)
7. Booth, K.S.: Lexicographically least circular substrings. Inf. Process. Lett. **10**(4–5), 240–242 (1980)
8. Breslauer, D.: An on-line string superprimitivity test. Inf. Process. Lett. **44**(6), 345–347 (1992)
9. Breslauer, D., Galil, Z.: Real-time streaming string-matching. ACM Trans. Algorithms **10**(4), 1–12 (2014). https://doi.org/10.1145/2635814
10. Černý, A.: Lyndon factorization of generalized words of Thue. Discrete Math. Theor. Comput. Sci. **5**, 17–46 (2002)
11. Conte, A., Grossi, R., Marino, A.: Large-scale clique cover of real-world networks. Inf. Comput. **270**, 104464 (2020)
12. Crochemore, M., et al.: Shortest covers of all cyclic shifts of a string. Theor. Comput. Sci. **866**, 70–81 (2021)
13. Crochemore, M., Rytter, W.: Jewels of Stringology: Text Algorithms. World Scientific, Singapore (2002)
14. Duval, J.P.: Factorizing words over an ordered alphabet. J. Algorithms **4**(4), 363–381 (1983)
15. Fine, N.J., Wilf, H.S.: Uniqueness theorems for periodic functions. Proc. Am. Math. Soc. **16**(1), 109–114 (1965). https://doi.org/10.2307/2034009
16. Fredricksen, H., Maiorana, J.: Necklaces of beads in k colors and k-ary de Bruijn sequences. Discret. Math. **23**(3), 207–210 (1978)
17. Galil, Z.: Optimal parallel algorithms for string matching. Inf. Control. **67**(1–3), 144–157 (1985). https://doi.org/10.1016/S0019-9958(85)80031-0
18. Gusfield, D.: Algorithms on strings, trees, and sequences: computer science and computational biology. ACM Sigact News **28**(4), 41–60 (1997)
19. Kärkkäinen, J., Sanders, P.: Simple linear work suffix array construction. In: Baeten, J.C.M., Lenstra, J.K., Parrow, J., Woeginger, G.J. (eds.) ICALP 2003. LNCS, vol. 2719, pp. 943–955. Springer, Heidelberg (2003). https://doi.org/10.1007/3-540-45061-0_73
20. Kociumaka, T.: Efficient data structures for internal queries in texts. Ph.D. thesis, University of Warsaw, October 2018 (2018). https://www.mimuw.edu.pl/kociumaka/files/phd.pdf
21. Kociumaka, T., Kubica, M., Radoszewski, J., Rytter, W., Waleń, T.: A linear time algorithm for seeds computation. In: Proceedings of the 23rd Annual ACM-SIAM Symposium on Discrete Algorithms, pp. 1095–1112. SIAM (2012)
22. Kociumaka, T., Pissis, S.P., Radoszewski, J., Rytter, W., Waleń, T.: Fast algorithm for partial covers in words. Algorithmica **73**(1), 217–233 (2014). https://doi.org/10.1007/s00453-014-9915-3
23. Kociumaka, T., Radoszewski, J., Rytter, W., Waleń, T.: Internal pattern matching queries in a text and applications. In: Proceedings of the 26th Annual ACM-SIAM Symposium on Discrete Algorithms, pp. 532–551. SIAM (2014)
24. Li, Y., Smyth, W.F.: Computing the cover array in linear time. Algorithmica **32**(1), 95–106 (2002). https://doi.org/10.1007/s00453-001-0062-2
25. Lothaire, M.: Applied combinatorics on words. Encyclopedia of Mathematics and its Applications, Cambridge University Press (2005). https://doi.org/10.1017/CBO9781107341005
26. Lothaire, M.: Algebraic Combinatorics on Words, vol. 90. Cambridge University Press, New York (2002)

27. Melançon, G.: Lyndon factorization of infinite words. In: Puech, C., Reischuk, R. (eds.) STACS 1996. LNCS, vol. 1046, pp. 147–154. Springer, Heidelberg (1996). https://doi.org/10.1007/3-540-60922-9_13

28. Melançon, G.: Lyndon factorization of Sturmian words. Discret. Math. **210**(1–3), 137–149 (2000)

29. Miyazaki, M., Shinohara, A., Takeda, M.: An improved pattern matching algorithm for strings in terms of straight-line programs. In: Apostolico, A., Hein, J. (eds.) CPM 1997. LNCS, vol. 1264, pp. 1–11. Springer, Heidelberg (1997). https://doi.org/10.1007/3-540-63220-4_45

30. Moore, D., Smyth, W.F.: Computing the covers of a string in linear time. In: Proceedings of the 5th Annual ACM-SIAM Symposium on Discrete Algorithms, pp. 511–515. SODA'94, Society for Industrial and Applied Mathematics, USA (1994)

31. Norman, R.Z., Rabin, M.O.: An algorithm for a minimum cover of a graph. Proc. Am. Math. Soc. **10**(2), 315–319 (1959)

32. Robin, G.: Grandes valeurs de la fonction somme des diviseurs et hypothese de Riemann. J. Math. Pures Appl. **63**, 187–213 (1984)

33. Shiloach, Y.: Fast canonization of circular strings. J. Algorithms **2**(2), 107–121 (1981)

34. Tisza, M.J., et al.: Discovery of several thousand highly diverse circular DNA viruses. Elife **9**, e51971 (2020)

35. Wagner, E.K., Hewlett, M.J., Bloom, D.C., Camerini, D.: Basic Virology, vol. 3. Blackwell Science, Malden, MA (1999)

Efficient Non-isomorphic Graph Enumeration Algorithms for Subclasses of Perfect Graphs

Jun Kawahara[1] , Toshiki Saitoh[2(✉)] , Hirokazu Takeda[2], Ryo Yoshinaka[3] ,
and Yui Yoshioka[2]

[1] Kyoto University, Kyoto, Japan
[2] Kyushu Institute of Technology, Kitakyushu, Japan
toshikis@ai.kyutech.ac.jp
[3] Tohoku University, Sendai, Japan

Abstract. Intersection graphs are well-studied in the area of graph algorithms. Some intersection graph classes are known to have algorithms enumerating all unlabeled graphs by reverse search. Since these algorithms output graphs one by one and the numbers of graphs in these classes are vast, they work only for a small number of vertices. Binary decision diagrams (BDDs) are compact data structures for various types of data and useful for solving optimization and enumeration problems. This study proposes enumeration algorithms for five intersection graph classes, which admit $O(n)$-bit string representations for their member graphs. Our algorithm for each class enumerates all unlabeled graphs with n vertices over BDDs representing the binary strings in time polynomial in n. Moreover, our algorithms are extended to enumerate those with constraints on the maximum (bi)clique size and/or the number of edges.

Keywords: Enumeration · Binary decision diagrams · Graph isomorphism · Graph classes · String representation

1 Introduction

This paper is concerned with efficient enumeration of unlabeled intersection graphs. An intersection graph has a geometric representation such that each vertex of the graph corresponds to a geometric object and the intersection of two objects represents an edge between the two vertices in the graph. Intersection graphs are well-studied for their practical and theoretical applications [2,17]. For example, interval graphs, which are represented by intervals on a real line, are applied in bioinformatics, scheduling, and so on [5]. Proper interval graphs are a subclass of interval graphs with interval representations where no interval is properly contained to another. These graph classes are related to important graph parameters: The bandwidth of a graph G is equal to the smallest value of the maximum clique sizes in proper interval graphs that extend G [6].

The literature has considered the enumeration problems for many of the intersection graph classes. The graph enumeration problem is to enumerate all the graphs with n vertices in a specified graph class. If it requires not enumerating two isomorphic graphs, it is called *unlabeled*. Otherwise, it is called *labeled*. Unlabeled enumeration algorithms based on reverse search [1] have been proposed for subclasses of interval graphs and permutation graphs [14,15,18,19]. Those algorithms generate graphs in time polynomial in the number of vertices per graph. In this regard, those algorithms are considered to be fast in theory. However, since those algorithms output graphs one by one and the numbers of graphs in these classes are vast, the total running time will be impractically long, and storing the output graphs requires a large amount of space.

The idea of using *binary decision diagrams (BDDs)* has been studied to overcome the difficulty of the high complexity of enumeration. BDDs can be seen as indexing and compressed data structures for various types of data, including graphs, via reasonable encodings. The technique so-called *frontier-based search*, given an arbitrary graph, efficiently constructs a BDD which represents all subgraphs satisfying a specific property [7,10,16]. Among those, Kawahara et al. [8] proposed enumeration algorithms for several sorts of intersection graphs, e.g., chordal and interval graphs. Using the obtained BDD, one can easily count the number of those graphs, generate a graph uniformly at random, and find an optimal one under some measurement, like the minimum weight. However, the enumeration by those algorithms is labeled. In other words, the obtained BDDs by those algorithms may have many isomorphic graphs. Hence, the technique cannot be used, for example, for generating a graph at uniformly random when taking isomorphism into account.

This paper proposes polynomial-time algorithms for unlabeled intersection graph enumeration using BDDs. The five intersection graph classes in concern are those of proper interval, cochain, bipartite permutation, (bipartite) chain, and threshold graphs. It is known that the unlabeled graphs with n vertices of these classes have natural $O(n)$-bit string encodings: We require $2n$ bits for proper interval and bipartite permutation graphs [14,15] and n bits for chain, cochain, and threshold graphs [11,13]. It may be a natural idea for enumerating those graphs to construct a BDD that represents those encoding strings. Here, we remark that there are different strings that represent isomorphic graphs, and we need to keep only a "canonical" one among those strings. Actually, if we make a BDD naively represent those canonical strings, the resultant BDD will be exponentially large. To solve the problem, we introduce new string encodings of intersection graphs of the respective classes so that the sizes of the BDDs representing canonical strings are polynomial in n. Our encodings are still natural enough to extend the enumeration technique to more elaborate tasks: namely, enumerating graphs with bounded maximum (bi)clique size and/or with maximum number of edges. One application of enumerating proper interval graphs with maximum clique size k is, for example, to enumerate graphs with the bandwidth at most k. Recall that the bandwidth of a graph is the minimum size of the maximum cliques in the proper interval graphs obtained by adding edges. Thus, conversely, we can obtain graphs of bandwidth at most k by removing edges from the enumerated graphs.

2   Preliminary

Graphs. Let $G = (V, E)$ be a simple graph with n vertices and m edges. A sequence $P = (v_1, v_2, \ldots, v_k)$ of vertices is a *path from v_1 to v_k* if v_i and v_j are distinct for $i \neq j$ and $(v_i, v_{i+1}) \in E$ for $i \in \{1, \ldots, k - 1\}$. The graph G is *connected* if for every two vertices $v_i, v_j \in V$, there exists a path from v_i to v_j. The neighbor set of a vertex v is denoted by $N(v)$, and the closed neighbor set of v is denoted by $N[v] = N(v) \cup \{v\}$. A vertex v is *universal* if $|N(v)| = n - 1$ and a vertex v is *isolate* if $|N(v)| = 0$. For $V' \subseteq V$ and $E' \subseteq E$ such that the endpoints of every edge in E' are in V', $G' = (V', E')$ is a *subgraph* of G. The graph G is *complete* if every vertex is universal. If a subgraph $G' = (V', E')$ of G is a complete graph, V' is called a *clique* of G. A clique C is *maximum* if for any clique C' in G, $|C| \geq |C'|$. A vertex set S is called an *independent* set if for each $v \in S$, $N(v) \cap S = \emptyset$. The *complement* of $G = (V, E)$ is the graph $\overline{G} = (V, \overline{E})$ where $\overline{E} = \{(u, v) \mid (u, v) \notin E\}$.

For a graph $G = (V, E)$, let (X, Y) be a partition of V; that is, $V = X \cup Y$ and $X \cap Y = \emptyset$. A graph $G = (X \cup Y, E)$ is *bipartite* if for every edge $(u, v) \in E$, either $u \in X$ and $v \in Y$ or $u \in Y$ and $v \in X$ holds. The bipartite graph G is *complete bipartite* if $E = \{(x, y) \mid x \in X, y \in Y\}$. For a subgraph $G' = (X' \cup Y', E')$ of G, $X' \cup Y'$ is called *biclique* if G' is complete bipartite. A biclique B is *maximum* if for any biclique B' in G, $|B| \geq |B'|$. Note that we here say that a biclique has the "maximum" size if the number of not edges but vertices of it is maximum. For a bipartite graph $G = (X \cup Y, E)$, $\overline{G}$ is called *cobipartite*. Note that X and Y are cliques in $\overline{G}$. An ordering $x_1, x_2, \ldots, x_{|X|}$ on X is an *inclusion ordering* if $N(x_i) \cap Y \subseteq N(x_j) \cap Y$ for every i, j with $i < j$.

Binary Strings. We use the binary alphabet $\Sigma = \{\text{L}, \text{R}\}$ in this paper. Let $s = c_1 c_2 \ldots c_n$ be a binary string on Σ^*. The length of s is n and we denote it by $|s|$. Let $\overline{\text{L}} = \text{R}$ and $\overline{\text{R}} = \text{L}$. For a string $s = c_1 c_2 \ldots c_n$, we define $\overline{s} = \overline{c_n}\, \overline{c_{n-1}} \ldots \overline{c_1}$. The *height* $h_s(i)$ of s at $i \in \{0, 1, \ldots, n\}$ is defined by $h_s(i) = |c_1 \ldots c_i|_\text{L} - |c_1 \ldots c_i|_\text{R}$, where $|t|_c$ denotes the number of occurrences of c in a string t. The string s is *balanced* if $h_s(n) = 0$; that is, the number of L is equal to that of R in s. The *height* of s is the maximum value in the height function for s and denoted by $h(s)$; that is, $h(s) = \max_i h_s(i)$. We say s is *larger* than a string s' with length n if there exists an index $i \in \{1, \ldots, n\}$ such that $h_s(i') = h_{s'}(i')$ for any $i' < i$ and $h_s(i) > h_{s'}(i)$, and we denote it by $s > s'$. The *alternate* string $\alpha(s)$ of s is obtained by reordering the characters of s from outside to center, alternately; that is, $\alpha(s) = c_1 c_n c_2 c_{n-1} \ldots c_{\lceil n/2 \rceil}$ if n is odd and $\alpha(s) = c_1 c_n c_2 c_{n-1} \ldots c_{n/2} c_{n/2+1}$ otherwise.

Binary Decision Diagrams. A *binary decision diagram (BDD)* is an edge labeled directed acyclic graph $D = (N, A)$ that classifies strings over a binary alphabet Σ of a fixed length n. To distinguish BDDs from the graphs we enumerate, we call elements of N *nodes* and those of A *arcs*. The nodes are partitioned into $n + 1$ groups: $N = N_1 \cup \cdots \cup N_{n+1}$. Nodes in N_i are said to be at *level i* for $1 \leq i \leq n + 1$. There is just one node at level 1, called the *root*.

Level $(n+1)$ nodes are only two: the 0-terminal
node and the 1-terminal node. Each node in N_i for $i \leq$
n has two outgoing arcs pointing at nodes in $N_{i+1} \cup$
N_{n+1}. Thus, the length of every path from the root to
a node in N_i is just $i-1$ for $i \leq n$. The terminal nodes
have no outgoing arcs. The two arcs from a node have
different labels from Σ. We call those arcs L-arc and
R-arc. When a string $s = c_1 \ldots c_n$ is given, we follow
the arcs labeled $c_1, \ldots, c_n$ from the root node. If we
reach the 1-terminal, then the input is accepted. If we
reach the 0-terminal, it is rejected. One may reach a

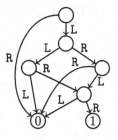

Fig. 1. An example BDD.

terminal node before reading the whole string. In that case, we do not care the
rest unread suffix of the string, and classify the whole string in accordance with
the terminal node. Figure 1 shows an example BDD, where LRLR and LLRR are
accepted and LLRL and RLRL are rejected.

3 Algorithms

3.1 Proper Interval Graphs and Cochain Graphs

Definition and Properties of Proper Interval Graphs. A graph $G = (V, E)$
with $V = \{v_1, \ldots, v_n\}$ is an *interval* graph if there exists a set of n intervals
$\mathcal{I} = \{I_1, \ldots, I_n\}$ such that $(v_i, v_j) \in E$ iff $I_i \cap I_j \neq \emptyset$ for $i, j \in \{1, \ldots, n\}$. The
set $\mathcal{I}$ of intervals is called an *interval representation* of G. For an interval I, we
denote the left and right endpoints of I by $l(I)$ and $r(I)$, respectively. Without
loss of generality, we assume that any two endpoints in $\mathcal{I}$ are distinct. An interval
representation $\mathcal{I}$ is *proper* if there are no two distinct intervals I_i and I_j in $\mathcal{I}$
such that $l(I_i) < l(I_j) < r(I_j) < r(I_i)$ or $l(I_j) < l(I_i) < r(I_i) < r(I_j)$. A graph
G is *proper interval* if it has a proper interval representation (Fig. 2).

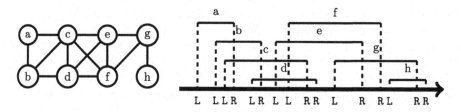

Fig. 2. Proper interval graph and its proper interval representation. The string representation of the proper interval representation is LLLRLRLLRRLRRLRR.

Proper interval graphs can be represented by binary strings as follows. Let G
be a proper interval graph with n vertices and $\mathcal{I}$ be a proper interval representation of G. We can represent $\mathcal{I}$ as a string by sweeping $\mathcal{I}$ from left to right and

encoding $l(I)$ by L and $r(I)$ by R, respectively. We denote the obtained string by $s(\mathcal{I})$ and call it the *string representation* of $\mathcal{I}$. The length of $s(\mathcal{I})$ is $2n$.

Lemma 1 ([15]). *Let $s(\mathcal{I}) = c_1 c_2 \ldots c_{2n}$ be a string representation of a connected proper interval graph G with n vertices.*

1. $c_1 = $ L *and* $c_{2n} = $ R,
2. $s(\mathcal{I})$ *is balanced; that is, the number of* L *is same as that of* R *in* $s(\mathcal{I})$, *and*
3. $h_{s(\mathcal{I})}(i) > 0$ *for* $i \in \{1, \ldots, 2n - 1\}$.

A connected proper interval graph has at most two string representations [4]. More strictly, for any two string representations s and s' of a connected proper interval graph G, $s = s'$ or $s = \overline{s'}$. The string representation is said to be *canonical* if $s > \overline{s}$ or $s = \overline{s}$. Thus, the canonical string representations have one-to-one correspondence to the proper interval graphs up to isomorphism [15].

Algorithm for n Vertices. We here present an enumeration algorithm of all connected proper interval graphs with n vertices up to isomorphism. We would like to construct a BDD representing all canonical string representations of proper interval graphs. However, for the efficiency of the BDD construction as described later, we instead construct a BDD representing alternate strings of all canonical representations of proper interval graphs.

We describe an overview of our algorithm. We construct the BDD in a breadth-first manner in the direction from the root node to the terminals. We create the root node in N_1, and for each node in N_i ($i \in \{1, \ldots, 2n\}$), we create its L and R-arcs and make each arc point at one of the existing nodes in N_{i+1} or N_{2n+1} or a newly created node. We call making an arc point at 0-terminal node *pruning*. For each node ν, we store into ν information on the paths from the root to ν as a tuple, which we call *state*. Two nodes having the same state never exist. When creating an (L or R) arc of a node, we compute the state of the destination from the state of the original node. If there is an existing node having the same state as the computed one, we make the arc point at the existing node, which we call *(node) sharing*.

Consider deciding whether a string s in Σ^{2n} is canonical or not; that is, $s > \overline{s}$ or $s = \overline{s}$ holds. Suppose that $s = c_1 c_2 \ldots c_{2n}$ and we have $\overline{s} = \overline{c_{2n}}\, \overline{c_{2n-1}} \ldots \overline{c_1}$. This can be done by comparing c_i with $\overline{c_{2n-i+1}}$ for $i = 1, \ldots, 2n$. When creating a node ν in the BDD construction process, we would like to conduct pruning early if we can determine that all the path labels from the root via ν will not be canonical. That is the reason we adopt alternate string representations. A node in level i ($\in \{1, \ldots, 2n\}$) corresponds to the $\lceil i/2 \rceil$th character in the string representation if i is odd, and the $(2n + 1 - i/2)$th one otherwise. For example, consider the path LRLRRR. Any path extending LRLRRR will represent a string of the form $s = $ LLRtRRR for which $\overline{s} = $ LLL$\bar{t}$LRR for some $t \in \Sigma^*$ and $s < \overline{s}$ holds. This implies s cannot be canonical. The path goes to the 0-terminal.

We make each node, say ν, maintain state $(i, h_\mathrm{L}, h_\mathrm{R}, F)$. The first element i is the level where ν is. We take an arbitrary path from the root node to ν, say $c_1 c_{2n} c_2 c_{2n-1} \ldots c_{\lceil i/2 \rceil - 1} c_{2n+2-\lceil i/2 \rceil}$ (the case where i is odd) or $c_1 c_{2n} c_2 c_{2n-1} \ldots c_{2n+2-\lceil i/2 \rceil} c_{\lceil i/2 \rceil}$ (the case where i is even). The second and third elements $h_\mathrm{L}, h_\mathrm{R}$

represent the heights of the sequences $c_1 c_2 \ldots c_{\lfloor i/2 \rfloor}$ and $\overline{c_{2n}}\,\overline{c_{2n-1}} \ldots \overline{c_{2n+2-\lceil i/2 \rceil}}$, respectively. Note that we must design an algorithm so that it is well-defined; that is, the values of the sequences obtained from all the paths from the root node to ν are the same. F represents whether ($\star$) $c_{\hat{i}} = \overline{c_{2n+1-\hat{i}}}$ holds for all $\hat{i} = 1, \ldots, \lceil i/2 \rceil - 1$. If $F = \top$, ($\star$) does not hold; that is, there exists i' such that $c_{i'} \neq \overline{c_{2n+1-i'}}$. If $c_{i'} = $ R and $\overline{c_{2n+1-i'}} = $ L, the canonicity condition does not meet. As shown later, such a node never exists because we conduct the pruning. Therefore, $F = \top$ means that $c_{i'} = $ L, $\overline{c_{2n+1-i'}} = $ R and $c_{i''} = \overline{c_{2n+1-i''}}$ holds for all $i'' \leq i' - 1$, which implies that the canonicity condition is satisfied whatever the other characters are. $F = \bot$ means that ($\star$) holds.

We discuss how to store states and conduct pruning in the process of the BDD construction. We make the root node have the state $(1, 0, 0, \bot)$. Let ν be a node that has the state (i, h_L, h_R, F) and ν_L and ν_R be nodes pointed at by L-arc and R-arc of ν. If $i = 1$, ν_R is 0-terminal, and if $i = 2$, ν_L is 0-terminal because of the condition (i) in Lemma 1. First, we consider the case where i is odd. L-arc and R-arc of ν mean that the $\lceil i/2 \rceil$th character is L and R, respectively. We make ν_L have state $(i + 1, h_L + 1, h_R, F)$. As for R-arc, if $h_L - 1 \leq 0$, we make R-arc of ν point at 0-terminal because R-arc means $c_{\lceil i/2 \rceil} = $ R and the height of $c_1 c_2 \ldots c_{\lfloor (i+1)/2 \rfloor}$ violates the condition of (iii) in Lemma 1. Otherwise, we make ν_R have state $(i + 1, h_L - 1, h_R, F)$. Next, we consider the case where i is even. L-arc and R-arc of ν mean that the $(2n + 1 - \lceil i/2 \rceil)$th character is L and R, respectively. If $F = \top$, we make ν_L and ν_R maintain states $(i + 1, h_L, h_R - 1, \top)$ and $(i + 1, h_L, h_R + 1, \top)$, respectively. (Recall that since $F = \top$ means that the canonicity condition has already been satisfied, we need not update F.) We conduct pruning for L-arc if $h_R - 1 \leq 0$. Let us consider the case where $F = \bot$. Recall that ($\star$) holds. Although we want to compare the $\lceil i/2 \rceil$th and $(2n+1-\lceil i/2 \rceil)$th characters to decide whether the canonicity condition holds or not, ν does not have the information on the $\lceil i/2 \rceil$th character. Instead, ν has h_L and h_R. We consider two cases (i) and (ii): (i) If $h_L - 1 = h_R$, it means that the $\lceil i/2 \rceil$th character is L. In this case, R-arc of ν means that the $(2n + 1 - \lceil i/2 \rceil)$th character is R, which implies that ($\star$) still holds. Therefore, we make ν_R maintain state $(i+1, h_L, h_R+1, \bot)$. L-arc of ν means that the $(2n+1-\lceil i/2 \rceil)$th character is L, which implies that ($\star$) no longer holds and the canonicity condition is satisfied. Therefore, we make ν_L maintain state $(i + 1, h_L, h_R - 1, \top)$. (ii) If $h_L - 1 \neq h_R$, it means that the $\lceil i/2 \rceil$th character is R. In this case, R-arc of ν means that the $(2n + 1 - \lceil i/2 \rceil)$th character is R, which violates the canonicity condition. We make R-arc of ν point at 0-terminal. L-arc of ν means that the $(2n+1-\lceil i/2 \rceil)$th character is L, which implies that ($\star$) still holds. Therefore, we make ν_L maintain state $(i + 1, h_L, h_R - 1, \bot)$.

Consider the case where $i = 2n$ (final level). Let the computed state as the destination of L- or R-arc of a node in N_{2n} be $(2n + 1, h'_L, h'_R, F')$. If $h'_L \neq h'_R$, the destination is pruned (0-terminal) because it violates the condition of (ii) in Lemma 1. Otherwise, we make the arc point at 1-terminal.

Theorem 1. *Our algorithm constructs a BDD representing all canonical string representations of connected proper interval graphs in* $\mathrm{O}(n^3)$ *time and space.*

Proof. We here analyze the complexity of the algorithm. For each level $i \in \{0, 1, \ldots, 2n\}$, the number of nodes in N_i is $O(n^2)$ because $0 \leq h_L, h_R \leq n$ and $F \in \{\bot, \top\}$. Thus, the total size of BDD is $O(n^3)$. The computation of the next state for each node can be run in constant time because it has only increment and we can access the nodes in constant time by using $O(n^2)$ pointers.   □

Algorithm for Maximum Clique Size k. We here present an algorithm that given natural numbers n and k, enumerates all proper interval graphs with n vertices and the maximum clique size at most k. It is well known that a clique of an interval graph G corresponds to overlap intervals of a point in an interval representation of G [3]. The number of overlapping intervals is same as the height of string representation of a proper interval graph. Thus, the enumeration of all proper interval graphs with the maximum clique size at most k can be seen as that of all canonical string representations with the height at most k. We modify the algorithm for n vertices by adding one pruning for the case when either of the heights h_L or h_R becomes larger than k. Therefore, our extended algorithm runs in $O(k^2 n)$ time and space since the ranges of h_L and h_R become k from n.

Algorithm for m Edges. To extend the algorithm for n vertices and m edges, we here show how to count the number of edges from the string representation. Let s be a string representation of a proper interval graph with m edges. Sweeping the string representation from left to right, for each $i \in \{1, \ldots, 2n\}$ with $c_i = L$, the height $h_s(i)$ is the number of intervals I_j with $j < i$ that overlap with i. This means that the vertex v corresponding to c_i is incident to $h_s(i)$ edges in G. Thus, we obtain the number of edges from the string representation as follows.

Lemma 2. *Let $s = c_1 \ldots c_{2n}$ be a string representation of a connected proper interval graph G with m edges and J be the set of indices i of s such that $c_i = L$. The summation of heights in J is equal to m; i.e., $\sum_{i \in J} h_s(i) = m$.*

In the construction of a BDD, each node stores the value to maintain the number of edges m'. The state of each node is now a quintuple (i, h_L, h_R, F, m'). For the L-arc of a node ν, the number of edges m' is updated to $m' + h_L$ if i is odd and to $m' + h_R - 1$ otherwise. When either i is odd and $m' + h_L > m$ or i is even and $m' + h_R - 1 > m$ holds, we make the L-arc of ν point at the 0-terminal since the number of edges is larger than m. We make each arc point at the 1-terminal if it gives a state $(2n + 1, h, h, F, m)$ for some h and F based on the state updating rule. Otherwise, it must point at the 0-terminal. For each $i \in \{1, \ldots, 2n\}$, the number of nodes in N_i is $O(n^2 m)$ since $0 \leq h_L, h_R \leq n$ and $0 \leq m' \leq m$ and the number of levels is $2n$. Therefore, the algorithm runs in $O(n^3 m)$ time.

Theorem 2. *A BDD representing all connected proper interval graphs with n vertices and maximum clique size k and with n vertices and m edges can be constructed in $O(k^2 n)$ time and $O(n^3 m)$ time, respectively.*

Cochain Graphs. A graph $G = (X \cup Y, E)$ is a *cochain* graph if G is cobipartite and each of X and Y has an inclusion ordering. In other words, X

and Y are cliques in G and we have two orderings over $X = \{x_1, \ldots, x_{n_X}\}$ and $Y = \{y_1, \ldots, y_{n_Y}\}$ such that $(x_i, y_j) \in E$ implies $(x_{i'}, y_{j'}) \in E$ for any $i \leq i'$ and $j \leq j'$. It is well-known [2] that cochain graphs are a subclass of proper interval graphs. Here, we give a concrete proper interval representation $\{I_1, \ldots, I_{n_X}, J_1, \ldots, J_{n_Y}\}$ of G, where x_i and y_j correspond to I_i and J_j, respectively, by

- $l(I_1) < \cdots < l(I_{n_X}) < r(I_1) < \cdots < r(I_{n_X}) < r(J_{n_Y})$,
- $l(I_{n_X}) < l(J_{n_Y}) < \cdots < l(J_1) < r(J_{n_Y}) < \cdots < r(J_1)$,
- $l(J_j) < r(I_i)$ iff $(x_i, y_j) \in E$ for $1 \leq i \leq n_X$ and $1 \leq j \leq n_Y$.

The inclusion ordering constraint guarantees that the above is well-defined and gives a proper interval representation. Therefore, one can specify a cochain graph as a proper interval graph by a $2n$-bit string representation. Moreover, the strong restriction of cochain graphs allows us to reduce the number of bits to specify a cochain graph. Obviously, the first n_X bits of the proper interval string representation of a cochain graph are all L and the last n_Y bits are all R. Thus, those $n = n_X + n_Y$ bits are redundant and removable. Indeed, one can recover the numbers n_X and n_Y from the remaining n bits. Since every surviving bit of R corresponds to $r(I_i)$ for some i, the number of those bits is just n_X. Similarly, n_Y is the number of bits of L in the new n-bit representation. Conversely, every n-bit string s can be seen as the string representation of a cochain graph with n vertices. However, the n-bit strings are not in one-to-one correspondence to the cochain graphs because universal vertices in the cochain graphs can be seen in either X or Y. To avoid the duplication, we assume that all universal vertices are in Y, so we only consider n-bit strings without R as a suffix. Using this n-bit string representation, we obtain an enumeration algorithm for cochain graph, and it runs in $O(n)$ time.

For the constraint problems, we use $2n$-bit strings because we need to compute the size of cliques or the number of edges. Our algorithms with constraints for cochain graphs are similar to that of proper interval graphs and need to recognize whether the strings represent cochain graphs.

Theorem 3. *A BDD representing all canonical string representations of cochain graphs with n vertices, n vertices and maximum clique size k, and n vertices and m edges can be constructed in $O(n)$, $O(k^2 n)$, and $O(n^3 m)$ time, respectively.*

3.2 Bipartite Permutation Graphs and Chain Graphs

Definition and Properties of Bipartite Permutation Graphs. Let π be a permutation on V; that is, π is a bijection from V to $\{1, \ldots, n\}$. We define $\overline{\pi}$ as $\overline{\pi}(v) = n + 1 - \pi(v)$ for all $v \in V$. We denote by π^{-1} the inverse of π.

A graph $G = (V, E)$ is *permutation* if it has a pair (π_1, π_2) of two permutations on V such that there exists an edge $(u, v) \in E$ iff $(\pi_1(u) - \pi_1(v))(\pi_2(u) - \pi_2(v)) < 0$. The pair $\mathcal{P} = (\pi_1, \pi_2)$ can be seen as the following intersection model on two parallel horizontal lines L_1 and L_2: the vertices in V are arranged on the line L_1 (resp. line L_2) according to π_1 (resp. π_2). Each vertex w corresponds to a line segment l_w, which joins w on L_1 and w on L_2. An edge (u, v) is in E iff l_u and l_v intersects, which is equivalent to $(\pi_1(u) - \pi_1(v))(\pi_2(u) - \pi_2(v)) < 0$. The model $\mathcal{P} = (\pi_1, \pi_2)$ is called a *permutation diagram*. A graph G is *bipartite permutation* if G is bipartite and permutation.

Let $\mathcal{P} = (\pi_1, \pi_2)$ be a permutation diagram of a connected bipartite permutation graph $G = (V, E)$. Let us observe properties of π_1 and π_2, which are discussed in [14]. First, there is no vertex $u \in V$ such that $\pi_1(u) = \pi_2(u)$ unless $n = 1$. Secondly, for all vertices $u, v \in V$ such that $\pi_1(u) < \pi_2(u)$, $\pi_1(v) < \pi_2(v)$ and $\pi_1(u) < \pi_1(v)$ hold, $\pi_2(u) > \pi_2(v)$ does not hold; that is, l_u and l_v never intersects. Therefore, $X = \{u \mid \pi_1(u) < \pi_2(u)\}$ and $Y = \{u \mid \pi_1(u) > \pi_2(u)\}$ give the vertex partition of G. By expressing the above observation with the intersection model, the line segments are never straight vertical and classified into X and Y depending on their tilt directions: lines in X go from upper left to lower right and those in Y go from lower left to upper right.

Based on the above discussion, let us give a string representation $s(\mathcal{P})$ of the permutation diagram $\mathcal{P}$. We define $s_x(\mathcal{P}) = x_1 \ldots x_n$ and $s_y(\mathcal{P}) = y_1 \ldots y_n$ as follows: For $i = 1, \ldots, n$, $x_i = \mathtt{L}$ if $\pi_1(\pi_1^{-1}(i))(= i) < \pi_2(\pi_1^{-1}(i))$, and $x_i = \mathtt{R}$ otherwise. Similarly, for $i = 1, \ldots, n$, $y_i = \mathtt{R}$ if $\pi_2(\pi_2^{-1}(i))(= i) > \pi_1(\pi_2^{-1}(i))$, and $y_i = \mathtt{L}$ otherwise. In other words, $x_i = \mathtt{L}$ iff the ith intersection point of L_1 is with a line segment from X in the intersection model. On the other hand, $y_i = \mathtt{L}$ iff the ith intersection point of L_2 is with a line segment from Y. We define the string representation $s(\mathcal{P})$ of $\mathcal{P}$ by $s(\mathcal{P}) = x_1 y_1 x_2 y_2 \ldots x_n y_n$. The string representation $s(\mathcal{P})$ has the following properties [14].

Lemma 3. *Let $s = c_1 c_2 \ldots c_{2n}$ be a string representation of a connected bipartite permutation graph G with n vertices. Then,*

(i) $c_1 = \mathtt{L}$ and $c_{2n} = \mathtt{R}$,
(ii) s is balanced; that is, the number of $\mathtt{L}$ is the same as that of $\mathtt{R}$ in s, and
(iii) $h_s(i) > 0$ for $i \in \{1, \ldots, 2n - 1\}$.

By horizontally, vertically, and rotationally flipping $\mathcal{P}$, we obtain essentially equivalent diagrams $\mathcal{P}^{\mathrm{V}} = (\pi_2, \pi_1)$, $\mathcal{P}^{\mathrm{H}} = (\overline{\pi_1}, \overline{\pi_2})$, and $\mathcal{P}^{\mathrm{R}} = (\overline{\pi_2}, \overline{\pi_1})$ of G, respectively.

Lemma 4 ([14]). *Let $\mathcal{P}_1$ and $\mathcal{P}_2$ be permutation diagrams of a connected bipartite permutation graph. At least one of the equations $s(\mathcal{P}_1) = s(\mathcal{P}_2)$, $s(\mathcal{P}_1) = s(\mathcal{P}_2^{\mathrm{V}})$, $s(\mathcal{P}_1) = s(\mathcal{P}_2^{\mathrm{H}})$, or $s(\mathcal{P}_1) = s(\mathcal{P}_2^{\mathrm{R}})$ holds.*

A string representation $s(\mathcal{P})$ is said to be *canonical* if all the inequalities $s(\mathcal{P}) \geq s(\mathcal{P}^{\mathrm{V}})$, $s(\mathcal{P}) \geq s(\mathcal{P}^{\mathrm{H}})$, and $s(\mathcal{P}) \geq s(\mathcal{P}^{\mathrm{R}})$ hold.

Algorithm for n Vertices. We construct the BDD representing the set of bipartite permutation graphs using the alternate strings of the canonical representation strings. Each BDD node is identified with a state tuple $(i, h_{\mathrm{L}}, h_{\mathrm{R}}, c_{\mathrm{L}}, c_{\mathrm{R}}, F_{\mathrm{V}}, F_{\mathrm{H}}, F_{\mathrm{R}})$. The integer i is the level where the node is. The heights h_L and h_R are those of $x_1 x_2 \dots x_n$ and $\overline{y_n} \, \overline{y_{n-1}} \dots \overline{y_1}$, respectively, the purpose of which is the same as in Sect. 3.1.

Let us describe F_V, F_H and F_R. F_R is $\perp$ or $\top$, which is used for deciding whether $s(\mathcal{P}) \geq s(\mathcal{P}^R)$ holds or not. Recall that if $s(\mathcal{P}) = x_1 y_1 x_2 y_2 \dots x_n y_n$, $s(\mathcal{P}^R) = \overline{y_n}\,\overline{x_n}\,\overline{y_{n-1}}\,\overline{x_{n-1}} \dots \overline{y_1}\,\overline{x_1}$. According to the variable order $\alpha(s(\mathcal{P}))$, we can decide whether $s(\mathcal{P}) > s(\mathcal{P}^R)$ holds or not using the heights h_L and h_R by the way described in Sect. 3.1. Then, F_R has the same role as F in Sect. 3.1. Next, we consider F_V, which is used for deciding the canonicity of $s(\mathcal{P}) \geq s(\mathcal{P}^V)$. Recall that if $s(\mathcal{P}) = x_1 y_1 x_2 y_2 \dots x_n y_n$, $s(\mathcal{P}^V) = y_1 x_1 y_2 x_2 \dots y_n x_n$. We need to compare x_1 with y_1, y_1 with $x_1, \dots$, and y_n with x_n in order. Recall that on the BDD, the value of y_i is represented by arcs of each node in level $4i - 1$. The value of x_i has already been determined by arcs of a node in level $4i - 3$. Therefore, to compare x_i with y_i, we store the value of x_i into nodes. Strictly speaking, if i is odd, then, $c_L = x_{\lceil i/2 \rceil - 1}$ and $c_R = y_{2n - \lceil i/2 \rceil + 2}$. If i is even, then, $c_L = x_{i/2}$ and $c_R = y_{2n - i/2 + 2}$. The stored values c_L and c_R are also used for deciding whether $s(\mathcal{P}) \geq s(\mathcal{P}^H)$ holds or not in a similar way.

We estimate the number of BDD nodes by counting the possible values of a state $(i, h_L, h_R, c_L, c_R, F_V, F_H, F_R)$. Since $1 \leq i \leq 2n$, $0 \leq h_L \leq n$, $0 \leq h_R \leq n$, and the number of possible states of c_L, c_R, F_V, F_H, F_R are two, the number of possible values of tuples is $2n \times (n+1)^2 \times 2^5 = O(n^3)$.

Algorithm for m edges. We present an algorithm that constructs the BDD representing the set of (string representations of) bipartite permutation graphs with n vertices and m edges when n and m are given. The number of edges of a bipartite permutation graph G is that of intersections of the permutation diagram of G. We use the following lemma.

Lemma 5. *The number of edges is* $\sum_{i=1}^{n} h_{s(\mathcal{P})}(2i)$.

We can easily obtain

$$\sum_{i=1}^{n} h_{s(\mathcal{P})}(2i) = \sum_{i=1}^{\lceil n/2 \rceil} h_{s(\mathcal{P})}(2i) + \sum_{i=1}^{\lfloor n/2 \rfloor} h_{s(\mathcal{P}^R)}(2i). \tag{1}$$

To count the number of edges, we store this value into each BDD node. Let us describe the detail. We make each BDD node maintain a tuple $(i, h_L, h_R, c_L, c_R, F_V, F_H, F_R, m')$. The first eight elements are the same as the ones described above. The last element m' is the current value of (1). Thus, the running time of the algorithm is $O(n^3 m)$.

Theorem 4. *A BDD representing all connected bipartite permutation graphs with n vertices, and n vertices and m edges can be constructed in $O(n^3)$ and $O(n^3 m)$ time, respectively.*

Chain Graphs. A graph $G = (X \cup Y, E)$ is a *chain* graph if G is bipartite and each of X and Y has an inclusion ordering. Let $(x_1, x_2, \dots, x_{|X|})$ and $(y_1, y_2, \dots, y_{|Y|})$ be an inclusion ordering of X and Y, respectively. Chain graphs are known to be a subclass of bipartite permutation graphs [2] and have the following permutation diagrams $\mathcal{P} = (\pi_1, \pi_2)$ [12]:

- $\pi_1 = (x_1, x_2, \ldots, x_{|X|}, y_{|Y|}, y_{|Y|-1}, \ldots, y_1)$,
- for $i, j \in \{1, \ldots, |X|\}$ with $i < j$, $\pi_2(x_i) < \pi_2(x_j)$,
- for $i, j \in \{1, \ldots, |Y|\}$ with $i < j$, $\pi_2(y_j) < \pi_2(y_i)$.

Chain graphs as bipartite permutation graphs have $2n$-bit string representations based on the permutation diagrams. From the diagram and Lemma 4, we observe that the string of π_1 is uniquely determined except for exchanging X and Y. Since π_1 can be fixed as above, any chain graph can be represented using an n-bit string by sweeping π_2: The ith element of π_2 is encoded as L if $\pi_2^{-1}(i) \in X$ and is encoded as R if $\pi_2^{-1}(i) \in Y$. If a chain graph G is disconnected, G consists of two parts: a connected chain graph component and a set of isolated vertices [9]. We observe that the connected chain graphs have a one-to-one correspondence with the string representations up to reversal [13]. On the other hand, isolated vertices may arbitrarily belong to X or Y. To determine a unique string representation, we assume that isolated vertices are all in X, where the representation strings must not end with R. Thus, we obtain an algorithm to construct a BDD representing all canonical n-bit string representations of chain graphs and it runs in $O(n)$. For the restriction problems, we adopt $2n$-bit strings defined as representations of bipartite permutation graphs instead of n-bit representations to compute the number of edges or the size of bicliques. In the algorithms, we need to check whether the constructed strings represent chain graphs satisfying the conditions described above.

Theorem 5. *A BDD representing all chain graphs with n vertices, n vertices and maximum biclique size k, and n vertices and m edges can be constructed in $O(n)$, $O(k^2 n)$, $O(n^3 m)$ time, respectively.*

3.3 Threshold Graphs

A graph G is a *threshold* graph if the vertex set of G can be partitioned into X and Y such that X is a clique and Y is an independent set and each of X and Y has an inclusion ordering. Threshold graphs are a subclass of interval graphs, and any threshold graph can be constructed by the following process [2,11]. First, if the size of the vertex set is one, the graph is threshold. Then, for a threshold graph G, (1) the graph by adding an isolated vertex to G is also threshold, and (2) the graph adding a universal vertex to G is also threshold. The sequence of the two operations (1) and (2) to construct a threshold graph is called a *construction sequence*. It is easy to see that the two threshold graphs G_1 and G_2 are not isomorphic if the construction sequences of (1) and (2) of G_1 and G_2 are different. From this characterization of threshold graphs, we obtain algorithms to construct a BDD representing all unlabeled threshold graphs by encoding the construction sequences of the operation (1) to L and (2) to R.

Theorem 6. *A BDD representing all threshold graphs with n vertices, n vertices and maximum clique size k, and n vertices and m edges can be constructed in $O(n)$ time, $O(kn)$ time, and $O(nm)$ time, respectively.*

Acknowledgments. The authors are grateful for the helpful discussions of this work with Ryuhei Uehara. This work was supported in part by JSPS KAKENHI Grant Numbers JP18H04091, JP19K12098, JP20H05794, and JP21H05857.

References

1. Avis, D., Fukuda, K.: Reverse search for enumeration. Discret. Appl. Math. **65**(1–3), 21–46 (1996)
2. Brandstädt, A., Le, V.B., Spinrad, J.P.: Graph Classes: A Survey. Society for Industrial and Applied Mathematics (1999)
3. Cormen, T.H., Leiserson, C.E., Rivest, R.L., Stein, C.: Introduction to Algorithms, 3rd edn. MIT Press, Cambridge (2009)
4. Deng, X., Hell, P., Huang, J.: Linear-time representation algorithms for proper circular-arc graphs and proper interval graphs. SIAM J. Comput. **25**(2), 390–403 (1996)
5. Golumbic, M.C.: Algorithmic Graph Theory and Perfect Graphs (Annals of Discrete Mathematics, vol. 57). Elsevier (2004)
6. Kaplan, H., Shamir, R.: Pathwidth, bandwidth, and completion problems to proper interval graphs with small cliques. SIAM J. Comput. **25**(3), 540–561 (1996)
7. Kawahara, J., Inoue, T., Iwashita, H., Minato, S.: Frontier-based search for enumerating all constrained subgraphs with compressed representation. IEICE Trans. Fundam. Electron. Commun. Comput. Sci. **100**(9), 1773–1784 (2017)
8. Kawahara, J., Saitoh, T., Suzuki, H., Yoshinaka, R.: Colorful frontier-based search: implicit enumeration of chordal and interval subgraphs. In: Kotsireas, I., Pardalos, P., Parsopoulos, K.E., Souravlias,' D., Tsokas, A. (eds.) SEA 2019. LNCS, vol. 11544, pp. 125–141. Springer, Cham (2019). https://doi.org/10.1007/978-3-030-34029-2_9
9. Kijima, S., Otachi, Y., Saitoh, T., Uno, T.: Subgraph isomorphism in graph classes. Discret. Math. **312**(21), 3164–3173 (2012)
10. Knuth, D.: The Art of Computer Programming, Volume 4A: Combinatorial Algorithms. No. Part 1, Pearson Education, London (2014)
11. Mahadev, N., Peled, U.: Threshold Graphs and Related Topics. Elsevier, Amsterdam (1995)
12. Okamoto, Y., Uehara, R., Uno, T.: Counting the number of matchings in chordal and chordal bipartite graph classes. In: Paul, C., Habib, M. (eds.) WG 2009. LNCS, vol. 5911, pp. 296–307. Springer, Heidelberg (2010). https://doi.org/10.1007/978-3-642-11409-0_26
13. Peled, U.N., Sun, F.: Enumeration of difference graphs. Discret. Appl. Math. **60**(1–3), 311–318 (1995)
14. Saitoh, T., Otachi, Y., Yamanaka, K., Uehara, R.: Random generation and enumeration of bipartite permutation graphs. J. Discrete Algorithms **10**, 84–97 (2012)
15. Saitoh, T., Yamanaka, K., Kiyomi, M., Uehara, R.: Random generation and enumeration of proper interval graphs. IEICE Trans. Inf. Syst. **93**(7), 1816–1823 (2010)
16. Sekine, K., Imai, H., Tani, S.: Computing the Tutte polynomial of a graph of moderate size. In: Staples, J., Eades, P., Katoh, N., Moffat, A. (eds.) ISAAC 1995. LNCS, vol. 1004, pp. 224–233. Springer, Heidelberg (1995). https://doi.org/10.1007/BFb0015427
17. Spinrad, J.P.: Efficient Graph Representations. American Mathematical Society, Providence, RI, Fields Institute monographs (2003)

18. Yamazaki, K., Qian, M., Uehara, R.: Efficient enumeration of non-isomorphic distance-hereditary graphs and ptolemaic graphs. In: Uehara, R., Hong, S.-H., Nandy, S.C. (eds.) WALCOM 2021. LNCS, vol. 12635, pp. 284–295. Springer, Cham (2021). https://doi.org/10.1007/978-3-030-68211-8_23
19. Yamazaki, K., Saitoh, T., Kiyomi, M., Uehara, R.: Enumeration of nonisomorphic interval graphs and nonisomorphic permutation graphs. Theor. Comput. Sci. **806**, 310–322 (2020)

Optimization

Better Hardness Results for the Minimum Spanning Tree Congestion Problem

Huong Luu$^{(\boxtimes)}$ and Marek Chrobak

Department of Computer Science, University of California at Riverside,
Riverside, USA
`hluu008@ucr.edu`

Abstract. In the spanning tree congestion problem, given a connected graph G, the objective is to compute a spanning tree T in G for which the maximum edge congestion is minimized, where the congestion of an edge e of T is the number of vertex pairs adjacent in G for which the path connecting them in T traverses e. The problem is known to be NP-hard, but its approximability is still poorly understood, and it is not even known whether the optimum can be efficiently approximated with ratio $o(n)$. In the decision version of this problem, denoted $K-\mathsf{STC}$, we need to determine if G has a spanning tree with congestion at most K. It is known that $K-\mathsf{STC}$ is NP-complete for $K \geq 8$, and this implies a lower bound of 1.125 on the approximation ratio of minimizing congestion. On the other hand, $3-\mathsf{STC}$ can be solved in polynomial time, with the complexity status of this problem for $K \in \{4, 5, 6, 7\}$ remaining an open problem. We substantially improve the earlier hardness result by proving that $K-\mathsf{STC}$ is NP-complete for $K \geq 5$. This leaves only the case $K = 4$ open, and improves the lower bound on the approximation ratio to 1.2.

1 Introduction

Problems involving constructing a spanning tree that satisfies certain requirements are among the most fundamental tasks in graph theory and algorithmics. One such problem is the *spanning tree congestion problem*, STC for short, that has been studied extensively for many years. Roughly, in this problem we seek a spanning tree T of a given graph G that approximates the connectivity structure of G in the following sense: Embed G into T by replacing each edge (u, v) of G by the unique u-to-v path in T. Define the *congestion of an edge e of T* as the number of such paths that traverse e. The objective of STC is to find a spanning tree T that minimizes the maximum edge congestion.

The general concept of edge congestion was first introduced in 1986, under the name of *load factor*, as a measure of quality of an embedding of one graph into another [3] (see also the survey in [20]). The problem of computing trees with low congestion was studied by Khuller *et al.* [12] in the context of solving commodities network routing problems. The trees considered there were not required to be

M. Chrobak—Research partially supported by National Science Foundation grant CCF-2153723.

C.-C. Lin et al. (Eds.): WALCOM 2023, LNCS 13973, pp. 167–178, 2023.
https://doi.org/10.1007/978-3-031-27051-2_15

spanning subtrees, but the variant involving spanning trees was also mentioned. In 2003, Ostrovskii provided independently a formal definition of STC and established some fundamental properties of spanning trees with low congestion [17]. Since then, many combinatorial and algorithmic results about this problem have been reported in the literature — we refer the readers to the survey paper by Otachi [18] for complete information, most of which is still up-to-date.

As established by Löwenstein [15], STC is NP-hard. As usual, this is proved by showing NP-completeness of its decision version, where we are given a graph G and an integer K, and we need to determine if G has a spanning tree with congestion at most K. Otachi et al. [19] strengthened this by proving that the problem remains NP-hard even for planar graphs. In [16], STC is proven to be NP-hard for chain graphs and split graphs. On the other hand, computing optimal solutions for STC can be achieved in polynomial time for some special classes of graphs: complete k-partite graphs, two-dimensional tori [14], outerplanar graphs [5], and two-dimensional Hamming graphs [13].

In our paper, we focus on the decision version of STC where the bound K on congestion is a fixed constant. We denote this variant by $K-$STC. Several results on the complexity of $K-$STC were reported in [19]. For example, the authors show that $K-$STC is decidable in linear time for planar graphs, graphs of bounded treewidth, graphs of bounded degree, and for all graphs when $K = 1, 2, 3$. On the other hand, they show that the problem is NP-complete for any fixed $K \geq 10$. In [4], Bodlaender et al. proved that $K-$STC is linear-time solvable for graphs in apex-minor-free families and chordal graphs. They also show an improved hardness result of $K-$STC, namely that it is NP-complete for $K \geq 8$, even in the special case of apex graphs that only have one unbounded degree vertex. As stated in [18], the complexity status of $K-$STC for $K \in \{4, 5, 6, 7\}$ remains an open problem.

Little is known about the approximability of STC. The trivial upper bound for the approximation ratio is $n/2$ [18]. As a direct consequence of the NP-completeness of $8-$STC, there is no polynomial-time algorithm to approximate the optimum spanning tree congestion with a ratio better than 1.125 (unless $\mathbb{P} = \mathbb{NP}$).

Our Contribution. Addressing an open question in [18], we provide an improved hardness result for $K-$STC:

Theorem 1. *For any fixed integer $K \geq 5$, $K-$STC is NP-complete.*

The proof of this theorem is given in Sect. 3. Combined with the results in [19], Theorem 1 leaves only the status of $4-$STC open. Furthermore, it also immediately improves the lower bound on the approximation ratio for STC:

Corollary 1. *For $c < 1.2$ there is no polynomial-time c-approximation algorithm for STC, unless $\mathbb{P} = \mathbb{NP}$.*

We remark that this hardness result remains valid even if an additive constant is allowed in the approximation bound. This follows by an argument in [4]. (In essence, the reason is that assigning a positive integer weight β to each edge increases its congestion by a factor β.)

Other Related Work. The spanning tree congestion problem is closely related to the tree spanner problem in which the objective is to find a spanning tree T of G that minimizes the stretch factor, defined as the maximum ratio, over all vertex pairs, between the length of the path in T and the length of the shortest path in G connecting these vertices. In fact, for any planar graph, its spanning tree congestion is equal to its dual's minimum stretch factor plus one [10,19]. This direction of research has been extensively explored, see [6,8,9]. As an aside, we remark that the complexity of the tree 3-spanner problem has been open since its first introduction in 1995 [6].

STC is also intimately related to problems involving cycle bases in graphs. As each spanning tree identifies a fundamental cycle basis of a given graph, a spanning tree with low congestion yields a cycle basis for which the edge-cycle incidence matrix is sparse. Sparsity of such matrices is desirable in linear-algebraic approaches to solving some graph optimization problems, for example analyses of distribution networks such as in pipe flow systems [1].

STC can be considered as an extreme case of the graph sparsification problem, where, given a graph G, the objective is to compute a sparse graph H that captures connectivity properties of G. Such H can be used instead of G for the purpose of various analyses, to improve efficiency. See [2,11,21] (and the references therein) for some approaches to graph sparsification.

2 Preliminaries

Let $G = (V, E)$ be a simple graph with vertex set V and edge set E. Consider a spanning tree $T \subseteq E$ of G. If $e = (u, v) \in T$, removing e from T splits T into two components. We denote by $T_{u,v}$ the component that contains u and by $T_{v,u}$ the component that contains v. Let the *cross-edge set* of e, denoted $\partial_{G,T}(e)$, be the set of edges in E that have one endpoint in $T_{u,v}$ and the other in $T_{v,u}$. In other words, $\partial_{G,T}(e)$ consists of the edges $(u', v') \in E$ for which the unique (simple) path in T from u' to v' goes through e. Note that $e \in \partial_{G,T}(e)$. The *congestion of* e, denoted by $\mathrm{cng}_{G,T}(e)$, is the cardinality of $\partial_{G,T}(e)$. The *congestion of tree* T is $\mathrm{cng}_G(T) = \max_{e \in T} \mathrm{cng}_{G,T}(e)$. Finally, the *spanning tree congestion of graph* G, denoted by $\mathrm{stc}(G)$, is defined as the minimum value of $\mathrm{cng}_G(T)$ over all spanning trees T of G.

The concept of the spanning tree congestion extends naturally to multi-graphs. For multigraphs, only one edge between any two given vertices can be in a spanning tree, but all of them belong to the cross-edge set $\partial_{G,T}(e)$ of any edge $e \in T$ whose removal separates these vertices in T (and thus all contribute to $\mathrm{cng}_{G,T}(e)$). As observed in [19], edge subdivision does not affect the spanning tree congestion of a graph. Therefore any multigraph can be converted into a simple graph by subdividing all multiple edges, without changing its minimum congestion. We use positive integer weights to represent edge multiplicities: an edge (u, v) with weight ω represents a bundle of ω edges connecting u to v. While we state our results in terms of simple graphs, we use weighted graphs in our proofs, with the understanding that they actually represent the corresponding simple graphs. As all weights used in the paper are constant, the computational complexity of $K-\mathsf{STC}$ is not affected.

In fact, it is convenient to generalize this further by introducing edges with *double weights*. A double weight of an edge e is denoted $\omega:\omega'$, where ω and ω' are positive integers such that $\omega \leq \omega'$, and its interpretation in the context of $K-\mathsf{STC}$ is as follows: given a spanning tree T, if $e \in E\backslash T$ then e contributes ω to the congestion $\mathrm{cng}_{G,T}(f)$ of any edge f for which $e \in \partial_{G,T}(f)$, and if $e \in T$ then e contributes ω' to its own congestion, $\mathrm{cng}_{G,T}(e)$. The lemma below implies that including edges with double weights that add up to at most K does not affect the computational complexity of $K-\mathsf{STC}$, and therefore we can formulate our proofs in terms of graphs where some edges have double weights.

Lemma 1. *Let (u,v) be an edge in G with double weight $\omega:\omega'$, where $1 \leq \omega \leq \omega'$ and $\omega + \omega' \leq K$ for some integer K. Consider another graph G' with vertex set $V' = V \cup \{w\}$ and edge set $E' = E \cup \{(u,w),(w,v)\}\backslash\{(u,v)\}$, in which the weight of (u,w) is ω and the weight of (w,v) is ω'. Then, $\mathrm{stc}(G) \leq K$ if and only if $\mathrm{stc}(G') \leq K$.*

Proof. ($\Rightarrow$) Suppose that G has a spanning tree T with $\mathrm{cng}_G(T) \leq K$. We will show that there exists a spanning tree T' of G' with $\mathrm{cng}_{G'}(T') \leq K$. We break the proof into two cases, in both cases showing that $\mathrm{cng}_{G',T'}(e) \leq K$ for each edge $e \in T'$.

<u>Case 1</u>: $(u,v) \in T$. Let $T' = T \cup \{(u,w),(w,v)\}\backslash\{(u,v)\}$. T' is clearly a spanning tree of G'. If $(x,y) \in E'\backslash\{(u,w),(w,v)\}$, the x-to-y paths in T and T' are the same, except that if the x-to-y path in T traverses edge (u,v) then the x-to-y path in T' will traverse (u,w) and (w,v) instead. Therefore, if $e \in T'\backslash\{(u,w),(w,v)\}$, $\partial_{G',T'}(e) = \partial_{G,T}(e)$, so $\mathrm{cng}_{G',T'}(e) = \mathrm{cng}_{G,T}(e) \leq K$. On the other hand, if $e \in \{(u,w),(w,v)\}$, $\partial_{G',T'}(e) = \partial_{G,T}(u,v)\backslash\{(u,v)\}\cup\{e\}$. Then, edge e contributes ω or ω' to $\mathrm{cng}_{G',T'}(e)$, while (u,v), by the definition of double weights, contributes $\omega' \geq \omega$ to $\mathrm{cng}_{G,T}(u,v)$. Hence, $\mathrm{cng}_{G',T'}(e) \leq \mathrm{cng}_{G,T}(u,v) \leq K$.

<u>Case 2</u>: $(u,v) \notin T$. Let $T' = T \cup \{(w,v)\}$, which is a spanning tree of G'. If $e \in T'\backslash\{(w,v)\}$, we have two subcases. If e is not on the u-to-v path in T', $\partial_{G',T'}(e) = \partial_{G,T}(e)$, so $\mathrm{cng}_{G',T'}(e) = \mathrm{cng}_{G,T}(e) \leq K$. If e is on the u-to-v path in T', $\partial_{G',T'}(e) = \partial_{G,T}(e)\cup\{(u,w)\}\backslash\{(u,v)\}$. As (u,w) contributes ω to $\mathrm{cng}_{G',T'}(e)$ and, by the definition of double weights, (u,v) contributes ω to $\mathrm{cng}_{G,T}(e)$, we obtain that $\mathrm{cng}_{G',T'}(e) = \mathrm{cng}_{G,T}(e) \leq K$. In the remaining case, for $e = (w,v)$, we have $\partial_{G',T'}(e) = \{(u,w),(w,v)\}$, so $\mathrm{cng}_{G',T'}(e) = \omega + \omega' \leq K$.

($\Leftarrow$) Let T' be the spanning tree of G' with congestion $\mathrm{cng}_{G'}(T') \leq K$. We will show that there exists a spanning tree T of G with $\mathrm{cng}_G(T) \leq K$. Note that at least one of edges (u,w) and (v,w) has to be in T'. We now consider three cases, in each case showing that $\mathrm{cng}_{G,T}(e) \leq K$ for each edge $e \in T$.

<u>Case 1</u>: $(u,w),(v,w) \in T'$. Let $T = T'\cup\{(u,v)\}\backslash\{(u,w),(w,v)\}$. T is clearly a spanning tree of G. The argument for this case is similar to Case 1 in the proof

for the ($\Rightarrow$) implication. For each edge $e \in T \setminus \{(u,v)\}$, its congestion in T is the same as in T'. The congestion of (u,v) in T is bounded by the congestion of (w,v) in T', which is at most K.

<u>Case 2</u>: $(v,w) \in T'$ and $(u,w) \notin T'$. Let $T = T' \setminus \{(w,v)\}$. T is a spanning tree of G. Here again, the argument is similar to the proof for Case 2 in the ($\Rightarrow$) implication. For each edge $e \in T$, if e is not on the u-to-v path in T, its congestion in T and T' is the same. If e is on the u-to-v path in T, the contributions of (u,v) and (u,w) to the congestion of e in T and T' are the same.

<u>Case 3</u>: $(u,w) \in T'$ and $(v,w) \notin T'$. Consider $T'' = T' \cup \{(v,w)\} \setminus \{(u,w)\}$, which is a different spanning tree of G'. It is sufficient to show that $\mathrm{cng}_{G'}(T'') \leq \mathrm{cng}_{G'}(T')$ because it will imply $\mathrm{cng}_{G'}(T'') \leq K$, and then we can apply Case 2 to T''. We examine the congestion values of each edge $e \in T''$. Suppose first that $e \neq (u,w)$. If e is not on the u-to-v path in T', $\partial_{G',T''}(e) = \partial_{G',T'}(e)$, so $\mathrm{cng}_{G',T''}(e) = \mathrm{cng}_{G',T'}(e)$. If e is on the u-to-v path in T', $\partial_{G',T''}(e) = \partial_{G',T'}(e) \cup \{(u,w)\} \setminus \{(v,w)\}$, so $\mathrm{cng}_{G',T''}(e) = \mathrm{cng}_{G',T'}(e) + \omega - \omega' \leq \mathrm{cng}_{G',T'}(e)$. In the last case when $e = (v,w)$, $\mathrm{cng}_{G',T''}(e) = \omega + \omega' \leq K$.

3 NP-Completeness Proof of $K-$STC for $K \geq 5$

In this section we prove our main result, the NP-completeness of $K-$STC. Our proof uses an NP-complete variant of the satisfiability problem called (2P1N)-SAT [7,22]. An instance of (2P1N)-SAT is a boolean expression ϕ in conjunctive normal form, where each variable occurs exactly three times, twice positively and once negatively, and each clause contains exactly two or three literals of different variables. The objective is to decide if ϕ is satisfiable, that is if there is a satisfying assignment that makes ϕ true.

For each constant K, $K-$STC is clearly in NP. We will present a polynomial-time reduction from (2P1N)-SAT. In this reduction, given an instance ϕ of (2P1N)-SAT, we construct in polynomial time a graph G with the following property:

($*$) ϕ has a satisfying truth assignment if and only if $\mathrm{stc}(G) \leq K$.

Throughout the proof, the three literals of x_i in ϕ will be denoted by x_i, x_i', and $\bar{x}_i$, where x_i, x_i' are the two positive occurrences of x_i and $\bar{x}_i$ is the negative occurrence of x_i. We will also use notation $\tilde{x}_i$ to refer to an unspecified literal of x_i, that is $\tilde{x}_i \in \{x_i, x_i', \bar{x}_i\}$.

We now describe the reduction. Set $k_i = K - i$ for $i = 1, 2, 3, 4$. (In particular, for $K = 5$, we have $k_1 = 4$, $k_2 = 3$, $k_3 = 2$, $k_4 = 1$.) G will consist of gadgets corresponding to variables, with the gadget corresponding to x_i having three vertices x_i, x_i', and $\bar{x}_i$, that represent its three occurrences in the clauses. G will also have vertices representing clauses and edges connecting literals with the clauses where they occur (see Fig. 1b for an example). As explained in Sect. 2, without any loss of generality we can allow edges in G to have constant-valued weights, single or double. Specifically, starting with G empty, the construction of G proceeds as follows:

- Add a *root vertex* r.
- For each variable x_i, construct the x_i-gadget (see Fig. 1a). This gadget has three vertices corresponding to the literals: a *negative literal vertex* $\bar{x}_i$ and two *positive literal vertices* x_i, x_i', and two auxiliary vertices y_i and z_i. Its edges and their weights are given in the table below:

edge	$(\bar{x}_i, z_i)$	(z_i, x_i)	(x_i, x_i')	(r, x_i')	(r, y_i)	(y_i, z_i)	$(y_i, \bar{x}_i)$
weight	$1:k_3$	$1:k_3$	$1:k_2$	k_3	k_4	k_4	$1:k_2$

- For each clause c, create a *clause vertex* c. For each literal $\tilde{x}_i$ in c, add the corresponding *clause-to-literal edge* $(c, \tilde{x}_i)$ of weight $1:k_2$. Importantly, as all literals in c correspond to different variables, these edges will go to different variable gadgets.
- For each two-literal clause c, add a *root-to-clause* edge (r, c) of weight $1:k_1$.

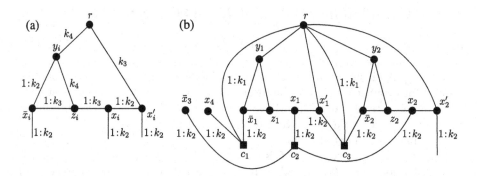

Fig. 1. (a)The x_i-gadget. (b) An example of a partial graph G for the boolean expression $\phi = (\bar{x}_1 \lor x_4) \land (x_1 \lor x_2 \lor \bar{x}_3) \land (x_1 \lor \bar{x}_2) \land \cdots$. Here, $c_1 = \bar{x}_1 \lor x_4$, $c_2 = x_1 \lor x_2 \lor \bar{x}_3$, and $c_3 = x_1 \lor \bar{x}_2$.

We now show that G has the required property ($*$), proving the two implications separately.

($\Rightarrow$) Suppose that ϕ has a satisfying assignment. Using this assignment, we construct a spanning tree T of G as follows:

- For every x_i-gadget, include in T edges (r, x_i'), (r, y_i), and (y_i, z_i). If $x_i = 0$, include in T edges $(\bar{x}_i, z_i)$ and (x_i, x_i'), otherwise include in T edges $(y_i, \bar{x}_i)$ and (z_i, x_i).
- For each clause c, include in T one clause-to-literal edge that is incident to any literal vertex that satisfies c in our chosen truth assignment for ϕ.

By routine inspection, T is indeed a spanning tree: Each x_i-gadget is traversed from r without cycles, and all clause vertices are leaves of T. Figures 2 and 3 show how T traverses an x_i-gadget in different cases, depending on whether

$x_i = 0$ or $x_i = 1$ in the truth assignment for ϕ, and on which literals are chosen to satisfy each clause. Note that the edges with double weights satisfy the assumption of Lemma 1 in Sect. 2, that is each such weight $1 : \omega'$ satisfies $1 \leq \omega'$ and $1 + \omega' \leq K$.

We need to verify that each edge in T has congestion at most K. All the clause vertices are leaves in T, thus the congestion of each clause-to-literal edge is $k_2 + 2 = K$; this holds for both three-literal and two-literal clauses. To analyze the congestion of the edges inside an x_i-gadget, we consider two cases, depending on the value of x_i in our truth assignment.

When $x_i = 0$, we have two sub-cases as shown in Fig. 2. The congestions of the edges in the x_i-gadget are as follows:

- In both cases, $\mathrm{cng}_{G,T}(r, x_i') = k_3 + 3$.
- In case (a), $\mathrm{cng}_{G,T}(r, y_i) = k_4 + 3$. In case (b), it is $k_4 + 2$.
- In case (a), $\mathrm{cng}_{G,T}(y_i, z_i) = k_4 + 4$. In case (b), it is $k_4 + 3$.
- In case (a), $\mathrm{cng}_{G,T}(\bar{x}_i, z_i) = k_3 + 3$. In case (b), it is $k_3 + 2$.
- In both cases, $\mathrm{cng}_{G,T}(x_i, x_i') = k_2 + 2$.

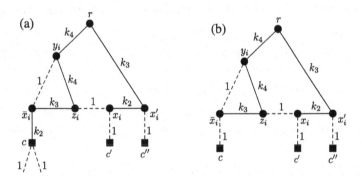

Fig. 2. The traversal of the x_i-gadget by T when $x_i = 0$. Solid lines are tree edges, dashed lines are non-tree edges. (a) $\bar{x}_i$ is chosen by clause c. (b) $\bar{x}_i$ is not chosen by clause c.

On the other hand, when $x_i = 1$, we have four sub-cases. Figure 2 illustrates cases (a)–(c). In case (d) (not shown in Fig. 2), none of the positive literal vertices x_i, x_i' is chosen to satisfy their corresponding clauses. The congestions of the edges in the x_i-gadget are as follows:

- In cases (a) and (b), $\mathrm{cng}_{G,T}(r, x_i') = k_3 + 3$. In cases (c) and (d), it is $k_3 + 2$.
- In cases (a) and (c), $\mathrm{cng}_{G,T}(r, y_i) = k_4 + 4$. In cases (b) and (d), it is $k_4 + 3$.
- In cases (a) and (c), $\mathrm{cng}_{G,T}(y_i, z_i) = k_4 + 4$. In cases (b) and (d), it is $k_4 + 3$.
- In cases (a) and (c), $\mathrm{cng}_{G,T}(z_i, x_i) = k_3 + 3$. In cases (b) and (d), it is $k_3 + 2$.
- In all cases, $\mathrm{cng}_{G,T}(y_i, \bar{x}_i) = k_2 + 2$.

In summary, the congestion of each edge of T is at most K. Thus $\mathrm{cng}_G(T) \leq K$; in turn, $\mathrm{stc}(G) \leq K$, as claimed.

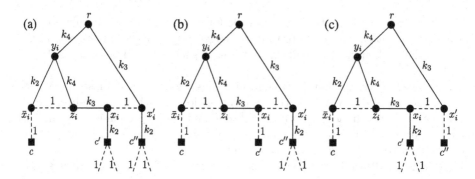

Fig. 3. The traversal of the x_i-gadget by T when $x_i = 1$. By c, c' and c'' we denote the clauses that contain literals $\bar{x}_i$, x_i and x'_i, respectively. (a) x_i and x'_i are chosen by clauses c' and c''. (b) x'_i is chosen by clause c''. (c) x_i is chosen by clause c'.

($\Leftarrow$) We now prove the other implication in (∗). We assume that G has a spanning tree T with $\text{cng}_G(T) \leq K$. We will show how to convert T into a satisfying assignment for ϕ. The proof consists of a sequence of claims showing that T must have a special form that will allow us to define this truth assignment.

Claim 1. *Each x_i-gadget satisfies the following property: for each literal vertex $\tilde{x}_i$, if some edge e of T (not necessarily in the x_i-gadget) is on the r-to-$\tilde{x}_i$ path in T, then $\partial_{G,T}(e)$ contains at least two distinct edges from this gadget other than (y_i, z_i).*

This claim is straightforward: it follows directly from the fact that there are two edge-disjoint paths from r to any literal vertex $\tilde{x}_i \in \{\bar{x}_i, x_i, x'_i\}$ that do not use edge (y_i, z_i).

Claim 2. *For each two-literal clause c, edge (r, c) is not in T.*

For each literal $\tilde{x}_i$ of clause c, there is an r-to-c path via the x_i-gadget, so, together with edge (r, c), G has three disjoint r-to-c paths. Thus, if (r, c) were in T, its congestion would be at least $k_1 + 2 > K$, proving Claim 2.

Claim 3. *All clause vertices are leaves in T.*

To prove Claim 3, suppose there is a clause c that is not a leaf. Then, by Claim 2, c has at least two clause-to-literal edges in T, say $(c, \tilde{x}_i)$ and $(c, \tilde{x}_j)$. We can assume that the last edge on the r-to-c path in T is $e = (c, \tilde{x}_i)$. Clearly, $r \in T_{\tilde{x}_i, c}$ and $\tilde{x}_j \in T_{c, \tilde{x}_i}$. By Claim 1, at least two edges of the x_j-gadget are in $\partial_{G,T}(e)$, and they contribute at least 2 to $\text{cng}_{G,T}(e)$. We now have some cases to consider.

If c is a two-literal clause, its root-to-clause edge (r, c) is also in $\partial_{G,T}(e)$, by Claim 2. Thus, $\text{cng}_{G,T}(e) \geq k_2 + 3 > K$ (see Fig. 4a). So assume now that c is a three-literal clause, and let $\tilde{x}_l \neq \tilde{x}_i, \tilde{x}_j$ be the third literal of c. If T contains $(c, \tilde{x}_l)$, the x_l-gadget would also contribute at least 2 to $\text{cng}_{G,T}(e)$, so

$\operatorname{cng}_{G,T}(e) \geq k_2 + 4 > K$ (see Fig. 4b). Otherwise, $(c, \tilde{x}_l) \notin T$, and $(c, \tilde{x}_l)$ itself contributes 1 to $\operatorname{cng}_{G,T}(e)$, so $\operatorname{cng}_{G,T}(e) \geq k_2 + 3 > K$ (see Fig. 4c).

We have shown that if a clause vertex c is not a leaf in T, then in all cases the congestion of T would exceed K, completing the proof of Claim 3.

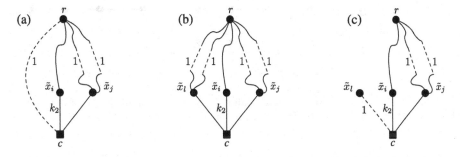

Fig. 4. Illustration of the proof of Claim 3. In (a) c is a two-literal clause; in (b) and (c), c is a three-literal clause.

Claim 4. *For each x_i-gadget, edge (r, x_i') is in T.*

Towards contradiction, suppose that (r, x_i') is not in T. Let (x_i', c) be the clause-to-literal edge of x_i'. If only one of the two edges $(x_i', x_i), (x_i', c)$ is in T, making x_i' a leaf, then the congestion of that edge is $k_3 + k_2 + 1 > K$. Otherwise, both $(x_i', x_i), (x_i', c)$ are in T. Because c is a leaf in T by Claim 3, $e = (x_i, x_i')$ is the last edge on the r-to-x_i' path in T. As shown in Fig. 5a, $\operatorname{cng}_{G,T}(e) \geq k_3 + k_2 + 2 > K$. This proves Claim 4.

Claim 5. *For each x_i-gadget, edge (r, y_i) is in T.*

To prove this claim, suppose (r, y_i) is not in T. We consider the congestion of the first edge e on the r-to-y_i path in T. By Claims 3 and 4, we have $e = (r, x_i')$, all vertices of the x_i-gadget have to be in $T_{x_i',r}$, and $T_{x_i',r}$ does not contain literal vertices of another variable $x_j \neq x_i$. For each literal $\tilde{x}_i$ of x_i, if a clause-to-literal edge $(c, \tilde{x}_i)$ is in T, then the two other edges of c contribute 2 to $\operatorname{cng}_{G,T}(e)$, otherwise $(c, \tilde{x}_i)$ contributes 1 to $\operatorname{cng}_{G,T}(e)$. Then, $\operatorname{cng}_{G,T}(e) \geq k_4 + k_3 + 3 > K$ (see Fig. 5b), proving Claim 5.

Claim 6. *For each x_i-gadget, exactly one of edges (z_i, x_i) and (x_i, x_i') is in T.*

By Claims 4 and 5, edges (r, y_i) and (r, x_i') are in T. Since the clause neighbor c' of x_i is a leaf of T, by Claim 3, if none of (z_i, x_i), (x_i, x_i') were in T, x_i would not be reachable from r in T. Thus, at least one of them is in T. Now, assume both (z_i, x_i) and (x_i, x_i') are in T (see Fig. 6a). Then, edge (y_i, z_i) is not in T, as otherwise we would create a cycle. Let us consider the congestion of edge $e = (r, x_i')$. Clearly, x_i and x_i' are in $T_{x_i',r}$. The edges of the two clause neighbors

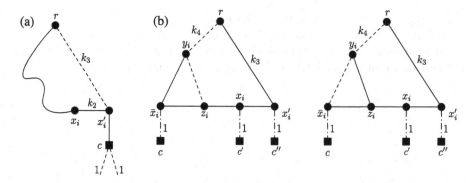

Fig. 5. (a) Illustration of the proof of Claim 4. (a) Illustration of the proof of Claim 5. Dot-dashed lines are edges that may or may not be in T.

c' and c'' of x_i and x_i' contribute at least 2 to $\mathrm{cng}_{G,T}(e)$, by Claim 3. In addition, by Claim 1, besides e and (y_i, z_i), $\partial_{G,T}(e)$ contains another edge of the x_i-gadget which contributes at least another 1 to $\mathrm{cng}_{G,T}(e)$. Thus, $\mathrm{cng}_{G,T}(e) \geq k_4 + k_3 + 3 > K$ — a contradiction. This proves Claim 6.

Claim 7. *For each x_i-gadget, edge (y_i, z_i) is in T.*

By Claims 4 and 5, the two edges (r, x_i') and (r, y_i) are in T. Now assume, towards contradiction, that (y_i, z_i) is not in T (see Fig. 6b). By Claim 6, only one of (z_i, x_i) and (x_i, x_i') is in T. Furthermore, the clause neighbor c' of x_i is a leaf of T, by Claim 3. As a result, (z_i, x_i) cannot be on the y_i-to-z_i path in T. To reach z_i from y_i, the two edges $(y_i, \bar{x}_i), (\bar{x}_i, z_i)$ have to be in T. Let us consider the congestion of $e = (y_i, \bar{x}_i)$. The edges of the clause neighbor c of $\bar{x}_i$ contribute at least 1 to the congestion of e, by Claim 3. Also, by Claim 1, besides e and (y_i, z_i), $\partial_{G,T}(e)$ contains another edge of the x_i-gadget which contributes at least 1 to $\mathrm{cng}_{G,T}(e)$. In total, $\mathrm{cng}_{G,T}(e) \geq k_4 + k_2 + 2 > K$, reaching a contradiction and completing the proof of Claim 7.

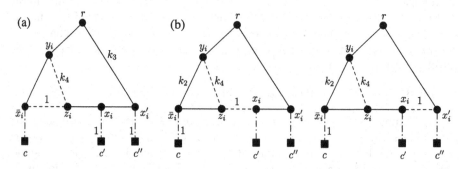

Fig. 6. (a) Illustration of the proof of Claim 6. (b) Illustration of the proof of Claim 7. Dot-dashed lines are edges that may or may not be in T.

Claim 8. *For each x_i-gadget, if its clause-to-literal edge $(\bar{x}_i, c)$ is in T, then its other two clause-to-literal edges (x_i, c') and (x'_i, c'') are not in T.*

Assume the clause-to-literal edge $(\bar{x}_i, c)$ of the x_i-gadget is in T. By Claim 7, edge (y_i, z_i) is in T. If $(y_i, \bar{x}_i)$ is also in T, edge $(\bar{x}_i, z_i)$ cannot be in T, and it contributes 1 to $\mathrm{cng}_{G,T}(y_i, \bar{x}_i)$. As shown in Fig. 7a, $\mathrm{cng}_{G,T}(y_i, \bar{x}_i) = k_2 + 3 > K$. Thus, $(y_i, \bar{x}_i)$ cannot be in T. Since c is a leaf of T, edge $(\bar{x}_i, z_i)$ has to be in T, for otherwise $\bar{x}_i$ would not be reachable from r. By Claim 6, one of edges (z_i, x_i) and (x_i, x'_i) is in T. If (z_i, x_i) is in T (see Fig. 7b), $\mathrm{cng}_{G,T}(y_i, z_i) \geq k_4 + 5 > K$. Hence, (z_i, x_i) is not in T, which implies that (x_i, x'_i) is in T.

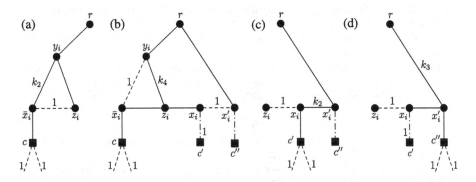

Fig. 7. Illustration of the proof of Claim 8. Dot-dashed lines are edges that may or may not be in T.

Now, we proceed by contradiction assuming that at least one other clause-to-literal edge of the x_i-gadget is in T. If edge (x_i, c') is in T, $\mathrm{cng}_{G,T}(x_i, x'_i) \geq k_2 + 3 > K$, as shown in Fig. 7c. Similarly, if (x'_i, c'') is in T, $\mathrm{cng}_{G,T}(r, x'_i) \geq k_3 + 4 > K$ (see Fig. 7d). So we reach a contradiction in both cases, thus proving Claim 8.

We are now ready to complete the proof of the ($\Leftarrow$) implication in the equivalence (∗). We use our spanning tree T of congestion at most K to create a truth assignment for ϕ by setting $x_i = 0$ if the clause-to-literal edge of $\bar{x}_i$ is in T, otherwise $x_i = 1$. By Claim 8, this truth assignment is well-defined. Each clause has one clause-to-literal edge in T which ensures that all clauses are indeed satisfied.

References

1. Alvarruiz Bermejo, F., Martínez Alzamora, F., Vidal Maciá, A.M.: Improving the efficiency of the loop method for the simulation of water distribution networks. J. Water Resour. Plan. Manag. **141**(10), 1–10 (2015)
2. Benczúr, A.A., Karger, D.R.: Approximating $s - t$ minimum cuts in $\tilde{O}(n^2)$ time. In: Proceedings of the 28th Annual ACM Symposium on Theory of Computing, pp. 47–55 (1996)

3. Bhatt, S., Chung, F., Leighton, T., Rosenberg, A.: Optimal simulations of tree machines. In: Proceedings of the 27th Annual Symposium on Foundations of Computer Science, pp. 274–282 (1986)
4. Bodlaender, H., Fomin, F., Golovach, P., Otachi, Y., Leeuwen, E.: Parameterized complexity of the spanning tree congestion problem. Algorithmica **64**, 1–27 (2012)
5. Bodlaender, H.L., Kozawa, K., Matsushima, T., Otachi, Y.: Spanning tree congestion of k-outerplanar graphs. Discret. Math. **311**(12), 1040–1045 (2011)
6. Cai, L., Corneil, D.G.: Tree spanners. SIAM J. Discret. Math. **8**(3), 359–387 (1995)
7. Dahlhaus, E., Johnson, D.S., Papadimitriou, C.H., Seymour, P.D., Yannakakis, M.: The complexity of multiterminal cuts. SIAM J. Comput. **23**(4), 864–894 (1994)
.8. Dragan, F.F., Fomin, F.V., Golovach, P.A.: Spanners in sparse graphs. J. Comput. Syst. Sci. **77**(6), 1108–1119 (2011)
9. Emek, Y., Peleg, D.: Approximating minimum max-stretch spanning trees on unweighted graphs. SIAM J. Comput. **38**(5), 1761–1781 (2009)
10. Fekete, S.P., Kremer, J.: Tree spanners in planar graphs. Discret. Appl. Math. **108**(1), 85–103 (2001)
11. Fung, W.S., Hariharan, R., Harvey, N.J., Panigrahi, D.: A general framework for graph sparsification. In: Proceedings of the 43rd Annual ACM Symposium on Theory of Computing, pp. 71–80 (2011)
12. Khuller, S., Raghavachari, B., Young, N.: Designing multi-commodity flow trees. Inf. Process. Lett. **50**(1), 49–55 (1994)
13. Kozawa, K., Otachi, Y.: Spanning tree congestion of rook's graphs. Discuss. Math. Graph Theory **31**(4), 753–761 (2011)
14. Kozawa, K., Otachi, Y., Yamazaki, K.: On spanning tree congestion of graphs. Discret. Math. **309**(13), 4215–4224 (2009)
15. Löwenstein, C.: In the Complement of a Dominating Set. Ph.D. thesis, Technische Universitat Ilmenau (2010)
16. Okamoto, Y., Otachi, Y., Uehara, R., Uno, T.: Hardness results and an exact exponential algorithm for the spanning tree congestion problem. In: Ogihara, M., Tarui, J. (eds.) TAMC 2011. LNCS, vol. 6648, pp. 452–462. Springer, Heidelberg (2011). https://doi.org/10.1007/978-3-642-20877-5_44
17. Ostrovskii, M.: Minimal congestion trees. Discret. Math. **285**(1), 219–226 (2004)
18. Otachi, Y.: A survey on spanning tree congestion. In: Fomin, F.V., Kratsch, S., van Leeuwen, E.J. (eds.) Treewidth, Kernels, and Algorithms. LNCS, vol. 12160, pp. 165–172. Springer, Cham (2020). https://doi.org/10.1007/978-3-030-42071-0_12
19. Otachi, Y., Bodlaender, H.L., van Leeuwen, E.J.: Complexity results for the spanning tree congestion problem. In: Thilikos, D.M. (ed.) WG 2010. LNCS, vol. 6410, pp. 3–14. Springer, Heidelberg (2010). https://doi.org/10.1007/978-3-642-16926-7_3
20. Rosenberg, A.L.: Graph embeddings 1988: recent breakthroughs, new directions. In: Reif, J.H. (ed.) AWOC 1988. LNCS, vol. 319, pp. 160–169. Springer, New York (1988). https://doi.org/10.1007/BFb0040384
21. Spielman, D.A., Teng, S.H.: Spectral sparsification of graphs. SIAM J. Comput. **40**(4), 981–1025 (2011)
22. Yoshinaka, R.: Higher-order matching in the linear lambda calculus in the absence of constants is NP-complete. In: Giesl, J. (ed.) RTA 2005. LNCS, vol. 3467, pp. 235–249. Springer, Heidelberg (2005). https://doi.org/10.1007/978-3-540-32033-3_18

Energy Efficient Sorting, Selection and Searching

Varunkumar Jayapaul[1], Seungbum Jo[2(✉)], Krishna Palem[3], and Srinivasa Rao Satti[4]

[1] Indian Institute of Technology Mandi, Kamand, India
varunkumar@iitmandi.ac.in
[2] Chungnam National University, Daejeon, South Korea
sbjo@cnu.ac.kr
[3] Rice University, Houston, USA
Krishna.V.Palem@rice.edu
[4] Norwegian University of Science and Technology, Trondheim, Norway
srinivasa.r.satti@ntnu.no

Abstract. In this paper, we introduce a model for studying energy efficient algorithms by extending the well-studied comparison model. In our model, the result of a comparison is determined based on two parameters: (i) the energy used to perform a comparison, and (ii) the absolute difference between the two values being compared – thus introducing an energy-accuracy trade-off. This model also extends the ideas presented by Geissmann and Penna [SOFSEM 2018] and Funke et al. [Comput. Geom. 2005] wherein they use two distinct types of comparisons namely low and full-energy (cheap and expensive) comparisons, by introducing multiple types of comparisons. In this extension, the accuracy of a comparison becomes a function of the energy used. We consider the fundamental problems of (i) sorting (ii) selection (iii) searching, and design efficient algorithms for these problems in the new model. We also present lower bounds on the energy usage for some of these problems, showing that some of our algorithms are asymptotically optimal with respect to the energy usage.

Keywords: Energy-efficiency · Sorting · Selection · Searching · Binary search tree

1 Introduction

Various studies have captured the relationship between energy consumption and computational capability. One direction to explore such a relationship is by considering models motivated by Landauer's Principle [4,17,21]. In these models, energy usage is proportional to the ratio of the input and output sizes. If one can reconstruct the input from the output, no energy is considered necessary for the computation to produce the output and in this case, the computation is called *reversible* computation. Demaine et al. [7] give three different models based on the energy cost of irreversible operations and give a time-space-energy trade-off for some fundamental sorting and graph algorithms.

On the other hand, one can consider a trade-off between the energy usage and its accuracy for each step of the computation (see [20] for details). This is motivated by the fact that probabilistic CMOS switches [16] give significant energy gains compared to the deterministic CMOS switches. For example, Arumugam et al. [3] proposed a probabilistic memory model which has a fixed probability of incorrectness for read/write operations on each memory cell and designed energy-accuracy trade-off algorithms for radix sort and pattern matching problems under the model. Geissmann and Penna introduced a comparison model in which the accuracy of a comparison depends on an energy parameter λ. They analyzed the performance of sorting algorithms by considering them as Markov processes [14].

The concept of incorrectness in computations has been well-studied for comparison-based models with several variants (without considering the energy consumption as a parameter). Feige et al. [8] proposed the model where each comparison has a fixed error probability and showed how many comparisons are necessary (and sufficient) for obtaining a correct output with a small probability of failure, and showed that there exists an asymptotically optimal algorithm for sorting, merging, and selection problems under this model, in the worst case. Under the same model, Leucci and Liu [18] considered the selection problem and the approximate minimum selection problem in the average and the worst case, respectively. Alonso et al. [2] showed that the number of inversions after performing Quicksort on an array of size n converges to $\Theta(n^2 p)$ when each comparison has error probability p. Geissmann et al. [11] recently proposed an $O(n \log n)$-time sorting algorithm with $O(\log n)$ maximum dislocation and $O(n)$ total dislocation with high probability for an array of size n when each comparison has a fixed error probability, and the answer does not change by repeating the same comparison. In this model, the dislocation is defined as a gap between the actual rank and its final position on the input elements. Without considering the probability of errors while performing a comparison as a parameter, Finocchi and Italiano [9] considered the comparison-based algorithms when some elements of the input can be read incorrectly. Huang et al. [15] considered the sorting problem under the model where one cannot compare each pair of elements with a fixed probability. Ajtai et al. [1] proposed a model of imprecise comparisons where two elements are compared correctly only if their value difference is at least δ (otherwise, the output is unpredictable) and designed a sorting algorithm of permutation where any two inversions have the value difference at most $k\delta$ for any constant $k > 1$ after the sorting. Geissmann and Penna [13] proposed an inexact comparison-based model motivated from the energy-accuracy trade-off mentioned in [20]. Their model has two types of comparisons *low-energy comparisons* and *high-energy comparisons*, wherein each comparison has a non-zero error probability (based on the value difference) with low-energy comparison, whereas a high-energy comparison has zero error probability. They considered the number of expected inversions after Insertion sort and Quicksort only with low-energy comparisons (without repeating the same comparison). Thus, by combining with any sorting algorithm parameterized by the number of inversions (e.g., Insertion sort with $(2, 4)$-finger search tree [19]) with high-energy comparisons, one can use less high-energy comparisons compared to performing sorting algorithms only with high-energy comparisons. The model with low and high-energy comparisons is also used in other problems as optimization [12] and clustering [5]. Note that the similar model was proposed by Funke et al. [10] which uses *cheap* and *expensive* comparisons, to improve the practical performance of geometric computations.

In this paper, we propose a new comparison model based on the energy-accuracy trade-off with distance-based errors, which was considered in the model of Ajtai et al. [1] and the model with low and high-energy comparisons [5,12,13]. Our model's primary motivation is to measure the total energy usage with various energy-accuracy trade-off models. More precisely, our model can use multiple amounts of energy where the energy usage for each comparison is decided by the *threshold function*. The value of the threshold function only depends on the maximum difference of two elements to be compared without any probability of incorrectness. If we compare two elements with energy less than the minimum energy to be compared correctly, the answer is unpredictable (i.e., the adversary can choose the answer arbitrarily). Note that our model can be applied to the various scenarios by defining the threshold function appropriately. For example, in the RAM model with word size $\Theta(\lg n)$ bits[1], our model with the threshold function $f(k) = O(\lg k)$ corresponds to the model when one needs to use the same energy to read each bit in the word (see Sect. 2 for a detailed definition of the threshold function f). Also for example, suppose there is a preference list of a set of objects investigated by multiple experts, and the experts have different levels of expertise in ranking the objects (see [6]). For simplicity, assume that more experienced experts can rank the objects whose ranks are closer; and that the cost of querying experts is proportional to their expertise. Then our model can be applied to the problem of obtaining a complete sorted list of preferences from the experts using minimum cost.

Our model extends the model with low and high-energy comparisons [5,12,13] in the sense that our model can use multiple types of energy with some unknown error probability, while giving a ratio of the energy consumption among them. For example, when the ratio between the low and high energy comparison is almost 1, sorting only with high energy comparisons might use less energy. Also, compared to the model of Ajtai et al. [1], our model can use multiple types of imprecise comparisons by increasing (or decreasing) the energy used for a comparison.

We consider the following fundamental problems (i) sorting, (ii) selection and (iii) searching under our model, and describe energy-efficient algorithms for the above three problems for a wide variety of threshold functions. Most of our new algorithms use less energy than the trivial algorithms, which perform the well-known optimal algorithms while using the full-energy (i.e., high-energy

---

[1] Throughout this paper, lg denotes the logarithm to the base 2, and we ignore ceiling and floors which do not affect our results asymptotically.

comparison in the model of low- and high-energy comparisons [13]) for each comparison operation. We summarize our results for each problems in Sect. 2. The basic idea of our algorithms is to use comparisons with low energy to obtain an approximated solution that only requires less number of comparisons with the full-energy. Note that the similar idea was used in [10,12,13] for sorting and optimization problems under low and high-energy comparisons model.

The paper is organized as follows. Section 2 describes our energy-accuracy trade-off comparison model, and gives the summarization of our results. Section 3 and 4, we consider sorting and finding the minimum under our model and give the algorithms which take less energy than the trivial algorithms, with some threshold functions.

2   Model

In this section, we introduce our energy-accuracy trade-off comparison model. In this model the cost of performing a comparison is measured by the *energy* used for comparing any two operands a and b. Note that, as is the case for the standard comparison model, any computation other than comparison (i.e., read/write and movement of data within the memory) is considered free and costs no energy. In this paper, we only consider inputs consisting of integers.

Now we define the energy to perform a single comparison. Let E be the minimum energy needed to compare any two elements from the input correctly (i.e., if we use energy E to compare two elements, then we are assured that the answer is correct). We refer to a comparison that uses energy E as a *full-energy* comparison, and a comparison that uses energy strictly less than E as a *partial-energy* comparison. Although one can simply choose E to be 1 to remove this parameter from the energy terms. We choose not to do so, so that it is clear from the complexity terms that we are referring to energy and not the number of comparisons.

The answer to a partial-energy comparison is based on following two parameters: (i) the (absolute) difference between the operands, and (ii) the threshold function, $f : \mathbb{Z}^+ \to \mathbb{R}^+$ which is a non-decreasing function with $f(1) = 1$. More specifically, when we compare two distinct elements a and b using a partial-energy comparison which consumes at least $E/f(|a - b|)$ energy, the answer is correct; otherwise (i.e., if we use energy less than $E/f(|a-b|)$), the answer could be incorrect (i.e., the adversary can choose the answer arbitrarily). Note that the case of energy $E/f(1)$ (i.e., full-energy comparison) is considered as the special case which allows to check the equality of two elements although $|a-b| < 1$ when $a = b$. In our model, the energy used by a comparison is fixed by choosing an argument for the threshold function. We refer to this argument as the *threshold*.

Since any threshold function is a non-decreasing function, it is clear that more energy is necessary to compare two values correctly with a smaller difference, regardless of which threshold function is used in our model. Also, any comparison-based algorithm with K comparisons can be performed under our model with energy $E \cdot K$ by trivially performing every comparison with full-energy. In the rest of the paper, given two integers a, b, and a positive integer

x, we use the notation $a \prec_x b$ (resp. $a \succ_x b$) to mean that a is smaller (resp. larger) than b according to the answer of the comparison with threshold x, and $a < b$ (resp. $a > b$) means that a is smaller (resp. larger) than b according to their actual values. Thus, $a \prec_x b$ implies that $a + (x - 1) < b$. The equality of two elements ($a = b$) can only be ascertained by using a full energy comparison to compare the two elements. In the rest of the paper, we study threshold functions in general, but we also showcase a few special threshold functions.

Summary of the Results. When a sequence P of n distinct integers are given, we obtain the following results:

- **Sorting**: We first show that at least energy $E \cdot (n - 1)$ is necessary to sort P (Theorem 5). We then give two sorting algorithms (i) an algorithm that runs in two phases, where every comparison within each phase uses the same amount of energy. We refer to this as a two-level sorting algorithm; and (ii) a multi-level sorting algorithm, which consists of several phases. Both the algorithms use energy $o(En \lg n)$ when the threshold function is non-trivial i.e. in $\omega(1)$. Furthermore, when the threshold function is $O(k^c)$ for $c > 1$, we show that the multi-level algorithm uses asymptotically optimal energy in the worst case (Theorem 7).
- **Selection**: We show that there exists a algorithm for (i) finding the minimum of P using energy $o(En)$ when the threshold function is in $\Omega(\lg^c k)$ for any constant $c > 1$ in the worst case (Theorem 9), and (ii) finding an element with rank r in P using energy $o(En)$ when the threshold function is in $\Omega(k^c)$ for any constant $c > 0$ in the worst case (Theorem 1). For the selection problem, we also give a randomized algorithm which uses energy $O(E \lg n)$ in average if the threshold function is in $\Omega(k)$ (Theorem 2). Note that when $f(k)$ is $O(k)$, our randomized algorithm uses asymptotically optimal energy on average.
- **Searching**: We consider two problems as (i) searching in a sorted array and (ii) searching in a balanced binary search tree with $o(n)$ additional space (for both of the cases, we assume that the input is already given). For (i), we give an algorithm that uses energy $o(E \lg n)$ in the worst case when the threshold function is in $\omega(1)$ (Theorem 3). When the threshold function is in $\Omega(\lg^c k)$ for any constant $c > 1$, the algorithm uses energy $\Theta(E)$, which is asymptotically optimal in energy usage. For (ii), We give an algorithm that uses energy $o(E \lg n)$ in the worst case when the threshold function is in $\omega(1)$ (Theorem 4). Again when the threshold function is in $\Omega(k^c)$ for any constant $c > 0$, the algorithm uses energy $\Theta(E)$.

Due to the page limit, the details of the results other than sorting and finding the minimum are omitted. The following theorems describe the results on selection and searching problems.

Theorem 1. *Given a sequence P containing n distinct integers, there exists an algorithm which can find the r-th smallest element in P using energy $o(En)$ if the threshold function $f(k)$ is in $\Omega(k^c)$ for any constant $c > 0$.*

Theorem 2. *Given a set P containing n distinct elements, there exists an algorithm which can find the r-th smallest element in P using expected energy $O(E \lg n)$ if the threshold function $f(k)$ is in $\Omega(k)$.*

Theorem 3. *Given a sorted sequence of n distinct integers, there exists an algorithm to search an element in P using energy $o(E \lg n)$ if the threshold function $f(k)$ is in $\omega(1)$. Furthermore, the algorithm uses energy $\Theta(E)$ if $f(k)$ is in $\Omega(\lg^c k)$ for some constant $c > 1$,*

Theorem 4. *Given a balanced BST T of n nodes consisting of distinct integers, there exists an algorithm to search a key in T energy $o(E \lg n)$ if the threshold function is $\omega(1)$, and using energy $O(E)$ if the threshold function $f(k)$ is in $\Omega(k^c)$ for some constant $c > 0$.*

3   Sorting

In this section, we consider the problem of sorting a sequence $P = P(1), P(2), \ldots, P(n)$ of n distinct integers. Regardless of the threshold function, one can obtain a simple algorithm that uses energy $O(En \lg n)$ by simply running any worst-case $O(n \lg n)$-time comparison-based sorting algorithm using only full-energy comparisons. In this section, we focus on how to minimize the total energy consumption under our model. We start by giving a simple lower bound on the energy consumption for sorting an arbitrary sequence of integers.

Theorem 5. *Any sorting algorithm that sorts a sequence of n distinct integers requires energy at least $E \cdot (n - 1)$.*

Proof. Let P be the input sequence consisting of n distinct integers that is already sorted. In order to verify that P is sorted, the algorithm needs to compare every adjacent pair of elements to make sure that they are in the correct order. Otherwise, the adversary can put one pair of adjacent elements in wrong order, while keeping the rest of elements in the correct order. Now, for any pair of adjacent elements $P(i)$ and $P(i + 1)$, if $|P(i) - P(i + 1)| = 1$ then any number of partial-energy comparisons cannot infer the correct ordering between $P(i)$ and $P(i + 1)$. Thus, the algorithm is forced to use full-energy comparisons (with energy E) to verify that each pair of adjacent elements is in the correct order. This forces every correct algorithm to perform at least $n - 1$ full-energy comparisons using energy $E \cdot (n - 1)$.

In the following sections, an *inversion* in P is defined as a pair of positions (i, j) in P, where $i < j$ and $P(i) > P(j)$, or vice versa. For an inversion (i, j), we say that i (resp. j) has an inversion with j (resp. i).

3.1   Two-Level Algorithm

Our two-level sorting algorithm is based on the two-level algorithm of Funke et al. [10]. Their notion of *cheap* and *expensive* comparisons is analogous to our concept of partial-energy and full-energy comparisons respectively. More specifically, the model of Funke et al. [10] is a special case of our model which only allows two types of comparisons: compare two values with threshold k for some fixed $k > 1$ (*cheap* comparison), and with threshold 1 (*expensive* comparison)[2]. The two-level algorithm of Funke et al. [10] to sort P is as follows:

1. (First level:) Perform any comparison-based sorting algorithm on P using cheap comparisons only. Let P' be the resulting sequence with I inversions in it.
2. (Second level:) Sort P' with $O(n \lg (2 + I/n))$ expensive comparisons using (2,4)-finger search tree [19].

Now we analyze the above two-level algorithm under our model. We will use a threshold k which will be decided later. For the first level of the algorithm, we choose Mergesort which gives at most $kn \lg n$ inversions after sorting (only with cheap comparisons) [10]. The energy usage is at most $O(\frac{E}{f(k)} \cdot n \lg n)$ for the first level of the algorithm, and $O(E \cdot n \lg(2 + \frac{kn \lg n}{n})) = O(En(\lg k + \lg \lg n))$ for the second level. Thus, we can sort P by using energy at most $O(En(\lg n/f(k) + \lg k + \lg \lg n))$ in total. The threshold functions f can be broadly categorized into three different classes and we shall list out the energy requirement in each of those classes.

1. When $f(k)$ is in $\Omega(k)$, let $k \in \{1, 2, \ldots, n\}$ be the minimum value which satisfies $f(k) \geq \lg n$. By choosing this k as our threshold, the total energy required to sort P would be $O(En \lg \lg n)$.
2. When $f(k)$ is in $\omega(1)$, we simply choose $k = \lg n$. Then since $f(\lg n) = \omega(1)$, the total energy required to sort is $O(En(\lg n/f(\lg n) + \lg \lg n)) = o(En \lg n)$.
3. When $f(k)$ is in $O(1)$, the cost of any partial-energy comparison asymptotically matches the cost of a full-energy comparison. So, there is no advantage in using partial-energy comparisons, and the energy required to sort is $\Theta(En \lg n)$.

The third case $(f(k) = O(1))$ is not interesting because the optimal strategy is the trivial strategy of using only full-energy comparisons. This case is only considered for the sake of completeness and we shall not be discussing the third case in the rest of the paper. Thus, as long as the threshold function is a super constant function, the two-level algorithm uses energy $o(En \lg n)$ to sort P. We summarize the results in the following theorem.

Theorem 6. *Given a sequence P of n distinct integers and a threshold function $f(x)$, there exists an algorithm which can sort P using energy at most (i) $O(En \lg \lg n)$ if $f(k)$ is in $\Omega(k)$, and (ii) $o(En \lg n)$ if $f(k)$ is in $\omega(1)$.*

---

[2] In [10], they defined the cheap comparison based on the absolute difference between the *rank* of operands. This corresponds to the case when input is a permutation over integers 1 to n.

Remark. *By using Quicksort instead of Mergesort during the first level of the two-level algorithm, we can get better results on average. Suppose we perform Quicksort using $O(n \lg n)$ cheap comparisons on average. Since there exist at most kn inversions after Quicksort with cheap comparisons [13], we can sort using additional $O(n \lg(2+k))$ expensive comparisons. Thus in this case, the two-level algorithm uses energy $O(En(\lg n/f(k)+\lg k))$ on average. This is especially useful when the threshold function is a super polynomial function since we can choose k to satisfy $\lg n/f(k) \leq 1$ in this case (thus, $O(En \lg k)$ energy is enough on average). For example, if $f(k)$ is in $2^{\sqrt{k}}$, we can choose k to be $(\lg \lg n)^2$ and the average energy consumption would be $O(En \lg \lg \lg n)$ which is $o(En \lg \lg n)$. Furthermore, the worst case energy usage in this case is $O(En(n/f(k) + \lg k))$ – this would be $O(En \lg \lg n)$ if we choose $k = (\lg n)^2$.*

3.2   Multi-level Algorithm

In this section, we describe a multi-level sorting algorithm to sort a sequence P of n distinct integers, which generalizes the two-level algorithm in Sect. 3.1. We prove that by using distinct threshold values at each level, we can obtain an algorithm whose energy usage matches asymptotically the lower bound of Theorem 5 when the threshold function $f(k)$ is in $\Omega(k^c)$ for any fixed constant $c > 0$. We first introduce a lemma which is a reformulation of Lemma 3 of Funke et al. [10].

Lemma 1. *([10]). Let P' be a sequence after performing Mergesort on a sequence P of n distinct integers using partial-energy comparisons only (with threshold k). Then for any two positions $i < j$ on P', $P'(i) \leq P'(j) + k \lg n$.*

Proof. We follow the proof in Funke et al. [10, Lemma 3] which bounds the maximum number of inversions after Mergesort with partial-energy (i.e., cheap) comparisons. The proof uses induction on the number of merging levels. Level 0 (the case that there is one element in the input) is trivial. Now suppose we merge two lists $P(x_1)P(x_2) \ldots P(x_{n/2})$ and $P(x_{n/2+1})P(x_{n/2+2}) \ldots P(x_n)$. Then by induction hypothesis, for any positions in the first list with $i < j$, $P(x_i) \leq P(x_j)+k \lg(n/2)$. Since we only use partial-energy comparisons with threshold k, at most k elements in the second list can be placed between $P(i)$ and $P(j)$ after merging two lists. Thus, the largest element which is placed between $P(i)$ and $P(j)$ after merging lists has the value at most $k+P(x_j)+k \lg n/2 = P(x_j)+k \lg n$.

Next, we introduce the following lemma which gives an upper bound on the distance between any two positions i and j on P where (i,j) is an inversion, after Mergesort with partial-energy comparison is performed.

Lemma 2. *Let P' be a sequence after performing Mergesort on a sequence P of n distinct integers using partial-energy comparisons only (with threshold k). Then for any inversion (i,j), $|i - j|$ is at most $2k \lg n$.*

Proof. Let (i, j) be an inversion in P' after performing Mergesort, and without loss of generality suppose $i < j$ and $P'(i) > P'(j)$. Then by Lemma 1, there exists at most $k \lg n$ elements between $P'(i)$ and $P'(j)$ which are smaller than $P'(i)$. Next, for any position l with $i < l < j$ and $P'(i) < P'(l)$, (l, j) is an inversion since $P'(j) < P'(i) < P'(l)$. Thus, by the same property, there exists at most $k \lg n$ elements between $P'(i)$ and $P'(j)$ which are larger than $P'(i)$. Thus, the number of elements between the positions i and j is at most $2k \lg n$.

Now we describe our multi-level sorting algorithm to sort P. Let $n_1 = n$ for the sake of simplicity of notation. The threshold values (k_i at level i, for $i \geq 1$) in the following description are chosen later, based on the threshold function.

First level: We first perform Mergesort on P using energy $E/f(k_1)$ for each comparison, thus using energy $O((En_1 \lg n_1)/f(k_1))$ in total. Then the resulting sequence P_1 has at most $k_1 n_1 \lg n_1$ inversions. Also for any inversion (a_1, b_1) in P_1, $|a_1 - b_1|$ is at most $2k_1 \lg n_1$ by Lemma 2.

j-th level: For j-th ($j > 1$) level, we first divide P_{j-1} into *blocks* of size $n_j = 4k_{j-1} \lg n_{j-1}$ except the last block. Also for any $i \in \{1, 2, \ldots, \lceil n_{j-1}/n_j \rceil\}$, we define i-th *group* in P_{j-1} as (at most) three consecutive blocks composed of the i-th block along with the $(i-1)$-th and $(i+1)$-th block, if they exist. Then by the property of P_{j-1}, for any inversion (a, b) in P_{j-1}, both the positions a and b are in (a/n_j)-th group. Now let P_j be a sequence obtained by performing Mergesort on each group of P_{j-1} sequentially from left to right, using energy $E/f(k_j)$ for each comparison (thus, the total energy usage is $O(\frac{n_1}{n_j} \cdot \frac{En_j \lg n_j}{f(k_j)}) = O(\frac{En_1 \lg n_j}{f(k_j)})$). Then by Lemma 2, for any inversion (a, b) in P_j, $|a - b|$ is at most $4k_j \lg n_j$ (thus, P_j has $O(k_j n_1 \lg n_j)$ inversions in total). This is from the fact that $|a - b|$ can be greater than $2k_j \lg n_j$ only when $P_j(a)$ and $P_j(b)$ are placed in two consecutive blocks (let these blocks be the r-th and $(r+1)$-th block, respectively) just before sorting the group which contains the $(r + 1)$-th block, but not the r-th block. The worst case occurs when b is moved ($2k_j \lg n_j$) positions to the right of its previous position after sorting the group, which implies $|a - b| \leq 4k_j \lg n_j$ (note that $P_j(b)$ never moves beyond the $(r + 1)$-th block in this case).

Final level: We iteratively perform the above procedure up to ℓ-th level where $4k_\ell \lg n_\ell$ is $O(1)$. The total energy used in all the ℓ levels is $O(En_1 \cdot \sum_{j=1}^{\ell} \frac{\lg n_j}{f(k_j)})$. Finally, since the resulting sequence P_ℓ has $O(n)$ inversions, we sort P_ℓ using $O(En)$ energy, (using the same strategy which was used in the second level of the two-level algorithm in Sect. 3.1, using full-energy comparisons) and complete the sorting procedure.

The following theorem shows that for any constant $c > 0$ the multi-level sorting algorithm with a threshold function $f(k)$ is $\Omega(k^c)$ uses asymptotically optimal energy to sort P.

Theorem 7. *Given a sequence P of n distinct integers, one can sort P using at most energy $O(En)$ if the threshold function $f(k) = \Omega(k^c)$ for some constant $c > 0$.*

Proof. Let $k_i = (\lg n_i)^{c'}$ for all $i \geq 1$ where c' is $2/c$ when $0 < c \leq 1$, and 1 otherwise $(c > 1)$. Then the multi-level sorting algorithm terminates after $O(\lg^* n)$ levels (since n_j is $O(\lg^{1+c'} n_{j-1})$). Thus, the total energy usage is $O(En((\sum_{j=1}^{O(\lg^* n)} \frac{1}{\lg n_i}) + 1)) = O(En)$ since the terms in the summation can be upper bounded by the terms of a geometric series in which last term is $O(En)$.

Remark. *Note that for some threshold functions $f(k) = o(k^c)$, the multi-level sorting algorithm uses asymptotically less energy than the two-level sorting algorithm. For example when $f(k)$ is in $\Theta(\lg^2 k)$, the multi-level sorting algorithm terminates after $O(\lg \lg n)$ levels by setting $k_i = 2^{(\lg n_i / \sqrt{\lg \lg n_i})}$ for all $i \geq 1$ (note that $n_{i+1} = k_i \lg n_i = O(\sqrt{n_i})$). Thus, the multi-level sorting algorithm uses $O(En(\sum_{j=1}^{O(\lg \lg n)} \frac{\lg n_i}{\lg \lg n_i}) = o(En \lg \lg n)$ energy in total, whereas the two-level algorithm of Theorem 6 uses $\omega(En \lg \lg n)$ energy.*

4   Finding the Minimum Element

In this section, we study the problem of finding the minimum (i.e., the case when $r = 1$). We first prove the following theorem, which gives a lower bound on the energy usage for finding the minimum element (thus, the same lower bound holds for finding an element with rank r).

Theorem 8. *Any selection algorithm that finds the minimum element from a sequence of n distinct integers requires energy at least $\sum_{i=1}^{n-1} E/f(i)$.*

Proof. Suppose the sequence is a permutation sorted in ascending order. To verify that the smallest element is placed correctly, the verification process needs to compare the smallest element with the i-th smallest element in the permutation using a threshold at most i. Then the total energy required would be $\sum_{i=1}^{n-1} E/f(i)$.

A naive way to find the minimum would involve using $n-1$ comparisons with full-energy which takes energy $\Theta(En)$ in total. To improve the energy usage, we first divide P into blocks of size $b = 4k \lg k$ (k will be chosen based on the threshold function), and let B_i be the i-th block of P. Now let C be a sequence of elements in P initialized as B_1. Then for $2 \leq i \leq n/b$, we iteratively perform a following procedure:

1. Perform Mergesort on the sequence $C \cdot B_i$ using $E/f(k)$ energy for each comparison (for any two sequences A and B, $A \cdot B$ denotes the concatenation of A and B).
2. By Lemma 2, the smallest element of $C \cdot B_i$ should be in the $2k \lg 2b < b$ leftmost elements after the sequence is sorted in increasing order. Update C as the sequence of of $4k \lg k$ leftmost elements in $C \cdot B_i$.

After performing the above procedure, it is clear that we can find the smallest element in P by performing at most $O(k \lg k)$ full-energy comparisons on C. Thus, we can find the minimum element in P using energy $O(E((n/b) \cdot b \lg b / f(k) + k \lg k) = O(E(n \lg k / f(k) + k \lg k))$ in total. When the threshold function $f(k)$ is in $\Omega(\lg^c k)$ for any constant $c > 1$, we can find the minimum element in P using energy $O(Ek \lg k) + o(En) = o(En)$ by setting $k = n^{\frac{1}{1+c}}$. We summarize the results in the following theorem.

Theorem 9. *Given a sequence P containing n distinct elements, there exists an algorithm which can find the minimum in P using energy $o(En)$ if the threshold function $f(k)$ is in $\Omega(\lg^c k)$ for any constant $c > 1$.*

5 Conclusion

This paper proposes a new comparison model based on the energy-accuracy trade-off and gives some energy-efficient algorithms under this model with the various cases of threshold functions. Many of the algorithms that we propose are asymptotically optimal under a wide range of t threshold functions. Note that our model can be applied not only to implement energy-efficient exact problems but also to energy-efficient approximation problems. For example, considering the energy-efficient algorithms for imprecise sorting [1] or sorting with small maximum dislocation [11] under our model would be interesting. Also in this paper, we only consider the case when the input consists of distinct integers. Extending our results to more general inputs (e.g., multisets) is also an interesting open problem.

References

1. Ajtai, M., Feldman, V., Hassidim, A., Nelson, J.: Sorting and selection with imprecise comparisons. ACM Trans. Algorithms **12**(2), 19:1–19:19 (2016)
2. Alonso, L., Chassaing, P., Gillet, F., Janson, S., Reingold, E.M., Schott, R.: Quicksort with unreliable comparisons: a probabilistic analysis. Comb. Probab. Comput. **13**(4–5), 419–449 (2004). https://doi.org/10.1017/S0963548304006297
3. Arumugam, G.P., et al.: Novel inexact memory aware algorithm co-design for energy efficient computation: algorithmic principles. In: Proceedings of the 2015 Design, Automation & Test in Europe Conference & Exhibition, DATE 2015, pp. 752–757. ACM (2015)
4. Bennett, C.H.: Time/space trade-offs for reversible computation. SIAM J. Comput. **18**(4), 766–776 (1989)
5. Bianchi, E., Penna, P.: Optimal clustering in stable instances using combinations of exact and noisy ordinal queries. Algorithms **14**(2), 55 (2021)
6. David, H.A.: The Method of Paired Comparisons, 2nd edition, vol. 12. London (1988)
7. Demaine, E.D., Lynch, J., Mirano, G.J., Tyagi, N.: Energy-efficient algorithms. In: Sudan, M. (ed.) Proceedings of the 2016 ACM Conference on Innovations in Theoretical Computer Science, Cambridge, MA, USA, 14–16 January 2016, pp. 321–332. ACM (2016)

8. Feige, U., Raghavan, P., Peleg, D., Upfal, E.: Computing with noisy information. SIAM J. Comput. **23**(5), 1001–1018 (1994)
9. Finocchi, I., Italiano, G.F.: Sorting and searching in faulty memories. Algorithmica **52**(3), 309–332 (2008)
10. Funke, S., Mehlhorn, K., Näher, S.: Structural filtering: a paradigm for efficient and exact geometric programs. Comput. Geom. **31**(3), 179–194 (2005)
11. Geissmann, B., Leucci, S., Liu, C., Penna, P.: Optimal sorting with persistent comparison errors. In: 27th Annual European Symposium on Algorithms, ESA 2019. LIPIcs, vol. 144, pp. 49:1–49:14 (2018)
12. Geissmann, B., Leucci, S., Liu, C.H., Penna, P., Proietti, G.: Dual-mode greedy algorithms can save energy. In: 30th International Symposium on Algorithms and Computation (ISAAC 2019), vol. 149, pp. 64–1. Schloss Dagstuhl-Leibniz-Zentrum für Informatik (2019)
13. Geissmann, B., Penna, P.: Inversions from sorting with distance-based errors. In: Tjoa, A.M., Bellatreche, L., Biffl, S., van Leeuwen, J., Wiedermann, J. (eds.) SOFSEM 2018. LNCS, vol. 10706, pp. 508–522. Springer, Cham (2018). https://doi.org/10.1007/978-3-319-73117-9_36
14. Geissmann, B., Penna, P.: Sorting processes with energy-constrained comparisons. Phys. Rev. E **97**(5), 052108 (2018)
15. Huang, Z., Kannan, S., Khanna, S.: Algorithms for the generalized sorting problem. In: Ostrovsky, R. (ed.) IEEE 52nd Annual Symposium on Foundations of Computer Science, FOCS 2011, pp. 738–747. IEEE Computer Society (2011)
16. Korkmaz, P., Akgul, B.E.S., Palem, K.V.: Energy, performance, and probability tradeoffs for energy-efficient probabilistic CMOS circuits. IEEE Trans. Circuits Syst. I Regul. Pap. **55**(8), 2249–2262 (2008)
17. Landauer, R.: Irreversibility and heat generation in the computing process. IBM J. Res. Dev. **5**(3), 183–191 (1961). https://doi.org/10.1147/rd.53.0183
18. Leucci, S., Liu, C.H.: Approximate minimum selection with unreliable comparisons. Algorithmica **84**(1), 60–84 (2022)
19. Mehlhorn, K.: Data Structures and Algorithms 1: Sorting and Searching, EATCS Monographs on Theoretical Computer Science, vol. 1. Springer, Cham (1984). https://doi.org/10.1007/978-3-642-69672-5
20. Palem, K.V., Avinash, L.: Ten years of building broken chips: the physics and engineering of inexact computing. ACM Trans. Embed. Comput. Syst. **12**(2s), 87:1–87:23 (2013)
21. Zurek, W.H.: Thermodynamic cost of computation, algorithmic complexity and the information metric. Nature **341**, 119–124 (1989)

Reconfiguration of Vertex-Disjoint Shortest Paths on Graphs

Rin Saito[1]([⊠])[iD], Hiroshi Eto[2][iD], Takehiro Ito[1][iD], and Ryuhei Uehara[3][iD]

[1] Graduate School of Information Sciences, Tohoku University, Sendai, Japan
rin.saito@dc.tohoku.ac.jp, takehiro@tohoku.ac.jp
[2] School of Computer Science and Systems Engineering,
Kyushu Institute of Technology, Iizuka, Japan
eto@ai.kyutech.ac.jp
[3] School of Information Science, Japan Advanced Institute of Science
and Technology, Nomi, Japan
uehara@jaist.ac.jp

Abstract. We introduce and study reconfiguration problems for (internally) vertex-disjoint shortest paths: Given two tuples of internally vertex-disjoint shortest paths for fixed terminal pairs in an unweighted graph, we are asked to determine whether one tuple can be transformed into the other by exchanging a single vertex of one shortest path in the tuple at a time, so that all intermediate results remain tuples of internally vertex-disjoint shortest paths. We also study the shortest variant of the problem, that is, we wish to minimize the number of vertex-exchange steps required for such a transformation, if exists. These problems generalize the well-studied SHORTEST PATH RECONFIGURATION problem. In this paper, we analyze the complexity of these problems from the viewpoint of graph classes, and give some interesting contrast.

Keywords: Combinatorial reconfiguration · Graph algorithm ·
Vertex-disjoint paths

1 Introduction

Combinatorial reconfiguration [6] has been extensively studied in the field of theoretical computer science. One of the most well-studied problems is the *reachability variant*: we are given two feasible solutions of a combinatorial search problem, and are asked to determine whether we can transform one into the other by repeatedly applying a prescribed reconfiguration step so that all intermediate results are also feasible. This kind of problems has been studied intensively for several combinatorial search problems. (See surveys [5,8].)

This work is partially supported by JSPS KAKENHI Grant Numbers JP18H04091, JP19K11814, JP20H05793, JP20H05961, JP20H05964 and JP20K11673.

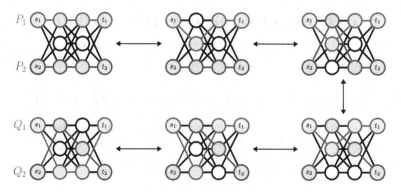

Fig. 1. Sequence of tuples of (internally) vertex-disjoint shortest paths for two terminal pairs (s_1, t_1) and (s_2, t_2).

For example, *the* SHORTEST PATH RECONFIGURATION (SPR) *problem* is defined as follows [7]: we are given two shortest paths between two specified vertices s and t (called *terminals*) in an unweighted graph, and are asked to determine whether or not we can transform one into the other by exchanging a single vertex in a shortest path at a time, so that all intermediate results remain shortest paths between s and t. Surprisingly, the problem is PSPACE-complete [1, 9], and polynomial-time algorithms have been developed for restricted graph classes [1–3].

1.1   Our Problems and Related Results

In this paper, as generalizations of the SPR problem, we introduce and study reconfiguration problems for (internally) vertex-disjoint shortest paths in an unweighted graph G. For k terminal pairs (s_i, t_i), $i \in \{1, 2, \ldots, k\}$, consider a tuple of k paths in G such that the i-th path in the tuple joins s_i and t_i. Then, the k paths in the tuple are said to be *internally vertex-disjoint* if the internal vertices of k paths are all distinct and do not contain any terminal.

We now define the REACHABILITY OF VERTEX-DISJOINT SHORTEST PATHS (RVDSP) problem, as follows. Suppose that we are given two tuples $\mathcal{P} = (P_1, P_2, \ldots, P_k)$ and $\mathcal{Q} = (Q_1, Q_2, \ldots, Q_k)$ of internally vertex-disjoint paths such that each of P_i and Q_i is a shortest path in an unweighted graph G joining two terminals s_i and t_i for all $i \in \{1, 2, \ldots, k\}$. Then, *the* RVDSP *problem* asks to determine whether or not one can transform $\mathcal{P}$ into $\mathcal{Q}$ by exchanging a single vertex of one shortest path in the tuple at a time, so that all intermediate results remain tuples of internally vertex-disjoint shortest paths for k terminal pairs. (See Fig. 1 as an example.) Thus, the RVDSP problem for $k = 1$ is equivalent to the SPR problem. We also study the *shortest variant, the* SHORTEST RECONFIGURATION OF VERTEX-DISJOINT SHORTEST PATHS (SRVDSP) *problem* which asks to determine whether or not there is a transformation between $\mathcal{P}$ and $\mathcal{Q}$ by at most ℓ vertex-exchange steps, for a given integer $\ell \geq 0$.

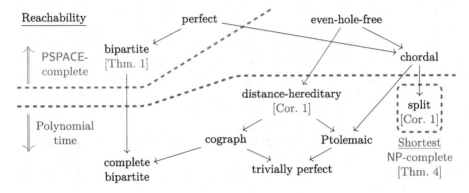

Fig. 2. Our results with respect to graph classes. Each arrow represents the inclusion relationship between graph classes: $A \rightarrow B$ means that the graph class B is a proper subclass of the graph class A. In addition, we prove that the RVDSP problem is PSPACE-complete for graphs with bounded bandwidth (Theorem 1), and the SRVDSP problem is solvable in polynomial time for distance-hereditary graphs and for split graphs if all k terminal pairs are the same (Theorem 3).

Kamiński et al. [7] introduced the SPR problem (i.e., the RVDSP problem for $k = 1$), and posed an open question of the complexity of the SPR problem. Bonsma [1] answered by proving that the SPR problem is PSPACE-complete for bipartite graphs. Since $P \subseteq NP \subseteq PSPACE$, this means that the problem admits no polynomial-time algorithm under the assumption of $P \neq PSPACE$, and furthermore implies that there is a yes-instance that requires super-polynomial steps for transforming one shortest path to the other under the assumption of $NP \neq PSPACE$. These are somewhat surprising because the problem of finding shortest paths (especially, in an unweighted graph) is easy. Bonsma [1] posed another open question whether the SPR problem can be solved in polynomial time for graphs with bounded treewidth. This question was answered negatively by Wrochna [9]: the SPR problem remains PSPACE-complete even for graphs with bounded bandwidth. Note that the bandwidth of a graph gives an upper bound on the pathwidth (and hence the treewidth) of the graph; and hence the PSPACE-completeness holds also for graphs with bounded treewidth.

On the positive side, the SPR problem has been shown to be solvable in polynomial time when restricted to graph classes, such as chordal graphs and claw-free graphs [1]; planar graphs [2]; circle graphs, permutation graphs, the Boolean hypercube, bridged graphs, and circular-arc graphs [3]. Furthermore, the shortest variant of the SPR problem (i.e., the SRVDSP problem for $k = 1$) is solvable in polynomial time for chordal graphs [1].

1.2   Our Contribution

In this paper, we study the computational complexity of the RVDSP and SRVDSP problems from the viewpoint of graph classes. Figure 2 summarizes our results. (Throughout the paper, k denotes the number of terminal pairs.)

We first observe that the RVDSP problem for every fixed $k \geq 1$ is PSPACE-complete for bipartite graphs and for graphs with bounded bandwidth. On the positive side, we give a polynomial-time algorithm to solve the RVDSP problem for distance-hereditary graphs and for split graphs. Interestingly, our algorithm for these two graph classes can be obtained as a corollary of a single theorem (Theorem 2) by introducing the concept of "st-completeness" of graphs for terminal pairs (s, t). Our algorithm is constructive, and finds an actual transformation (if exists) that requires polynomial number of vertex-exchange steps.

We then prove that the SRVDSP problem is NP-complete for split graphs. On the positive side, the problem is solvable in polynomial time for split graphs and for distance-hereditary graphs if all k terminal pairs are the same, that is, $(s_1, t_1) = (s_2, t_2) = \cdots = (s_k, t_k)$.

Our results give the following interesting contrast:

1. the RVDSP problem is PSPACE-complete for bipartite graphs (Theorem 1), while it is solvable in polynomial time for complete bipartite graphs (Corollary 1);
2. for split graphs, the RVDSP problem is solvable in polynomial time (Corollary 1), whereas the SRVDSP problem is NP-complete (Theorem 4); and
3. the SRVDSP problem for $k = 1$ is solvable in polynomial time for chordal graphs [1] (thus, for split graphs), while the SRVDSP problem for general k is NP-complete for split graphs (Theorem 4).

We omit the proofs for claims with ($*$) from this extended abstract.

2   Preliminaries

In this paper, we assume that graphs are simple and unweighted. For a graph G, we denote by $V(G)$ and $E(G)$ the vertex and edge sets of G, respectively. Let $n = |V(G)|$ and $m = |E(G)|$ throughout the paper. For $u, v \in V(G)$, a path in G joining u and v is called a uv-path. We denote by $d_G(u, v)$ the minimum number of edges in any uv-path in G; we sometimes omit the subscript G if it is clear from the context. The *diameter of G* is the maximum $d_G(u, v)$ among any two vertices u, v in G. For two sets A and B, we denote by $A \triangle B$ the *symmetric difference* of A and B, that is, $(A \setminus B) \cup (B \setminus A)$.

Let k be a positive integer, and let (s_i, t_i) be a pair of vertices in G, called *terminals*, for $i \in \{1, 2, \ldots, k\}$. Then, k paths $P_1, P_2, \ldots, P_k$ in G are said to be *internally vertex-disjoint* if P_i is an $s_i t_i$-path in G for each $i \in \{1, 2, \ldots, k\}$, and their internal vertices are all distinct and do not contain any terminal. Note that internally vertex-disjoint paths may share terminals. In the following, we call internally vertex-disjoint paths simply *vertex-disjoint* paths. For a tuple $\mathcal{P} = (P_1, P_2, \ldots, P_k)$ of vertex-disjoint paths, let $V(\mathcal{P}) = \bigcup_{i=1}^{k} V(P_i)$.

In this paper, we consider only *shortest $s_i t_i$-paths* in G, $i \in \{1, 2, \ldots, k\}$. For two tuples $\mathcal{P} = (P_1, P_2, \ldots, P_k)$ and $\mathcal{P}' = (P_1', P_2', \ldots, P_k')$ of vertex-disjoint shortest paths, we write $\mathcal{P} \leftrightarrow \mathcal{P}'$ if $\sum_{i=1}^{k} |V(P_i) \triangle V(P_i')| = 2$; in other words, $\mathcal{P}'$ can be obtained from $\mathcal{P}$ by exchanging a single (internal) vertex in some shortest path P_i with a vertex that is not contained in $V(\mathcal{P})$. A sequence $\langle \mathcal{P}_0, \mathcal{P}_1, \ldots, \mathcal{P}_\ell \rangle$ of tuples of vertex-disjoint shortest paths is called a *reconfiguration sequence* between $\mathcal{P}_0$ and $\mathcal{P}_\ell$ if $\mathcal{P}_{r-1} \leftrightarrow \mathcal{P}_r$ for all $r \in \{1, 2, \ldots, \ell\}$. The *length* of a reconfiguration sequence $\langle \mathcal{P}_0, \mathcal{P}_1, \ldots, \mathcal{P}_\ell \rangle$ is defined to be ℓ. We now define two following problems.

The REACHABILITY OF VERTEX-DISJOINT SHORTEST PATHS (RVDSP) problem
Input: An unweighted graph G, and two tuples $\mathcal{P}$ and $\mathcal{Q}$ of vertex-disjoint shortest paths for k terminal pairs (s_i, t_i).
Task: Determine if there is a reconfiguration sequence between $\mathcal{P}$ and $\mathcal{Q}$.

The SHORTEST RECONFIGURATION OF VERTEX-DISJOINT SHORTEST PATHS (SRVDSP) problem
Input: An unweighted graph G, two tuples $\mathcal{P}$ and $\mathcal{Q}$ of vertex-disjoint shortest paths for k terminal pairs (s_i, t_i), and an integer $\ell \geq 0$.
Task: Determine if there is a reconfiguration sequence between $\mathcal{P}$ and $\mathcal{Q}$ of length at most ℓ.

Note that both RVDSP and SRVDSP problems are decision problems, and do not ask for an actual reconfiguration sequence as an output. We sometimes denote simply by $(G, \mathcal{P}, \mathcal{Q})$ an instance of the RVDSP problem, and by $(G, \mathcal{P}, \mathcal{Q}, \ell)$ an instance of the SRVDSP problem.

Definitions of layers and st-completeness

For two distinct vertices $s, t \in V(G)$ and $j \in \{0, 1, \ldots, d(s,t)\}$, let $L_j = \{v \in V(G) \mid d(s,v) = j, d(s,v) + d(v,t) = d(s,t)\}$. We call L_j the j-th st-layer, that is, L_j is the set of vertices v such that $d(s,v) = j$ and v is contained in some shortest st-path. Note that $L_0 = \{s\}$ and $L_{d(s,t)} = \{t\}$. We denote by G_{st} the subgraph of G induced by all st-layers L_j, $j \in \{0, 1, \ldots, d(s,t)\}$. Then, any shortest st-path in G is contained in G_{st}. We say that G is st-*complete* if every vertex in L_j is adjacent in G to all vertices in L_{j+1} for all $j \in \{0, 1, \ldots, d(s,t) - 1\}$. Since G is an unweighted graph, we have the following lemma.

Lemma 1. *For two vertices $s, t \in V(G)$, one can construct G_{st} and check whether G is st-complete in $O(m + n)$ time.*

The st-completeness of a graph G is a useful property, because we can forget the structure of G in the following sense: if we choose exactly one vertex from each st-layer L_j, $j \in \{0, 1, \ldots, d(s,t)\}$, the set of the chosen vertices *always* forms a shortest st-path in G.

3   Reachability Variant

Recall that the RVDSP problem for $k = 1$ (i.e., the SPR problem) is PSPACE-complete for bipartite graphs [1], and for graphs with bounded bandwidth [9].

By introducing dummy vertex-disjoint shortest paths, one can observe the following hardness results.

Theorem 1. *For every fixed $k \geq 1$, the RVDSP problem is PSPACE-complete for bipartite graphs, and for graphs of bounded bandwidth.*

The main result of this section is the following theorem, whose proof will be given in Sects. 3.1 and 3.2.

Theorem 2. *Let G be a graph of diameter d as an input of the RVDSP problem such that G is st-complete for all k terminal pairs. Then, the RVDSP problem is solvable in $O(mk + ndk^2)$ time. Furthermore, in the same running time, one can find a reconfiguration sequence of length $O(d^2k^2)$ if exists.*

From Theorem 2, one can show that the RVDSP problem is solvable in polynomial time for some graph classes, as in the following corollary. A graph G is *split* if $V(G)$ can be partitioned into a clique and an independent set. A graph G is *distance hereditary* if $d_G(u, v) = d_{G'}(u, v)$ for every connected induced subgraph G' of G and all $u, v \in V(G')$.

Corollary 1 ($*$). *The RVDSP problem is solvable in polynomial time for split graphs, and for distance-hereditary graphs.*

Note that any split graph is of diameter at most 3, and hence we can drop the factor d in Theorem 2 for split graphs. On the other hand, the diameter of distance-hereditary graphs can be $\Omega(n)$. We also note that a complete bipartite graph is distance hereditary, and hence the RVDSP problem can be solved in polynomial time for complete bipartite graphs. Therefore, Theorem 1 and Corollary 1 give an interesting contrast of the complexity of the RVDSP problem. (See Fig. 2 again.)

3.1 Characterization of Reachability

Let $(G, \mathcal{P}, \mathcal{Q})$ be an instance of the RVDSP problem such that G is $s_i t_i$-complete for each $i \in \{1, 2, \ldots, k\}$. We denote by L_j^i the j-th $s_i t_i$-layer for each pair of integers $i \in \{1, 2, \ldots, k\}$ and $j \in \{0, 1, \ldots, d_G(s_i, t_i)\}$. For G and $\mathcal{P}$, we define a directed graph $G_{\mathcal{P}}$, called an *auxiliary graph for $\mathcal{P}$*, as follows: $V(G_{\mathcal{P}}) = V(G)$, and for each i, j, we add arcs (u, v) from $u \in V(P_i) \cap L_j^i$ to all vertices $v \in L_j^i$, that is, $A(G_{\mathcal{P}}) = \bigcup_{i,j} \{(u, v) \mid u \in V(P_i) \cap L_j^i, v \in L_j^i\}$.

For each pair of $i \in \{1, 2, \ldots, k\}$ and $j \in \{0, 1, \ldots, d_G(s_i, t_i)\}$, we place a (labeled) *token* t_j^i on the vertex $u \in V(P_i) \cap L_j^i$ in the auxiliary graph $G_{\mathcal{P}}$. Note that no two tokens are placed on the same vertex, because paths in $\mathcal{P}$ are vertex-disjoint. Conversely, the $s_i t_i$-completeness of G ensures that any placement of the token t_j^i to a vertex in L_j^i yields a shortest $s_i t_i$-path in G. The vertex on which t_j^i is placed is sometimes referred simply as t_j^i. We say that the token t_j^i is *$\mathcal{P}$-movable* if there exists a directed path in $G_{\mathcal{P}}$ from t_j^i to a vertex $v \notin V(\mathcal{P})$. We sometimes call such a $\mathsf{t}_j^i v$-path a *t_j^i-escape path under $\mathcal{P}$*. The $\mathcal{P}$-movable tokens have a good property, as follows.

Lemma 2 (∗). *Let $\mathcal{P}$ and $\mathcal{P}'$ be two tuples of vertex-disjoint shortest paths in G such that $\mathcal{P} \leftrightarrow \mathcal{P}'$. Then, a token t_j^i is $\mathcal{P}$-movable if and only if t_j^i is $\mathcal{P}'$-movable.*

For each pair of integers $i \in \{1, 2, \ldots, k\}$ and $j \in \{0, 1, \ldots, d_G(s_i, t_i)\}$, the vertex in $V(Q_i) \cap L_j^i$ is called the *target position* for the token t_j^i. For a tuple $\mathcal{P}'$ of vertex-disjoint shortest paths, we denote by $T_{\mathcal{P}'}$ the set of all tokens that are *not* placed on their target positions in $\mathcal{P}'$. The following lemma is the key for the proof of Theorem 2.

Lemma 3. *$(G, \mathcal{P}, \mathcal{Q})$ is a yes-instance if and only if every token in $T_{\mathcal{P}}$ is $\mathcal{P}$-movable.*

Proof. We first prove the only-if direction. Suppose that $(G, \mathcal{P}, \mathcal{Q})$ is a yes-instance, and hence there is a reconfiguration sequence $\langle \mathcal{P}_0, \mathcal{P}_1, \ldots, \mathcal{P}_\ell \rangle$ between $\mathcal{P} = \mathcal{P}_0$ and $\mathcal{Q} = \mathcal{P}_\ell$. Because every token t_j^i in $T_{\mathcal{P}}$ is not placed on its target position in $\mathcal{P}$, the token must be moved at least once in the reconfiguration sequence. Assume that t_j^i was moved between $\mathcal{P}_r$ and $\mathcal{P}_{r+1}$, from a vertex $y_1 \in L_j^i \cap V(\mathcal{P}_r)$ to another vertex $y_2 \in L_j^i \setminus V(\mathcal{P}_r)$. Then, the auxiliary graph $G_{\mathcal{P}_r}$ has an arc (y_1, y_2); this arc forms a t_j^i-escape path under $\mathcal{P}_r$, and hence t_j^i is $\mathcal{P}_r$-movable. Since $\mathcal{P}_r \leftrightarrow \mathcal{P}_{r-1} \leftrightarrow \cdots \leftrightarrow \mathcal{P}_0 = \mathcal{P}$, Lemma 2 implies that t_j^i is $\mathcal{P}$-movable.

We then prove the if direction, by induction on $|T_{\mathcal{P}}|$. If $|T_{\mathcal{P}}| = 0$, then $\mathcal{P} = \mathcal{Q}$ and hence $(G, \mathcal{P}, \mathcal{Q})$ is a yes-instance. Thus, suppose that $|T_{\mathcal{P}}| \geq 1$ and every token in $T_{\mathcal{P}}$ is $\mathcal{P}$-movable. We consider two following cases.

Case (a): We first consider the case where there exists a token t_j^i placed on the vertex $y_1 \in V(P_i) \cap L_j^i$ such that its target position y_2 is not occupied by any token in $\mathcal{P}$, that is, $V(Q_i) \cap L_j^i = \{y_2\}$ and $y_2 \notin V(\mathcal{P})$. In this case, we can move t_j^i to y_2 directly, and obtain the tuple $\mathcal{P}'$ of vertex-disjoint shortest paths; notice that there is an arc (y_1, y_2) in $G_{\mathcal{P}}$ by the definition of $G_{\mathcal{P}}$. Since $\mathcal{P} \leftrightarrow \mathcal{P}'$, Lemma 2 says that every $\mathcal{P}$-movable token is $\mathcal{P}'$-movable. Since t_j^i reaches its target position y_2, we have $|T_{\mathcal{P}'}| = |T_{\mathcal{P}}| - 1$. Therefore, we can apply the induction hypothesis to $(G, \mathcal{P}', \mathcal{Q})$.

Case (b): We then consider the other case, that is, the target positions of all tokens are occupied by tokens in $\mathcal{P}$. In this case, we can find a directed cycle $C = x_1 x_2 \ldots x_\alpha$ in $G_{\mathcal{P}}$ such that x_{r+1} is the target position of the token placed on x_r for all $r \in \{1, 2, \ldots, \alpha\}$; for convenience, we regard $x_{\alpha+1} = x_1$. Since $|T_{\mathcal{P}}| \geq 1$, we can assume that $\alpha \geq 2$ (i.e., C is not a self-loop). Therefore, all tokens placed on $x_1, x_2, \ldots, x_\alpha$ belong to $T_{\mathcal{P}}$, and hence all of them are $\mathcal{P}$-movable. Then, there exists at least one token t such that t is placed on a vertex in C, say x_α, and $G_{\mathcal{P}}$ has a t-escape path $x_\alpha y_1 y_2 \ldots y_\beta$ with $y_1, y_2, \ldots, y_\beta \notin V(C)$. (See Fig. 3.) Then, we move tokens as follows:

1. move the token on y_r to y_{r+1} for each r, $\beta - 1 \geq r \geq 1$;
2. move the token on x_α to y_1 (now no token is placed on x_α);
3. move the token on x_r to x_{r+1} for each r, $\alpha - 1 \geq r \geq 1$;
4. move the token on y_1 (which was placed on x_α in $\mathcal{P}$) to x_1; and
5. move the token on y_r to y_{r-1} for each r, $2 \leq r \leq \beta$.

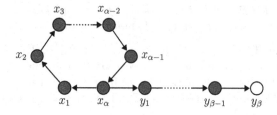

Fig. 3. Illustration for Case (b).

Note that, in Step 4, we can move the token on y_1 to x_1, because x_1 is the target position of the token which was placed on x_α in $\mathcal{P}$. After Step 5, each token on $y_1, y_2, \ldots, y_\beta$ are placed on the same vertex as in $\mathcal{P}$, and each token on $x_1, x_2, \ldots, x_\alpha$ reaches its target position. Let $\mathcal{P}'$ be the resulting tuple of vertex-disjoint shortest paths. Lemma 2 says that every $\mathcal{P}$-movable token is $\mathcal{P}'$-movable. Since $|T_{\mathcal{P}'}| = |T_\mathcal{P}| - \alpha$ and $\alpha \geq 2$, we can apply the induction hypothesis to $(G, \mathcal{P}', \mathcal{Q})$.   □

3.2   Proof of Theorem 2

Our proof of Lemma 3 naturally yields an algorithm which finds an actual reconfiguration sequence between $\mathcal{P}$ and $\mathcal{Q}$ if exists. Starting from any token t in $T_\mathcal{P}$, we traverse $G_\mathcal{P}$ by repeatedly visiting the target position of the currently visited token; if we reach a vertex which is not occupied by any token (that is, the vertex is not contained in any path in $\mathcal{P}$), then we apply Case (a) of the proof; otherwise we find a directed cycle $x_1, x_2, \ldots, x_\alpha$ to which we apply Case (b). In Case (b) we can find the path $x_\alpha y_1 y_2 \ldots y_\beta$ by a breadth-first search on $G_\mathcal{P}$ starting from any vertex in the directed cycle.

We now estimate the length of our reconfiguration sequence between $\mathcal{P}$ and $\mathcal{Q}$. Recall that there are $O(dk)$ tokens, where d is the diameter of G and k is the number of terminal pairs. In Case (a), one token in $T_\mathcal{P}$ reaches its target position by one step. In Case (b), α tokens in $T_\mathcal{P}$ reach their target positions by $(\beta - 1) + 1 + (\alpha - 1) + 1 + (\beta - 1) = \alpha + 2\beta - 1 = O(dk)$ steps. Therefore, in both cases, at least one token in $T_\mathcal{P}$ reaches its target position by $O(dk)$ steps. Since there are $O(dk)$ tokens, the length of our reconfiguration sequence between $\mathcal{P}$ and $\mathcal{Q}$ can be bounded by $O(dk) \times O(dk) = O(d^2 k^2)$ in total.

Finally, we estimate the running time of the algorithm. By Lemma 1 we can check if a given graph G is st-complete for all k terminal pairs, and construct layers in $O(k(n+m))$ time. Since $\sum_{j=1}^{d(s_i, t_i)} |L_j^i| \leq n$ for each $i \in \{1, 2, \ldots, k\}$, the auxiliary graph $G_\mathcal{P}$ has at most nk arcs. For each token in $T_\mathcal{P}$, we traverse $G_\mathcal{P}$ at most twice to apply Case (a) or (b). Thus, we can move at least one token in $T_\mathcal{P}$ to its target position in $O(nk) + O(dk) = O(nk)$ time. Since there are $O(dk)$ tokens, all tokens can be moved to their target positions in $O(nk) \times O(dk) = O(ndk^2)$ time. In total, our algorithm runs in $O(mk + ndk^2)$ time.

This completes the proof of Theorem 2.   □

4   Shortest Variant

Our polynomial-time algorithm in Sect. 3 for the RVDSP problem does not always return a reconfiguration sequence of the shortest length. Indeed, we will show in this section that the SRVDSP problem is NP-complete even for split graphs, while the RVDSP problem is solvable in polynomial time for split graphs.

4.1   Polynomial-Time Solvable Cases

We first give tractable cases of the SRVDSP problem, based on the algorithm in Sect. 3. We say that k terminal pairs are *identical* if $(s_1, t_1) = (s_2, t_2) = \cdots = (s_k, t_k)$. Then, we have the following theorem.

Theorem 3 ($*$). *Let $(G, \mathcal{P}, \mathcal{Q}, \ell)$ be an instance of the SRVDSP problem such that G is st-complete for all k terminal pairs, and k terminal pairs are identical. Then, the SRVDSP problem is solvable in polynomial time.*

Similarly as in Corollary 1, we have the following corollary from Theorem 3. Note that, if $k = 1$, then it is an identical terminal pair.

Corollary 2. *The SRVDSP problem is solvable in polynomial time for split graphs, and for distance-hereditary graphs, if k terminal pairs are identical. In particular, the shortest variant of the SPR problem is solvable in polynomial time for split graphs, and for distance-hereditary graphs.*

4.2   NP-Completeness

We finally prove the following theorem.

Theorem 4. *The SRVDSP problem is NP-complete for split graphs.*

By Theorem 2, recall that the RVDSP problem for split graphs can be solved in polynomial time, and admits a reconfiguration sequence of length $O(k^2)$ if exists. Therefore, the SRVDSP problem for split graphs belongs to the class NP. As a proof of Theorem 4, we will thus prove that the SRVDSP problem is NP-hard for split graphs, by giving a polynomial-time reduction from 3SAT [4].

Suppose that we are given a 3CNF formula ϕ, where each clause consists of exactly three literals. Let α and β be the numbers of variables and clauses in ϕ, respectively. We write $x_1, x_2, \ldots, x_\alpha$ for variables in ϕ, and $C_1, C_2, \ldots, C_\beta$ for clauses in ϕ. We will construct the corresponding graph G_ϕ which forms a split graph. Recall that a split graph G_ϕ can be partitioned into a clique and an independent set. In the following, we call a vertex in the clique a *clique vertex*, and call a vertex in the independent set an *independent vertex*. In our reduction, independent vertices will be terminals. As shown in Corollary 1, a split graph G_ϕ is st-complete for any terminal pair (s, t). Therefore, roughly speaking, we will focus on how to move tokens placed on clique vertices in G_ϕ.

Reduction

We first create a *variable gadget* G_{x_i} for each variable x_i in ϕ. The variable gadget G_{x_i} has five clique vertices $a_i, b_i, c_i, x_i^\top, x_i^\perp$, and eight independent vertices $s_{i1}, s_{i2}, s_{i3}, s_{i4}, t_{i1}, t_{i2}, t_{i3}, t_{i4}$. Then, we define four vertex sets $L^{i1}, L^{i2}, L^{i3}, L^{i4}$, as follows:

$$L^{i1} = \{a_i, b_i, x_i^\top\}, \ \ L^{i2} = \{a_i, b_i, x_i^\perp\}, \ \ L^{i3} = \{c_i, x_i^\top\}, \ \ L^{i4} = \{c_i, x_i^\perp\}.$$

We join the clique and independent vertices in G_{x_i} so that each L^{ir} forms the 1-st $s_{ir}t_{ir}$-layer for terminal pair (s_{ir}, t_{ir}), $r \in \{1, 2, 3, 4\}$; more specifically, for each $r \in \{1, 2, 3, 4\}$, we join each of s_{ir} and t_{ir} with all clique vertices in L^{ir}. For each $r \in \{1, 2, 3, 4\}$, we define shortest $s_{ir}t_{ir}$-paths P_{ir} and Q_{ir} as follows:

$$
\begin{aligned}
P_{i1} &= s_{i1}a_i t_{i1}, & Q_{i1} &= s_{i1}b_i t_{i1}, \\
P_{i2} &= s_{i2}b_i t_{i2}, & Q_{i2} &= s_{i2}a_i t_{i2}, \\
P_{i3} &= Q_{i3} = s_{i3}x_i^\top t_{i3}, & & \\
P_{i4} &= Q_{i4} = s_{i4}x_i^\perp t_{i4}. & &
\end{aligned}
$$

Notice that only the internal vertices of P_{i1} and P_{i2} are swapped in Q_{i1} and Q_{i2}. By the construction, it suffices to focus on which clique vertex is chosen as an internal vertex of a shortest $s_{ir}t_{ir}$-path.

We then create a *clause gadget* G_{C_j} for each clause C_j in ϕ. We assume that $C_j = (l_{jp} \vee l_{jq} \vee l_{jr})$, where l_{jh} is either x_h or $\neg x_h$ for each $h \in \{p, q, r\}$. Let $\sigma(j, h)$ denote $\top$ if $l_{jh} = x_h$, otherwise $\perp$. The clause gadget G_{C_j} has six new clique vertices a_{jh}, b_{jh} where $h \in \{p, q, r\}$, and three clique vertices $x_p^{\sigma(j,p)}, x_q^{\sigma(j,q)}, x_r^{\sigma(j,r)}$ which are already introduced in variable gadgets. In addition, G_{C_j} has 12 new independent vertices $s_{jh1}, s_{jh2}, t_{jh1}, t_{jh2}$ where $h \in \{p, q, r\}$. We define six vertex sets, as follows:

$$
\begin{aligned}
L^{jp1} &= \{x_p^{\sigma(j,p)}, \ b_{jp}, \ a_{jp}, \ b_{jq}\}, \\
L^{jq1} &= \{x_q^{\sigma(j,q)}, \ b_{jq}, \ a_{jq}, \ b_{jr}\}, \\
L^{jr1} &= \{x_r^{\sigma(j,r)}, \ b_{jr}, \ a_{jr}, \ b_{jp}\}, \\
L^{jh2} &= \{a_{jh}, \ b_{jh}\} \quad \text{for } h \in \{p, q, r\}.
\end{aligned}
$$

Similarly as in the variable gadgets, we join the clique and independent vertices in G_{C_j}, as follows: for each $h \in \{p, q, r\}$, we join each of s_{jh1} and t_{jh1} with all clique vertices in L^{jh1}, and also join each of s_{jh2} and t_{jh2} with all clique vertices in L^{jh2}. For each $h \in \{p, q, r\}$, we define shortest $s_{jh1}t_{jh1}$-paths P_{jh1} and Q_{jh1}, and shortest $s_{jh2}t_{jh2}$-paths P_{jh2} and Q_{jh2}, as follows:

$$
\begin{aligned}
P_{jh1} &= s_{jh1}a_{jh}t_{jh1}, & Q_{jh1} &= s_{jh1}b_{jh}t_{jh1}, \\
P_{jh2} &= s_{jh2}b_{jh}t_{jh2}, & Q_{jh2} &= s_{jh2}a_{jh}t_{jh2}.
\end{aligned}
$$

We next join all clique vertices in variable and clause gadgets so that they form a clique in G_ϕ. This completes the constructions of G_ϕ, $\mathcal{P}$, and $\mathcal{Q}$. Then, there are $k = 4\alpha + 6\beta$ terminal pairs, and we set $\ell = 5\alpha + 9\beta$. In this way, the corresponding instance $(G_\phi, \mathcal{P}, \mathcal{Q}, \ell)$ of the SRVDSP problem can be constructed in time polynomial in α and β.

Lemma 4 (∗). *ϕ is satisfiable if and only if $(G_\phi, \mathcal{P}, \mathcal{Q}, \ell)$ is a yes-instance.*

5 Conclusion

In this paper, we introduced two reconfiguration problems, the RVDSP and SRVDSP problems, as generalizations of the well-studied SPR problem. We studied the computational complexity of the RVDSP and SRVDSP problems from the viewpoint of graph classes, and gave some interesting contrast. (Recall Fig. 2 and the numbered list at the end of the introduction.)

It remains open to clarify the complexity status of the RVDSP problem for chordal graphs, and for planar graphs. Note that the SPR problem (i.e., the RVDSP problem for $k = 1$) is solvable in polynomial time for chordal graphs [1], and for planar graphs [2].

Acknowledgment. We are grateful to Kai Matsudate, Tohoku University, Japan, for valuable discussions with him.

References

1. Bonsma, P.S.: The complexity of rerouting shortest paths. Theor. Comput. Sci. **510**, 1–12 (2013). https://doi.org/10.1016/j.tcs.2013.09.012
2. Bonsma, P.S.: Rerouting shortest paths in planar graphs. Discret. Appl. Math. **231**, 95–112 (2017). https://doi.org/10.1016/j.dam.2016.05.024
3. Gajjar, K., Jha, A.V., Kumar, M., Lahiri, A.: Reconfiguring shortest paths in graphs. In: Proceedings of AAAI 2022, pp. 9758–9766. AAAI Press (2022). https://ojs.aaai.org/index.php/AAAI/article/view/21211
4. Garey, M.R., Johnson, D.S.: Computers and Intractability: A Guide to the Theory of NP-Completeness. W.H. Freeman, New York (1979)
5. van den Heuvel, J.: The complexity of change. In: Surveys in Combinatorics 2013, London Mathematical Society Lecture Note Series, vol. 409, pp. 127–160. Cambridge University Press (2013). https://doi.org/10.1017/CBO9781139506748.005
6. Ito, T., et al.: On the complexity of reconfiguration problems. Theor. Comput. Sci. **412**(12–14), 1054–1065 (2011). https://doi.org/10.1016/j.tcs.2010.12.005
7. Kamiński, M., Medvedev, P., Milanič, M.: Shortest paths between shortest paths. Theor. Comput. Sci. **412**(39), 5205–5210 (2011). https://doi.org/10.1016/j.tcs.2011.05.021
8. Nishimura, N.: Introduction to reconfiguration. Algorithms **11**(4), 52 (2018). https://doi.org/10.3390/a11040052
9. Wrochna, M.: Reconfiguration in bounded bandwidth and tree-depth. J. Comput. Syst. Sci. **93**, 1–10 (2018). https://doi.org/10.1016/j.jcss.2017.11.003

k-Transmitter Watchman Routes

Bengt J. Nilsson[1] and Christiane Schmidt[2]([⊠])

[1] Department of Computer Science and Media Technology, Malmö University, Malmö, Sweden
bengt.nilsson.TS@mau.se
[2] Department of Science and Technology, Linköping University, Norrköping, Sweden
christiane.schmidt@liu.se

Abstract. We consider the watchman route problem for a k-transmitter watchman: standing at point p in a polygon P, the watchman can see $q \in P$ if $\overline{pq}$ intersects P's boundary at most k times—q is k-visible to p. Traveling along the k-transmitter watchman route, either all points in P or a discrete set of points $S \subset P$ must be k-visible to the watchman. We aim for minimizing the length of the k-transmitter watchman route.

We show that even in simple polygons the shortest k-transmitter watchman route problem for a discrete set of points $S \subset P$ is NP-complete and cannot be approximated to within a logarithmic factor (unless P=NP), both with and without a given starting point. Moreover, we present a polylogarithmic approximation for the k-transmitter watchman route problem for a given starting point and $S \subset P$ with approximation ratio $O(\log^2(|S| \cdot n) \log \log(|S| \cdot n) \log |S|)$ (with $|P| = n$).

Keywords: Watchman route · k-Transmitter · k-Transmitter watchman route · NP-Hardness · Approximation algorithm · NP-completeness

1 Introduction

In the classical *Watchman Route Problem* (WRP)—introduced by Chin and Ntafos [1], we ask for the shortest (closed) route in an environment (usually a polygon P), such that a mobile guard traveling along this route sees all points of the environment. The WRP has mostly been studied for the "traditional" definition of visibility: a point $p \in P$ sees another point $q \in P$ if the line segment $\overline{pq}$ is fully contained in P. This mimics human vision, as this, e.g., does not allow looking through obstacles or around corners. In contrast to the classical guarding problems with stationary guards, the *Art Gallery Problem* (AGP)— where we aim to place a minimum number of non-moving guards that see the complete environment—the WRP is solvable in polynomial time in simple polygons with [2–4] and without [5,6] a given boundary start point. In polygons with holes, the WRP is NP-hard [1,7].

Supported by grants 2018-04001 (Nya paradigmer för autonom obemannad flygledning) and 2021-03810 (Illuminate: bevisbart goda algoritmer för bevakningsproblem) from the Swedish Research Council (Vetenskapsrådet).

However, we may also have other vision types for the watchman. If we, for example, consider a mobile robot equipped with a laser scanner, then the scanning creates point clouds, which are easier to map afterwards when the robot was immobile while taking a single scan. This results in the model of "discrete vision": information on the environment can be acquired only at discrete points, at other times the watchman is blind. Carlsson et al. [8] showed that the problem of finding the minimum number of vision points—the discrete set of points at which the vision system is active—along a given path (e.g., the shortest watchman route) is NP-hard in simple polygons. Carlsson et al. [9] also presented an efficient algorithm to solve the problem of placing vision points along a given watchman route in streets. Another natural restriction for a mobile robot equipped with laser scanners is a limited visibility range (resolution degrades with increasing distance), see [10–12].

Another type of visibility is motivated by modems: When we try to connect to a modem, we observe that one wall will not prevent this connection (i.e., obstacles are not always a problem). However, many walls separating our location from the modem result in a failed connection. This motivates studying so-called *k*-transmitters: $p \in P$ sees $q \in P$ if the line segment $\overline{pq}$ intersects P's boundary at most k times. If more than k walls are intersected, we no longer "see" an object—the connection is not established. Different aspects of guarding with *k*-transmitters (the AGP with *k*-transmitters) have been studied. First, the focus was on worst-case bounds, so-called Art Gallery theorems. Aichholzer et al. [13] presented tight bounds on the number of *k*-transmitters in monotone and monotone orthogonal polygons. Other authors explored *k*-transmitter coverage of regions other than simple polygons, such as coverage of the plane in the presence of lines or line segment obstacles [14,15]. Ballinger et al. [14] also presented a tight bound for a very special class of polygons: spiral polygons, so-called *spirangles*. Moreover, for simple *n*-gons they provided a lower bound of $\lfloor n/6 \rfloor$ 2-transmitters. Cannon et al. [16] showed that it is NP-hard to compute a minimum cover of point 2-transmitters, point *k*-transmitters, and edge 2-transmitters (where a guard is considered to be the complete edge) in a simple polygon. The point 2-transmitter result extends to orthogonal polygons. Moreover, they gave upper and lower bounds for the number of edge 2-transmitters in general, monotone, orthogonal monotone, and orthogonal polygons; and improved the bound from [14] for simple *n*-gons to $\lfloor n/5 \rfloor$ 2-transmitters. For the AGP with *k*-transmitters, no approximation algorithms have been obtained so far, but Biedl et al. [17] recently presented a first constant-factor approximation result for so-called sliding *k*-transmitters (traveling along an axis-parallel line segment *s* in the polygon, covering all points *p* of the polygon for which the perpendicular from *p* onto *s* intersects at most *k* edges of the polygon).

Of course, *k*-transmitters do not have to be stationary (or restricted to travel along a special structure as in [17]): we might have to find a shortest tour such that a mobile *k*-transmitter traveling along this route can establish a connection with all (or a discrete subset of the) points of an environment, the *WRP with a k-transmitter*. This problem is the focus of this paper and to the best of our knowledge it has not been studied before. Given that our watchman moves inside the polygon, we consider even values for k only—while odd numbers of k can be interesting when we, for example, want to monitor the plane in presence of line-segment or line obstacles, or when we want to monitor parts of a polygon's exterior.

For the original WRP, we know that an optimal tour must visit all essential cuts: the non-dominated extensions of edges incident to a reflex vertex. However, already if we want to see a discrete set of points with a mobile k-transmitter, for $k \geq 2$, we do not have such a structure: the region visible to a k-transmitter point can have $O(n)$ connected components, to see the point, the mobile k-transmitter can visit any of these.

Guarding a discrete set of points—though with stationary guards—is, e.g., considered in the problem of guarding treasures in an art gallery: Deneen and Joshi [18] presented an approximation algorithm for finding the minimum number of guards that monitors a discrete set of treasure points, Carlsson and Jonsson [19] added weights to the treasures and aimed for placing a single guard maximizing the total value of the guarded treasures.

Roadmap. In Sect. 2, we introduce notation; in Sect. 3, we detail some special properties of k-transmitters. In Sect. 4, we show that the WRP with k-transmitters monitoring a discrete set of points is NP-complete and cannot be approximated to within a logarithmic factor in simple polygons even for $k = 2$. In Sect. 5, we present an approximation algorithm for the WRP with k-transmitters monitoring a discrete set of points and has a given starting point.

2   Notation and Preliminaries

We let P be a polygon, in general, we are interested in P being simple. We define $\partial(P)$ as the boundary of P, and let n denote the number of vertices of P.

A point $q \in P$ is k-*visible* to a k-transmitter $p \in \mathbb{R}^2$ if $\overline{qp}$ intersects P's boundary in at most k connected components. This includes "normal" guards for $k = 0$. For a point $p \in P$, we define the k-*visibility region* of p, $k\mathrm{VR}(p)$, as the set of points in P that are k-visible from p, see Fig. 1(a). For a set $X \subseteq P$: $k\mathrm{VR}(X) = \bigcup_{p \in X} k\mathrm{VR}(p)$. A k-visibility region can have $O(n)$ connected components (CCs), see [16] and Observation 2 in Sect. 3, we denote these components by $k\mathrm{VR}^j(p), j = 1, \ldots, J_p$, with $J_p \in O(n)$.

The boundary of each CC of $k\mathrm{VR}(p)$ contains edges that coincide with (parts of) edges of $\partial(P)$, and so-called *windows*. Cutting P along a window w partitions it into two subpolygons. We denote by $P_s(w)$ the subpolygon that contains a given point $s \notin w$ and consider the window w to belong to $\partial(P_s(w))$. A window w_1 *dominates* another window w_2 if $P_s(w_2) \subset P_s(w_1)$. A window w is *essential* if it is not dominated by any other window.

For a given point s in a simple polygon P there exists one window w per CC $k\mathrm{VR}^j(p)$, such that any path from s to a window $w' \neq w, w' \in k\mathrm{VR}^j(p)$ intersects w, that is, $P_s(w) \subseteq P_s(w')$. We denote this window as the *cut* of $k\mathrm{VR}^j(p)$ w.r.t. s, see Fig. 1(a).

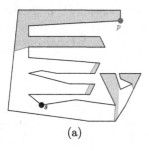

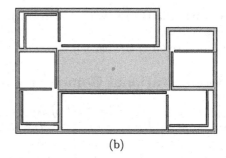

(a)   (b)

Fig. 1. (a): Point p with its 2-visibility region shown in light blue, $2\mathrm{VR}(p)$ has five CCs. The cuts of these CCs w.r.t. s are shown in red. (b): The complete boundary of this polygon is visible from the pink 2-transmitter watchman route. In particular, this holds for the red part of the polygon's boundary (seen, e.g., from the four marked pink points). However, the turquoise point is not 2-visible from that route. Thus, not all of P is 2-visible from that route. (Color figure online)

We aim to find shortest watchman routes for k-transmitters. In particular, we aim to find a route R, such that either all points of a polygon P or a set of points $S \subset P$ is k-visible for the watchman following R, that is, $k\mathrm{VR}(R) = P$ or $S \subset k\mathrm{VR}(R)$. We define the *$k$-Transmitter WRP for $X \subseteq P$* and P, k-TrWRP(X, P), possibly with a given starting point $s \in P$, k-TrWRP(X, P, s), as the problem of finding the shortest route for a k-transmitter within P, starting at s, from which every point in X is k-visible. Let OPT(S, P) and OPT(S, P, s) be *optimal* w.r.t. k-TrWRP(S, P) and k-TrWRP(S, P, s), respectively.

In Sect. 5, we use an approximation algorithm by Garg et al. [20] for the group Steiner tree problem. The *group Steiner tree problem* was introduced by Reich and Widmayer [21]: given a graph $G = (V, E)$ with cost function $c : E \to \mathbb{R}^+$ and subsets of vertices $\gamma_1, \gamma_2, \dots, \gamma_Q \subseteq V$, so-called *groups*, we aim to find the minimum-cost subtree T of G that contains at least one vertex from each of the groups, that is, a connected subgraph $T = (V', E'), V' \subseteq V, E' \subseteq E$ that minimizes $\sum_{e \in E'} c_e$ such that $V' \cap \gamma_q \neq \emptyset$, $\forall q \in \{1, \dots, Q\}$. For $|V| = m$, Garg et al. [20] obtained a randomized algorithm with an approximation ratio of $O(\log^2 m \log \log m \log Q)$.

3 Special Observations for k-Transmitters

For 0-transmitter watchmen guarding a simple polygon's boundary is enough to guard all of the polygon, this does not hold for k-transmitters with $k \geq 2$:

Observation 1. *For a simple polygon P and $k \geq 2$: $\partial(P)$ being k-visible to a k-transmitter watchman route is not a sufficient condition for P being k-visible to that k-transmitter watchman route, see Fig. 1(b).*

The visibility region $0\mathrm{VR}(p)$ for any point $p \in P$ has a single connected component (and is also a simple polygon). This does not hold for larger k, as already for $k = 2$ we have:

Observation 2 (Observation 1 in [16]). *In a simple polygon P, the 2-visibility region of a single guard can have $O(n)$ connected components. [More precise: The 2-visibility region of a single guard can have at most n connected components.]*

4   Computational Complexity

Theorem 1. *For a discrete set of points $S \subset P$ and a simple polygon P, the k-Transmitter WRP for S and P, k-TrWRP(S, P), does not admit a polynomial-time approximation algorithm with approximation ratio $\alpha \cdot \ln |S|$ for a constant $\alpha > 0$ unless P=NP, even for $k = 2$.*

Proof. We give a gap-preserving reduction from Set Cover (SC):
Set Cover (SC):
Input: A set system $(\mathcal{U}, \mathcal{C})$, with $\cup_{C \in \mathcal{C}} C = \mathcal{U}$.
Output: Minimum cardinality sub-family $\mathcal{B} \subseteq \mathcal{C}$ that covers $\mathcal{U}$, i.e., $\cup_{B \in \mathcal{B}} B = \mathcal{U}$.

Given an instance of the Set Cover problem, we construct a polygon P with $S = \mathcal{U} \cup \{v\}$. For that construction, we build a bipartite graph G with vertex set $V(G) = \mathcal{U} \cup \mathcal{C}$ and edge set $E(G) = \{e = (u, c) \mid u \in \mathcal{U}, c \in \mathcal{C}, u \in c\}$. See Fig. 2(a) for an example of this graph G.

We start the construction of P, see Fig. 2(b), with a spiral structure with $v \in S$ located in the center of the spiral, to its end we attach $|\mathcal{C}|$ spikes (narrow polygonal corridors of four vertices each), each ending at the same y-coordinate (each $C \in \mathcal{C}$ corresponds to the tip of one spike). Let the length of the longest spike be $\ell_{\mathcal{C}}$, and let the length of the spikes differ by $\varepsilon' \ll \ell_{\mathcal{C}}$ only. All points $u \in \mathcal{U}$ are located in a long horizontal box to which T-shaped structures are attached, such that the crossbeams leave gaps only where an edge in $E(G)$ connects a C from a spike to a u in the horizontal box. These two polygon parts are connected by a very long vertical polygonal corridor. Let the length of this corridor be $\ell_{\text{vert}} = 4 \cdot |\mathcal{C}| \cdot \ell_{\mathcal{C}}$.

Because of the placement of v, any 2-transmitter needs to enter the spiral structure to reach a point in 2VR(v) (indicated in light green in Fig. 2(c)). All $u \in \mathcal{U}$ are visible only to points in the horizontal box, in the T-shaped structures, to points at the bottom of the long vertical corridor (shown in light pink in Fig. 2(d)) and from the tips of the spikes (representing the $C \in \mathcal{C}$). Covering the vertical corridor twice to reach any of the light pink points from 2VR(v) is more expensive than even visiting all tips of the spikes. Hence, any optimal k-transmitter watchman route must visit spike tips to see all $u \in \mathcal{U}$. To obtain the shortest k-transmitter watchman route we must visit as few spike tips as possible: we must visit the minimum number of spike tips, such that all pink points $u \in \mathcal{U}$ are covered. This is exactly the solution to the Set Cover problem.

Set Cover cannot be approximated in polynomial time to within a factor $(1 - o(1)) \ln |\mathcal{U}|$, where $|S| = |\mathcal{U}| + 1$; [22].

For each $s_i \in S$, we can compute the 2-visibility region of s_i and check whether the given route intersects it, thus, k-TrWRP(S, P) is in NP.

By choosing s to be located on the window of 2VR(v) in the above construction and using $S = \mathcal{U}$, we obtain:

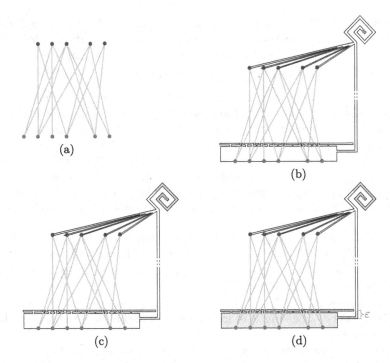

Fig. 2. Example construction for the SC instance $(\mathcal{U}, \mathcal{C})$ with $\mathcal{U} = \{1, 2, 3, 4, 5, 6\}$, $\mathcal{C} = \{\{2, 4\}, \{1, 3, 5\}, \{1, 2, 5, 6\}, \{2, 4, 6\}, \{4, 5\}\}$. (a) Graph G, (b) polygon P with ϵ shown in green, (c) $k\text{VR}(v)$ shown in light green, (d) set of points that are not located in the $|\mathcal{C}|$ spikes and see all $u \in \mathcal{U}$ (all points in the horizontal box, in the T-shaped structures, and points at the bottom of the long vertical corridor). (Color figure online)

Corollary 1. *For S, P, and α as in Theorem 1, $s \in P$, k-TrWRP(S, P, s) does not admit a polynomial-time approximation algorithm with approximation ratio $\alpha \cdot \ln|S|$, for $k \geq 2$.*

We can generalize the construction by replacing the elements of U in Fig. 2 with small almost horizontal spikes that need to be covered by the tour visiting the minimum number of set spikes at the top of the figures. We claim:

Corollary 2. *For a simple polygon P and $s \in P$, k-TrWRP(P, P, s) does not admit a polynomial-time approximation algorithm with approximation ratio $\alpha \cdot \ln n$ for a constant $\alpha > 0$, for $k \geq 4$.*

5   Approximation Algorithm for k-TrWRP(S, P, s)

In this section, we develop an approximation algorithm ALG(S, P, s) for the k-transmitter watchman route problem for a simple polygon P, a discrete set S of points in P, and a given starting point s. We prove:

Theorem 2. *Let P be a simple polygon, $n = |P|$. Let $\mathrm{OPT}(S, P, s)$ be the optimal solution for the k-$\mathrm{TrWRP}(S, P, s)$ and let R be the solution output by our algorithm $\mathrm{ALG}(S, P, s)$. Then R has length within $O\big(\log^2(|S| \cdot n) \log\log(|S| \cdot n) \log |S|\big)$ of $\mathrm{OPT}(S, P, s)$.*

The basic idea of our approximation algorithm is to create a candidate point for each connected component of the k-visibility region of each point in S. These candidate points are defined by the intersections of geodesics from the starting point s to the cuts and the cuts themselves. We then build a complete graph on these candidate points (using the length of geodesics in P between two points as the edge length in the graph). Finally, we group all candidate points that belong to the same point in S and construct a group Steiner tree; by doubling this tree, we obtain a route. Our approximation algorithm performs the following steps:

1. For each $s_i \in S$, we compute the k-visibility region within P, $k\mathrm{VR}(s_i)$—and say that all CCs $k\mathrm{VR}^j(s_i)$ have "color" s_i. We denote $k\mathrm{VR}^j(s_i)$ as k_i^j. Let the cut of each k_i^j be denoted by $c_{i,j}$, and let $\mathcal{C}^{all}$ denote the set of all cuts in P. See Fig. 3(a) for an example of this step.
2. We compute a geodesic $g_{i,j}$ from s to each cut $c_{i,j}$. Let $p_{i,j}$ be the point where $g_{i,j}$ intersects $c_{i,j}$. See Fig. 3(b) for an example of this step, the $p_{i,j}$ are shown in light green.
3. We build the complete graph on the $p_{i,j}$ and s: for an edge $\{x, y\}$, we have $\mathrm{cost}(\{x, y\}) = \mathrm{geodesic}_P(x, y)$. We introduce further vertices and edges: one vertex $\hat{c}_{i,j}$ per cut $c_{i,j} \in \mathcal{C}^{all}$. We add edges $\{p_{i,j}, \hat{c}_{i,j}\}$ with edge cost 0, and edges $\{p_{i,j}, \hat{c}_{i',j'}\}$ with edge cost 0 for all cuts $c_{i',j'}$ that $g_{i,j}$ intersects. (Rationale: any path or tour visiting $p_{i,j}$ must visit $c_{i',j'}/c_{i,j}$.) Let the resulting graph be denoted as $G = (V, E)$. See Fig. 3(c)/(d): the points of type $\hat{c}_{i,j}$ are shown in green. We have $|V(G)| = O(n \cdot |S|)$.
4. With $\gamma_i = \bigcup_{j=1}^{J_i} p_{i,j} \cup \bigcup_{j=1}^{J_i} \hat{c}_{i,j}$, $\gamma_0 = s$, $Q = |S| + 1$—that is, each group γ_i contains all vertices in $V(G)$ of color s_i, γ_0 contains the starting point that we must visit—we approximate the group Steiner tree problem on G, using the approximation by Garg et al. [20], the approximation ratio is $O\big(f(|V(G)|, |S|)\big)$, where $f(N, M) = \log^2 N \log\log N \log M$, e.g., polylogarithmic in $|V(G)|$ and $|S|$.
5. We double the resulting tree to obtain a route R; it visits at least one vertex per color (one point in each $k\mathrm{VR}(s_i)$). Thus, R is a feasible solution for k-$\mathrm{TrWRP}(S, P, s)$ visiting one point per γ_i. R is a polylog-approximation to the best tour that is feasible for k-$\mathrm{TrWRP}(S, P, s)$, visiting one point per γ_i using edges in G (denoted by $\mathrm{OPT}_G(S, P, s)$).

To prove that R is indeed an approximation with the claimed approximation factor, we alter the optimum k-transmitter watchman route, $\mathrm{OPT}(S, P, s)$, (which we of course do not know in reality) to pass points that represent vertices of $V(G)$, and show that this new tour is at most 3 times as long as the optimum route. The visited points are intersection points of independent geodesics and cuts (we obtain independent geodesics by ordering the geodesics to essential cuts

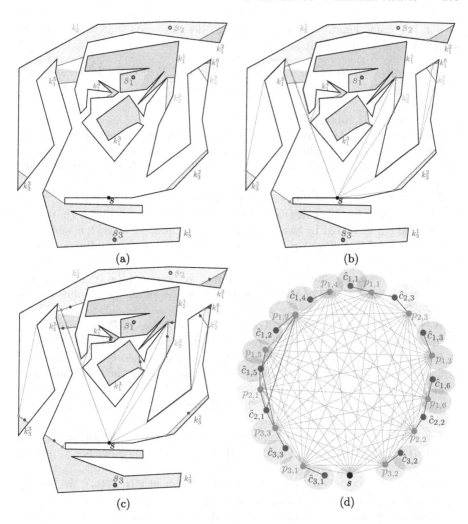

Fig. 3. Example for the idea of our approximation algorithm, $S = \{s_1, s_2, s_3\}$. (a)–(c): The cuts $c_{i,j}$ are shown in red, geodesics and all $p_{i,j}$ are shown in light green, all $\hat{c}_{i,j}$ are shown in dark green. The CCs of the visibility region of a point s_i are colored in a lighter shade of the same color as the point itself (orange for s_1, turquoise for s_2, and pink for s_3). Line segments are slightly offset to enhance visibility in case they coincide with polygon boundary. (d) Resulting graph G. Edges with edge cost 0 are shown in dark green, edges with edge cost of the length of the geodesic between the two points in P are shown in gray. We highlight each vertex of G with the color of the point s_i to which it belongs. All colored in the same color constitute the set γ_i (γ_1 highlighted in light orange, γ_2 highlighted in light turquoise, γ_3 highlighted in light pink, γ_0 highlighted in yellow). (Color figure online)

by non-increasing length and filtering out geodesics to cuts that were visited by longer geodesics). The basic idea is:

a. We identify all cuts of the $k\text{VR}(s_i)$ that $\text{OPT}(S, P, s)$ visits, let these be the set $\mathcal{C}$ ($\mathcal{C} \subseteq \mathcal{C}^{all}$). Let $o_{i,j}$ denote the point where $\text{OPT}(S, P, s)$ visits $c_{i,j}$ (for the first time).

b. We identify the subset of essential cuts $\mathcal{C}' \subseteq \mathcal{C}$.

c. We order the geodesics to the essential cuts $\mathcal{C}'$ by decreasing length: $\ell(g_1) \geq \ell(g_2) \geq \ldots \geq \ell(g_{|\mathcal{C}'|})$, where $\ell(\cdot)$ defines the Euclidean length.

d. $\mathcal{C}'' \leftarrow \mathcal{C}'$; FOR $t = 1$ TO $|\mathcal{C}'|$, we identify all $\mathcal{C}_t \subset \mathcal{C}'$ that g_t intersects, and set $\mathcal{C}'' \leftarrow \mathcal{C}'' \setminus \mathcal{C}_t$.
$\mathcal{C}'' \subseteq \mathcal{C}'$. We let $\mathcal{G}_{\mathcal{C}''}$ be the set of geodesics that end at cuts in $\mathcal{C}''$.

e. The geodesics in $\mathcal{G}_{\mathcal{C}''}$ constitute a set of *independent* geodesics, that is, no essential cut is visited by two of these geodesics. Moreover, each essential cut visited by $\text{OPT}(S, P, s)$—each cut in $\mathcal{C}'$—is touched by exactly one of the geodesics.

f. The geodesics in $\mathcal{G}_{\mathcal{C}''}$ intersect the cuts in $\mathcal{C}''$ in points of the type $p_{i,j}$, points that represent vertices of $V(G)$. We denote the set of all these points as $\mathcal{P}_{\mathcal{C}''}$ ($\mathcal{P}_{\mathcal{C}''} \subseteq \{p_{i,j} \mid i = 1, \ldots, |S|, j = 1, \ldots, J_i\}$).

g. We build the relative convex hull of all $o_{i,j}$ and all points in $\mathcal{P}_{\mathcal{C}''}$ (relative w.r.t. the polygon P). We denote this relative convex hull by $\text{CH}_P(\text{OPT}, \mathcal{P}_{\mathcal{C}''})$.

h. Because we have a set of independent geodesics, no geodesic can intersect $\text{CH}_P(\text{OPT}, \mathcal{P}_{\mathcal{C}''})$ between a point $o_{i,j}$ and a point $p_{i,j}$ on the same cut. Thus, between any pair of points of the type $o_{i,j}$ on $\text{CH}_P(\text{OPT}, \mathcal{P}_{\mathcal{C}''})$, we have at most two points of $\mathcal{P}_{\mathcal{C}''}$. We show that $\text{CH}_P(\text{OPT}, \mathcal{P}_{\mathcal{C}''})$ has length of at most three times $\|\text{OPT}(S, P, s)\|$.

i. The relative convex hull of the points in $\mathcal{P}_{\mathcal{C}''}$, $\text{CH}_P(\mathcal{P}_{\mathcal{C}''})$, is not longer than $\text{CH}_P(\text{OPT}, \mathcal{P}_{\mathcal{C}''})$, and we show that $\text{CH}_P(\mathcal{P}_{\mathcal{C}''})$ visits one point per γ_i (except for γ_0).

j. Because s ($= \gamma_0$) might be located in the interior of $\text{CH}_P(\mathcal{P}_{\mathcal{C}''})$, we need to connect s to $\text{CH}_P(\mathcal{P}_{\mathcal{C}''})$. This costs at most $\|\text{OPT}(S, P, s)\|$.

k. Thus, we obtain (note $f(N, M) = \log^2 N \log \log N \log M$):

$$\|R\| \leq \alpha_1 \cdot f(|V(G)|, |S|)\|\text{OPT}_G(S, P, s)\| \leq \alpha_2 \cdot f(n|S|, |S|)\|\text{CH}_P(\mathcal{P}_{\mathcal{C}''})\|$$
$$\leq \alpha_3 \cdot f(n|S|, |S|)\|\text{CH}_P(\text{OPT}, \mathcal{P}_{\mathcal{C}''})\| \leq \alpha_4 \cdot f(n|S|, |S|)\|\text{OPT}(S, P, s)\|$$

for suitable constants $\alpha_1, \ldots, \alpha_4$.

Hence, to show Theorem 2, we need to prove steps e, h, and i. We show step e using Lemma 1; step h using Lemmas 2, 3, and 4; and step i using Lemmas 5, 6, 7, and 8. For the proofs of Lemmas 1, 4, and 6 we refer to the paper's full version.

Lemma 1. *For the geodesics in $\mathcal{G}_{\mathcal{C}''}$, we have:*

I. *The geodesics in $\mathcal{G}_{\mathcal{C}''}$ are independent, that is, no cut in $\mathcal{C}'$ is visited by two of these geodesics.*

II. *Each cut in $\mathcal{C}'$ is visited by a geodesic in $\mathcal{G}_{\mathcal{C}''}$.*

Lemma 2. *Consider a cut $c \in \mathcal{C}''$, from CC j of a k-visibility region for $s_i \in S$, $kVR^j(s_i)$, for which both the point $o_{i,j}$ and the point $p_{i,j}$ are on $\mathrm{CH}_P(\mathrm{OPT}, \mathcal{P}_{\mathcal{C}''})$. No geodesic in $\mathcal{G}_{\mathcal{C}''}$ intersects c between $o_{i,j}$ and $p_{i,j}$.*

Proof. Assume that there exists a geodesic $g_{c'} \in \mathcal{G}_{\mathcal{C}''}$ to a cut $c' \neq c, c' \in \mathcal{C}''$ that intersects c between $o_{i,j}$ and $p_{i,j}$. Let c' be the cut of $kVR^{j'}(s_{i'})$. Let p_c denote the point in which $g_{c'}$ intersects c. If $\ell(g_{c'}) > \ell(g_c)$, we would have deleted g_c in step d, hence $c \notin \mathcal{C}''$. If $\ell(g_{c'}) < \ell(g_c)$, the geodesic to c' restricted to the part between s and p_c, $g_{c'[s;p_c]}$, is shorter than g_c, a contradiction to g_c being the geodesic to c. If $\ell(g_{c'}) = \ell(g_c)$, either $\ell(g_{c'[s;p_c]}) < \ell(g_{c'}) = \ell(g_c)$ or (if p_c on c') $p_{i,j} = p_c$ and the claim holds.

Lemma 3. *Between any pair of points of the type $o_{i,j}$ on $\mathrm{CH}_P(\mathrm{OPT}, \mathcal{P}_{\mathcal{C}''})$, we have at most two points in $\mathcal{P}_{\mathcal{C}''}$.*

Proof. Let $o_{i,j}$ and $o_{i',j'}$ be two consecutive points from OPT on $\mathrm{CH}_P(\mathrm{OPT}, \mathcal{P}_{\mathcal{C}''})$, see Fig. 4 for an example of this proof construction. By Lemma 2, $p_{i,j}$ and $p_{i',j'}$ can lie between $o_{i,j}$ and $o_{i',j'}$, but we can have no point $p_{\kappa,\lambda}$ between $o_{i,j}$ and $p_{i,j}$ or between $o_{i',j'}$ and $p_{i',j'}$. Assume that there exists a point $p_{\kappa,\lambda}$ between $p_{i,j}$ and $p_{i',j'}$ on $\mathrm{CH}_P(\mathrm{OPT}, \mathcal{P}_{\mathcal{C}''})$. Moreover, let $p_{i,j}, p_{i',j'}$, and $p_{\kappa,\lambda}$ be on cuts c, c' and c'', respectively. OPT visits $o_{\kappa,\lambda}$ on c''. As $o_{i,j}$ and $o_{i',j'}$ are consecutive points from OPT on $\mathrm{CH}_P(\mathrm{OPT}, \mathcal{P}_{\mathcal{C}''})$, OPT visits the three points either in order $o_{i,j}, o_{i',j'}, o_{\kappa,\lambda}$ or $o_{\kappa,\lambda}, o_{i,j}, o_{i',j'}$. W.l.o.g., assume the order $o_{i,j}, o_{i',j'}, o_{\kappa,\lambda}$. The cut c'' is a straight-line segment. Consider the convex polygon $P_\triangle$ with vertices $o_{i,j}, p_{i,j}, p_{\kappa,\lambda}, o_{\kappa,\lambda}, o_{i',j'}, o_{i,j}$. The point $p_{i',j'}$ must lie in $P_\triangle$'s interior. Moreover, $o_{i',j'}$ cannot lie on $\mathrm{CH}_P(\mathrm{OPT}, \mathcal{P}_{\mathcal{C}''})$; a contradiction.

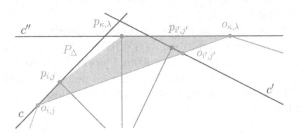

Fig. 4. Example for the proof of Lemma 3. Cuts are shown in red, points of the type $p_{i,j}$ in green, and the optimal route and points of the type $o_{i,j}$ in orange. The convex polygon $P_\triangle$ is shown in turquoise. (Color figure online)

Lemma 4. $\|\mathrm{CH}_P(OPT, \mathcal{P}_{\mathcal{C}''})\| \leq 3 \cdot \|\mathrm{OPT}(S, P, s)\|$

Lemma 5. $\|\mathrm{CH}_P(\mathcal{P}_{\mathcal{C}''})\| \leq \|\mathrm{CH}_P(\mathrm{OPT}, \mathcal{P}_{\mathcal{C}''})\|.$

Proof. We have $\mathcal{P}_{\mathcal{C}''} \subseteq \mathrm{OPT} \cup \mathcal{P}_{\mathcal{C}''}$, hence, the claim follows trivially.

Lemma 6. *All points in $\mathcal{P}_{\mathcal{C}''}$ lie on their relative convex hull $\mathrm{CH}_P(\mathcal{P}_{\mathcal{C}''})$.*

Lemma 7. $CH_P(\mathcal{P}_{\mathcal{C}''})$ *visits all cuts in* $\mathcal{C}$.

Proof. We have $\mathcal{C} = \mathcal{C}'' \cup \{\mathcal{C}' \setminus \mathcal{C}''\} \cup \{\mathcal{C} \setminus \mathcal{C}'\}$. Cuts in $\{\mathcal{C} \setminus \mathcal{C}'\}$ are dominated by cuts in $\mathcal{C}'$. Thus, any tour visiting all cuts in $\mathcal{C}'$ must visit all cuts in $\{\mathcal{C} \setminus \mathcal{C}'\}$. Cuts in $\mathcal{C}''$ are visited by Lemma 6.

Assume that there is a cut $c \in \{\mathcal{C}' \setminus \mathcal{C}''\}$ not visited by $CH_P(\mathcal{P}_{\mathcal{C}''})$. Then the cut c is not in $\mathcal{C}''$ (we filtered g_c out in step d), hence, there exists a geodesic $g_{c_{i,j}} \in \mathcal{G}_{\mathcal{C}''}$ with $\ell(g_{c_{i,j}}) \geq \ell(g_c)$ that intersects c; $g_{c_{i,j}}$ visits $c_{i,j}$ in the point $p_{i,j}$. Thus, any tour that visits both s and $p_{i,j}$ must intersect c. By Lemma 6, $CH_P(\mathcal{P}_{\mathcal{C}''})$ is such a tour; a contradiction.

Lemma 8. $CH_P(\mathcal{P}_{\mathcal{C}''})$ *visits one point per* γ_i, *except for* γ_0.

Proof. Because $OPT(S, P, s)$ is feasible, the set $\mathcal{C}$ does include at least one cut colored in $s_i, \forall i$. By Lemma 7, $CH_P(\mathcal{P}_{\mathcal{C}''})$ visits all cuts in $\mathcal{C}$. Hence, it visits at least one point per γ_i.

This concludes the proof of Theorem 2.

6   Conclusion

We proved that even in simple polygons the k-transmitter watchman route problem for $S \subset P$ cannot be approximated to within a logarithmic factor (unless $P = NP$)—both the variant with a given starting point and the floating watchman route. Moreover, we provided an approximation algorithm for k-TrWRP(S, P, s), that is, the variant where we need to see a discrete set of points $S \subset P$ with a given starting point. The approximation ratio of our algorithm is $O(\log^2(|S| \cdot n) \log \log(|S| \cdot n) \log |S|)$.

Obvious open questions concern approximation algorithms for the other versions of the WRP for k-transmitters: k-TrWRP(P, P, s), k-TrWRP(P, P), and k-TrWRP(S, P). Moreover, for 0-transmitters ("normal" guards), we have a very clear structure: any watchman route must visit all non-dominated extensions of edges incident to a reflex vertex. Any structural analogue for k-transmitters ($k \geq 2$) would be of great interest.

References

1. Chin, W.-P., Ntafos, S.: Optimum watchman routes. In: Proceedings of the Second Annual Symposium on Computational Geometry, SCG 1986, pp. 24–33. ACM, New York (1986). ISBN 0-89791-194-6
2. Chin, W.-P., Ntafos, S.: Shortest watchman routes in simple polygons. Discret. Comput. Geom. **6**(1), 9–31 (1991). https://doi.org/10.1007/BF02574671
3. Tan, X., Hirata, T., Inagaki, Y.: Corrigendum to "an incremental algorithm for constructing shortest watchman routes". Int. J. Comput. Geom. Appl. **9**(3), 319–323 (1999)

4. Dror, M., Efrat, A., Lubiw, A., Mitchell, J.S.B.: Touring a sequence of polygons. In: Proceedings of the 35th Annual ACM Symposium Theory Computing, pp. 473–482. ACM Press (2003)
5. Carlsson, S., Jonsson, H., Nilsson, B.J.: Finding the shortest watchman route in a simple polygon. In: Ng, K.W., Raghavan, P., Balasubramanian, N.V., Chin, F.Y.L. (eds.) ISAAC 1993. LNCS, vol. 762, pp. 58–67. Springer, Heidelberg (1993). https://doi.org/10.1007/3-540-57568-5_235
6. Tan, X.: Fast computation of shortest watchman routes in simple polygons. Inf. Process. Lett. **77**(1), 27–33 (2001)
7. Dumitrescu, A., Tóth, C.D.: Watchman tours for polygons with holes. Comput. Geom. **45**(7), 326–333 (2012). ISSN 0925-7721
8. Carlsson, S., Nilsson, B.J., Ntafos, S.C.: Optimum guard covers and *m*-watchmen routes for restricted polygons. Int. J. Comput. Geom. Appl. **3**(1), 85–105 (1993)
9. Carlsson, S., Nilsson, B.J.: Computing vision points in polygons. Algorithmica **24**(1), 50–75 (1999)
10. Bhattacharya, A., Ghosh, S.K., Sarkar, S.: Exploring an unknown polygonal environment with bounded visibility. In: Alexandrov, V.N., Dongarra, J.J., Juliano, B.A., Renner, R.S., Tan, C.J.K. (eds.) ICCS 2001. LNCS, vol. 2073, pp. 640–648. Springer, Heidelberg (2001). https://doi.org/10.1007/3-540-45545-0_74
11. Fekete, S., Mitchell, J.S.B., Schmidt, C.: Minimum covering with travel cost. J. Comb. Optim. 1–20 (2010). ISSN 1382-6905
12. Schmidt, C.: Algorithms for mobile agents with limited capabilities. Ph.D. thesis, Braunschweig Institute of Technology (2011)
13. Aichholzer, O., et al.: Modem illumination of monotone polygons. Comput. Geom.: Theory Appl. SI: in memoriam Ferran Hurtado (2018)
14. Ballinger, B., et al.: Coverage with *k*-transmitters in the presence of obstacles. In: Wu, W., Daescu, O. (eds.) COCOA 2010. LNCS, vol. 6509, pp. 1–15. Springer, Heidelberg (2010). https://doi.org/10.1007/978-3-642-17461-2_1 ISBN 978-3-642-17460-5
15. Fabila-Monroy, R, Vargas, A.R., Urrutia, J.: On modem illumination problems. In: XIII Encuentros de Geometria Computacional, Zaragoza (2009)
16. Cannon, S., Fai, T.G., Iwerks, J., Leopold, U., Schmidt, C.: Combinatorics and complexity of guarding polygons with edge and point 2-transmitters. Comput. Geom.: Theory Appl. SI: in memory of Ferran Hurtado **68**, 89–100 (2018). ISSN 0925-7721
17. Biedl, T., et al.: Guarding orthogonal art galleries with sliding k-transmitters: hardness and approximation. Algorithmica **81**(1), 69–97 (2019). ISBN 1432-0541
18. Deneen, L.L., Joshi, S.: Treasures in an art gallery. In: Proceedings of the 4th Canadian Conference on Computational Geometry, pp. 17–22 (1992)
19. Carlsson, S., Jonsson, H.: Guarding a treasury. In: Proceedings of the 5th Canadian Conference on Computational Geometry, pp. 85–90 (1993)
20. Garg, N., Konjevod, G., Ravi, R.: A polylogarithmic approximation algorithm for the group Steiner tree problem. J. Algorithms **37**(1), 66–84 (2000)
21. Reich, G., Widmayer, P.: Beyond Steiner's problem: a VLSI oriented generalization. In: Nagl, M. (ed.) WG 1989. LNCS, vol. 411, pp. 196–210. Springer, Heidelberg (1990). https://doi.org/10.1007/3-540-52292-1_14
22. Feige, U.: A threshold of ln n for approximating set cover. J. ACM **45**, 314–318 (1998)

Graph Algorithm

Splitting Plane Graphs to Outerplanarity

Martin Gronemann, Martin Nöllenburg, and Anaïs Villedieu[(✉)]

Algorithms and Complexity Group, TU Wien, Vienna, Austria
{mgronemann,noellenburg,avilledieu}@ac.tuwien.ac.at

Abstract. Vertex splitting replaces a vertex by two copies and partitions its incident edges amongst the copies. This problem has been studied as a graph editing operation to achieve desired properties with as few splits as possible, most often planarity, for which the problem is NP-hard. Here we study how to minimize the number of splits to turn a plane graph into an outerplane one. We tackle this problem by establishing a direct connection between splitting a plane graph to outerplanarity, finding a connected face cover, and finding a feedback vertex set in its dual. We prove NP-completeness for plane biconnected graphs, while we show that a polynomial-time algorithm exists for maximal planar graphs. Finally, we provide upper and lower bounds for certain families of maximal planar graphs.

Keywords: Vertex splitting · Outerplanarity · Feedback vertex set

1 Introduction

Graph editing problems are fundamental problems in graph theory. They define a set of basic operations on a graph G and ask for the minimum number of these operations necessary in order to turn G into a graph of a desired target graph class $\mathcal{G}$ [24,29,34,42]. For instance, in the Cluster Editing problem [38] the operations are insertions or deletions of individual edges and the target graph class are cluster graphs, i.e., unions of vertex-disjoint cliques. In graph drawing, a particularly interesting graph class are planar graphs, for which several related graph editing problems have been studied, e.g., how many vertex deletions are needed to turn an arbitrary graph into a planar one [32] or how many vertex splits are needed to obtain a planar graph [16,23]. In this paper, we are interested in the latter operation: vertex splitting. A *vertex split* creates two copies of a vertex v, distributes its edges among these two copies and then deletes v from G.

Further, we are translating the graph editing problem into a more geometric or topological drawing editing problem. This means that we apply the splitting operations not to the vertices of an abstract graph, but to the vertices of a planar graph drawing, or more generally to a planar embedded (or *plane*) graph. In a plane graph, each vertex has an induced cyclic order of incident edges, which needs to be respected by any vertex split in the sense that we must split its

Anaïs Villedieu is supported by the Austrian Science Fund (FWF) under grant P31119.

C.-C. Lin et al. (Eds.): WALCOM 2023, LNCS 13973, pp. 217–228, 2023.
https://doi.org/10.1007/978-3-031-27051-2_19

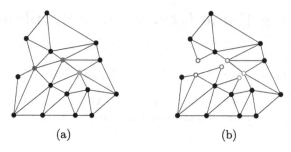

Fig. 1. (a) An instance of OUTERPLANE SPLITTING NUMBER, where the colored vertices will be split; (b) resulting outerplane graph after the minimum 3 splits.

cyclic order into two contiguous intervals, one for each of the two copies. From a different perspective, the two faces that serve as the separators of these two edge intervals are actually merged into a single face by the vertex split.

Finally, we consider outerplanar graphs as the target graph class. Thus, we want to apply a minimum number of vertex splits to a plane graph G, which merge a minimum number of faces in order to obtain an outerplanar embedded graph G', where all vertices are incident to a single face, called the *outer face*. We denote this minimum number of splits as the *outerplane splitting number* osn(G) of G (see Fig. 1). Outerplanar graphs are a prominent graph class in graph drawing (see, e.g., [7,17,27,28]) as well as in graph theory and graph algorithms more generally (e.g., [10,18,31]). For instance, outerplanar graphs admit planar circular layouts or 1-page book embeddings [5]. Additionally, outerplanar graphs often serve as a simpler subclass of planar graphs with good algorithmic properties. For instance, they have treewidth 2 and their generalizations to k-outerplanar graphs still have bounded treewidth [6,9], which allows for polynomial-time algorithms for NP-complete problems that are tractable for such bounded-treewidth graphs. This, in turn, can be used to obtain a PTAS for these problems on planar graphs [4].

We are now ready to define our main computational problem as follows.

Problem 1 (OUTERPLANE SPLITTING NUMBER). Given a plane biconnected graph $G = (V, E)$ and an integer k, can we transform G into an outerplane graph G' by applying at most k vertex splits to G?

Contributions. In this paper, we introduce the above problem OUTERPLANE SPLITTING NUMBER. WE START BY SHOWING the key property for our subsequent results, namely that (minimum) sets of vertex splits to turn a plane biconnected graph G into an outerplane one correspond to (minimum) connected face covers in G (Sect. 2), which in turn are equivalent to (minimum) feedback vertex sets in the dual graph of G. Using this tool we then show that for general plane biconnected graphs OUTERPLANE SPLITTING NUMBER is NP-complete (Sect. 3), whereas for maximal planar graphs we can solve it in polynomial time (Sect. 4).

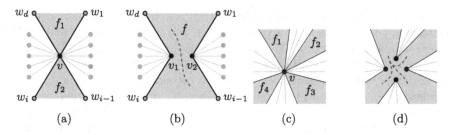

Fig. 2. (a) Two touching faces f_1, f_2 with a common vertex v on their boundary. (b) Result of the split of v with respect to f_1, f_2 joining them into a new face f. (c-d) Merging 4 faces $f_1, \ldots, f_4$ covering a single vertex v with 3 splits.

Finally, we provide upper and lower bounds on the outerplane splitting number for maximal planar graphs (Sect. 5).

Related Work. Splitting numbers have been studied mostly for abstract (non-planar) graphs with the goal of turning them into planar graphs. The PLANAR SPLITTING NUMBER problem is NP-complete in general [16], but exact splitting numbers are known for complete and complete bipartite graphs [21,23], as well as for the 4-cube [15]. For two-layer drawings of general bipartite graphs, the problem is still NP-complete, but FPT [2] when parametrized by the number of split vertices. It has also been studied for other surfaces such as the torus [19] and the projective plane [20]. Another related concept is the split thickness of a graph G (or its folded covering number [26]), which is the smallest k such that G can be transformed into a planar graph by applying at most k splits per vertex. Recognizing graphs with split thickness 2 is NP-hard, but there is a constant-factor approximation algorithm and a fixed-parameter algorithm for graphs of bounded treewidth [14]. Recently, the complexity of the embedded splitting number problem of transforming non-planar graph drawings into plane ones has been investigated [35]. Beyond the theoretical investigations of splitting numbers and planarity, there are also applied work in graph drawing making use of vertex splitting to untangle edges [41] or to improve layout quality for community exploration [3,22].

Regarding vertex splitting for achieving graph properties other than planarity, Trotter and Harary [39] studied vertex splitting to turn a graph into an interval graph. Paik et al. [36] considered vertex splitting to remove long paths in directed acyclic graphs and Abu-Khzam et al. [1] studied heuristics using vertex splitting for a cluster editing problem.

Preliminaries. The key concept of our approach is to merge a set of faces of a given plane graph $G = (V, E)$ with vertex set $V = V(G)$ and edge set $E = E(G)$ into one big face which is incident to all vertices of G. Hence, the result is outerplanar. The idea is that if two faces f_1 and f_2 share a vertex v on their boundary (we say f_1 and f_2 *touch*, see Fig. 2a), then we can split v into two new vertices v_1, v_2. In this way, we are able to create a narrow gap, which merges

f_1, f_2 into a bigger face f (see Fig. 2b). With this in mind, we formally define an *embedding-preserving split* of a vertex v w.r.t. two incident faces f_1 and f_2. We construct a new plane graph $G' = (V', E')$ with $V' = V \setminus \{v\} \cup \{v_1, v_2\}$. Consider the two neighbors of v both incident to f_1 and let w_1 be the second neighbor in clockwise order. Similarly, let w_i be the second vertex adjacent to v and incident to f_2. We call w_d the vertex preceding w_1 in the cyclic ordering or the neighbors, with d being the degree of v, see Fig. 2a. Note that while $w_1 = w_{i-1}$ and $w_i = w_d$ is possible, $w_d \neq w_1$ and $w_{i-1} \neq w_i$. For the set of edges, we now set $E' = E \setminus \{(v, w_1), \ldots, (v, w_d)\} \cup \{(v_2, w_1), \ldots, (v_2, w_{i-1})\} \cup \{(v_1, w_i), \ldots, (v_1, w_d)\}$ and assume that they inherit their embedding from G. From now on we refer to this operation simply as a *split* or when f_1, f_2 are clear from the context, we may refer to *merging* the two faces at v. The vertices v_1, v_2 introduced in place of v are called *copies* of v. If a copy v_i of a vertex v is split again, then any copy of v_i is also called a copy of the original vertex v.

We can now reformulate the task of using as few splits as possible. Our objective is to find a set of faces S that satisfies two conditions. (1) Every vertex in G has to be on the boundary of at least one face $f \in S$, that is, the faces in S *cover* all vertices in V.[1] And (2) for every two faces $f, f' \in S$ there exists a set of faces $\{f_1, \ldots, f_k\} \subseteq S$ such that $f = f_1, \ldots, f_k = f'$, and f_i touches f_{i+1} for $1 \leq i < k$. In other words, S is connected in terms of touching faces. We now introduce the main tool in our constructions that formalizes this concept.

2   Face-Vertex Incidence Graph

Let $G = (V, E)$ be a plane biconnected graph and F its set of faces. The *face-vertex incidence graph* is defined as $H = (V \cup F, E_H)$ and contains the edges $E_H = \{(v, f) \in V \times F : v$ is on the boundary of $f\}$. Graph H is by construction bipartite and we assume that it is plane by placing each vertex $f \in F$ into its corresponding face in G.

Definition 1. *Let G be a plane biconnected graph, let F be the set of faces of G, and let H be its face-vertex incidence graph. A* face cover *of G is a set $S \subseteq F$ of faces such that every vertex $v \in V$ is incident to at least one face in S. A face cover S of G is a* connected face cover *if the induced subgraph $H[S \cup V]$ of $S \cup V$ in H is connected.*

We point out that the problem of finding a connected face cover is not equivalent to the Connected Face Hitting Set Problem [37], where a connected set of vertices incident to every face is computed. We continue with two lemmas that are concerned with merging multiple faces at the same vertex (Fig. 2c).

Lemma 1. *Let G be a plane biconnected graph and $S \subseteq F$ a subset of the faces F of G that all have the vertex $v \in V$ on their boundary. Then $|S| - 1$ splits are sufficient to merge the faces of S into one.*

---

[1] Testing whether such S with $|S| \leq k$ exists, is the NP-complete problem FACE COVER [8].

Proof. Let $f_1, \ldots, f_k$ with $k = |S|$ be the faces of S in the clockwise order as they appear around v (f_1 chosen arbitrarily). We iteratively merge f_1 with f_i for $2 \leq i \leq k$, which requires in total $|S| - 1$ splits (see Fig. 2c and Fig. 2d).   □

Lemma 2. *Let G be a plane biconnected graph and let S be a connected face cover of G. Then $|S| - 1$ splits are sufficient to merge the faces of S into one.*

Proof. Let $H' = H[S \cup V]$ and compute a spanning tree T in H'. For every vertex $v \in V(T) \cap V(G)$, we apply Lemma 1 with the face set $F'(v) = \{f \in S \cap V(T) \mid (v, f) \in E(T)\}$. We root the tree at an arbitrary face $f' \in S$, which provides a hierarchy on the vertices and faces in T. Every vertex $v \in V(T) \cap V(G)$ requires by Lemma 1 $|F'(v)| - 1$ splits. Note that that for all leaf vertices in T, $|F'(v)| = 1$, i.e., they will not be split. Each split is charged to the children of v in T. Since H is bipartite, so is T. It follows that every face $f \in S \setminus \{f'\}$ is charged exactly once by its parent, thus $|S| - 1$ splits suffice.   □

Lemma 3. *Let G be a plane biconnected graph and σ a sequence of k splits to make G outerplane. Then G has a connected face cover of size $k + 1$.*

Proof. Since by definition applying σ to G creates a single big face that is incident to all vertices in $V(G)$ by iteratively merging pairs of original faces defining a set $S \subseteq F$, it is clear that S is a face cover of G and since the result of the vertex splits and face merges creates a single face, set S must also be connected.   □

As a consequence of Lemmas 2 and 3 we obtain that OUTERPLANE SPLITTING NUMBER and computing a minimum connected face cover are equivalent.

Theorem 1. *Let G be a plane biconnected graph. Then G has outerplane splitting number k if and only if it has a connected face cover of size $k + 1$.*

3   NP-Completeness

In this section, we prove that finding a connected face cover of size k (and thus OUTERPLANE SPLITTING NUMBER) is NP-complete. The idea is to take the dual of a planar biconnected VERTEX COVER instance and subdivide every edge once (we call this an *all-1-subdivision*). Note that the all-1-subdivision of a graph G corresponds to its vertex-edge incidence graph and the all-1-subdivision of the dual of G corresponds to the face-edge incidence graph of G. A connected face cover then corresponds to a vertex cover in the original graph, and vice versa. The following property greatly simplifies the arguments regarding Definition 1.

Property 1. *Let G' be an all-1-subdivision of a biconnected planar graph G and S a set of faces that cover $V(G')$. Then S is a connected face cover of G'.*

Proof. Let H be the all-1-subdivision of the dual of G, and assume to the contrary that the induced subgraph $H' = H[S \cup V(G)]$ is not connected. Then there exists an edge $(u, v) \in E(G)$ such that u and v are in different connected components in H'. Let w be the subdivision vertex of (u, v) in G'. As a subdivision

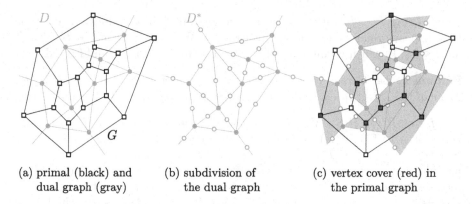

(a) primal (black) and   (b) subdivision of   (c) vertex cover (red) in
    dual graph (gray)   the dual graph   the primal graph

Fig. 3. Link between the primal graph G, its vertex cover, the dual D and its subdivision D^*.

vertex, w is incident to only two faces, one of which, say f, must be contained in S. But f is also incident to u and v and hence u and v are in the same component of H' via face f, a contradiction. Hence H' is connected and S is a connected face cover of G'.   □

The proof of the next theorem is very similar to the reduction of Bienstock and Monma to show NP-completeness of FACE COVER [8]; due to differences in the problem definitions, such as the connectivity of the face cover and whether the input graph is plane or not, we provide the full reduction for the sake of completeness.

Theorem 2. *Deciding whether a plane biconnected graph G has a connected face cover of size at most k is* NP-*complete.*

Proof. Clearly the problem is in NP. To prove hardness, we first introduce some notation. Let G be a plane biconnected graph and D the corresponding dual graph. Furthermore, let D^* be the all-1-subdivision of D. We prove now that a connected face cover S^* of size k in D^* is in a one-to-one correspondence with a vertex cover S of size k in G (see Fig. 3). More specifically, we show that the dual vertices of the faces of S^* that form a connected face cover in D^*, are a vertex cover for G and vice versa. The reduction is from the NP-complete VERTEX COVER problem in biconnected planar graphs in which all vertices have degree 3 (cubic graphs) [33].

CONNECTED FACE COVER $\Rightarrow$ VERTEX COVER : Let G be such a biconnected plane VERTEX COVER instance. Assume we have a connected face cover S^* with $|S^*| = k$ for D^*. Note that the faces of D^* correspond to the vertices in G. We claim that the faces S^*, when mapped to the corresponding vertices $S \subseteq V(G)$ are a vertex cover for G. Assume otherwise, that is, there exists an edge $e^* \in E(G)$ that has no endpoints in S. However, e^* has a dual edge $e \in E(D)$ and therefore a subdivision vertex $v_e \in V(D^*)$. Hence, there is a face

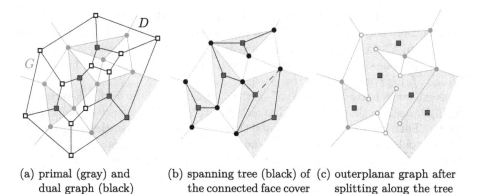

(a) primal (gray) and
dual graph (black)

(b) spanning tree (black) of
the connected face cover

(c) outerplanar graph after
splitting along the tree

Fig. 4. The connected face cover (blue) is a feedback vertex set (red) in the dual. (Color figure online)

$f \in S^*$ that has v_e on its boundary by definition of connected face cover. And when mapped to D, f has e on its boundary, which implies that the primal edge e^* has at least one endpoint in S^*; a contradiction.

VERTEX COVER $\Rightarrow$ CONNECTED FACE COVER : To prove that a vertex cover S induces a connected face cover S^* in D^*, we have to prove that S^* covers all vertices and the induced subgraph in the face-vertex incidence graph H is connected. We proceed as in the other direction. S covers all edges in $E(G)$, thus every edge $e \in E(D)$ is bounded by at least one face of S^*. Hence, every subdivision vertex in $V(D^*)$ is covered by a face of S^*. Furthermore, every vertex in D^* is adjacent to a subdivision vertex, thus, also covered by a face in S^*. Since S^* is covering all vertices, we obtain from Property 1 that S^* is a connected face cover. □

4   Feedback Vertex Set Approach

A *feedback vertex set* $S^\circ \subset V(G)$ of a graph G is a vertex subset such that the induced subgraph $G[V(G) \setminus S^\circ]$ is acyclic. We show here that finding a connected face cover S of size k for a plane biconnected graph G is equivalent to finding a feedback vertex set $S^\circ \subset V(D)$ of size k in the dual graph D of G. The *weak dual*, i.e., the dual without a vertex for the outer face, of an outerplanar graph is a forest. Thus we must find the smallest number of splits in G which transform D into a forest. In other words, we must must break all the cycles in D, and hence all of the vertices in the feedback vertex set S° of D correspond to the faces of G that should be merged together (see Fig. 4).

Property 2. *Let H be the face-vertex incidence graph of a plane biconnected graph G and let S° be a feedback vertex set in the dual D of G. Then S° induces a connected face cover S in G.*

Proof. We need to show that S° is a face cover and that it is connected. First, assume there is a vertex $v \in V(G)$ of degree $\deg(v) = d$ that is not incident to a vertex in S°, i.e., a face of G. Since G is biconnected, v is incident to d faces $f_1, \ldots, f_d$, none of which is contained in S°. But then $D[V(D) \setminus S^\circ]$ has a cycle $(f_1, \ldots, f_d)$, a contradiction.

Next, we define $\overline{S^\circ} = V(D) \setminus S^\circ$ as the complement of the feedback vertex set S° in D. Assume that $H[V \cup S^\circ]$ has at least two separate connected components C_1, C_2. Then there must exist a closed curve in the plane separating C_1 from C_2, which avoids the faces in S° and instead passes through a sequence $(f_1, \ldots, f_\ell)$ of faces in $\overline{S^\circ}$, where each pair (f_i, f_{i+1}) for $i \in \{1, \ldots, \ell - 1\}$ as well as (f_ℓ, f_1) are adjacent in the dual D. Again this implies that there is a cycle in $D[V(D) \setminus S^\circ]$, a contradiction. Thus S° is a connected face cover. $\square$

Theorem 3. *A plane biconnected graph G has outerplane splitting number k if and only if its dual D has a minimum feedback vertex set of size $k + 1$.*

Proof. Let S° be a minimum feedback vertex set of the dual D of G with cardinality $|S^\circ| = k + 1$ and let H be the face-vertex incidence graph of G. We know from Property 2 that $H' = H[V(G) \cup S^\circ]$ is connected and hence S° induces a connected face cover S with $|S| = k + 1$. Then by Lemma 2 G has $\mathrm{osn}(G) \leq k$.

Let conversely σ be a sequence of k vertex splits that turn G into an outerplane graph G' and let F be the set of faces of G. By Lemma 3 we obtain a connected face cover S of size $k + 1$ consisting of all faces that are merged by σ. The complement $\overline{S} = F \setminus S$ consists of all faces of G that are not merged by the splits in σ and thus the remaining (inner) faces of the outerplane graph G'. Since G' is outerplane and biconnected, $\overline{S}$ is the vertex set of the weak dual of G', which must be a tree. Hence S is a feedback vertex set in D of size $k + 1$ and the minimum feedback vertex set in D has size at most $k + 1$. $\square$

Since all faces in a maximal planar graph are triangles, the maximum vertex degree of its dual is 3. Thus, we can apply the polynomial-time algorithm of Ueno et al. [40] to this dual, which computes the minimum feedback vertex set in graphs of maximum degree 3 by reducing the instance to polynomial-solvable matroid parity problem instance, and obtain

Corollary 1. *We can solve* OUTERPLANE SPLITTING NUMBER *for maximal planar graphs in polynomial time.*

Many other existing results for feedback vertex set extend to OUTERPLANE SPLITTING NUMBER, e.g., it has a kernel of size $13k$ [11] and admits a PTAS [13].

5   Lower and Upper Bounds

In this section we provide some upper and lower bounds on the outerplane splitting number in certain maximal planar graphs.

5.1   Upper Bounds

Based on the equivalence of Theorem 3 we obtain upper bounds on the outer-plane splitting number from suitable upper bounds on the feedback vertex set problem, which has been studied for many graph classes, among them cubic graphs [12]. Liu and Zhao [30] showed that cubic graphs $G = (V, E)$ of girth at least four (resp., three) have a minimum feedback vertex set of size at most $\frac{|V|}{3}$ (resp., $\frac{3|V|}{8}$). Kelly and Liu [25] showed that connected planar subcubic graphs of girth at least five have a minimum feedback vertex set of size at most $\frac{2|V|+2}{7}$. Recall that the girth of a graph is the length of its shortest cycle.

Proposition 1. *The outerplane splitting number of a maximal planar graph $G = (V, E)$ of minimum degree (i) 3, (ii) 4, and (iii) 5, respectively, is at most (i) $\frac{3|V|-10}{4}$, (ii) $\frac{2|V|-7}{3}$, and (iii) $\frac{4|V|-13}{7}$, respectively.*

Proof. Maximal planar graphs with $n = |V|$ vertices have $2n - 4$ faces. So the corresponding dual graphs have $2n - 4$ vertices. Moreover, since the degree of a vertex in G corresponds to the length of a facial cycle in the dual, graphs with minimum vertex degree 3, 4, or 5 have duals with girth 3, 4, or 5, respectively. So if the minimum degree in G is 3, we obtain an upper bound on the feedback vertex set of $(3n - 6)/4$; if the minimum degree is 4, the bound is $(2n - 4)/3$; and if the minimum degree is 5, the bound is $(4n - 6)/7$. The claim then follows from Theorem 3. □

5.2   Lower Bounds

We first provide a generic lower bound for the outerplane splitting number of maximal planar graphs. Let G be an n-vertex maximal planar graph with $2n - 4$ faces. Each face is a triangle incident to three vertices. In a minimum-size connected face cover S^*, the first face covers three vertices. Due to the connectivity requirement, all other faces can add at most two newly covered vertices. Hence we need at least $\frac{n-1}{2}$ faces in any connected face cover. By Theorem 1 this implies that $osn(G) \geq \frac{n-3}{2}$.

Proposition 2. *Any maximal planar graph G has outerplane splitting number at least $\frac{|V(G)|-3}{2}$.*

   Next, towards a better bound, we define a family of maximal planar graphs $T_d = (V_d, E_d)$ of girth 3 for $d \geq 0$ that have outerplane splitting number at least $\frac{2|V_d|-8}{3}$. The family are the complete planar 3-trees of depth d, which are defined recursively as follows. The graph T_0 is the 4-clique K_4. To obtain T_d from T_{d-1} for $d \geq 1$ we subdivide each inner triangular face of T_{d-1} into three triangles by inserting a new vertex and connecting it to the three vertices on the boundary of the face.

Proposition 3. *The complete planar 3-tree T_d of depth d has outerplane splitting number at least $\frac{2|V_d|-8}{3}$.*

Proof. Each T_d is a maximal planar graph with $n_d = 3 + \sum_{i=0}^{d} 3^i = \frac{3^{d+1}+5}{2}$ vertices. All 3^d leaf-level vertices added into the triangular faces of T_{d-1} in the last step of the construction have degree 3 and are incident to three exclusive faces, i.e., there is no face that covers more than one of these leaf-level vertices. This immediately implies that any face cover of T_d, connected or not, has size at least 3^d. From $n_d = \frac{3^{d+1}+5}{2}$ we obtain $d = \log_3 \frac{2n_d-5}{3}$ and $3^d = \frac{2n_d-5}{3}$. Theorem 1 then implies that $\text{osn}(T_d) \geq \frac{2n_d-8}{3}$.   □

6   Open Problems

We have introduced the OUTERPLANE SPLITTING NUMBER problem and established its complexity for plane biconnected graphs. The most important open question revolves around the embedding requirement. Splitting operations can be defined more loosely and allow for any new embedding and neighborhood of the split vertices. In general, it is also of interest to understand how the problem differs when the input graph does not have an embedding at all, as in the original splitting number problem. Since OUTERPLANE SPLITTING NUMBER can be solved in polynomial time for maximal planar graphs but is hard for plane biconnected graphs, there is a complexity gap to be closed when faces of degree more than three are involved. Vertex splitting in graph drawings has so far been studied to achieve planarity and outerplanarity. A natural extension is to study it for other graph classes or graph properties.

References

1. Abu-Khzam, F.N., Barr, J.R., Fakhereldine, A., Shaw, P.: A greedy heuristic for cluster editing with vertex splitting. In: Proceedings of the 4th International Conference on Artificial Intelligence for Industries (AI4I), pp. 38–41. IEEE (2021). https://doi.org/10.1109/AI4I51902.2021.00017
2. Ahmed, R., Kobourov, S., Kryven, M.: An FPT algorithm for bipartite vertex splitting. In: Angelini, P., von Hanxleden, R. (eds.) GD 2022. LNCS, vol. 13764, pp. 261–268. Springer, Cham (2023). https://doi.org/10.1007/978-3-031-22203-0_19
3. Angori, L., Didimo, W., Montecchiani, F., Pagliuca, D., Tappini, A.: Hybrid graph visualizations with ChordLink: algorithms, experiments, and applications. IEEE Trans. Vis. Comput. Graph. **28**(2), 1288–1300 (2022). https://doi.org/10.1109/TVCG.2020.3016055
4. Baker, B.S.: Approximation algorithms for NP-complete problems on planar graphs. J. ACM **41**(1), 153–180 (1994). https://doi.org/10.1145/174644.174650
5. Bernhart, F., Kainen, P.C.: The book thickness of a graph. J. Comb. Theor. Ser. B **27**(3), 320–331 (1979). https://doi.org/10.1016/0095-8956(79)90021-2
6. Biedl, T.: On triangulating k-outerplanar graphs. Discrete Appl. Math. **181**, 275–279 (2015). https://doi.org/10.1016/j.dam.2014.10.017
7. Biedl, T.: Small drawings of outerplanar graphs, series-parallel graphs, and other planar graphs. Discrete Comput. Geom. **45**(1), 141–160 (2010). https://doi.org/10.1007/s00454-010-9310-z

8. Bienstock, D., Monma, C.L.: On the complexity of covering vertices by faces in a planar graph. SIAM J. Comput. **17**(1), 53–76 (1988). https://doi.org/10.1137/0217004

9. Bodlaender, H.L.: A partial k-arboretum of graphs with bounded treewidth. Theor. Comput. Sci. **209**(1–2), 1–45 (1998). https://doi.org/10.1016/S0304-3975(97)00228-4

10. Bodlaender, H.L., Fomin, F.V.: Approximation of pathwidth of outerplanar graphs. J. Algorithms **43**(2), 190–200 (2002). https://doi.org/10.1016/S0196-6774(02)00001-9

11. Bonamy, M., Kowalik, L.: A 13k-kernel for planar feedback vertex set via region decomposition. Theor. Comput. Sci. **645**, 25–40 (2016). https://doi.org/10.1016/j.tcs.2016.05.031

12. Bondy, J.A., Hopkins, G., Staton, W.: Lower bounds for induced forests in cubic graphs. Can. Math. Bull. **30**(2), 193–199 (1987). https://doi.org/10.4153/CMB-1987-028-5

13. Demaine, E., Hajiaghayi, M.: Bidimensionality: new connections between FPT algorithms and PTASs. In: Proceedings of the 37th Annual ACM Symposium on Theory of Computing (STOC), pp. 590–601 (2005). https://doi.org/10.1145/1070432.1070514

14. Eppstein, D.: On the planar split thickness of graphs. Algorithmica **80**(3), 977–994 (2017). https://doi.org/10.1007/s00453-017-0328-y

15. Faria, L., de Figueiredo, C.M.H., de Mendonça Neto, C.F.X.: The splitting number of the 4-cube. In: Lucchesi, C.L., Moura, A.V. (eds.) LATIN 1998. LNCS, vol. 1380, pp. 141–150. Springer, Heidelberg (1998). https://doi.org/10.1007/BFb0054317

16. Faria, L., de Figueiredo, C.M., de Mendonça, C.F.X.: Splitting number is NP-complete. Discrete Appl. Math. **108**(1), 65–83 (2001). https://doi.org/10.1016/S0166-218X(00)00220-1

17. Frati, F.: Planar rectilinear drawings of outerplanar graphs in linear time. Comput. Geom. **103**, 101854 (2022). https://doi.org/10.1016/j.comgeo.2021.101854

18. Frederickson, G.N.: Searching among intervals and compact routing tables. Algorithmica **15**(5), 448–466 (1996). https://doi.org/10.1007/BF01955044

19. Hartsfield, N.: The toroidal splitting number of the complete graph k_n. Discrete Math. **62**(1), 35–47 (1986). https://doi.org/10.1016/0012-365X(86)90039-7

20. Hartsfield, N.: The splitting number of the complete graph in the projective plane. Graphs Comb. **3**(1), 349–356 (1987). https://doi.org/10.1007/BF01788557

21. Hartsfield, N., Jackson, B., Ringel, G.: The splitting number of the complete graph. Graphs Comb. **1**(1), 311–329 (1985). https://doi.org/10.1007/BF02582960

22. Henry, N., Bezerianos, A., Fekete, J.: Improving the readability of clustered social networks using node duplication. IEEE Trans. Vis. Comput. Graph. **14**(6), 1317–1324 (2008). https://doi.org/10.1109/TVCG.2008.141

23. Jackson, B., Ringel, G.: The splitting number of complete bipartite graphs. Arch. Math. **42**(2), 178–184 (1984). https://doi.org/10.1007/BF01772941

24. Kant, G.: Augmenting outerplanar graphs. J. Algorithms **21**(1), 1–25 (1996). https://doi.org/10.1006/jagm.1996.0034

25. Kelly, T., Liu, C.: Minimum size of feedback vertex sets of planar graphs of girth at least five. Eur. J. Comb. **61**, 138–150 (2017). https://doi.org/10.1016/j.ejc.2016.10.009

26. Knauer, K.B., Ueckerdt, T.: Three ways to cover a graph. Discrete Math. **339**(2), 745–758 (2016). https://doi.org/10.1016/j.disc.2015.10.023

27. Lazard, S., Lenhart, W.J., Liotta, G.: On the edge-length ratio of outerplanar graphs. Theor. Comput. Sci. **770**, 88–94 (2019). https://doi.org/10.1016/j.tcs.2018.10.002

28. Lenhart, W., Liotta, G.: Proximity drawings of outerplanar graphs (extended abstract). In: North, S. (ed.) GD 1996. LNCS, vol. 1190, pp. 286–302. Springer, Heidelberg (1997). https://doi.org/10.1007/3-540-62495-3_55

29. Lewis, J.M., Yannakakis, M.: The node-deletion problem for hereditary properties is NP-complete. J. Comput. Syst. Sci. **20**(2), 219–230 (1980). https://doi.org/10.1016/0022-0000(80)90060-4

30. Liu, J., Zhao, C.: A new bound on the feedback vertex sets in cubic graphs. Discrete Math. **148**(1–3), 119–131 (1996). https://doi.org/10.1016/0012-365X(94)00268-N

31. Maheshwari, A., Zeh, N.: External memory algorithms for outerplanar graphs. In: ISAAC 1999. LNCS, vol. 1741, pp. 307–316. Springer, Heidelberg (1999). https://doi.org/10.1007/3-540-46632-0_31

32. Marx, D., Schlotter, I.: Obtaining a planar graph by vertex deletion. Algorithmica **62**(3–4), 807–822 (2012). https://doi.org/10.1007/s00453-010-9484-z

33. Mohar, B.: Face covers and the genus problem for apex graphs. J. Comb. Theor. Ser. B **82**(1), 102–117 (2001). https://doi.org/10.1006/jctb.2000.2026

34. Natanzon, A., Shamir, R., Sharan, R.: Complexity classification of some edge modification problems. Discrete Appl. Math. **113**(1), 109–128 (2001). https://doi.org/10.1016/S0166-218X(00)00391-7

35. Nöllenburg, M., Sorge, M., Terziadis, S., Villedieu, A., Wu, H.Y., Wulms, J.: Planarizing graphs and their drawings by vertex splitting. In: Angelini, P., von Hanxleden, R. (eds.) GD 2022. LNCS, vol. 13764, pp. 232–246. Springer, Cham (2023). https://doi.org/10.1007/978-3-031-22203-0_17

36. Paik, D., Reddy, S.M., Sahni, S.: Vertex splitting in dags and applications to partial scan designs and lossy circuits. Int. J. Found. Comput. Sci. **9**(4), 377–398 (1998). https://doi.org/10.1142/S0129054198000301

37. Schweitzer, P., Schweitzer, P.: Connecting face hitting sets in planar graphs. Inf. Process. Lett. **111**(1), 11–15 (2010). https://doi.org/10.1016/j.ipl.2010.10.008

38. Shamir, R., Sharan, R., Tsur, D.: Cluster graph modification problems. Discrete Appl. Math. **144**(1–2), 173–182 (2004). https://doi.org/10.1016/j.dam.2004.01.007

39. Trotter, W.T., Harary, F.: On double and multiple interval graphs. J. Graph. Theor. **3**(3), 205–211 (1979). https://doi.org/10.1002/jgt.3190030302

40. Ueno, S., Kajitani, Y., Gotoh, S.: On the nonseparating independent set problem and feedback set problem for graphs with no vertex degree exceeding three. Discrete Math. **72**(1–3), 355–360 (1988). https://doi.org/10.1016/0012-365X(88)90226-9

41. Wu, H.Y., Nöllenburg, M., Viola, I.: Multi-level area balancing of clustered graphs. IEEE Trans. Vis. Comput. Graph. **28**(7), 2682–2696 (2022). https://doi.org/10.1109/TVCG.2020.3038154

42. Yannakakis, M.: Node-and edge-deletion NP-complete problems. In: Proceedings of the 10th Annual ACM Symposium on Theory of Computing (STOC), pp. 253–264. ACM (1978). https://doi.org/10.1145/800133.804355

Certifying Induced Subgraphs
in Large Graphs

Ulrich Meyer[1], Hung Tran[1(✉)], and Konstantinos Tsakalidis[2]

[1] Goethe University Frankfurt, Frankfurt, Germany
{umeyer,htran}@ae.cs.uni-frankfurt.de
[2] University of Liverpool, Liverpool, UK
K.Tsakalidis@liverpool.ac.uk

Abstract. We introduce I/O-effiient certifying algorithms for bipartite graphs, as well as for the classes of split, threshold, bipartite chain, and trivially perfect graphs. When the input graph is a member of the respective class, the certifying algorithm returns a certificate that characterizes this class. Otherwise, it returns a forbidden induced subgraph as a certificate for non-membership. On a graph with n vertices and m edges, our algorithms take $O(\text{SORT}(n + m))$ I/Os in the worst case for split, threshold and trivially perfect graphs. In the same complexity bipartite and bipartite chain graphs can be certified with high probability. We provide implementations for split and threshold graphs and provide a preliminary experimental evaluation.

Keywords: Certifying algorithm · Graph algorithm · External memory

1 Introduction

Certifying algorithms [13] ensure the correctness of an algorithm's output without having to trust the algorithm itself. The user of a certifying algorithm inputs x and receives the output y with a *certificate* or *witness* w that proves that y is a correct output for input x. In a subsequent step, the certificate can be inspected using an authentication algorithm that considers the input, output and certificate and returns whether the output is indeed correct. Certifying the bipartiteness of a graph is a textbook example where the returned witness w is a bipartition of the vertices (YES-certificate) or an induced *odd-length cycle* subgraph, i.e. a cycle of vertices with an odd number of edges (NO-certificate).

Emerging big data applications need to process large graphs efficiently. Standard models of computation in internal memory (RAM, pointer machine) do not capture the algorithmic complexity of processing graphs with size that exceed the main memory. The *I/O model* by Aggarwal and Vitter [1] is suitable for studying large graphs stored in an external memory hierarchy, e.g. comprised of cache, RAM and hard disk memories. The input data elements are stored in *external memory* (EM) packed in *blocks* of at most B elements and computation is free

C.-C. Lin et al. (Eds.): WALCOM 2023, LNCS 13973, pp. 229–241, 2023.
https://doi.org/10.1007/978-3-031-27051-2_20

in *main memory* for at most M elements. The *I/O-complexity* is measured in *I/O-operations* (*I/Os*) that transfer a block from external to main memory and vice versa. *I/O-optimal* external memory algorithms for sorting n elements take $\text{SORT}(n) = O((n/B)\log_{M/B}(n/B))$ I/Os and reading or writing n contiguous items (which is referred to as scanning) requires $\text{SCAN}(n) = O(n/B)$ I/Os.

1.1   Previous Work

Certifying bipartiteness in internal memory takes linear time in the number of edges by any traversal of the graph. However, all known external memory breadth-first search [2] and depth-first search [4] traversal algorithms take suboptimal $\omega(\text{SORT}(n+m))$ I/Os for an input graph with n vertices and m edges.

Heggernes and Kratsch [10] present optimal internal memory algorithms for certifying whether a graph belongs to the classes of split, threshold, bipartite chain, and trivially perfect graphs. They return in linear time a YES-certificate characterizing the corresponding class or a forbidden induced subgraph of the class (NO-certificate). The YES- and NO-certificates are authenticated in linear and constant time, respectively. A straightforward application to the I/O model leads to suboptimal certifying algorithms since graph traversal algorithms in external memory are much more involved and no worst-case efficient algorithms are known.

1.2   Our Results

We present I/O-efficient certifying algorithms for *bipartite, split, threshold, bipartite chain*, and *trivially perfect* graphs. All algorithms return in the membership case, a YES-certificate w characterizing the graph class, or a $O(1)$-size NO-certificate in the non-membership case. All YES-certificates can be authenticated using $O(\text{SORT}(n+m))$ I/Os as detailed in the full version of the paper [14]. Additionally, we perform experiments for split and threshold graphs showing scaling well beyond the size of main memory.

2   Preliminaries and Notation

For a graph $G = (V, E)$, let $n = |V|$ and $m = |E|$ denote the number of vertices V and edges E, respectively. For a vertex $v \in V$ we denote by $N(v)$ the *neighborhood* of v and by $N[v] = N(v) \cup \{v\}$ the *closed neighborhood* of v. The *degree* $\deg(v)$ of a vertex v is given by $\deg(v) = |N(v)|$. A vertex v is called *simplicial* if $N(v)$ is a clique and *universal* if $N[v] = V$.

Graph Subgraphs and Orderings. The subgraph of G that is induced by a subset $A \subseteq V$ of vertices is denoted by $G[A]$. The *substructure* (subgraph) of a cycle on k vertices is denoted by C_k and of a path on k vertices is denoted by P_k. The $2K_2$ is a graph that is isomorphic to the following constant size graph: $(\{a, b, c, d\}, \{ab, cd\})$.

Henceforth we refer to different types of orderings of vertices: an ordering $(v_1, \ldots, v_n)$ is a (i) *perfect elimination ordering* (*peo*) if v_i is simplicial in

$G[\{v_i, v_{i+1}, \ldots, v_n\}]$ for all $i \in \{1, \ldots, n\}$, and a (ii) *universal-in-a-component-ordering* (*uco*) if v_i is universal in its connected component in $G[\{v_i, v_{i+1}, \ldots, v_n\}]$ for all $i \in \{1, \ldots, n\}$. For a subset $X = \{v_1, \ldots, v_k\}$, we call $(v_1, \ldots, v_k)$ a *nested neighborhood ordering* (*nno*) if $(N(v_1) \setminus X) \subseteq (N(v_2) \setminus X)) \subseteq \ldots \subseteq (N(v_k) \setminus X)$. Finally for any ordering, we partition $N(v_i)$ into lower and higher ranked neighbors, respectively, $L(v_i) = \{x \in N(v_i) : v_i$ is ranked higher than $x\}$ and $H(v_i) = \{x \in N(v_i) : v_i$ is ranked lower than $x\}$.

Graph Representation. We assume an *adjacency array representation* [15] where the graph $G = (V, E)$ is represented by two arrays $P = [\ P_i\ ]_{i=1}^n$ and $E = [\ u_i\ ]_{i=1}^m$. The neighbors of a vertex v_i are then given by the vertices at position $P[v_i]$ to $P[v_{i+1}] - 1$ in E. We use the adjacency array representation to straightforwardly allow for efficient scanning of G: (i) scanning k consecutive adjacency lists consisting of m edges requires $O(\text{SCAN}(m))$ I/Os and (ii) computing and scanning the degrees of k consecutive vertices requires $O(\text{SCAN}(k))$ I/Os.

Time-Forward Processing. *Time-forward processing* (*TFP*) is a generic technique to manage data dependencies of external memory algorithms [12]. These dependencies are typically modeled by a directed acyclic graph $G = (V, E)$ where every vertex $v_i \in V$ models the computation of z_i and an edge $(v_i, v_j) \in E$ indicates that z_i is required for the computation of z_j.

Computing a solution then requires the algorithm to traverse G according to some topological order $\prec_T$ of the vertices V. The TFP technique achieves this in the following way: after z_i has been calculated, the algorithm inserts a message $\langle v_j, z_i \rangle$ into a minimum priority-queue data structure for every successor $(v_i, v_j) \in E$ where the items are sorted by the recipients according to $\prec_T$. By construction, v_j receives all required values z_i of its predecessors $v_i \prec_T v_j$ as messages in the data structure. Since these predecessors already removed their messages from the data structure, items addressed to v_j are currently the smallest elements in the data structures and thus can be dequeued with a delete-minimum operation. By using suitable external memory priority-queues [3], TFP incurs $O(\text{SORT}(k))$ I/Os, where k is the number of messages sent.

3 Certifying Graph Classes in External Memory

3.1 Certifying Split Graphs in External Memory

A split graph is a graph that can be partitioned into two sets of vertices (K, I) where K and I induce a clique and an independent set, respectively. The partition (K, I) is called the *split partition*. They are additionally characterized by the forbidden induced subgraphs $2K_2$, C_4 and C_5, meaning that any vertex subset of a split graph cannot induce these structures [9]. Since split graphs are a subclass of chordal graphs, there exists a peo of the vertices for every split graph. In fact, any non-decreasing degree ordering of a split graph is a peo [10].

Our algorithm adapts the internal memory certifying algorithm of Heggernes and Kratsch [10] to external memory by adopting TFP. As output it either returns the split partition (K, I) as a YES-certificate or one of the forbidden subgraphs C_4, C_5 or $2K_2$ as a NO-certificate. We present the algorithm as a whole and refer to details in Proposition 1 and Proposition 2 at the end of the subsection.

First, we compute a non-decreasing degree ordering $\alpha = (v_1, \ldots, v_n)$ and relabel[1] the graph according to α. Thereafter we check whether α is a peo in $O(\text{SORT}(n + m))$ I/Os by Proposition 1. In the non-membership case, the algorithm returns three vertices v_j, v_k, v_i where $\{v_i, v_j\}, \{v_i, v_k\} \in E$ but $\{v_j, v_k\} \notin E$ and $i < j < k$, violating that v_i is simplicial in $G[\{v_i, \ldots, v_n\}]$. In order to return a forbidden subgraph we find additional vertices that complete the induced subgraphs. Note that (v_k, v_i, v_j) already forms a P_3 and may extend to a C_4 if $N(v_k) \cap N(v_j)$ contains a vertex $z \neq v_i$ that is not adjacent to v_i. Computing $(N(v_k) \cap N(v_j)) \setminus N(v_i)$ requires scanning the adjacencies of $O(1)$ many vertices totaling to $O(\text{SCAN}(n))$ I/Os. If $(N(v_k) \cap N(v_j)) \setminus N(v_i)$ is empty we try to extend the P_3 to a C_5 or output a $2K_2$ otherwise. To do so, we find vertices $x \neq v_i$ and $y \neq v_i$ for which $\{x, v_j\}, \{y, v_k\} \in E$ but $\{x, v_k\}, \{y, v_j\} \notin E$ that are also not adjacent to v_i, i.e. $\{x, v_i\}, \{y, v_i\} \notin E$. Both x and y exist due to the ordering α [10] and are found using $O(1)$ scanning steps requiring $O(\text{SCAN}(n))$ I/Os. If $\{x, y\} \in E$ then (v_j, v_i, v_k, y, x) is a C_5, otherwise $G[\{v_j, x, v_k, y\}]$ constitutes a $2K_2$. Determining whether $\{x, y\} \in E$ requires scanning $N(x)$ and $N(y)$ using $O(\text{SCAN}(n))$ I/Os.

In the membership case, α is a peo and the algorithm proceeds to verify first the clique K and then the independent set I of the split partition (K, I). Note that for a split graph the maximum clique of size k must consist of the k-highest ranked vertices in α [10] where k can be computed using $O(\text{SORT}(m))$ I/Os by Proposition 2. Therefore, it suffices to verify for each of the k candidates v_i whether it is connected to $\{v_{i+1}, \ldots, v_n\}$ since the graph is undirected. For a sorted sequence of edges relabeled by α, we check this property using $O(\text{SCAN}(m))$ I/Os. If we find a vertex $v_i \in \{v_{n-k+1}, \ldots, v_n\}$ where $\{v_i, v_j\} \notin E$ with $i < j$ then $G[\{v_i, \ldots, v_n\}]$ already does not constitute a clique and we have to return a NO-certificate. Since the maximum clique has size k, there are k vertices with degree at least $k - 1$. By these degree constraints there must exist an edge $\{v_i, x\} \in E$ where $x \in \{v_1, \ldots, v_{i-1}\}$ [10]. Additionally, it holds that $\{x, v_j\} \notin E$ and there exists an edge $\{z, v_j\} \in E$ where $z \in \{v_1, \ldots, v_{i-1}\}$ that cannot be connected to x, i.e. $\{x, z\} \notin E$ [10]. Thus, we first scan the adjacency lists of v_i and v_j to find x and z in $O(\text{SCAN}(n))$ I/Os and return $G[\{v_i, v_j, x, z\}]$ as the $2K_2$ NO-certificate. Otherwise let $K = \{v_{n-k+1}, \ldots, v_n\}$.

Lastly, the algorithm verifies whether the remaining vertices form an independent set. We verify that each candidate v_i is not connected to $\{v_{i+1}, \ldots, v_{n-k}\}$, since the graph is undirected. For this, it suffices to scan over $n - k$ consecutive adjacency lists in $O(\text{SCAN}(m))$ I/Os. More precisely, we scan the adjacency lists from v_{n-k} to v_1 and in case an edge $\{v_i, v_j\}$ where $i < j \leq n - k$ is

---

[1] If a vertex v_i has rank k in α it will be relabeled to v_k. The relabeling results in an adjacency array representation of the relabeled graph requiring $O(\text{SORT}(n + m))$ I/Os.

Algorithm 1: Recognizing Perfect Elimination in EM

Data: edges E of graph G, non-decreasing degree ordering $\alpha = (v_1, \ldots, v_n)$
Output: bool whether α is a peo, three invalidating vertices $\{v_i, v_j, v_k\}$ if not

1 Sort E and relabel according to α
2 **for** $i = 1, \ldots, n$ **do**
3   Retrieve $H(v_i)$ from E
4   **if** $H(v_i) \neq \emptyset$ **then**
5   Let u be the smallest successor of v_i in $H(v_i)$
6   **for** $x \in H(v_i) \setminus \{u\}$ **do**
7   PQ.push($\langle u, x, v_i \rangle$)   // inform u of x coming from v_i
8   **while** $\langle v, v_k, v_j \rangle \leftarrow$ PQ.top() *where* $v = v_i$ **do** // for each message to v_i
9   **if** $v_k \notin H(v_i)$ **then**   // v_i does not fulfill peo property
10   **return** FALSE, $\{v_i, v_j, v_k\}$
11   PQ.pop()

12 **return** TRUE

---

found we find two more vertices to again complete a $2K_2$. For the first occurrence of such a vertex v_i, we remark that $\{v_{i+1}, \ldots, v_{n-k}\}$ and $\{v_{n-k+1}, \ldots, v_n\}$ form an independent set and a clique, respectively. Therefore there exists a vertex $y \in K$ that is adjacent to x but not to v_i [10]. We find y by scanning $N(x)$ and $N(v_i)$ in $O(\mathrm{SCAN}(n))$ I/Os. To complete the $2K_2$ we similarly find $z \in N(y) \setminus (N(x) \cup N(y_i))$ in $O(\mathrm{SCAN}(n))$ I/Os which is guaranteed to exist [10].

Proposition 1. *Verifying that a non-decreasing degree ordering $\alpha = (v_1, \ldots, v_n)$ of a graph G is a perfect elimination ordering requires $O(\mathrm{SORT}(n + m))$ I/Os.*

Proof. We follow the approach of [8, Theorem 4.5] and adapt it to the external memory using TFP, see Algorithm 1.

After relabeling and sorting the edges by α we iterate over the vertices in the order given by α. For a vertex v_i the set of neighbors $N(v_i)$ needs to be a clique. In order to verify this for all vertices, for a vertex v_i we first retrieve $H(v_i)$. Then let $u \in H(v_i)$ be the smallest ranked neighbor according to α. In order for v_i to be simplicial, u needs to be adjacent to all vertices of $H(v_i) \setminus \{u\}$. In TFP-fashion we insert a message $\langle u, w \rangle$ into a priority-queue where $w \in H(v_i) \setminus \{u\}$ to inform u of every vertex it should be adjacent to. Conversely, after sending all adjacency information, we retrieve for v_i all messages $\langle v_i, - \rangle$ directed to v_i and check that all received vertices are indeed neighbors of v_i.

Relabeling and sorting the edges takes $O(\mathrm{SORT}(m))$ I/Os. Every vertex v_i inserts at most all its neighbors into the priority-queue totaling up to $O(m)$ messages which requires $O(\mathrm{SORT}(m))$ I/Os. Checking that all received vertices are neighbors only requires a scan over all edges since vertices are handled in non-descending order by α. $\qquad\square$

Algorithm 2: Maximum Clique Size for Chordal Graphs in EM

Data: edges E of input graph G, peo $\alpha = (v_1, \ldots, v_n)$
Output: maximum clique size χ

1 Sort E and relabel according to α
2 $\chi \leftarrow 0$
3 **for** $i = 1, \ldots, n$ **do**
4   Retrieve $H(v_i)$ from E `// scan E`
5   **if** $H(v_i) \neq \emptyset$ **then**
6   Let u be the smallest successor of v_i in $H(v_i)$
7   PQ.push($\langle u, |H(v_i)| - 1 \rangle$)   `// `$v_i$` simplicial `$\Rightarrow$` G[N(`$v_i$`)] is clique`
8       $S(v_i) \leftarrow -\infty$
9   **while** $\langle v, S \rangle \leftarrow$ PQ.top() *where* $v = v_i$ **do**
10          $S(v_i) \leftarrow \max\{S(v_i), S\}$ `// compute maximum over all`
11   PQ.pop()
12      $\chi \leftarrow \max\{\chi, S(v_i)\}$
13 **return** χ

Proposition 2. *Computing the size of a maximum clique in a split graph requires $O(\text{SORT}(m))$ I/Os.*

Proof. Note that split graphs are both chordal and co-chordal [9]. For chordal graphs, computing the size of a maximum clique in internal memory takes linear time [8, Theorem 4.17] and is easily convertible to an external memory algorithm using $O(\text{SORT}(m))$ I/Os. To do so, we simulate the data accesses of the internal memory variant using priority-queues to employ TFP, see Algorithm 2. Instead of updating each $S(v_i)$ value immediately, we delay its consecutive computation by sending a message $\langle v_i, S \rangle$ to v_i to inform v_i, that v_i is part of a clique of size S. After collecting all messages, the overall maximum is computed and the global value of the currently maximum clique is updated if necessary.   □

By the above description it follows that split graphs can be certified using $O(\text{SORT}(n + m))$ I/Os which we summarize in Theorem 1.

Theorem 1. *A graph G can be certified whether it is a split graph or not in $O(\text{SORT}(n + m))$ I/Os. In the membership case the algorithm returns the split partition (K, I) as the YES-certificate, and otherwise it returns an $O(1)$-size NO-certificate.*

3.2 Certifying Threshold Graphs in External Memory

Threshold graphs [6,8,11] are split graphs with the additional property that the independent set I of the split partition (K, I) has an nno. Its corresponding forbidden subgraphs are $2K_2$, P_4 and C_4. Alternatively, threshold graphs can be characterized by a graph generation process: repeatedly add universal or isolated vertices to an initially empty graph. Conversely, by repeatedly removing universal and isolated vertices from a threshold graph the resulting graph must be the empty graph. In comparison to certifying split graphs, threshold graphs thus require additional steps.

First, the algorithm certifies whether the input is a split graph. In the non-membership case, if the returned NO-certificate is a C_5 we extract a P_4 otherwise we return the subgraph immediately. For the membership case, we recognize whether the input is a threshold graph by repeatedly removing universal and isolated vertices using the previously computed peo α in $O(\text{SORT}(m))$ I/Os by Proposition 3 (see below). If the remaining graph is empty, we return the independent set I with its non-decreasing degree ordering. Note that after removing a universal vertex v_i, vertices with degree one become isolated. Since low-degree vertices are at the front of α, an I/O-efficient algorithm cannot determine them on-the-fly after removing a high-degree vertex. Therefore pre-processing is required. For every vertex v_i we compute the number of vertices $S(v_i)$ that become isolated after the removal of $\{v_i, \ldots, v_n\}$. To do so, we iterate over α in non-descending order and check for v_i with $L(v_i) = \emptyset$. Since v_i has no lower ranked neighbors, it would become isolated after removing all vertices in $H(v_i)$, in particular when the successor with smallest index $v_j \in H(v_i)$ is removed. We save v_j in a vector S and sort S in non-ascending order. The values $S(v_n), \ldots, S(v_1)$ are now accessible by a scan over S to count the occurrences of each v_j in $O(\text{SCAN}(m))$ I/Os.

In the non-membership case, there must exist a P_4 since the input is split and cannot contain a C_4 or a $2K_2$. We can delete further vertices from the remaining graph that cannot be part of a P_4. For this, let $K' \subset K$ and $I' \subset I$ be the remaining vertices of the split partition. Any $v \in K'$ where $N(v) \cap I' = \emptyset$ and any $v \in I'$ where $N(v) \cap K' = K'$ cannot be part of a P_4 [10] and can therefore be deleted. We proceed by considering and removing vertices of K by non-descending degree and vertices of I by non-ascending degree. After this process, we retrieve the highest-degree vertex v in I for which there exists $\{v, y\} \notin E$ and $\{y, z\} \in E$ where $y \in K$ and $z \in I$ [10]. Additionally, there is a neighbor $w \in K$ of v for which $\{w, z\} \notin E$ [10] and we return the P_4 given by $G[\{v, w, y, z\}]$. Finding the P_4 therefore only requires $O(\text{SCAN}(n + m))$ I/Os.

Proposition 3. *Verifying that a non-decreasing degree ordering $\alpha = (v_1, \ldots, v_n)$ of a graph G emits an empty graph after repeatedly removing universal and isolated vertices requires $O(\text{SORT}(n) + \text{SCAN}(m))$ I/Os.*

Proof. Generating the values $S(v_n), \ldots, S(v_1)$ requires a scan over all adjacency lists in non-descending order and sorting S which takes $O(\text{SORT}(n) + \text{SCAN}(m))$ I/Os. Afte pre-processing, the algorithm only requires a reverse scan over the degrees $d_n, \ldots, d_1$. We iterate over α in reverse order, where for each v_i we check whether $L(v_i) = \emptyset$. If v_i is not isolated it must be universal. Therefore we compare its current degree $\deg(v_i)$ with the value $(n-1) - n_{\text{del}}$ where $n_{\text{del}} = \sum_{j=j+1}^{n} S(v_j)$. All operations take $O(\text{SCAN}(m))$ I/Os in total. □

We summarize our findings for threshold graphs in Theorem 2.

Theorem 2. *A graph G can be certified whether it is a threshold graph or not in $O(\text{SORT}(n+m))$ I/Os. In the membership case the algorithm returns a nested neighborhood ordering β as the YES-certificate, and otherwise it returns an $O(1)$-size NO-certificate.*

Proof. Certifying that the input graph is a split graph requires $O(\text{SORT}(n+m))$ I/Os by Theorem 1. If it is, we check if the input is a threshold graph directly by checking whether the graph is empty after repeatedly removing universal and isolated vertices in $O(\text{SORT}(m))$ I/Os by Proposition 3. Otherwise we have to find a P_4, since the input is a split but not a threshold graph. Hence, this step requires $O(\text{SCAN}(n+m))$ I/Os and the total I/Os are $O(\text{SORT}(n+m))$. □

3.3   Certifying Trivially Perfect Graphs in External Memory

Trivially perfect graphs have no vertex subset that induces a P_4 or a C_4 [8]. In contrast to split graphs, any non-increasing degree ordering of a trivially perfect graph is a uco [10]. In fact, this is a one-to-one correspondence: a non-increasing sorted degree sequence of a graph is a uco iff the graph is trivially perfect [10].

In external memory this can be verified using TFP by adapting the algorithm in [10]. After computing a non-increasing degree ordering γ the algorithm relabels the edges of the graph according to γ and sorts them. Now we iterate over the vertices in non-descending order of γ, process for each vertex v_i its received messages and relay further messages forward in time.

Initially all vertices are labeled with 0. Then, at step i vertex v_i checks that all adjacent vertices $N(v_i)$ have the same label as v_i. After this, v_i relabels each vertex $u \in N(v_i)$ with its own index i and is then removed from the graph. In the external memory setting we cannot access labels of vertices and relabel them on-the-fly but rather postpone the comparison of the labels to the adjacent vertices instead. To do so, v_i forwards its own label $\ell(v_i)$ to $u \in H(v_i)$ by sending two messages $\langle u, v_i, \ell(v_i) \rangle$ and $\langle u, v_i, i \rangle$ to u, signaling that u should compare its own label to v_i's label $\ell(v_i)$ and then update it to i. Since the label of any adjacent vertex is changed after processing a vertex, when arriving at vertex v_j an odd number of messages will be targeted to v_j, where the last one corresponds to its actual label at step j. Then, after collecting all received labels, we compare disjoint consecutive pairs of labels and check whether they match. In the membership case, we do not find any mismatch and return γ as the YES-certificate. Otherwise, we have to return a P_4 or C_4.

In the description of [10] the authors stop at the first anomaly where v_i detects a mismatch in its own label and one of its neighbors. We simulate the same behavior by writing out every anomaly we find, e.g. that v_j does not have the expected label of v_i via an entry $\langle v_i, v_j, k \rangle$ where k denotes the label of v_j. After sorting the entries, we find the earliest anomaly $\langle v_i, v_j, k \rangle$ with the largest label k of v_i's neighbors. Since v_j received the label k from v_k, but v_i did not, it is clear that v_k is not universal in its connected component in $G[\{v_k, v_{k+1}, \ldots, v_n\}]$ and we thus will return a P_4 or C_4. Note that (v_k, v_j, v_i) already constitutes a P_3 where $\deg(v_k) \geq \deg(v_j)$, because v_j received the label k. Since v_j is adjacent to both v_k and v_i and $\deg(v_k) \geq \deg(v_j)$, there must exist a vertex $x \in N(v_k)$ where $\{v_j, x\} \notin E$. Thus, $G[\{v_k, v_j, v_i, x\}]$ is a P_4 if $\{v_i, x\} \notin E$ and a C_4 otherwise. Finding x and determining whether the forbidden subgraph is a P_4 or a C_4 requires scanning $O(1)$ adjacency lists in $O(\text{SCAN}(n))$ I/Os.

Proposition 4. *Verifying that a non-increasing degree ordering* $\gamma = (v_1, \ldots, v_n)$ *of a graph G with n vertices and m edges is a universal-in-a-component-ordering requires $O(\text{SORT}(m))$ I/Os.*

Proof. Every vertex v_i receives exactly two messages per neighbor in $L(v_i)$ and verifies that all consecutive pairs of labels match. Then, either the label i is sent to each higher ranked neighbor of $H(v_i)$ via TFP or it is verified that γ is not a uco. Since at most $O(m)$ messages are inserted, the resulting overall complexity is $O(\text{SORT}(m))$ I/Os. Correctness follows from [10] since the adapted algorithm performs the same operations but only delays the label comparisons. $\qquad\square$

We again summarize our results in Theorem 3.

Theorem 3. *A graph G can be certified whether it is a trivially perfect graph or not in $O(\text{SORT}(n + m))$ I/Os. In the membership case the algorithm returns the universal-in-a-component ordering γ as the YES-certificate, and otherwise it returns an $O(1)$-size NO-certificate.*

3.4 Certifying Bipartite Chain Graphs in External Memory

Bipartite chain graphs are bipartite graphs where one part of the bipartition has an nno [16] similar to threshold graphs. Its forbidden induced subgraphs are $2K_2, C_3$ and C_5. By definition, bipartite chain graphs are bipartite graphs which therefore requires I/O-efficient bipartiteness testing.

We follow the linear time internal memory approach of [10] with slight adjustments to accommodate the external memory setting. First, we check whether the input is indeed a bipartite graph. Instead of using breadth-first search which is very costly in external memory, even for constrained settings [2], we can use a more efficient approach with spanning trees which is presented in the following in Lemma 1. Note that, computing a spanning forest only requires $O(\text{SORT}(n + m))$ I/Os with high probability [5] and is therefore no real restriction to Lemma 1. In case the input is not connected, we simply return two edges of two different components as the $2K_2$. If the graph is connected, we proceed to verify that the graph is bipartite and return a NO-certificate in the form of a C_3, C_5 or $2K_2$ in case it is not. In order to find a C_3, C_5 or $2K_2$ some modifications to Lemma 1 are necessary. Essentially, the algorithm instead returns a minimum odd cycle that is built from T and a single non-tree edge. Due to minimality we can then find a $2K_2$. The result is summarized in Corollary 1 and proven in the full version of the paper [14].

Then, it remains to show that each side of the bipartition has an nno. Let U be the larger side of the partition. By [11] it suffices to show that the input is a chain graph if and only if the graph obtained by adding all possible edges with both endpoints in U is a threshold graph. Instead of materializing the mentioned threshold graph, we implicitly represent the adjacencies of vertices in U to retain the same I/O-complexity and apply Theorem 2 using $O(\mathrm{SORT}(n+m))$ I/Os. If the input is bipartite but not chain, we repeatedly delete vertices that are connected to all other vertices of the other side and the resulting isolated vertices, similar to Subsect. 3.3 and [10]. After this, the vertex v with highest degree has a non-neighbor y in the other partition. By similar arguments y is adjacent to another vertex z that is adjacent to a vertex x where $\{v, x\} \notin E$ [10]. As such $G[\{v, y, z, x\}]$ is a $2K_2$ and can be found in $O(\mathrm{SCAN}(n))$ I/Os and returned as the NO-certificate.

Lemma 1. *A graph G can be certified whether it is a bipartite graph or not in $O(\mathrm{SORT}(n+m))$ I/Os, given a spanning forest of the input graph. In the membership case the algorithm returns a bipartition $(U, V \setminus U)$ as the YES-certificate, and otherwise it returns an odd-length cycle as the NO-certificate.*

Proof. In case there are multiple connected components, we operate on each individually and thus assume that the input is connected. Let T be the edges of the spanning tree and $E \setminus T$ the non-tree edges. Any edge $e \in E \setminus T$ may produce an odd cycle by its addition to T. In fact, the input is bipartite if and only if $T \cup \{e\}$ is bipartite for all $e \in E \setminus T^2$. We check whether an edge $e = \{u, v\}$ closes an odd cycle in T by computing the distance $d_T(u, v)$ of its endpoints in T. Since this is required for every non-tree edge $E \setminus T$, we resort to batch-processing. Note that T is a tree and hence after choosing a designated root $r \in V$ it holds that $d_T(u, v) = d_T(u, \mathrm{LCA}_T(u, v)) + d_T(v, \mathrm{LCA}_T(u, v))$ where $\mathrm{LCA}_T(u, v)$ is the lowest common ancestor of u and v in T. Therefore for every edge $E \setminus T$ we compute its lowest common ancestor in T using $O((m/n) \cdot \mathrm{SORT}(n)) = O(\mathrm{SORT}(m))$ I/Os [5].

Additionally, for each vertex $v \in V$ we compute its depth in T in $O(\mathrm{SORT}(m))$ I/Os using Euler Tours [5] and inform each incident edge of this value by a few scanning and sorting steps. Similarly, each edge $e = \{u, v\}$ is provided of the depth of $\mathrm{LCA}_T(u, v)$. Then, after a single scan over $E \setminus T$ we compute $d_T(u, v)$ and check if it is even. If any value is even, we return the odd cycle as a NO-certificate or a bipartition in T as the YES-certificate. Both can be computed using Euler Tours in $O(\mathrm{SORT}(m))$ I/Os. □

Corollary 1. *If a connected graph G contains a C_3, C_5 or $2K_2$ then any of these subgraphs can be found in $O(\mathrm{SORT}(n+m))$ I/Os given a spanning tree of G.*

We summarize our findings for bipartite chain graphs in Theorem 4.

Theorem 4. *A graph G can be certified whether it is a bipartite chain graph or not in $O(\mathrm{SORT}(n+m))$ I/Os with high probability. In the membership case the algorithm returns the bipartition $(U, V \setminus U)$ and nested neighborhood orderings of both partitions as the YES-certificate, and otherwise it returns an $O(1)$-size NO-certificate.*

---

[2] Since T is bipartite, one can think of T as a representation of a 2-coloring on T.

Proof. Computing a spanning tree T requires $O(\text{SORT}(n + m))$ I/Os with high probability by an external memory variant of the Karger, Klein and Tarjan minimum spanning tree algorithm [5]. By Corollary 1 we find a C_3, C_5 or $2K_2$ if the input is not bipartite or not connected. We proceed by checking the nno's of both partitions in $O(\text{SORT}(n + m))$ I/Os using Theorem 2. $\square$

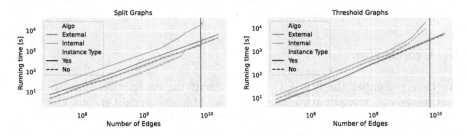

Fig. 1. Running times of the certifying algorithms for split (left) and threshold graphs (right) for different random graph instances. The black vertical lines depict the number of elements that can concurrently be held in internal memory.

4 Experimental Evaluation

We implemented our external memory certifying algorithms for split and threshold graphs in C++ using the STXXL library [7]. To provide a comparison of our algorithms, we also implemented the internal memory state-of-the-art algorithms by Heggernes and Kratsch [10]. STXXL offers external memory versions of fundamental algorithmic building blocks like scanning, sorting and several data structures. Our benchmarks are built with GNU g++-10.3 and executed on a machine equipped with an AMD EPYC 7302P processor and 64 GB RAM running Ubuntu 20.04 using six 500 GB solid-state disks.

In order to validate the predicted scaling behaviour we generate our instances parameterized by n. For YES-instances of split graphs we generate a split partition (K, I) with $|K| = n/10$ and add each possible edge $\{u, v\}$ with probability $1/4$ for $u \in I$ and $v \in K$. Analogously, YES-instances of threshold graphs are generated by repeatedly adding either isolated or universal vertices with probability $9/10$ and $1/10$, respectively. We additionally attempt to generate NO-instances by adding $O(1)$ many random edges to the YES-instances. In a last step, we randomize the vertex indices to remove any biases of the generation process.

In Fig. 1 we present the running times of all algorithms on multiple YES- and NO-instances. It is clear that the performance of both external memory algorithms is not impacted by the main memory barrier while the running time of their internal memory counterparts already increases when at least half the main memory is used. This effect is amplified immensely after exceeding the size of main memory for split graphs, Fig. 1.

Certifying the produced NO-instances of split graphs seems to require less time than their corresponding unmodified YES-instances as the algorithm typically stops early. Furthermore, due to the low data locality of the internal memory variant it is apparent that the external memory algorithm is superior for the YES-instances. The performance on both YES- and NO-instances is very similar in external memory. This is in part due to the fact that the common relabeling step is already relatively costly. For threshold graphs, however, the external memory variant outperforms the internal memory variant due to improved data locality.

5   Conclusions

We have presented the first I/O-efficient certifying recognition algorithms for split, threshold, trivially perfect, bipartite and bipartite chain graphs. Our algorithms require $O(\text{SORT}(n + m))$ I/Os matching common lower bounds for many algorithms in external memory. In our experiments we show that the algorithms perform well even for graphs exceeding the size of main memory.

Further, it would be interesting to extend the scope of certifying recognition algorithms to more graph classes for the external memory regime.

Acknowledgements. This work is partially supported by the International Exchanges Grant IES\R3\203041 of the Royal Society and by the Deutsche Forschungsgemeinschaft (DFG) under grant ME 2088/5-1 (FOR 2975 | Algorithms, Dynamics, and Information Flow in Networks).

References

1. Aggarwal, A., Vitter, J.S.: The input/output complexity of sorting and related problems. Commun. ACM **31**(9), 1116–1127 (1988)
2. Ajwani, D., Meyer, U.: Design and engineering of external memory traversal algorithms for general graphs. In: Lerner, J., Wagner, D., Zweig, K.A. (eds.) Algorithmics of Large and Complex Networks. LNCS, vol. 5515, pp. 1–33. Springer, Heidelberg (2009). https://doi.org/10.1007/978-3-642-02094-0_1
3. Arge, L.: The buffer tree: a technique for designing batched external data structures. Algorithmica **37**(1), 1–24 (2003)
4. Buchsbaum, A.L., Goldwasser, M.H., Venkatasubramanian, S., Westbrook, J.R.: On external memory graph traversal. In: SODA, pp. 859–860. ACM/SIAM (2000)
5. Chiang, Y., Goodrich, M.T., Grove, E.F., Tamassia, R., Vengroff, D.E., Vitter, J.S.: External-memory graph algorithms. In: SODA, pp. 139–149. ACM/SIAM (1995)
6. Chvátal, V.: Set-packing and threshold graphs. Res. Rep. Comput. Sci. Dept., Univ. Waterloo, 1973 (1973)
7. Dementiev, R., Kettner, L., Sanders, P.: STXXL: standard template library for XXL data sets. Softw. Pract. Exp. **38**(6), 589–637 (2008)
8. Golumbic, M.C.: Algorithmic Graph Theory and Perfect Graphs. Elsevier, Amsterdam (2004)
9. Hammer, P.L., Földes, S.: Split graphs. Congr. Numer. **19**, 311–315 (1977)
10. Heggernes, P., Kratsch, D.: Linear-time certifying recognition algorithms and forbidden induced subgraphs. Nord. J. Comput. **14**(1–2), 87–108 (2007)

11. Mahadev, N.V., Peled, U.N.: Threshold Graphs and Related Topics. Elsevier, Amsterdam (1995)
12. Maheshwari, A., Zeh, N.: A survey of techniques for designing I/O-efficient algorithms. In: Meyer, U., Sanders, P., Sibeyn, J. (eds.) Algorithms for Memory Hierarchies. LNCS, vol. 2625, pp. 36–61. Springer, Heidelberg (2003). https://doi.org/10.1007/3-540-36574-5_3
13. McConnell, R.M., Mehlhorn, K., Näher, S., Schweitzer, P.: Certifying algorithms. Comput. Sci. Rev. **5**(2), 119–161 (2011)
14. Meyer, U., Tran, H., Tsakalidis, K.: Certifying induced subgraphs in large graphs. CoRR abs/2210.13057 (2022)
15. Sanders, P., Mehlhorn, K., Dietzfelbinger, M., Dementiev, R.: Sequential and Parallel Algorithms and Data Structures - The Basic Toolbox. Springer, Cham (2019). https://doi.org/10.1007/978-3-030-25209-0
16. Yannakakis, M.: Node-deletion problems on bipartite graphs. SIAM J. Comput. **10**(2), 310–327 (1981)

Some Algorithmic Results for Eternal Vertex Cover Problem in Graphs

Kaustav Paul and Arti Pandey[✉]

Department of Mathematics, Indian Institute of Technology Ropar,
Nangal Road, Rupnagar 140001, Punjab, India
{kaustav.20maz0010,arti}@iitrpr.ac.in

Abstract. Eternal vertex cover problem is a variant of the vertex cover problem. It is a two player (attacker and defender) game in which given a graph $G = (V, E)$, the defender needs to allocate guards at some vertices so that the allocated vertices form a vertex cover. Attacker can attack one edge at a time and the defender needs to move the guards along the edges such that at least one guard moves through the attacked edge and the new configuration still remains a vertex cover. The attacker wins if no such move exists for the defender. The defender wins if there exists a strategy to defend the graph against infinite sequence of attacks. The minimum number of guards with which the defender can form a winning strategy is called the *eternal vertex cover number* of G, and is denoted by $evc(G)$. Given a graph G, the problem of finding the eternal vertex cover number is NP-hard for general graphs, and remains NP-hard even for bipartite graphs. We have given a polynomial time algorithm to find the Eternal vertex cover number in chain graphs and cographs. We have also given a linear-time algorithm to find the eternal vertex cover number for split graphs, an important subclass of chordal graphs.

Keywords: Eternal vertex cover · Chain graphs · Split graphs · Cographs

1 Introduction

In 2009, Klostermeyer and Mynhardt introduced the Eternal vertex cover problem [8], which is a dynamic variant of the vertex cover problem. The problem is a two player (attacker and defender) game such that given a graph $G = (V, E)$, the defender is permitted to allocate guards in some vertices of G so that the vertices, where guards are allocated form a vertex cover. The attacker can attack one edge at a time. Now for each guard, the defender can either move the guard to one of its neighbour or can keep it untouched, such that at least one guard from any of the endpoint of the attacked edge move through the edge to settle at the other end point. So, the new allocation should also remain a vertex cover to defend the next attack. If no such configuration exists then the attacker wins. If the allocation can defend infinite sequence of attacks, then the defender wins.

C.-C. Lin et al. (Eds.): WALCOM 2023, LNCS 13973, pp. 242–253, 2023.
https://doi.org/10.1007/978-3-031-27051-2_21

The minimum number of guards with which a winning strategy for the defender can be formed is known as the *eternal vertex cover number* of G, and is denoted by $evc(G)$. In this paper, we are assuming that at most one guard can be allocated to each vertex. If C_i be the allocation of the guards before the i-th attack, then after defending the i-th attack by moving the guards to configuration C_{i+1}, C_{i+1} needs to be a vertex cover (for each $i \in \mathbb{N}$), to form a winning strategy for the defender. If it is not then the $(i + 1)$-th attack will be on the edge which is not covered by C_{i+1} and the attacker will win. So after reconfiguring at each step, the vertices where the guards are allocated should form a vertex cover. This implies $evc(G) \geq mvc(G)$, where $mvc(G)$ denotes the size of the minimum vertex cover of G. It is also known that twice as many guards as $mvc(G)$ can form an eternal vertex cover by placing the guards at both end points of a maximum matching. So, for any graph G, we have

$$mvc(G) \leq evc(G) \leq 2mvc(G)$$

Klostermeyer and Mynhardt have also given a characterization of the graphs for which $evc(G) = 2mvc(G)$ is attained [8]. Babu et al. have given some special graph classes for which it attains the lower bound [2].

Fomin et al. have shown that the problem is NP-hard [6]. Fomin et al. have also presented a 2-approximation algorithm based on the endpoints of the matching [6]. Babu et al. proved that the problem remains NP-hard even for locally connected graphs which includes all biconnected internally triangulated planar graphs [2]. Babu et al. recently proved that the problem remains NP-hard for bipartite graphs [3]. Babu et al. proposed polynomial-time algorithms for cactus graphs and chordal graphs [4,5]. Babu et al. proved that the problem can also be solved in polynomial time for co-bipartite graphs [3]. In this paper, we further extend the algorithmic study of the problem by proposing polynomial-time algorithms for some special graph classes. Araki et al. have given the $evc(G)$ for generalized trees where each edge of the tree is replaced by some elementary bipartite graphs [1].

The rest of the paper is organized as follows: In Sect. 2.1, all notations and definitions used in the paper are presented. In Sect. 2.2, some theorems from existing literature are stated, which are used in the proofs presented in this paper. In Sect. 2.3, eternal vertex cover number is provided for some special subclasses of bipartite graphs. In Sect. 3, a linear-time algorithm is given to compute $evc(G)$ in chain graphs. In Sect. 4, a linear-time algorithm to compute $evc(G)$ in split graphs is presented. In Sect. 5, a polynomial time algorithm to compute $evc(G)$ in cographs is presented. Finally, Sect. 6 concludes the paper.

2 Preliminaries

2.1 Definitions and Notations

All graphs considered in this paper are finite, undirected and simple. Let $G = (V, E)$ is a graph. The set of neighbours of a vertex v in G is denoted by $N(v)$.

A set $I \subseteq V$ is called an independent set of G if for all $u, v \in I$, $\{u, v\} \notin E$. Degree of a vertex $v \in V$ is the number of neighbours of v in G and it is denoted as $deg(v)$. Given a subset V' of V, the number of neighbours of v in V' is denoted by $deg_{V'}(v)$. A vertex $v \in V$ is said to be a cut vertex if $G[V \setminus \{v\}]$ is not connected. The join of two graphs H_1 and H_2 is a graph formed from disjoint copies of H_1 and H_2 by connecting each vertex of $V(H_1)$ to each vertex of $V(H_2)$.

A vertex cover S of $G = (V, E)$ is subset of V, which contains at least one end point from each edge in E. If S is a vertex cover then $V \setminus S$ is an independent set. A vertex cover of minimum cardinality is called a minimum vertex cover. Cardinality of minimum vertex cover is denoted as minimum vertex cover number or $mvc(G)$. Given $B \subseteq V$, the cardinality of the minimum vertex cover containing B is denoted as $mvc_B(G)$. If the induced graph on S, i.e. $G[S]$ is connected, S is called a connected vertex cover. The cardinality of minimum vertex cover is denoted as $cvc(G)$. The independent set of maximum cardinality is called maximum independent set of G and its cardinality is denoted as $mis(G)$.

Consider a graph $G = (V, E)$ with $|V| = n$ and $|E| = m$. The guards are needed to be allocated in order to protect against infinite sequence of attacks. One edge can be attacked at a time and each guard either moves to a neighbour vertex or stays on the same vertex.

A hamiltonian cycle of a graph $G = (V, E)$ is a cycle in G, that visits each $v \in V$ exactly once. A graph possessing a hamiltonian cycle is known as hamiltonian graph. A graph $G = (V, E)$ is said to be k-regular if $deg(v) = k$, for each $v \in V$.

Let $G = (X \cup Y, E)$ be a bipartite graph. G is said to be a chain graph if vertices in X can be ordered $\{x_1, x_2, \ldots, x_{|X|}\}$, such that $N(x_1) \subseteq N(x_2) \subseteq \ldots \subseteq N(x_{|X|})$. Similarly vertices of Y can be ordered $\{y_1, y_2, \ldots, y_{|Y|}\}$, such that $N(y_1) \supseteq N(y_2) \supseteq \ldots \supseteq N(y_{|Y|})$. The cardinality of X and Y are denoted by p and q respectively, in this paper.

A graph $G = (V, E)$ is called a split graph if V can be partitioned in K and I, such that K is clique and I is an independent set. The class of split graphs is an important subclass of chordal graphs.

A graph $G = (V, E)$ is called a cograph if it can be generated from K_1 by complementation and disjoint union. Recursively, the class of cographs can be defined as follows

1. K_1 is a cograph.
2. Complement of a cograph is a cograph.
3. G_1 and G_2 are cographs, then $G_1 \cup G_2$ is a cograph.

Cographs can be represented as join of k graphs, $G_1, G_2, \ldots, G_k$ where G_i is either K_1 or disconnected graph.

2.2 Existing Results Used in This Paper

For the sake of convenience, we are stating some important theorems, which will be used in the proofs presented in our paper.

Theorem 1. *[2] Let $G = (V, E)$ be a graph with no isolated vertex for which every minimum vertex cover is connected. If for every vertex $v \in V$, there exists a minimum vertex cover S_v of G such that $v \in S_v$, then $evc(G) = mvc(G)$. Otherwise, $evc(G) = mvc(G) + 1$.*

Theorem 2. *[8] Let $G = (V, E)$ be a nontrivial, connected graph and let D be a minimum connected vertex cover of G. Then $evc(G) \leq |D| + 1$.*

Theorem 3. *[2] Let $G = (V, E)$ be a graph with at least 2 vertices and X be the set of cut vertices of G. If every minimum vertex cover S of G with $X \subseteq S$ is connected, then the following characterization holds: $evc(G) = mvc(G)$ if and only if for every vertex $v \in V \setminus X$, there exists a minimum vertex cover S_v of G such that $X \cup \{v\} \subseteq S_v$.*

Theorem 4. *[2] Let $G = (V, E)$ be a graph with no isolated vertices. If $evc(G) = mvc(G)$, then for every vertex $v \in V$, there is some minimum vertex cover of G containing v.*

2.3 Eternal Vertex Cover Number for Some Subclasses of Bipartite Graph

For a k-regular bipartite graph, the following observation can be made.

Observation 1. *Given a k-regular bipartite graph $G = (X \cup Y, E)$, for each $e \in E$, there exists a perfect matching that contains e.*

Note that, if the initial guard allocation is X (or Y), then attack on any edge e can be defended by moving the guards to Y (or X) through the perfect matching that contains e. So, from the Observation 1 it can be concluded that for a k-regular bipartite graph G, $evc(G) = mvc(G) = |X| = |Y|$.

For a hamiltonian bipartite graph $G = (X \cup Y, E)$ (with $|X| = |Y| = n$), suppose a hamiltonian cycle of G is $v_1 v_2 \cdots v_{2n} v_1$, where $X = \{v_1, v_3, \ldots, v_{2n-1}\}$ and $Y = \{v_2, v_4, \ldots, v_{2n}\}$. Then, we have the following observation.

Observation 2. *Given a hamiltonian bipartite graph $G = (X \cup Y, E)$ and a hamiltonian cycle $v_1 v_2 \cdots v_{2n} v_1$ of G, X and Y are the only two possible minimum vertex covers of G.*

From Observation 2, it can be concluded that for each $e \in E$, there exists a perfect matching that contains e, implying $evc(G) = mvc(G) = |X| = |Y|$.

3 A Polynomial Time Algorithm for Chain Graphs

In this section, we present a linear-time algorithm to compute the $evc(G)$ of a given chain graph G. We also show that for a chain graph G, $evc(G) \in \{mvc(G), mvc(G) + 1, mvc(G) + 2\}$.

For a chain graph $G = (X \cup Y, E)$, we assume that it is connected and $|X| \leq |Y|$. The eternal vertex cover problem in the class of chain graphs are studied in 2 exhaustive cases: (i) chain graphs having pendent vertices only in Y, and (ii) chain graphs having pendant vertices both in X and Y or only in X.

3.1 For Chain Graphs Where only Y Can Have Pendant Vertices

In this section we will assume that either there exists no pendant vertex in the graph or only Y contains pendant vertices. Note that a minimum vertex cover of a chain graph can be computed in linear time [9]. Let S be a minimum vertex cover G. If $|S| < min\{|X|, |Y|\}$, then $|X \cap S| \neq \phi$ and $|Y \cap S| \neq \phi$. First, we state the following observation.

Observation 3. *Given a chain graph $G = (X \cup Y, E)$ and a minimum vertex cover S of G; if $x_i \in S$, then $x_j \in S$, for each $i \leq j \leq p$ and if $y_i \in S$, then $y_j \in S$, for each $1 \leq j \leq i$.*

Lemma 1. *For a chain graph $G = (X \cup Y, E)$, if $mvc(G) < min\{|X|, |Y|\}$, then $evc(G) = mvc(G) + 1$.*

Proof. From Observation 3, if $|S| < min\{|X|, |Y|\}$, then $y_1, x_p \in S$. This implies that S is a connected vertex cover, and hence $mvc(G) = cvc(G)$. Also, each vertex cover of size $mvc(G)$ is connected, as it always contain y_1 and x_p. But there does not exist any minimum vertex cover S' that contains x_1 (If $x_1 \in S'$, then by Observation 3, $X \subseteq S'$, which implies that $mvc(G) \geq |X| > |S|$, a contradiction). So, by Theorem 1, if for a chain graph G, $mvc(G) < min\{|X|, |Y|\}$, then $evc(G) = mvc(G) + 1$ and the initial configuration of guards is $\{x_1\} \cup S$.   □

Now we consider the case when $mvc(G) = min\{|X|, |Y|\}$. Again two cases may arise, one is $|X| < |Y|$ and the another is $|X| = |Y|$.

Claim 1. *For a chain graph $G = (X \cup Y, E)$, if $|X| < |Y|$ and $mvc(G) = min\{|X|, |Y|\}$, then $mvc(G) \neq evc(G)$.*

Proof. Let $evc(G) = mvc(G)$, then $x_p \in S$, for any minimum vertex cover S of G (by Observation 3). If the attacker attacks $\{x_p, y_q\}$, then the guard at x_p moves to y_q and rest of the guards are adjusted so that the new configuration remains a vertex cover. Since in the new configuration, $y_q \in S'$, (where S' is a minimum vertex cover), by Observation 3, $Y \subseteq S'$. Which leads to a contradiction since $mvc(G) < |Y|$. Hence $mvc(G) \neq evc(G)$.   □

Lemma 2. *For a chain graph $G = (X \cup Y, E)$, if $|X| < |Y|$, $mvc(G) = min\{|X|, |Y|\}$, and there exists a minimum vertex cover containing x_p, y_1, then $mvc(G) = evc(G) + 1$.*

Proof. If for a given chain graph G, there exists a minimum vertex cover that contains x_p, y_1, then $cvc(G) = mvc(G)$. Since $evc(G) \neq mvc(G)$ and by Theorem 2, $evc(G) \leq cvc(G) + 1$, we may conclude that $evc(G) = mvc(G) + 1$.   □

Now let us consider the case when there does not exist any minimum vertex cover that contains x_p, y_1, $mvc(G) = min\{|X|, |Y|\}$ and $|X| < |Y|$. In this case, X is the only minimum vertex cover.

Lemma 3. *For a given chain graph $G = (X \cup Y, E)$, if $mvc(G) = min\{|X|, |Y|\}$ and $|Y| = |X| + 1$, and X is the only minimum vertex cover of G, then $evc(G) = mvc(G) + 1$.*

Proof. Let $|N(x_1)| > 2$ or $y_{q-1} \notin N(x_1)$. If the initial configuration is $\{x_1, x_2, \ldots, x_p, y_q\}$, attack any edge $\{x_i, y_j\}$ $(y_j \neq y_q)$; by Hall's Theorem there exists a perfect matching from $X \setminus \{x_i\}$ to $Y \setminus \{y_j, y_q\}$, since $| \cup_{j=1}^{k} N(x_j) | \geq k+1$, for each $k \in [p]$. So all the guards can be moved from $X \cup \{y_q\}$ to Y.

Now if Y is the guard allocation configuration and $\{y_j, x_i\}(y_j \neq y_q)$ is attacked then the next configuration will be $X \cup \{y_q\}$. If $y_j = y_q$ then the configuration will be $X \cup \{y_{q-1}\}$. Thus any infinite sequence of attack can be defended using $mvc(G) + 1$ guards. So $evc(G) = mvc(G) + 1$. If $|N(x_1)| \leq 2$ and $y_{q-1} \in N(x_1)$, then it is easy to observe $evc(G) = mvc(G) + 1$. □

Observation 4. *Let $G = (X \cup Y, E)$ is a chain graph with $|Y| > |X| + 1$ for which the only minimum vertex cover is X and S be a vertex cover of size $mvc(G) + 1$. If $| S \cap Y | \geq 2$ and $y_i \in S$, then $y_j \in S$, for each $j \in [i]$. We may also conclude that there exists two kinds of vertex covers of size $mvc(G) + 1$*

i. $X \cup \{y_i\}; i \in [q]$.
ii. $\{y_1, \ldots, y_{i+1}, x_{i+1}, \ldots, x_p\}; i \in [p-2]$.

Let $k = min\{i \mid \{x_i, y_q\} \in E\}$.

Lemma 4. *For a given chain graph $G = (X \cup Y, E)$ with $mvc(G) = min\{|X|, |Y|\}$ and $|Y| > |X| + 1$, if X is the only minimum vertex cover of G and $| \cup_{j=1}^{k-1} N(x_j) | = k$, then $evc(G) = mvc(G) + 1$.*

Proof. By above definition $k = min\{i \mid \{x_i, y_q\} \in E\}$, if $| \cup_{j=1}^{k-1} N(x_j) | = k$. Then any attack can be defended by moving the guards from the configuration $X \cup \{y_q\}$ to configuration $\{y_1, \ldots, y_k, x_k, \ldots, x_p\}$ (or from $\{y_1, \ldots, y_k, x_k, \ldots, x_p\}$ to $X \cup \{y_q\}$). So, in this case $evc(G) = mvc(G) + 1$. □

Let $V' = \{i \mid | \cup_{j=1}^{i} N(x_j)| = i + 1\}$.

Lemma 5. *For a given chain graph $G = (X \cup Y, E)$ with $mvc(G) = min\{|X|, |Y|\}$ and $|Y| > |X| + 1$, if X is the only minimum vertex cover of G and $| \cup_{j=1}^{k-1} N(x_j)| > k$, then $evc(G) = mvc(G) + 2$.*

Proof. If $| \cup_{j=1}^{k-1} N(x_j)| > k$ and $V' \neq \phi$, then let $l = max\{i \mid i \in V'\}$. If the initial configuration is of type-ii, then attack $\{x_p, y_q\}$ and make the configuration $X \cup \{y_q\}$, if possible. Then attack $\{x_{l+1}, y_{l+1}\}$, the guard at x_{l+1} moves to y_{l+1} and since $\{y_q, x_{l+1}\} \notin E$, so there does not exist any guard which can move to x_{l+1}, hence no defending move exists, hence $evc(G) = mvc(G) + 2$.

If the set $V' = \phi$, then $| \cup_{j=1}^{i} N(x_j) | > i + 2$, which implies all vertex covers of size $mvc(G) + 1$ are of type-i. Now whatever the initial configuration be attack $\{x_p, y_q\}$. The configuration after defending this should be $X \cup \{y_q\}$. Now attack $\{x_{k-1}, y_{k-1}\}$, the guard at x_{k-1} moves to y_{k-1} now there is no guard which can move to x_{k-1} and form a vertex cover. So $evc(G) = mvc(G) + 2$. □

Now, consider the case when $|X| = |Y|$.

Lemma 6. *For a given chain graph $G = (X \cup Y, E)$, if $mvc(G) = min\{|X|, |Y|\}$, $|X| = |Y|$ and there exists a minimum vertex cover containing y_1 and x_p, then $evc(G) = mvc(G) + 1$.*

Proof. There exists a minimum vertex cover of G that contains both y_1 and x_p. This implies there exists $i \in [p]$, such that $\cup_{j=1}^{i} N(x_j) = \cup_{j=1}^{i} \{y_j\}$ and $evc(G) \in \{mvc(G), mvc(G) + 1\}$. If $evc(G) = mvc(G)$, then the initial configuration can be of 3 types: (i) X, (ii) Y and (iii) $\{y_1, \ldots, y_i, x_{i+1}, \ldots, x_p\}$, $i \in [p]$.

If the initial configuration is of type-iii, then attack $\{x_1, y_1\}$ and change it to X if possible. Then attack $\{y_i, x_{i+1}\}$, so the guard at x_{i+1} moves to y_i and i guards at $x_1, x_2, \ldots, x_i$ have $i - 1$ places, i.e. $y_1, y_2, \ldots, y_{i-1}$ to move. Hence no new configuration can be made which will form a vertex cover.

If the initial configuration is Y, then attack $\{y_i, x_{i+1}\}$. The guard at y_i moves to x_{i+1} and $p - i$ guards at $y_{i+1}, \ldots, y_p$ have $p - i - 1$ places, i.e. $x_{i+2}, \ldots, x_p$ to move. Hence no new configuration can be made which will form a vertex cover.

This implies G can not be defended with $mvc(G)$ guards. So, $evc(G) = mvc(G) + 1$. $\qquad \square$

Lemma 7. *For a given chain graph $G = (X \cup Y, E)$, with $mvc(G) = min\{|X|, |Y|\}$ and $|Y| = |X|$, if the only minimum vertex covers are X and Y, then $evc(G) = mvc(G)$.*

Proof. The only type of minimum vertex covers are X and Y. This implies $|\cup_{j=1}^{l} N(x_j)| \geq l + 1$, for all $l \in [p-1]$. Now if the initial configuration is X, then attack on any edge $\{x_i, y_j\}$ can be defended by moving all the guards to Y, this can be done since by Hall's Theorem there exists a perfect matching in $(X \setminus \{x_i\}, Y \setminus \{y_j\})$. Similarly, if the initial configuration is Y, then attack on any edge $\{x_i, y_j\}$ can be defended by moving all the guards to X, this can also be done since by Hall's Theorem there exists a perfect matching in $(X \setminus \{x_i\}, Y \setminus \{y_j\})$. So, $evc(G) = mvc(G)$. $\qquad \square$

3.2   For Chain Graphs with Pendant Vertices in X or in X, Y both

If y_1 and x_p both have pendant vertices attached (consider that the graph is not K_2; for K_2, $evc(G) = mvc(G) = 1$), then there exists a minimum vertex cover that contains x_p and y_1, which implies $evc(G) \in \{mvc(G), mvc(G) + 1\}$. Now if $evc(G) = mvc(G)$, then there exists a configuration such that a guard is allocated at the pendant vertex x_1 (if not then we can attack the edge $\{y_1, x_1\}$ and shift the guard at y_1 to x_1). This implies that there is no guard in y_1. Now attack $\{x_p, y_1\}$, then the guard at x_p moves to y_1 and the guard at x_1 has to stay at x_1. So in this new configuration, x_1 and y_1 both have guards allocated, a contradiction since no minimum vertex cover can contain the pendant vertex and its respective stem. So, $evc(G) = mvc(G) + 1$.

Now consider the case when only X has pendant vertices, that is, only y_1 is the stem. If $mvc(G) < min\{|X|, |Y|\}$, then $evc(G) = mvc(G) + 1$. If $mvc(G) =$

$|X|$, then y_1 has only one pendant neighbour (otherwise $mvc(G) < |X|$, leading to a contradiction). Since $\{y_1, x_2, \ldots, x_p\}$ forms a minimum vertex cover and it is connected, $mvc(G) = cvc(G)$. This implies that $evc(G) \in \{mvc(G), mvc(G)+1\}$ Further, two cases may arise.

Case 1: $|X| < |Y|$
If $evc(G) = mvc(G)$, the initial guard allocation can be of 2 types; X and $\{y_1, \ldots, y_i, x_{i+1}, \ldots, x_p\}$.

   If the initial configuration is X, then if $\{x_p, y_1\}$ is attacked then the guard at x_1 can not move anywhere, failing to produce a valid defending move.

   If the initial configuration is $\{y_1, \ldots, y_i, x_{i+1}, \ldots, x_p\}$ then attack $\{x_1, y_1\}$, the only configuration it can form is X. But then, attacking $\{x_p, y_1\}$ will lead to a win for the attacker.

   So, $evc(G) \neq mvc(G)$. This implies $evc(G) = mvc(G) + 1$.

Case 2: $|X| = |Y|$
If the initial guard allocation is X or Y, then attacking $\{x_p, y_1\}$ will lead to a win for the attacker.

   If the initial configuration is $\{y_1, \ldots, y_i, x_{i+1}, \ldots, x_p\}$ then attack $\{x_1, y_1\}$, the only configuration it can form is X. But then, attacking $\{x_p, y_1\}$ will lead to a win for the attacker.

   So, $evc(G) \neq mvc(G)$. This implies that $evc(G) = mvc(G) + 1$.

   The above characterization is done by observing a property that for a given chain graph $G = (X \cup Y, E)$, whether there exists a minimum vertex cover S that contains both x_p and y_1 or not. This property can be checked in polynomial time for a given chain graph. Before starting the process of the algorithm, by preprocessing, an array $A[1, 2, \ldots, p]$ can be formed, where i^{th} cell contains the degree of x_i. If there exists a $j \in [p-1]$, such that $A[j] \leq j$, then there exists a minimum vertex cover of G that contains both x_p and y_1. If there does not exist such j, then the only vertex covers are of the form X or Y.

   From the above lemmas and results, we can conclude the following theorem.

Theorem 5. *Given a connected chain graph $G = (V, E)$, $evc(G)$ can be computed in $O(n + m)$ time.*

4   A Linear Time Algorithm for Split Graphs

In this section, we present a linear-time algorithm to compute the eternal vertex cover number for split graphs. Note that, there already exists a quadratic time algorithm to compute $evc(G)$ for chordal graphs. Since the class of split graphs is a subclass of chordal graphs, we also have a quadratic time algorithm to compute $evc(G)$ for split graphs. But, in this section, we present a linear-time algorithm to compute $evc(G)$ for any split graph G.

   The following result is already known regarding the eternal vertex cover number of chordal graphs.

Theorem 6. *[4] Given a connected chordal graph $G = (V, E)$ and the set of all cut vertices X of G, $evc(G) = mvc_X(G)$ if and only if for every vertex $v \in V(G) \backslash X$, we have $mvc_{X \cup \{v\}}(G) = mvc_X(G)$; otherwise $evc(G) = mvc_X(G) + 1$.*

Since split graphs are chordal graphs, for any split graph G we have $evc(G) \in \{mvc_X(G), mvc_X(G) + 1\}$.

Let $G = (K \cup I, E)$ be a connected split graph, where K is a clique and I is an independent set. Without loss of generality, we may assume that K is a maximal clique of G. Let X denote the set of cut vertices of G. Now, we first prove the following lemmas.

Lemma 8. *If for each $x \in K$, $|N(x)| > |K| - 1$, then $mvc(G) = mvc_X(G) = |K|$. Otherwise $mvc(G) = mvc_X(G) = |K| - 1$.*

Proof. If for each $x \in K$, $|N(x)| > |K| - 1$, then each $x \in K$ has at least one neighbour in I. Note that any minimum vertex cover must contain at least $|K| - 1$ vertices from K. If there exists a minimum vertex cover S that contains only $|K| - 1$ vertices from K. Then there exists a vertex $v \in K$, such that v does not belong to S. So, S must contain all neighbours of v from I, implying that $|S| \geq |K|$. Since K is itself a vertex cover of size $|K|$, if v has more than one neighbour in I, then $|S| > |K|$, a contradiction. So, K always form a minimum vertex cover in this case. Since $X \subseteq K$, it can be concluded that $mvc(G) = mvc_X(G) = |K|$.

Now if there exists $x \in K$, such that $|N(x)| = |K| - 1$, then $K \setminus \{x\}$ forms a minimum vertex cover of cardinality $|K| - 1$. Note that x cannot be a cut vertex (as it has no neighbour in I). So, $X \subseteq K \setminus \{x\}$ and $K \setminus \{x\}$ forms a minimum vertex cover, implying that $mvc(G) = mvc_X(G) = |K| - 1$. □

Lemma 9. *$evc(G) \in \{mvc(G), mvc(G) + 1\}$.*

Proof. The proof follows from the fact that $evc(G) \in \{mvc_X(G), mvc_X(G) + 1\}$ and $mvc(G) = mvc_X(G)$. □

Lemma 10. *Let $mvc(G) = |K| - 1$. Then $evc(G) = mvc(G) + 1$ if $I \neq \phi$ and $evc(G) = mvc(G)$ if $I = \emptyset$.*

Proof. If $I \neq \phi$, then consider a vertex $y \in I$. By Theorem 4, if $evc(G) = mvc(G) = |K| - 1$, then there exists a minimum vertex cover S that contains y, which implies $|S \cap K| \leq |K| - 2$, leading to a contradiction. Hence $evc(G) = mvc(G) + 1$. If $I = \phi$, then G is a complete graph, implying $evc(G) = mvc(G)$. □

Lemma 11. *Let $mvc(G) = |K|$ and there exists at least one pendant vertex $y_i \in I$, then $evc(G) = mvc(G) + 1$.*

Proof. Let x_j be the only neighbour of the pendant vertex y_i, then $x_j \in X$. On contrary assume that $evc(G) = mvc(G)$, then by Theorem 6 there exists a minimum vertex cover S that contains both X and y_i. Hence $x_j \in S$ as $y_i \in S$. Then there must be a vertex $x_k \in K$, which does not belong to S.

Since $mvc(G) = |K|$, by Lemma 8, $N(x) \cap I \neq \phi$ for all $x \in K$. Hence, if x_k is not in S then all of its neighbours should be in S. Since y_i is not a neighbour of x_k, no neighbour of x_k in I belongs to S. Hence contradiction arises. So, $evc(G) = mvc(G) + 1$.   □

Lemma 12. *Let $mvc(G) = |K|$, G has no pendant vertices and for each $x \in K$, $deg(x) \geq |K| + 1$. Then, $evc(G) = mvc(G) + 1$.*

Proof. Note that $mvc_X(G) = mvc(G)$. On contrary assume that $evc(G) = mvc(G)$. Then, by Theorem 4, for any $y_i \in I$, there exists a minimum vertex cover S that contains y_i. Then $|K \cap S| = |K| - 1$. Let $x_j \in K$ be the vertex which is not in S. Since $|N_I(x_j)| \geq 2$, S contains at least 2 vertices from I. But, then $|S| \geq |K| + 1$, a contradiction arises. Hence, $evc(G) = mvc(G) + 1$.   □

Lemma 13. *Let G does not has any pendant vertex with $mvc(G) = |K|$ and $X_1 = \{x \in K : deg_I(x) = 1\}$. If $N(X_1) \cap I = I$, then $evc(G) = mvc(G)$, otherwise if $N(X_1) \cap I$ is properly contained in I, then $evc(G) = mvc(G) + 1$.*

Proof. Proof of the Lemma 13 has been omitted due to space constraint.

So, by the above lemmas we can conclude the following theorem.

Theorem 7. *For a connected split graph $G(K \cup I, E)$, $evc(G)$ can be computed in time $O(n + m)$.*

Proof. The proof of the theorem is straightforward from the above lemmas. Before starting the algorithm, by preprocessing, an array $A[1, 2, \ldots, n]$ can be formed, such that $A[i]$ stores the degree of the vertex v_i. By help of this array the algorithm can run in $O(n + m)$ time.   □

5   A Polynomial Time Algorithm for Cographs

As mentioned earlier, any connected cograph can be written as join of k graphs, $G_1, G_2, \ldots, G_k$ where each G_i is either K_1 or a disconnected graph. Note that for a connected cograph $G = (V, E)$, the maximum independent set of G is a subset of $V(G_i)$, for some $i \in [k]$. So, any minimum vertex cover of G contains vertices from at least $k - 1$ number of G_i's.

By Theorem 1, given a connected cograph $G = (V, E)$, for which each minimum vertex cover is connected, $evc(G)$ can be calculated by checking $mvc_v(G) = mvc(G)$ for each $v \in V$. To check this condition for any $v \in V$, a new graph $G' = (V', E')$ can be formed from G, where $V' = V \cup \{u\}$, $E' = E \cup \{uv\}$; then we can check whether $mvc(G) = mvc(G')$. The class of cographs is not closed under pendant vertex addition. But cographs are also weakly chordal graphs, which are closed under pendant vertex addition. So, we are giving a polynomial time algorithm $EVC\_CHECK(G)$ for connected cographs $G = (V, E)$ for which every minimum vertex cover is connected, to compute $evc(G)$. For this, we are using the algorithm given in [7] to compute minimum vertex cover for weakly chordal graphs. So, for each vertex of the cograph $G = (V, E)$ (for which every minimum vertex cover is connected) we will add a pendent vertex (only one at each step and we will delete the previous pendent vertex while adding pendent to the next vertex) and use the algorithm in [7] to compute whether the mvc is same for the old and new graph. If it is same for each vertex of $G = (V, E)$, then $evc(G) = mvc(G)$; and $evc(G) = mvc(G) + 1$ otherwise.

Since the algorithm to find minimum vertex cover in weakly chordal graphs mentioned in [7] runs in $O(nm)$ time, we may conclude the following theorem from the above discussion.

Theorem 8. *Given a connected cograph $G = (V, E)$, for which every minimum vertex cover is connected, $evc(G)$ can be calculated in time $O(nm)$.*

The algorithm mentioned in Theorem 8 will be called as EVC_CHECK from here on.

When $k = 2$, the graph G is join of 2 subgraphs G_1 and G_2. Here we are assuming $|G_1| \leq |G_2|$ and both G_1 and G_2 are non-empty. If $mis(G) > min\{|G_1|, |G_2|\}$, then maximum independent set I of G is a subset of G_2. If $I \subset G_2$, then each minimum vertex cover S is connected, since $G_2 \cap S \neq \phi$ and $G_1 \cap S \neq \phi$. So $evc(G)$ can be computed by $EVC\_CHECK(G)$. If $I = G_2$, then G_1 is the only minimum vertex cover and there does not exist any minimum vertex cover S that contains any vertex of G_2, so by Theorem 4, $evc(G) \neq mvc(G)$. In this case, $G_1 \cup \{u\}$, such that $u \in G_2$, forms an initial configuration of guards, as G_2 is independent, implying $evc(G) = mvc(G) + 1$.

So, the case remains to be observed is, when $mis(G) \leq min\{|G_1|, |G_2|\}$. If $mis(G) < min\{|G_1|, |G_2|\}$, then any minimum vertex cover S is connected, since $G_2 \cap S \neq \phi$ and $G_1 \cap S \neq \phi$. So, $evc(G)$ can be calculated by $EVC\_CHECK(G)$.

Now for the case when $mis(G) = min\{|G_1|, |G_2|\}$ and by previous assumption, $|G_1| = min\{|G_1|, |G_2|\}$. If $|G_1| = |G_2|$, then G_1 or G_2 is an independent set.

If both are independent then G is $K_{|G_1|, |G_1|}$, and $evc(G) = mvc(G)$.

If G_2 is not independent, then G_1 is independent. If there exists a minimum vertex cover S that contains at least one vertex of G_1, then it contains all vertex of G_1, so it contains no vertex from G_2. But since G_2 is not independent, then there exists at least one edge in $E(G_2)$ for which no endpoint is in S. Hence no minimum vertex cover contains any vertex from G_1. So, by Theorem 4, $evc(G) \neq mvc(G)$. So, $evc(G) = mvc(G) + 1$ and $G_2 \cup \{u\}$ where $u \in G_1$, forms an initial guard allocation configuration.

Lemma 14. *Given a connected cograph $G = (V, E)$ which is a join of G_1 and G_2. If $mis(G) = min\{|G_1|, |G_2|\}$ and $|G_1| < |G_2|$, then $evc(G)$ can be calculated in polynomial time.*

Proof. Proof of Lemma 14 is omitted due to space constraint.

Observation 5. *Given a connected cograph $G = (V, E)$ and $k \geq 3$, every minimum vertex cover of G is connected.*

Note that, for $k \geq 3$, $evc(G)$ can be computed in $O(nm)$ time using $EVC\_CHECK(G)$. So, from the above observations and lemma the following theorem can be concluded.

Theorem 9. *Given a connected cograph $G = (V, E)$, $evc(G)$ can be computed in $O(nm)$-time.*

6 Conclusion and Future Aspects

In this paper we have given polynomial time algorithms for three restricted subclasses of perfect graphs, i.e. chain graphs, split graphs and cographs. For split graphs, running time of our algorithm is linear. The class of split graphs is an important subclass of chordal graphs, for which a quadratic time algorithm was already known in the literature. It will also be interesting to try for linear-time algorithms for eternal vertex cover problem for chordal graphs, or some other important subclasses of chordal graphs. The eternal vertex cover problem is NP-hard for bipartite graphs and the class of chain graphs is the largest class of bipartite graphs for which linear time algorithm has been found. The complexity status of the eternal vertex cover problem is still unknown for other important subclasses of bipartite graphs. Here we have solved the complexity status of eternal vertex cover problem for cographs, but for larger graph classes like distance hereditary graphs, it is yet to be solved.

References

1. Araki, H., Fujito, T., Inoue, S.: On the eternal vertex cover numbers of generalized trees. IEICE Trans. Fundam. Electron. Commun. Comput. Sci. **98-A**(6), 1153–1160 (2015)
2. Babu, J., Chandran, L.S., Francis, M., Prabhakaran, V., Rajendraprasad, D., Warrier, N.J.: On graphs whose eternal vertex cover number and vertex cover number coincide. Discret. Appl. Math. **319**, 171–182 (2022)
3. Babu, J., Misra, N., Nanoti, S.G.: Eternal vertex cover on bipartite graphs. In: Kulikov, A.S., Raskhodnikova, S. (eds.) Computer Science–Theory and Applications. CSR 2022. LNCS, vol. 13296, pp. 64–76. Springer, Cham (2022). https://doi.org/10.1007/978-3-031-09574-0_5
4. Babu, J., Prabhakaran, V.: A new lower bound for the eternal vertex cover number of graphs. J. Comb. Optim. **06**, 2482–2498 (2021)
5. Babu, J., Prabhakaran, V., Sharma, A.: A substructure based lower bound for eternal vertex cover number. Theor. Comput. Sci. **890**, 87–104 (2021)
6. Fomin, F.V., Gaspers, S., Golovach, P.A., Kratsch, D., Saurabh, S.: Parameterized algorithm for eternal vertex cover. Inf. Process. Lett. **110**(16), 702–706 (2010)
7. Hayward, R.B., Spinrad, J., Sritharan, R.: Weakly chordal graph algorithms via handles. In: Proceedings of the Eleventh Annual ACM-SIAM Symposium on Discrete Algorithms, 9–11 January 2000, San Francisco, CA, USA, pp. 42–49. ACM/SIAM (2000)
8. Klostermeyer, W.F., Mynhardt, C.M.: Edge protection in graphs. Australas. J. Comb. **45**, 235–250 (2009)
9. McConnell, R.M., Spinrad, J.P.: Modular decomposition and transitive orientation. Discret. Math. **201**(1–3), 189–241 (1999)

On the Complexity of Distance-d Independent Set Reconfiguration

Duc A. Hoang$^{(\boxtimes)}$

Graduate School of Informatics, Kyoto University, Kyoto, Japan
hoang.duc.8r@kyoto-u.ac.jp

Abstract. For a fixed positive integer $d \geq 2$, a *distance-d independent set (DdIS)* of a graph is a vertex subset whose distance between any two members is at least d. Imagine that there is a token placed on each member of a DdIS. Two DdISs are adjacent under Token Sliding (TS) if one can be obtained from the other by moving a token from one vertex to one of its unoccupied adjacent vertices. Under Token Jumping (TJ), the target vertex needs not to be adjacent to the original one. The DISTANCE-d INDEPENDENT SET RECONFIGURATION (DdISR) problem under TS/TJ asks if there is a corresponding sequence of adjacent DdISs that transforms one given DdIS into another. The problem for $d = 2$, also known as the INDEPENDENT SET RECONFIGURATION problem, has been well-studied in the literature and its computational complexity on several graph classes has been known. In this paper, we study the computational complexity of DdISR on different graphs under TS and TJ for any fixed $d \geq 3$. On chordal graphs, we show that DdISR under TJ is in P when d is even and PSPACE-complete when d is odd. On split graphs, there is an interesting complexity dichotomy: DdISR is PSPACE-complete for $d = 2$ but in P for $d = 3$ under TS, while under TJ it is in P for $d = 2$ but PSPACE-complete for $d = 3$. Additionally, certain well-known hardness results for $d = 2$ on general graphs, perfect graphs, planar graphs of maximum degree three and bounded bandwidth can be extended for $d \geq 3$.

Keywords: Reconfiguration problem · Distance-d independent set · Computational complexity · Token sliding · Token jumping

1 Introduction

Recently, *reconfiguration problems* have attracted the attention from both theoretical and practical viewpoints. The input of a reconfiguration problem consists of two *feasible solutions* of some *source problem* (e.g., SATISFIABILITY, INDEPENDENT SET, VERTEX COVER, DOMINATING SET, etc.) and a *reconfiguration rule* that describes an adjacency relation between solutions. One of the primary goal is to decide whether one feasible solution can be transformed into the other via a sequence of adjacent feasible solutions where each intermediate member is

C.-C. Lin et al. (Eds.): WALCOM 2023, LNCS 13973, pp. 254–266, 2023.
https://doi.org/10.1007/978-3-031-27051-2_22

obtained from its predecessor by applying the given reconfiguration rule exactly once. Such a sequence, if exists, is called a *reconfiguration sequence*. Readers may recall the classic Rubik's cube puzzle as an example of a reconfiguration problem, where each configuration of the Rubik's cube corresponds to a feasible solution, and two configurations (solutions) are adjacent if one can be obtained from the other by rotating a face of the cube by either 90, 180, or 270 degrees. The question is whether one can transform an arbitrary configuration to the one where each face of the cube has only one color. For an overview of this research area, we refer readers to the surveys [7, 26, 27, 29].

Reconfiguration problems involving *vertex subsets* (e.g., clique, independent set, vertex cover, dominating set, etc.) of a graph have been extensively considered in the literature. In such problems, to make it more convenient for describing reconfiguration rules, one usually view a vertex subset of a graph as a set of tokens placed on its vertices. Some of the well-known reconfiguration rules in this setting are:

- Token Sliding (TS): a token can only move to one of the unoccupied adjacent vertices;
- Token Jumping (TJ): a token can move to any unoccupied vertex; and
- Token Addition/Removal (TAR(k)): a token can either be added to an unoccupied vertex or removed from an occupied one, such that the number of tokens is always (upper/lower) bounded by some given positive integer k.

Let G be a simple, undirected graph. An *independent set* of a graph G is a set of pairwise non-adjacent vertices. The INDEPENDENT SET (MAXIS) problem, which asks if G has an independent set of size at least some given positive integer k, is one of the fundamental NP-complete problems in the computational complexity theory [14]. Given an integer $d \geq 2$, a *distance-d independent set* (also known as *d-scattered set* or sometimes *d-independent set*[1]) of G is a set of vertices whose pairwise distance is at least d. This "distance-d independent set" concept generalizes "independent set": an independent set is also a distance-2 independent set but may not be a distance-d independent set for $d \geq 3$. Given a fixed integer $d \geq 2$, the DISTANCE-d INDEPENDENT SET (MAXDdIS) problem asks if there is a distance-d independent set of G whose size is at least some given positive integer k. Clearly, MAXD2IS is nothing but MAXIS. It is known that MAXDdIS is NP-complete for every fixed $d \geq 3$ on general graphs [21] and even on regular bipartite graphs when $d \in \{3, 4, 5\}$ [22]. Eto et al. [12] proved that MAXDdIS is NP-complete for every fixed $d \geq 3$ even for bipartite graphs and for planar bipartite graphs of maximum degree three. They also proved that on chordal graphs, MAXDdIS is in P for any even $d \geq 2$ and remains NP-complete for any odd $d \geq 3$. The complexity of MAXDdIS on several other graphs has

---

[1] In fact, the terminology *d-independent set* is sometimes used to indicate a vertex subset such that any two members are of distance at least $d + 1$ for $d \geq 1$. We note that sometimes a *d-independent set* is defined as a vertex subset S such that the maximum degree of the subgraph induced by S is at most d. In some other contexts, a *d-independent set* is nothing but an independent set of size d.

also been studied. Additionally, MaxDdIS and its variants have been studied extensively from different viewpoints, including exact exponential algorithms, approximability, and parameterized complexity. We refer readers to [12, 19, 32] and the references therein for more details.

In this paper, we take MaxDdIS as the source problem and initiate the study of DISTANCE-d INDEPENDENT SET RECONFIGURATION (DdISR) from the computational complexity viewpoint. The problem for $d = 2$, which is usually known as the INDEPENDENT SET RECONFIGURATION (ISR) problem, has been well-studied from both classic and parameterized complexity viewpoints. Readers are referred to [7, 27] for a complete overview of recent developments regarding ISR. We now briefly mention some known results regarding the computational complexity of ISR on different graph classes. ISR remains PSPACE-complete under any of TS, TJ, TAR on general graphs [17], planar graphs of maximum degree three and bounded bandwidth [15, 30, 31] and perfect graphs [18]. Under TS, the problem is PSPACE-complete even on split graphs [3]. Interestingly, on bipartite graphs, ISR under TS is PSPACE-complete while under any of TJ and TAR it is NP-complete [24]. On the positive side, ISR under any of TJ and TAR is in P on even-hole-free graphs [18] (which also contains chordal graphs, split graphs, interval graphs, trees, etc.), cographs [5], and claw-free graphs [6]. ISR under TS is in P on cographs [18], claw-free graphs [6], trees [10], bipartite permutation graphs and bipartite distance-hereditary graphs [13], and interval graphs [4, 8].

DdISR for $d \geq 3$ was first studied by Siebertz [28] from the parameterized complexity viewpoint. More precisely, in [28], Siebertz proved that DdISR under TAR is in FPT for every $d \geq 2$ on "nowhere dense graphs" (which generalized the previously known result for $d = 2$ of Lokshtanov et al. [25]) and it is W[1]-hard for some value of $d \geq 2$ on "somewhere dense graphs" that are closed under taking subgraphs.

Since the TJ and TAR rules are somewhat equivalent [18], i.e., any TJ-sequence between two size-k token-sets can be converted into a TAR-sequence between them whose members are token-sets of size at least $k-1$ and vice versa[2], in this paper, we consider DdISR ($d \geq 3$) under only TS and TJ. In short, we show the following results.

- It is worth noting that
  - Even though it is well-known that MaxDdIS on G and MaxIS on its $(d-1)$th power (this concept will be defined later) are equivalent, this does *not* necessarily holds for their reconfiguration variants. (Sect. 3.1)
  - The definition of DdIS implies the triviality of DdISR for large enough d on graphs whose (connected) components' diameters are bounded by some constant, including cographs and split graphs. (Sect. 3.2)
- On chordal graphs and split graphs, there are some interesting complexity dichotomies. (Sect. 4)
  - Under TJ on chordal graphs, DdISR is in P for even d and PSPACE-complete for odd d.

---

[2] They proved the result for ISR, but it is not hard to extend it for DdISR.

- On split graphs, DdISR under TS is PSPACE-complete for $d = 2$ [3] but in P for $d = 3$. Under TJ, it is in P for $d = 2$ [18] but PSPACE-complete for $d = 3$.
- There is a proof of the PSPACE-hardness of DdISR on general graphs under TJ for $d \geq 3$ which is not a direct extension of the corresponding result by Ito et al. [17] for $d = 2$. (Sect. 5) Additionally, several known results for $d = 2$, including the one by Ito et al., can be extended for $d \geq 3$. (Sect. 6)

Due to the space limitation, the proofs of some statements (marked by $*$) are omitted and can be found in the full version [16].

2   Preliminaries

For terminology and notation not defined here, see [11]. Let G be a simple, undirected graph with vertex-set $V(G)$ and edge-set $E(G)$. For two sets I, J, we sometimes use $I - J$ and $I + J$ to indicate $I \setminus J$ and $I \cup J$, respectively. Additionally, we simply write $I - u$ and $I + u$ instead of $I - \{u\}$ and $I + \{u\}$, respectively. The *neighbors* of a vertex v in G, denoted by $N_G(v)$, is the set $\{w \in V(G) : vw \in E(G)\}$. The *closed neighbors* of v in G, denoted by $N_G[v]$, is simply the set $N_G(v) + v$. Similarly, for a vertex subset $I \subseteq V(G)$, its *neighbor* $N_G(I)$ and *closed neighbor* $N_G[I]$ are respectively $\bigcup_{v \in I} N_G(v)$ and $N_G(I) + I$. The *degree* of a vertex v in G, denoted by $\deg_G(v)$, is $|N_G(v)|$. The *distance* between two vertices u, v in G, denoted by $\text{dist}_G(u, v)$, is the number of edges in a shortest path between them. For convenience, if there is no path between u and v then $\text{dist}_G(u, v) = \infty$. The *diameter* of G, denoted by $\text{diam}(G)$, is the largest distance between any two vertices. A *(connected) component* of G is a maximal subgraph in which there is a path connecting any pair of vertices. An *independent set (IS)* of G is a vertex subset I such that for any $u, v \in I$, we have $uv \notin E(G)$. A *distance-d independent set (DdIS)* of G for an integer $d \geq 2$ is a vertex subset $I \subseteq V(G)$ such that for any $u, v \in I$, $\text{dist}_G(u, v) \geq d$. We use $\alpha_d(G)$ to denote the *maximum* size of a distance-d independent set of G. When $d = 2$, we use the well-known notation $\alpha(G)$ instead of $\alpha_2(G)$.

Unless otherwise noted, we denote by (G, I, J, R, d) an instance of DdISR under $\mathsf{R} \in \{\mathsf{TS}, \mathsf{TJ}\}$ where I and J are two distinct DdISs of a given graph G, for some fixed $d \geq 2$. Since (G, I, J, R, d) is obviously a no-instance if $|I| \neq |J|$, from now on, we always assume that $|I| = |J|$. Imagine that a token is placed on each vertex in a DdIS of a graph G. A TS-*sequence* in G between two DdISs I and J is the sequence $\mathcal{S} = \langle I = I_0, I_1, \ldots, I_q = J \rangle$ such that for $i \in \{0, \ldots, q-1\}$, the set I_i is a DdIS of G and there exists a pair $x_i, y_i \in V(G)$ such that $I_i - I_{i+1} = \{x_i\}$, $I_{i+1} - I_i = \{y_i\}$, and $x_i y_i \in E(G)$. By simply removing the restriction $x_i y_i \in E(G)$, we immediately obtain the definition of a TJ-*sequence* in G. Depending on the considered rule $\mathsf{R} \in \{\mathsf{TS}, \mathsf{TJ}\}$, we can also say that I_{i+1} is obtained from I_i by *immediately sliding/jumping* a token from x_i to y_i and write $x_i \xrightarrow{G}_\mathsf{R} y_i$. As a result, we can also write $\mathcal{S} = \langle x_0 \xrightarrow{G}_\mathsf{R} y_0, \ldots, x_{q-1} \xrightarrow{G}_\mathsf{R} y_{q-1} \rangle$. In short, $\mathcal{S}$ can be viewed as a (ordered) sequence of either DdISs or token-moves. (Recall that we defined S as a sequence between I and J. As a result, when regarding S as a sequence of token-moves, we implicitly assume that the initial DdIS is I.) With respect to the latter viewpoint, we say that S *slides/jumps a token t from u to v in G* if t is originally placed on $u \in I_0 = I$ and finally on $v \in I_q = J$ after performing S. The *length* of a R-sequence is simply the number of times the rule R is applied.

We conclude this section with the following simple remark: since MAXDdIS is in NP, DdISR is always in PSPACE [17]. As a result, to show the PSPACE-completeness of DdISR, it is sufficient to construct a polynomial-time reduction from a known PSPACE-hard problem and prove its correctness.

3   Observations

3.1   Graphs and Their Powers

An extremely useful concept for studying distance-d independent sets is the so-called *graph power*. For a graph G and an integer $s \geq 1$, the *sth power of* G is the graph G^s whose vertices are $V(G)$ and two vertices u, v are adjacent in G^s if $\text{dist}_G(u, v) \leq s$. Observe that I is a distance-d independent set of G if and only if I is an independent set of G^{d-1}. Therefore, MAXDdIS in G is equivalent to MAXIS in G^{d-1}.

However, this may *not* apply for their reconfiguration variants. More precisely, the statement "the DdISR's instance (G, I, J, R, d) is a yes-instance if and only if the ISR's instance $(G^{d-1}, I, J, R, 2)$ is a yes-instance" holds for R = TJ but not for R = TS. The reason is that we do not care about edges when performing token-jumps (as long as they result new DdISs), therefore whatever token-jump we perform in G can also be done in G^{d-1} and vice versa. Therefore, we have

Proposition 1. *Let $\mathcal{G}$ and $\mathcal{H}$ be two graph classes and suppose that for every $G \in \mathcal{G}$ we have $G^{d-1} \in \mathcal{H}$ for some fixed integer $d \geq 2$. If ISR under TJ on $\mathcal{H}$ can be solved in polynomial time, so does DdISR under TJ on $\mathcal{G}$.*

Recall that the power of any interval graph is also an interval graph [1,9] and ISR under TJ on even-hole-free graphs (which contains interval graphs) is in P [18]. Along with Proposition 1, we immediately obtain the following corollary.

Corollary 2. *DdISR under TJ is in P on interval graphs for any $d \geq 2$.*

On the other hand, when using token-slides, we need to consider which edge can be used for moving tokens, and clearly G^{d-1} has much more edges than G, which means certain token-slides we perform in G^{d-1} cannot be done in G. Figure 1 describes an example of a DdISR's no-instance $(G, I, J, TS, 3)$ whose corresponding ISR's instance $(G^2, I, J, TS, 2)$ is a yes-instance. One can verify that in the former instance no token can move without breaking the "distance-3 restriction", while in the latter I can be transformed into J using exactly two token-slides. Moreover, these moves use edges that do not appear in G.

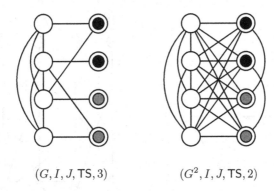

$(G, I, J, \mathsf{TS}, 3)$ $(G^2, I, J, \mathsf{TS}, 2)$

Fig. 1. A DdISR's no-instance $(G, I, J, \mathsf{TS}, 3)$ whose corresponding ISR's instance $(G^2, I, J, \mathsf{TS}, 2)$ is a yes-instance. Tokens in I (resp., J) are marked with black (resp., gray) color.

3.2 Graphs with Bounded Diameter Components

The following observation is straightforward.

Proposition 3. *Let $\mathcal{G}$ be a graph class such that there is some constant $c > 0$ satisfying* $\mathsf{diam}(C_G) \leq c$ *for any $G \in \mathcal{G}$ and any component C_G of G. Then, DdISR on $\mathcal{G}$ under $\mathsf{R} \in \{\mathsf{TS}, \mathsf{TJ}\}$ is in P for every $d \geq c + 1$.*

Proof. When $d \geq c+1$, any DdIS contains at most one vertex in each component of G, and the problem becomes trivial: under TJ, the answer is always "yes"; under TS, compare the number of tokens in each component. □

As a result, on cographs (a.k.a P_4-free graphs), one can immediately derive the following corollary.

Corollary 4. *DdISR on cographs under $\mathsf{R} \in \{\mathsf{TS}, \mathsf{TJ}\}$ is in P for any $d \geq 2$.*

Proof. It is well-known that the problems for $d = 2$ is in P [5,18]. Since a connected cograph has diameter at most two, Proposition 3 settles the case $d \geq 3$. □

4 Chordal Graphs and Split Graphs

In this section, we will focus on chordal graphs and split graphs. Recall that the odd power of a chordal graph is also chordal [1,2] and ISR under TJ on even-hole-free graphs (which contains chordal graphs) is in P [18]. Therefore, it follows from Proposition 1 that

Corollary 5. *DdISR is in P on chordal graphs under TJ for any even $d \geq 2$.*

In contrast, we have the following theorem.

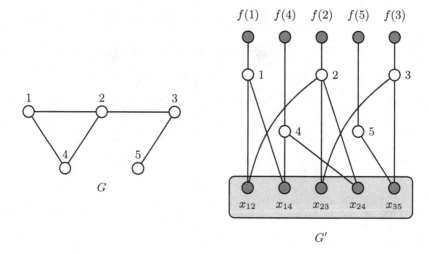

Fig. 2. An example of constructing a chordal graph G' from G for $d = 5$. Vertices in a light-gray box form a clique. New vertices in $V(G') - V(G)$ are marked with the gray color.

Theorem 6. *DdISR is PSPACE-complete on chordal graphs under TJ for any odd $d \geq 3$.*

Proof. We reduce from the ISR problem, which is known to be PSPACE-complete under any of TS and TJ [17]. Let $(G, I, J, \mathsf{TJ}, 2)$ be an ISR's instance. We construct a DdISR's instance $(G', I', J', \mathsf{TJ}, d)$ ($d \geq 3$ is odd) as follows. We note that the same reduction was used by Eto et al. [12] for showing the NP-completeness of MaxDdIS on chordal graphs for odd $d \geq 3$. We first describe how to construct G' from G. First, for each edge $uv \in V(G)$, add a new vertex x_{uv} and create a new edge between x_{uv} and both u and v. Next, we add an edge in G' between x_{uv} and $x_{u'v'}$ for any pair of distinct edges $uv, u'v' \in E(G)$. Finally, for each $v \in V(G)$ we add a new path P_v on $(d-3)/2$ vertices and then add a new edge between v and one of P_v's endpoints. Let the resulting graph be G'. One can verify that G' is indeed a chordal graph: it is obtained from a split graph by attaching new paths to certain vertices. Clearly this construction can be done in polynomial time. (For example, see Fig. 2.)

For each $u \in V(G)$, we define $f(u) \in V(G')$ to be the vertex whose distance in G' from u is largest among all vertices in $V(P_u) + u$. Let $f(X) = \bigcup_{x \in X}\{f(x)\}$ for a vertex subset $X \subseteq V(G)$. From the construction of G', note that if u and v are two vertices of distance 2 in G, one can always find a shortest path Q between $f(u)$ and $f(v)$ whose length is exactly d. Indeed, Q can be obtained by joining the paths from $f(u)$ to u, from u to x_{uw}, from x_{uw} to x_{wv}, from x_{wv} to v, and from v to $f(v)$, where $w \in N_G(u) \cap N_G(v)$. It follows that if I is an independent set of G then $f(I)$ is a distance-d independent set of G'. Therefore, we can set $I' = f(I)$ and $J' = f(J)$.

From the construction of G', note that for each $uv \in E(G)$, x_{uv} is of distance exactly one from each x_{wz} for $wz \in E(G) - uv$, at most two from each $v \in V(G)$, and at most $2 + (d - 3)/2 \leq d - 1$ from each vertex in P_v for $v \in V(G)$. It follows that any distance-d independent set of G' of size at least two must not contain any vertex in $\bigcup_{uv \in E(G)} \{x_{uv}\}$.

We now show that there is a TJ-sequence between I and J in G if and only if there is a TJ-sequence between I' and J' in G'. If $|I| = |J| = 1$, the claim is trivial. As a result, we consider the case $|I| = |J| \geq 2$. Since $f(I)$ is a distance-d independent set in G' if I is an independent set in G, the only-if direction is clear. It remains to show the if direction. Let $\mathcal{S}'$ be a TJ-sequence in G' between I' and J'. We modify $\mathcal{S}'$ by repeating the following steps:

- Let $x \xrightarrow{G'}_{\mathsf{TJ}} y$ be the first token-jump that move a token from $x \in f(V(G))$ to some $y \in V(P_u) + u - f(u)$ for some $u \in V(G)$. If no such token-jump exists, we stop. Let I_x and I_y be respectively the distance-d independent sets obtained before and after this token-jump. In particular, $I_y = I_x - x + y$.
- Replace $x \xrightarrow{G'}_{\mathsf{TJ}} y$ by $x \xrightarrow{G'}_{\mathsf{TJ}} f(u)$ and replace the first step after $x \xrightarrow{G'}_{\mathsf{TJ}} y$ of the form $y \xrightarrow{G'}_{\mathsf{TJ}} z$ by $f(u) \xrightarrow{G'}_{\mathsf{TJ}} z$. From the construction of G', note that any path containing $f(u)$ must also contains all vertices in $V(P_u) + u$. Additionally, $I_x \cap (V(P_u) + u) = \emptyset$, otherwise no token in I_x can jump to y. Therefore, the set $I_x - x + f(u)$ is also a distance-d independent set. Moreover, $\mathrm{dist}_{G'}(f(u), z) \geq \mathrm{dist}_{G'}(y, z)$ for any $z \in V(G') - V(P_u)$. Roughly speaking, this implies that no token-jump between $x \xrightarrow{G'}_{\mathsf{TJ}} y$ and $y \xrightarrow{G'}_{\mathsf{TJ}} z$ breaks the "distance-d restriction". Thus, after the above replacements, $\mathcal{S}'$ is still a TJ-sequence in G'.
- Repeat the first step.

After modification, the final resulting TJ-sequence $\mathcal{S}'$ in G' contains only token-jumps between vertices in $f(V(G))$. By definition of f, we can construct a TJ-sequence between I and J in G simply by replacing each step $x \xrightarrow{G'}_{\mathsf{TJ}} y$ in $\mathcal{S}'$ by $f^{-1}(x) \xrightarrow{G}_{\mathsf{TJ}} f^{-1}(y)$. Our proof is complete. □

Now, we consider the split graphs. Proposition 3 implies that on split graphs (where each component has diameter at most 3), DdISR is in P under $R \in \{\mathsf{TS}, \mathsf{TJ}\}$ for any $d \geq 4$. Interestingly, recall that when $d = 2$, the problem under TS is PSPACE-complete even on split graphs [3] while under TJ it is in P [18]. It remains to consider the case $d = 3$.

Observe that the constructed graph G' in the proof of Theorem 6 is indeed a split graph when $d = 3$. Therefore, we have the following corollary.

Corollary 7. *D3ISR is* PSPACE-*complete on split graphs under* TJ.

In contrast, under TS, we have the following proposition.

Proposition 8. *D3ISR is in* P *on split graphs under* TS.

Proof. Let $(G, I, J, \mathsf{TS}, 3)$ be an instance of D3ISR and suppose that $V(G)$ can be partitioned into a clique K and an independent set S. One can assume without loss of generality that G is connected, otherwise each component can be solved independently. If $|I| = |J| = 1$, the problem becomes trivial: $(G, I, J, \mathsf{R}, 3)$ is always a yes-instance. Thus, we now consider $|I| = |J| \geq 2$. Observe that for every $u \in V(G)$ and $v \in K$, we have $\mathrm{dist}_G(u, v) \leq 2$. Therefore, in this case, both I and J are subsets of S. Now, no token in $I \cup J$ can be slid, otherwise such a token must be slid to some vertex in K, and each vertex in K has distance at most two from any other token, which contradicts the restriction that tokens must form a D3IS. Hence, $(G, I, J, \mathsf{TS}, 3)$ is always a no-instance if $|I| = |J| \geq 2$. $\square$

5   A Reduction Under TJ on General Graphs

Recall that Ito et al. [17] proved the PSPACE-completeness of ISR under TJ/TAR by reducing from 3-SATISFIABILITY RECONFIGURATION (3SAT-R). In this section, we present a simple proof for the PSPACE-hardness of DdISR ($d \geq 3$) on general graphs under TJ by *reducing from* ISR *instead of* 3SAT-R.

Theorem 9 (*). *DdISR is* PSPACE-*complete under* TJ *for any* $d \geq 3$.

Proof (Sketch). Our goal is to construct a DdISR's instance $(G', I, J, \mathsf{TJ}, d)$ ($d \geq 3$) from a given ISR's instance $(G, I, J, \mathsf{TJ}, 2)$. Intuitively, we aim to construct G' from G such that two vertices of distance at least 2 in G would be of distance at least d in G'. We also need to slightly adjust G' to ensure that TJ-sequences in G between I and J can be converted to those in G' and vice versa. (This implies the correctness of our reduction.) More precisely, we adjust G' to avoid placing any token on a vertex in $V(G') - V(G)$ when reconfiguring in G'. $\square$

6   Extending Some Known Results for $d = 2$

In this section, we prove that several known results on the complexity of DdISR for the case $d = 2$ can be extended for $d \geq 3$.

6.1   General Graphs

Ito et al. [17] proved that ISR is PSPACE-complete on general graphs under TJ/TAR. Indeed, their proof uses only maximum independent sets, which implies that any token-jump is also a token-slide [6], and therefore the PSPACE-completeness also holds under TS. We will show that the reduction of Ito et al. [17] can be extended for showing the PSPACE-completeness of DdISR for $d \geq 3$.

Theorem 10 (*). *DdISR is* PSPACE-*complete under* $\mathsf{R} \in \{\mathsf{TS}, \mathsf{TJ}\}$ *for any* $d \geq 3$.

6.2   Perfect Graphs

In this section, we will show the PSPACE-completeness of DdISR ($d \geq 3$) on perfect graphs by extending the corresponding known result of Kamiński et al. [18] for ISR.

Theorem 11 (∗). *DdISR is* PSPACE-*complete on perfect graphs under* R $\in$ {TS, TJ} *for any $d \geq 3$.*

6.3   Planar Graphs

In this section, we claim that the PSPACE-hard reduction of Hearn and Demaine [15] for ISR under TS can be extended to DdISR ($d \geq 3$) under R $\in$ {TS, TJ}.

Theorem 12 (∗). *DdISR is* PSPACE-*complete under* R $\in$ {TS, TJ} *on planar graphs of maximum degree three and bounded bandwidth for any $d \geq 2$.*

7   Open Problem: Trees

Since the power of a tree is a (strongly) chordal graph [20, 23] and ISR on chordal graphs under TJ is in P [18], Proposition 1 implies that DdISR under TJ on trees is in P for any $d \geq 3$. On the other hand, the complexity of DdISR under TS for $d \geq 3$ remains unknown.

Conjecture 13. DdISR under TS on trees is in P for $d \geq 3$.

Demaine et al. [10] showed that the problem for $d = 2$ is in P. Their algorithm is based on the so-called *rigid tokens*. Given a tree T and a DdIS I of T ($d \geq 2$), a token t on $u \in I$ is (T, I, d)-*rigid* if there is no TS-sequence that slides t from u to some vertex $v \in N_T(u)$. We denote by $\mathcal{R}(T, I, d)$ the set of all vertices where (T, I, d)-rigid tokens are placed. Demaine et al. proved that $\mathcal{R}(T, I, 2)$ can be found in linear time. Clearly it holds for any $d \geq 2$ that every instance (T, I, J, TS, d) where $\mathcal{R}(T, I, d) \neq \mathcal{R}(T, J, d)$ is a no-instance. When $\mathcal{R}(T, I, 2) = \mathcal{R}(T, J, 2) = \emptyset$, they proved that $(T, I, J, \text{TS}, 2)$ is always a yes-instance. Based on these observations, one can derive a polynomial-time algorithm for solving ISR on trees under TS.

Indeed, for $d \geq 3$, even when $\mathcal{R}(T, I, d) = \mathcal{R}(T, J, d) = \emptyset$, (T, I, J, TS, d) may be a no-instance. An example of such instances is described in Fig. 3. As a result, Demaine et al.'s strategy cannot be directly applied and thus the problem for $d \geq 3$ becomes more challenging.

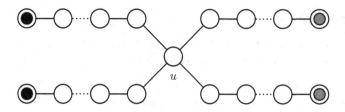

Fig. 3. A no-instance $(T, I, J, \mathsf{TS}, d)$ $(d \geq 3)$ with $\mathcal{R}(T, I, d) = \mathcal{R}(T, J, d) = \emptyset$. Tokens in I (resp., J) are marked with the black (resp. gray) color. All tokens are of distance $d - 1$ from u.

Acknowledgements. This research is partially supported by the Japan Society for the Promotion of Science (JSPS) KAKENHI Grant Number JP20H05964 (AFSA).

References

1. Agnarsson, G., Greenlaw, R., Halldórsson, M.M.: On powers of chordal graphs and their colorings. Congr. Numer. **144**, 41–65 (2000)
2. Balakrishnan, R., Paulraja, P.: Powers of chordal graphs. J. Aust. Math. Soc. **35**(2), 211–217 (1983). https://doi.org/10.1017/S1446788700025696
3. Belmonte, R., Kim, E.J., Lampis, M., Mitsou, V., Otachi, Y., Sikora, F.: Token sliding on split graphs. Theory Comput. Syst. **65**(4), 662–686 (2020). https://doi.org/10.1007/s00224-020-09967-8
4. Bonamy, M., Bousquet, N.: Token sliding on chordal graphs. In: Bodlaender, H.L., Woeginger, G.J. (eds.) WG 2017. LNCS, vol. 10520, pp. 127–139. Springer, Cham (2017). https://doi.org/10.1007/978-3-319-68705-6_10
5. Bonsma, P.S.: Independent set reconfiguration in cographs and their generalizations. J. Graph Theory **83**(2), 164–195 (2016). https://doi.org/10.1002/jgt.21992
6. Bonsma, P., Kamiński, M., Wrochna, M.: Reconfiguring independent sets in claw-free graphs. In: Ravi, R., Gørtz, I.L. (eds.) SWAT 2014. LNCS, vol. 8503, pp. 86–97. Springer, Cham (2014). https://doi.org/10.1007/978-3-319-08404-6_8
7. Bousquet, N., Mouawad, A.E., Nishimura, N., Siebertz, S.: A survey on the parameterized complexity of the independent set and (connected) dominating set reconfiguration problems. arXiv preprint arXiv:2204.10526 (2022)
8. Briański, M., Felsner, S., Hodor, J., Micek, P.: Reconfiguring independent sets on interval graphs. In: Bonchi, F., Puglisi, S.J. (eds.) Proceedings of MFCS 2021. LIPIcs, vol. 202, pp. 23:1–23:14. Schloss Dagstuhl - Leibniz-Zentrum für Informatik (2021). https://doi.org/10.4230/LIPIcs.MFCS.2021.23
9. Chen, M., Chang, G.J.: Families of graphs closed under taking powers. Graphs Comb. **17**(2), 207–212 (2001). https://doi.org/10.1007/PL00007241
10. Demaine, E.D., et al.: Linear-time algorithm for sliding tokens on trees. Theor. Comput. Sci. **600**, 132–142 (2015). https://doi.org/10.1016/j.tcs.2015.07.037
11. Diestel, R.: Graph Theory. GTM, vol. 173. Springer, Heidelberg (2017). https://doi.org/10.1007/978-3-662-53622-3

12. Eto, H., Guo, F., Miyano, E.: Distance-d independent set problems for bipartite and chordal graphs. J. Comb. Optim. **27**(1), 88–99 (2013). https://doi.org/10.1007/s10878-012-9594-4

13. Fox-Epstein, E., Hoang, D.A., Otachi, Y., Uehara, R.: Sliding token on bipartite permutation graphs. In: Elbassioni, K., Makino, K. (eds.) ISAAC 2015. LNCS, vol. 9472, pp. 237–247. Springer, Heidelberg (2015). https://doi.org/10.1007/978-3-662-48971-0_21

14. Garey, M.R., Johnson, D.S.: Computers and intractability: a guide to the theory of NP-completeness. W.H. Freeman and Company (1979)

15. Hearn, R.A., Demaine, E.D.: PSPACE-completeness of sliding-block puzzles and other problems through the nondeterministic constraint logic model of computation. Theor. Comput. Sci. **343**(1–2), 72–96 (2005). https://doi.org/10.1016/j.tcs.2005.05.008

16. Hoang, D.A.: On the complexity of distance-d independent set reconfiguration. arXiv preprint arXiv:2208.07199 (2022)

17. Ito, T., et al.: On the complexity of reconfiguration problems. Theor. Comput. Sci. **412**(12–14), 1054–1065 (2011). https://doi.org/10.1016/j.tcs.2010.12.005

18. Kamiński, M., Medvedev, P., Milanič, M.: Complexity of independent set reconfigurability problems. Theor. Comput. Sci. **439**, 9–15 (2012). https://doi.org/10.1016/j.tcs.2012.03.004

19. Katsikarelis, I., Lampis, M., Paschos, V.T.: Structurally parameterized d-scattered set. Discret. Appl. Math. **308**, 168–186 (2020). https://doi.org/10.1016/j.dam.2020.03.052

20. Kearney, P.E., Corneil, D.G.: Tree powers. J. Algorithms **29**(1), 111–131 (1998). https://doi.org/10.1006/jagm.1998.9999

21. Kong, M., Zhao, Y.: On computing maximum k-independent sets. Congr. Numer. **95**, 47–47 (1993)

22. Kong, M., Zhao, Y.: Computing k-independent sets for regular bipartite graphs. Congr. Numer. **143**, 65–80 (2000)

23. Lin, Y.L., Skiena, S.S.: Algorithms for square roots of graphs. SIAM J. Discret. Math. **8**(1), 99–118 (1995). https://doi.org/10.1137/S089548019120016X

24. Lokshtanov, D., Mouawad, A.E.: The complexity of independent set reconfiguration on bipartite graphs. ACM Trans. Algorithms **15**(1), 7:1–7:19 (2019). https://doi.org/10.1145/3280825

25. Lokshtanov, D., Mouawad, A.E., Panolan, F., Ramanujan, M.S., Saurabh, S.: Reconfiguration on sparse graphs. In: Dehne, F., Sack, J.-R., Stege, U. (eds.) WADS 2015. LNCS, vol. 9214, pp. 506–517. Springer, Cham (2015). https://doi.org/10.1007/978-3-319-21840-3_42

26. Mynhardt, C., Nasserasr, S.: Reconfiguration of colourings and dominating sets in graphs. In: Chung, F., Graham, R., Hoffman, F., Mullin, R.C., Hogben, L., West, D.B. (eds.) 50 years of Combinatorics, Graph Theory, and Computing, pp. 171–191. CRC Press, 1st edn. (2019). https://doi.org/10.1201/9780429280092-10

27. Nishimura, N.: Introduction to reconfiguration. Algorithms **11**(4), 52 (2018). https://doi.org/10.3390/a11040052

28. Siebertz, S.: Reconfiguration on nowhere dense graph classes. Electron. J. Comb. **25**(3), P3.24 (2018). https://doi.org/10.37236/7458

29. van den Heuvel, J.: The complexity of change. In: Surveys in Combinatorics, London Math. Soc. Lecture Note Ser., vol. 409, pp. 127–160. Cambridge University Press (2013). https://doi.org/10.1017/CBO9781139506748.005

30. van der Zanden, T.C.: Parameterized complexity of graph constraint logic. In: Proceedings of IPEC 2015. LIPIcs, vol. 43, pp. 282–293. Schloss Dagstuhl - Leibniz-Zentrum für Informatik (2015). https://doi.org/10.4230/LIPIcs.IPEC.2015.282
31. Wrochna, M.: Reconfiguration in bounded bandwidth and treedepth. J. Comput. Syst. Sci. **93**, 1–10 (2018). https://doi.org/10.1016/j.jcss.2017.11.003
32. Yamanaka, K., Kawaragi, S., Hirayama, T.: Exact exponential algorithm for distance-3 independent set problem. IEICE Trans. Inf. Syst. **102**(3), 499–501 (2019). https://doi.org/10.1587/transinf.2018FCL0002

On Star-Multi-interval Pairwise Compatibility Graphs

Angelo Monti[1] and Blerina Sinaimeri[2]([📧])

[1] Computer Science Departmente, Sapienza University of Rome, Rome, Italy
monti@di.uniroma1.it
[2] LUISS University, Rome, Italy
bsinaimeri@luiss.it

Abstract. A graph G is a star-k-PCG if there exists a non-negative edge weighted star tree S and k mutually exclusive intervals $I_1, I_2, \ldots, I_k$ of non-negative reals such that each vertex of G corresponds to a leaf of S and there is an edge between two vertices in G if the distance between their corresponding leaves in S lies in $I_1 \cup I_2 \cup \ldots \cup I_k$. These graphs are related to different well-studied classes of graphs such as PCGs and multithreshold graphs. It is well known that for any graph G there exists a k such that G is a star-k-PCG. Thus, for a given graph G it is interesting to know which is the minimum k such that G is a star-k-PCG.

In this paper, we focus on classes of graphs where k is constant and prove that circular graphs and two dimensional grid graphs are both star-2-PCGs and that they are not star-1-PCGs. Moreover we show that 4-dimensional grids are not star-2-PCG.

Keywords: Pairwise compatibility graph · Multithreshold graph · Graph theory · Grid graphs

1 Introduction

A graph G is a k-PCG (known also as multi-interval PCG) if there exists a non-negative edge weighted tree T and k mutually exclusive intervals $I_1, I_2, \ldots, I_k$ of non-negative reals such that each vertex of G corresponds to a leaf of T and there is an edge between two vertices in G if the distance between their corresponding leaves in T lies in $I_1 \cup I_2 \cup \ldots \cup I_k$ (see *e.g.* [1]). Such tree T is called the k-*witness* tree of G. Figure 1 depicts an example of a graph that is a 1-PCG graph of a star tree.

The class of 1-PCGs (simply known as PCGs) arose from the phylogenetic tree reconstruction problem [7]. Indeed, sampling leaves in a phylogenetic tree under distance constraints is related to sampling cliques in a PCG [7]. Later on, in [9], it was shown that PCGs could be applied to describe rare evolutionary events and scenarios with horizontal gene transfer.

To date, many graph classes have been proven to be 1-PCGs (see for a survey [12]). Not all graphs are 1-PCGs, while it is known that for each graph

C.-C. Lin et al. (Eds.): WALCOM 2023, LNCS 13973, pp. 267–278, 2023.
https://doi.org/10.1007/978-3-031-27051-2_23

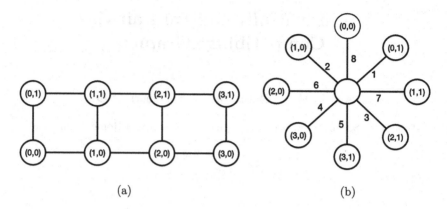

Fig. 1. (a) The grid $G_{4,2}$ (b) An edge-weighted star S such that $G_{4,2}$ is a star-1-PCG G of S for $I_1 = [8,10]$.

$G = (V, E)$ there is an integer $k \leq |E|$ such that G is a k-PCG [1,2]. Given a graph G, determining the minimum k for which G is k-PCG is a difficult problem. Indeed, it is not known whether it can be solved in polynomial time even for $k = 1$. Thus, in the hope to simplify the problem, the k-witness tree has been constrained to have a simple topology as a star or a caterpillar. This is furthermore motivated by the fact that, in the literature, most of the witness trees of k-PCGs have such simple structures. Nevertheless, the problem is not easy even for star trees. Only recently, a complete characterization of the 1-PCGs of a star was provided in [14]. This characterization gave a first polynomial-time algorithm for recognizing graphs that are 1-PCG of a star. This result has been further improved in [8] where, through a new characterization, a linear time algorithm has been provided. Motivated by these results, in this paper we consider k-PCGs having as witnesses star trees (in short star-k-PCGs). It is worth mentioning that star-k-PCGs are $2k$-threshold graphs, a class of graphs recently introduced in [5] and which has already drawn a lot of attention in the community [3,5,6,11]. Thus, the results presented in this paper can be also considered in terms of multithreshold graphs.

It is already known that every graph G is a star-k-PCG for some positive integer $k \leq |E(G)|$ [1]. Moreover, for each positive integer k there are graphs that are not star-k-PCGs and are star-$(k + 1)$-PCGs [3]. However, except for star-1-PCGs, a complete characterization of star-k-PCGs for $k \geq 2$ is still an open problem. In this paper we study the relation between the star-k-PCGs and circular and grid graphs. In [15] it is shown that every circular graph is a 1-PCG of a caterpillar tree. Here we show that they are not star-1-PCGs but are star-2-PCGs.

In [10] it was proved that 2-dimensional grid graphs are 2-PCGs of a caterpillar and this result was improved in [4] showing that 2-dimensional grid graphs are indeed 1-PCGs of a caterpillar. Here we prove that two dimensional grid graphs G_{n_1,n_2}, are star-1-PCG only if $\min\{n_1, n_2\} \leq 2$ and are star-2-PCGs otherwise. From the results in [5] it can be easily shown that d-dimensional grids are not star-$(d-4)$-PCGs. Here we improve this result for $d = 4$, by showing that 4-dimensional grids are not star-2-PCG. Finally, all our constructions can be obtained in linear time.

2   Preliminaries

For a graph $G = (V, E)$ and a vertex $u \in V$, the set $N(u) = \{v : \{u, v\} \in E\}$ is called the *neighborhood* of u.

Let S be an edge weighted star tree for each leaf v_i of S we denote by $w(v_i) = w_i$ the weight of the edge incident to v_i. For a graph G, *the weighted star tree of G* is a star whose leaves are the vertices of G.

A *circular* graph, denoted as C_n, $n \geq 3$, is a graph that consists of a single cycle of n vertices.

For any integer $n \geq 1$ we denote by $[n]$ the set $\{0, 1, 2, \ldots, n-1\}$. A d-*dimensional grid* graph $G_{n_1, \ldots, n_d}$, is a graph such that the vertex set is given by $[n_1] \times [n_2] \times \ldots \times [n_d]$ and there is an edge between two vertices if and only if they differ in exactly one coordinate and the difference is 1. More formally, a vertex u is described by its coordinates $(i_1, \ldots, i_d)$. For any dimension j we denote by u_j the coordinate of u in the j-th dimension. Two vertices u and u' are adjacent if there is a dimension i such that $|u_i - u'_i| = 1$ and for all $l \neq i$, $u_l = u'_l$ (see Fig. 1 for an example). Notice that in this paper we only consider bounded grid graphs.

Notice that every vertex of the d-dimensional grid has at most $2d$ neighbors. Two vertices u and u' are called *opposed* if there exists a dimension i for which $|u_i - u'_i| = 2$ and for all $j \neq i$, $u_j = u'_j$. Notice that in a d-dimensional grid for any vertex u, the set of its $2d$ neighbors can be partitioned in d pairs of opposed vertices. Consider a vertex u and any two of its neighbors v, v' that are not opposed. Then there exists exactly one vertex x different from u, denoted $Q^u_{v,v'}$ such that $N(u) \cap N(x) = \{v, v'\}$.

We need the following lemma which is inspired by [14].

Lemma 1. *Let G be a graph and let k be a positive integer. If for any weighted star S of G, there exist $x \in V(G)$, vertices $v_1, \ldots, v_{k+1}$ in $N(x)$ and $u_1, \ldots u_k$ not in $N(x) \cup \{x\}$, such that $w(v_1) \leq w(u_1) \leq \ldots \leq w(u_k) \leq w(v_{k+1})$, then G is not a star-k-PCG.*

Proof. Consider a graph G and let S be any of its star trees. Assume the assumptions of the lemma hold and thus there exist a vertex x with $v_1, \ldots, v_{k+1}$ in $N(x)$ and $u_1, \ldots u_k$ not in $N(x) \cup \{x\}$ such that the following holds

$$w(v_1) \leq w(u_1) \leq \ldots \leq w(u_k) \leq w(v_{k+1}).$$

We show that the edges $\{x, v_i\}$ must belong to $k + 1$ different intervals and thus S cannot be a k-witness tree. For any $1 \leq i < j \leq k + 1$ we consider the edges $e_i = \{x, v_i\}$, $e_j = \{x, v_j\}$ and the non-edge $\bar{e}_i = \{x, u_i\}$. We have that $w(x) + w(v_i) \leq w(x) + w(u_i) \leq w(x) + w(v_j)$ and thus if the edges e_i, e_j belong to the same interval so does the non-edge $\bar{e}_i$. This concludes the proof.   □

3   Circular Graphs

It is not hard to see that the circular graphs C_3 and C_4 are star-1-PCGs. For C_n with $n \geq 5$ we prove the following result.

Theorem 1. *For any $n \geq 5$ the graph C_n is not a star-1-PCG.*

Proof. Consider $C_n = v_1, v_2, \ldots, v_n, v_1$, and let S be any of its witness star trees. For $n \geq 5$, all the n vertices have distinct neighbourhoods. Consider the vertices of C_n ordered by weight of their incident edge in S. Thus, let $w(v_{i_1}) \leq \ldots \leq w(v_{i_n})$ and notice that there are exactly $n-1$ consecutive pairs $v_{i_j}, v_{i_{j+1}}$ which will correspond to at most $n-1$ different neighbourhoods. Thus, it must necessarily exists a vertex x with $N(x) = \{v, v'\}$ and a vertex u for which $w(v) \leq w(u) \leq w(v')$. Notice that if $x \neq u$ we are done as the conditions of Lemma 1 are satisfied. Otherwise let $x = u$, we have that $w(x) + w(v) \leq w(v') + w(v) \leq w(v') + w(x)$. Notice that since C_n is triangle free $\{v, v'\}$ is a non-edge. Hence, the edges $\{x, v\}$ and $\{x, v'\}$ cannot belong to the same interval and thus C_n is not a star-1-PCG.   □

We now prove that circular graphs are star-2-PCGs for $n \geq 5$. To this purpose we extend one construction for the 1-witness star of a path. Notice that it is already known that path graphs are star-1-PCGs [14] and two different ways to weight the witness star are presented in [10]. However, for circular graphs we need a slightly different construction.

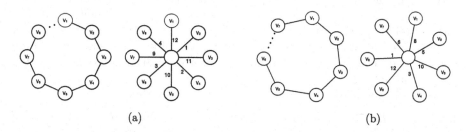

(a)   (b)

Fig. 2. (a) An edge-weighted star S such that C_8 is a star-2-PCG G of S for $I_1 = [12, 13]$ and $I_2 = [16]$. (b) An edge-weighted star S such that C_7 is a star-2-PCG of S for $I_1 = [13, 15]$ and $I_2 = [7]$. The normal and dotted edges in the cycle correspond to edges for which the sum of their endpoints falls in I_1 and I_2, respectively.

Theorem 2. *For any $n \geq 3$ the graph C_n is a star-2-PCG.*

Proof. The construction depends on the parity of n.

Case n even. Let $C_n = v_1, v_2, \ldots, v_n, v_1$, we construct a star tree S as follows: For every $1 \le i \le n$ we set w_i so that:

$$w_i = \begin{cases} n + \frac{n}{2} - \frac{i-1}{2}, & \text{if } i \text{ is odd} \\ \frac{i}{2}, & \text{if } i \text{ is even} \end{cases}$$

We define two intervals:

$$I_1 = \left[n + \frac{n}{2}, n + \frac{n}{2} + 1\right]$$
$$I_2 = [2n]$$

In Fig. 2a we depict an example of the witness star tree for C_8. Consider any two different vertices v_i, v_j. There are three cases to consider:

(a) Both i and j are odd. In this case $w_i + w_j = 3n - \frac{i+j}{2} + 1 > 2n + 1$ (where the last inequality follows from $\frac{i+j}{2} < n$). Thus, $w_i + w_j \notin I_1$ and $w_i + w_j \notin I_2$.
(b) Both i and j are even. In this case $w_i + w_j = \frac{i+j}{2} < n$. Thus, $w_i + w_j \notin I_1$ and $w_i + w_j \notin I_2$.
(c) The vertices i and j have different parity. W.l.o.g. let i be odd and j be even. In this case $w_i + w_j = n + \frac{n}{2} - \frac{i-j-1}{2}$. Clearly, $w_i + w_j \in I_1$ if and only if $\frac{i-j-1}{2} \in \{-1, 0\}$. If $\frac{i-j-1}{2} = -1$ then $i = j - 1$. Otherwise if $\frac{i-j-1}{2} = 0$ we have $i = j+1$. Thus $w_i + w_j \in I_1$ if and only if $|i-j| = 1$ which corresponds to an edge in C_n, more precisely in $P_n = v_1, v_2, \ldots, v_n$. Moreover, $w_i + w_j \in I_2$ if and only if $j - i = n - 1$. The latter holds only for $j = n$ and $i = 1$ as $1 \le i, j \le n$.

Thus, from points (a)–(c) we conclude that there is an edge among i and j if and only if $(i, j) \in C_n$.

Case n odd. For every $1 \le i \le n$ we set w_i so that:

$$w_i = \begin{cases} n - i, & \text{if } i \text{ is even} \\ n + i, & \text{if } i \text{ is odd and } i \ne n \\ n - 1, & \text{if } i = n \end{cases}$$

We define two intervals:

$$I_1 = [2n - 1, 2n + 1]$$
$$I_2 = [n]$$

In Fig. 2b we depict an example of the witness star tree for C_7. Consider any two different vertices v_i, v_j and there are three cases to consider:

(a) Both i and j are even. In this case $w_i + w_j = 2n - (i + j) < 2n - 1$ (where the last inequality follows from $i + j \ge 6$). Thus $w_i + w_j \notin I_1$. Moreover $i + j$ is even thus $w_i + w_j = 2n - (i + j)$ is even too. Hence $w_i + w_j \notin I_2$.

(b) Both i and j are odd. First assume i and j different from n. In this case $w_i + w_j = 2n + (i + j) > 2n + 1$ (where the last inequality follows from $i + j \geq 4$). Thus $w_i + w_j \notin I_1 \cup I_2$. Assume now that $i = n$, in this case we have $w_n + w_j = n - 1 + n + j = 2n + j - 1$. If $j = 1$ (*i.e.* we are considering the edge $\{v_n, v_1\}$) then $w_n + w_j = 2n \in I_1$. Otherwise, for $j \geq 3$, it holds that $w_n + w_j \geq 2n + 2$. Thus, $w_n + w_j \notin I_1 \cup I_2$.

(c) The vertices i and j have different parity and *w.l.o.g.* let i be even and j be odd. Assume first that $j \neq n$. In this case $w_i + w_j = 2n - i + j$. Clearly, $w_i + w_j \in I_1$ if and only if $-1 \leq j - i \leq 1$ and since $i \neq j$ it must be $i = j + 1$ or $i = j - 1$ which correspond to edges of $P_{n-1} = v_1, v_2, \ldots, v_{n-1}$. Consider now the case $j = n$. We have $w_i + w_j = 2n - i - 1$. Clearly, as $i \geq 2$, $w_i + w_j \leq 2n - 3 \notin I_1$. Finally, $w_i + w_j \in I_2$ if and only if $i = n - 1$. This corresponds to the edge $\{v_{n-1}, v_n\}$ of C_n.

Thus there is an edge among i and j if and only if $\{i, j\} \in C_n$. This concludes the proof. $\qquad\square$

4   Grid Graphs

In [5] it has been shown that every graph with minimum degree δ in which any two distinct vertices have at most c neighbors in common is not a star-$(\delta - c - 2)$-PCG. Thus, any d-dimensional grid is not a star-$(d - 4)$-PCG as the minimum degree is d and any two vertices can share at most 2 neighbors. Thus, the following theorem holds.

Theorem 3 ([5]). *For any integer $d \geq 5$, the d-dimensional grid $G_{n_1,\ldots,n_d}$ with $n_1, \ldots, n_d \geq 3$ is not a star-$(d - 4)$-PCG.*

In this section we analyse the cases $d = 2$ and $d = 4$.

Theorem 4. *A 2-dimensional grid G_{n_1,n_2} with $\min\{n_1, n_2\} \leq 2$ is a star-1-PCG.*

Proof. If $\min\{n_1, n_2\} = 1$ then the graph is a path and it is already known that it is a star-1-PCG [10,14]. Assume now $\min\{n_1, n_2\} = 2$ and *w.l.o.g.* let $n_2 = 2$. It is worth mentioning that it is already known that $G_{n_1,2}$ is a PCG but the witness tree is a caterpillar [13]. We define a witness star tree S for $G_{n_1,2}$ as follows: For each vertex (i, j) with $0 \leq i \leq n_1 - 1$ and $0 \leq j \leq 1$, we define its weight $w(i, j)$, as follows:

$$w(i, j) = \begin{cases} 2n_1 - i & \text{if } i + j \text{ is even} \\ i + 1 & \text{if } i + j \text{ is odd} \end{cases}$$

We define

$$I_1 = [2n_1, 2n_1 + 2]$$

See Fig. 1 for a construction of a 1-witness star tree for $G_{2,4}$. We now show that S is a 1-witness star tree for $G_{n_1,2}$. For this let (i, j) and (i', j') be two vertices and we consider the following three cases:

Case $j = j' = 0$. Notice that we must have $i \neq i'$ and $i + i' \leq n_1 - 1 + n_1 - 3 = 2n_1 - 4$. We consider the following three subcases:

- Both i and i' are odd. We have $w(i, 0) + w(i', 0) = i + i' + 2 \leq 2n_1 - 2 \notin I_1$.
- Both i and i' are even. We have $w(i, 0) + w(i', 0) = 4n_1 - (i + i') \geq 2n_1 + 4 \notin I_1$.
- i and i' have different parity. W.l.o.g. assume i odd and i' even. Then $w(i, 0) + w(i', 0) = 2n_1 + i - i' + 1$. Thus, as $I_1 = [2n_1, 2n_1 + 2]$ it must hold that either $2n_1 + i - i' + 1 = 2n_1$ or $2n_1 + i - i' + 1 = 2n_1 + 2$. Thus, $w(i, 1) + w(i', 1) \in I_1$ if and only if $|i - i'| = 1$, which corresponds to the edges of $G_{n_1,2}$ for which $|i - i'| = 1$ and $j = j' = 0$.

Case $j = j' = 1$. This case is symmetrical to the previous one. Thus, $w(i, 1) + w(i', 1) \in I_1$ if and only if $|i - i'| = 1$, which corresponds to the edges of $G_{n_1,2}$ for which $|i - i'| = 1$ and $j = j' = 1$.

Case j and j' are of different parity. W.l.o.g. assume $j = 0$ and $j' = 1$. Then we consider the following three subcases:

- Both i and i' are odd. We have $w(i, 0) + w(i', 1) = 2n_1 + i - i' + 1$. If $i \neq i'$ then $|i - i'| \geq 2$ and thus either $w(i, 0) + w(i', 1) \geq 2n_1 + 3 \notin I_1$ or $w(i, 0) + w(i', 1) \leq 2n_1 - 1 \notin I_1$. Otherwise, if $i = i'$ then $w(i, 0) + w(i', 1) = 2n_1 + 1 \in I_1$ which corresponds to the edges of $G_{n_1,2}$ for which $i = i'$ and $|j - j'| = 1$.
- Both i and i' are even. We have $w(i, 0) + w(i', 1) = 2n_1 - i + i' + 1$ and the case follows identical to the previous one. Thus, $w(i, 1) + w(i', 1) \in I_1$ if and only if $i = i'$, which corresponds to the edges of $G_{n_1,2}$ for which $i = i'$ and $|j - j'| = 1$.
- i and i' have different parity. Notice that we must have $i \neq i'$ and thus $i + i' \leq 2n_1 - 4$. Assume first i odd and i' even. Then $w(i, 0) + w(i', 1) = i + i' + 2 \leq 2n_1 - 2 \notin I_1$. Otherwise, if i even and i' odd. Then $w(i, 0) + w(i', 1) = 4n_1 - (i + i') \geq 2n_1 + 4 \notin I_1$.

Thus, S is a 1-witness star tree for $G_{n_1,2}$.   □

Theorem 5. *A 2-dimensional grid G_{n_1,n_2} with $n_1, n_2 \geq 3$ is not a star-1-PCG.*

Proof. Note that any induced subgraph of a star-1-PCG graph is also star-1-PCG graph. Thus, the theorem follows since the circular graph C_8 is an induced subgraph of G_{n_1,n_2} with $n_1, n_2 \geq 3$; by Theorem 1 C_8 is not a star-1-PCG.   □

We now prove that any 2-dimensional grid graph is a star-2-PCG.

Theorem 6. *For any two positive integers n_1, n_2, the 2-dimensional grid graph G_{n_1,n_2} is a star-2-PCG.*

Proof. Notice that if a graph G is a star-2-PCG, then so is any vertex induced subgraph of G. Hence, it is sufficient to focus on the case where $n_1 = n_2 = h$. Let $G = G_{h,h}$ and consider a star S on h^2 vertices of G. For each vertex (i, j), we define its weight $w(i, j)$, as follows:

$$w(i, j) = \begin{cases} \dfrac{(i + j - 1)h}{2} + i + 1, & \text{if } i + j \text{ is odd} \\ (2h - 1)h - \dfrac{(i + j)h}{2} - i, & \text{if } i + j \text{ is even} \end{cases}$$

We define two intervals:

$$I_1 = [2h(h-1), 2h(h-1)+1]$$
$$I_2 = [2h(h-1)+h+1, 2h(h-1)+h+2]$$

Thus, by construction two vertices (i,j) and (i',j') are adjacent if the sum of their weights is one of the 4 integer values in $S = \{2h(h-1), 2h(h-1)+1, 2h(h-1)+h+1, 2h(h-1)+h+2\}$ (see Fig. 3). Consider two vertices (i,j), (i',j') in the grid and we check under which conditions they are adjacent. Recall that $0 \le i, j, i', j' \le h-1$. There are two cases to consider.

Case $i+j$ and $i'+j'$ have the same parity. By definition of a 2-dimensional grid these vertices are not adjacent. Consider first the case where $i+j$ and $i'+j'$ are both odd. Notice that $w(i,j) = \frac{(i+j-1)h}{2} + i + 1$ is maximized for $i = h-1$ and $j = h-2$ (notice that as $i+j$ is odd we cannot have $i = j = h-1$). Hence, $w(i,j) + w(i',j') < (2h-4)h + 2h = (2h-2)h$ and thus is not in S where the last inequality holds as $(i,j) \ne (i',j')$.

Consider now the case $i+j$ and $i'+j'$ are both even. Notice that $w(i,j) = (2h-1)h - \frac{(i+j)h}{2} - i$ is minimized for $i = h-1$ and $j = h-1$. Hence, $w(i,j) + w(i',j') > 2h(h-1)+2$ and thus is not in S.

Case $i+j$ and $i'+j'$ have different parity. W.l.o.g. assume $i+j$ odd and $i'+j'$ even and thus:

$$w(i,j) + w(i',j') = (2h-1)h + (i-i'+j-j'-1)\frac{h}{2} + i - i' + 1$$

We consider now for what values of i, j, i', j' $w(i,j) + w(i',j') = s \in S$. To this purpose we solve the following equations for each possible value of s.

- In the case $s = 2h(h-1)$ we obtain $(i - i' + j - j' + 1)h + 2(i - i' + 1) = 0$. Let $c = i - i' + j - j' + 1$ and we consider for which values of c the equation has solutions. Notice first that for $c \le -3$ and $c \ge 3$ there are no solutions as $-2h \le 2(i - i' + 1) \le 2h$ (recall that $0 \le i, j, i', j' \le h-1$). Moreover, as $i + j$ is odd and $i' + j'$ is even we have that c must be even. Thus, the only possible cases that remain to consider are $c \in \{-2, 0, 2\}$. If $c = 2$ then $2h + 2(i - i' + 1) = 0$ and thus $i - i' + 1 = -h$. From this and $i - i' + j - j' + 1 = 2$ we have $j - j' = h + 2$ which is not possible since $j - j' \le h-1$. If $c = -2$ then $-2h + 2(i - i' + 1) = 0$ and thus $i - i' + 1 = h$. From this and $i - i' + j - j' + 1 = -2$ we have $j - j' = -(h + 2)$ which is not possible since $j - j' \ge -(h + 1)$. The only possibility is $c = 0$ and as a consequence $2(i - i' + 1) = 0$ from which we have $i = i' - 1$. Then as $i - i' + j - j' + 1 = 0$ we have $j = j'$.
- In the case $s = 2h(h-1)+1$ following a similar argument as in the previous point we have the only possibility is $i = i'$ and $j = j' - 1$.
- In the case $s = 2h(h-1)+h+1$ following a similar argument as in the previous point we have the only possibility is $i = i'$ and $j = j' + 1$.
- In the case $s = 2h(h-1)+h+2$ following a similar argument as in the previous point we have the only possibility is $i = i' + 1$ and $j = j'$.

From the previous four items we have that two vertices (i,j), (i',j') are adjacent if and only if $i = i'$ and $|j - j'| = 1$ or $j = j'$ and $|i - i'| = 1$ and $j = j'$. This concludes the proof. $\qquad\square$

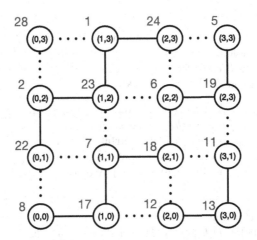

Fig. 3. The graph $G_{4,4}$ where the weight associated to each vertex v corresponds to the weight of the edge incident to v in the 2-witness tree star. The two intervals are $I_1 = [24, 25]$ and $I_2 = [29, 30]$. The normal and dotted edges correspond to edges for which the sum of their endpoints falls in I_1 and I_2, respectively.

We now consider 4-dimensional grids.

Theorem 7. *For any four positive integers $n_1, n_2, n_3, n_4 \geq 3$ the 4-dimensional grid G_{n_1,n_2,n_3,n_4} is not a star-2-PCG.*

Proof. Since $G_{3,3,3,3}$ is an induced subgraph of $G_{n_1,\ldots,n_4}$ and thus it is sufficient to show $G_{3,3,3,3}$ is not a star-2-PCG. Consider $G = G_{3,3,3,3}$, and let S be any of its star trees. Notice that due to Lemma 1 it is sufficient to prove that there exists a vertex x with v_1, v_2, v_3 in $N(x)$ and u_1, u_2 not in $N(x) \cup \{x\}$ such that $w(v_1) \leq w(u_1) \leq w(v_2) \leq w(u_k) \leq w(v_3)$. We now proceed to find this vertex x.

Consider the vertex $a = (1, 1, 1, 1)$. If a is our vertex x we are done. Otherwise we have $x \neq a$. Notice that a has exactly 8 neighbors and w.l.o.g. let $b_1, \ldots, b_8$ be its neighbors such that $w(b_1) < w(b_2) < \ldots < w(b_8)$ in S (notice that in d-dimensional grid graphs there are no two vertices with the same neighborhood and thus all the vertices have pairwise different weights). We show that there exist $b_i, b_j \in N(a)$ such that $x = Q^a_{b_i,b_j}$. There are two cases to consider:

Case I. For any $u \notin N(a)$ it holds $w(u) < w(b_1)$ or $w(u) > w(b_8)$. At least one between the pairs (b_2, b_4) and (b_2, b_5) is not an opposed pair. W.l.o.g. let (b_2, b_4) be such a pair. Let $y = Q^a_{b_2,b_4}$ and we show that $x = y$. To this purpose consider an arbitrary vertex $v \in N(y) - \{b_2, b_4\}$. As $v \notin N(a)$ we have that either $w(v) < w(b_1)$ or $w(v) > w(b_8)$. We consider each case separately.

- If $w(v) < w(b_1)$ we can set $v_1 = v$, $v_2 = b_2$, $v_3 = b_4$ and $u_1 = b_1$, $u_2 = b_3$ and notice that v_1, v_2, v_3 are in $N(y)$ and u_1, u_2 are not in $N(y)$ and moreover $w(v_1) < w(u_1) < w(v_2) < w(u_2) < w(v_3)$. Thus, we can take $x = y$.
- If $w(v) > w(b_8)$ we can set $v_1 = b_2$, $v_2 = b_4$, $v_3 = v$ and $u_1 = b_3$, $u_2 = b_5$ and we can take $x = y$.

Case II. There exists $1 \le i \le 7$ such that for any $u \notin N(a)$ at least one of the followings holds: (i) $w(u) < w(b_1)$, (ii) $w(b_i) < w(u) < w(b_{i+1})$, (iii) $w(u) > w(b_8)$. We will consider only the cases $1 \le i \le 4$ as the reasoning in the cases for $i > 4$ is identical to the cases $8 - i$.

Case II.a. $1 \le i \le 2$. At least one between the pairs (b_4, b_6) and (b_4, b_7) is not an opposed pair. W.l.o.g. let (b_4, b_6) be such a pair. Let $y = Q^a_{b_4, b_6}$. We show that $x = y$. To this purpose consider an arbitrary vertex $v \in N(y) - \{b_4, b_6\}$. As $v \notin N(a)$ we have that one of the followings must hold: (i) $w(v) < w(b_1)$, (ii) $w(b_i) < w(v) < w(b_{i+1})$, (iii) $w(v) > w(b_8)$. We consider each case separately.

(i) If $w(v) < w(b_1)$ we can set $v_1 = v$, $v_2 = b_4$, $v_3 = b_6$ and $u_1 = b_1$, $u_2 = b_5$ and thus we can take $x = y$.
(ii) If $w(b_i) < w(v) < w(b_{i+1})$ we can set $v_1 = v$, $v_2 = b_4$, $v_3 = b_6$ and $u_1 = b_3$, $u_2 = b_5$ and thus we can take $x = y$.
(iii) If $w(v) > w(b_8)$ we can $v_1 = b_4$, $v_2 = b_6$, $v_3 = v$ and $u_1 = b_5$, $u_2 = b_8$ and thus we can take $x = y$.

Case II.b. $3 \le i \le 4$. At least one between the pairs (b_2, b_6) and (b_2, b_7) is not an opposed pair. W.l.o.g. let (b_2, b_6) be such a pair. Let $y = Q^a_{b_2, b_6}$. We show that $x = y$. Similarly to the previous case let $v \in N(y) - \{b_2, b_6\}$. as $v \notin N(a)$ we have that one of the followings must hold: (i) $w(v) < w(b_1)$, (ii) $w(b_i) < w(v) < w(b_{i+1})$, (iii) $w(v) > w(b_8)$. We consider each case separately.

(i) If $w(v) < w(b_1)$ we can set $v_1 = v$, $v_2 = b_2$, $v_3 = b_6$ and $u_1 = b_1$, $u_2 = b_5$ and thus we can take $x = y$.
(ii) If $w(b_i) < w(v) < w(b_{i+1})$ we can set $v_1 = b_2$, $v_2 = v$, $v_3 = b_6$ and $u_1 = b_3$, $u_2 = b_5$ and thus we can take $x = y$.
(iii) If $w(v) > w(b_8)$ we can $v_1 = b_2$, $v_2 = b_6$, $v_3 = v$ and $u_1 = b_5$, $u_2 = b_8$ and thus we can take $x = y$.

This concludes the proof. $\qquad\qquad\qquad\qquad\qquad\qquad\qquad\qquad\qquad\quad$ □

5 Conclusions and Open Problems

In this paper we consider the problem of characterizing star-multi-interval pairwise compatibility graphs. This is particularly interesting as this class connects two important graph classes: the PCGs and multithreshold graphs for both of which a complete characterization is not yet known. Here we study the relation between the star-multi-interval pairwise compatibility graphs and circular and grid graphs. We show that circular graphs, C_n are star-1-PCGs for every $1 \leq n < 4$ and are star-2-PCGs for any $n \geq 5$. We then consider d-dimensional grids and for the specific cases $d = 2$ and $d = 4$ we improve on the existing general result stating that any d-dimensional grid is not a star-$(d-4)$-PCG [5]. More specifically, we show that 2-dimensional grid graphs G_{n_1,n_2}, are star-1-PCG only if $\min\{n_1, n_2\} \leq 2$ and they are star-2-PCGs otherwise. For 4-dimensional grids we show that they are not star-2-PCGs. It remains open to determine the minimum value k for which d-dimensional grids, $d \geq 3$, are star-k-PCGs.

Acknowledgments. The authors would like to thank Manuel Lafond for pointing out the connection of star-k-PCGs and $2k$-threshold graphs as well as the anonymous reviewers whose comments and suggestions helped us to improve the quality of the manuscript.

References

1. Ahmed, S., Rahman, M.S.: Multi-interval pairwise compatibility graphs. In: Gopal, T.V., Jäger, G., Steila, S. (eds.) TAMC 2017. LNCS, vol. 10185, pp. 71–84. Springer, Cham (2017). https://doi.org/10.1007/978-3-319-55911-7_6

2. Calamoneri, T., Lafond, M., Monti, A., Sinaimeri, B.: On generalizations of pairwise compatibility graphs (2021). https://doi.org/10.48550/ARXIV.2112.08503

3. Chen, G., Hao, Y.: Multithreshold multipartite graphs. J. Graph Theory 1–6 (2022). https://doi.org/10.1002/jgt.22805

4. Hakim, S.A., Papan, B.B., Rahman, M.S.: New results on pairwise compatibility graphs. Inf. Process. Lett. **178**, 106284 (2022). https://doi.org/10.1016/j.ipl.2022.106284

5. Jamison, R., Sprague, A.: Multithreshold graphs. J. Graph Theory **94**(4), 518–530 (2020)

6. Jamison, R.E., Sprague, A.P.: Double-threshold permutation graphs. J. Algebraic Combin. **56**, 23–41 (2021). https://doi.org/10.1007/s10801-021-01029-7

7. Kearney, P., Munro, J.I., Phillips, D.: Efficient generation of uniform samples from phylogenetic trees. In: Benson, G., Page, R.D.M. (eds.) WABI 2003. LNCS, vol. 2812, pp. 177–189. Springer, Heidelberg (2003). https://doi.org/10.1007/978-3-540-39763-2_14

8. Kobayashi, Y., Okamoto, Y., Otachi, Y., Uno, Y.: Linear-time recognition of double-threshold graphs. Algorithmica **84**(4), 1163–1181 (2022). https://doi.org/10.1007/s00453-021-00921-9

9. Long, Y., Stadler, P.F.: Exact-2-relation graphs. Discrete Appl. Math. **285**, 212–226 (2020). https://doi.org/10.1016/j.dam.2020.05.015, https://www.sciencedirect.com/science/article/pii/S0166218X20302638

10. Papan, B.B., Pranto, P.B., Rahman, M.S.: On 2-interval pairwise compatibility properties of two classes of grid graphs. Comput. J. (2022). https://doi.org/10.1093/comjnl/bxac011

11. Puleo, G.J.: Some results on multithreshold graphs. Graphs Combin. **36**(3), 913–919 (2020). https://doi.org/10.1007/s00373-020-02168-7

12. Rahman, M.S., Ahmed, S.: A survey on pairwise compatibility graphs. AKCE Int. J. Graphs Comb. **17**(3), 788–795 (2020). https://doi.org/10.1016/j.akcej.2019.12.011

13. Salma, S., Rahman, M., Hossain, M.: Triangle-free outerplanar 3-graphs are pairwise compatibility graphs. J. Graph Algorithms Appl. **17**(2), 81–102 (2013). https://doi.org/10.7155/jgaa.00286
14. Xiao, M., Nagamochi, H.: Characterizing star-PCGs. Algorithmica **82**(10), 3066–3090 (2020). https://doi.org/10.1007/s00453-020-00712-8
15. Yanhaona, M.N., H.K.R.M.: Pairwise compatibility graphs. J. Appl. Math. Comput. (30), 479–503 (2009). https://doi.org/10.1007/s12190-008-0204-7

Parameterized Complexity of Optimizing List Vertex-Coloring Through Reconfiguration

Yusuke Yanagisawa$^{(\boxtimes)}$, Akira Suzuki, Yuma Tamura, and Xiao Zhou

Graduate School of Information Sciences, Tohoku University, Sendai, Japan
yusuke.yanagisawa.r7@dc.tohoku.ac.jp, {akira,tamura,zhou}@tohoku.ac.jp

Abstract. In the combinatorial reconfiguration framework, we study the relationship between two feasible solutions of a combinatorial search problem under a prescribed reconfiguration rule. In this paper, we deal with the OPT-LIST COLORING RECONFIGURATION problem. Given a graph G, a list function L, a list L-coloring f_0 of G and an integer p, OPT-LIST COLORING RECONFIGURATION asks for finding a list L-coloring $f_{\mathtt{sol}}$ of G such that $f_{\mathtt{sol}}$ can be transformed from f_0 in a step-by-step fashion and $f_{\mathtt{sol}}$ uses at most p colors. We first observe that the problem remains NP-hard for empty graphs even if every vertex of a given graph has a list of size two. Moreover, we prove that the problem is PSPACE-complete for bipartite graphs with bounded bandwidth and pathwidth two, even if the number k of colors that can be used in the reconfiguration process is bounded by some constant. On the positive side, we give an FPT algorithm parameterized by k for graphs of pathwidth one. We also design an FPT algorithm parameterized by $k + \mathsf{vc}$, where vc is vertex cover number of a given graph.

1 Introduction

In the *combinatorial reconfiguration* framework, we study the relationship between two feasible solutions of a combinatorial search problem under a prescribed reconfiguration rule. In particular, a *reconfiguration problem* asks whether, given two feasible solutions of the problem, one solution can be transformed into the other step-by-step, so that each intermediate solution is also feasible. Combinatorial reconfiguration models "dynamic" transformations of real-world systems, such as communication networks and road networks, as well as topics in recreational mathematics such as the 15-puzzle and the block-pushing puzzle. Since both cases do not allow for drastic changes in a short period of time, we wish to transform the current configuration into a more desirable one

A. Suzuki—Partially supported by JSPS KAKENHI Grant Numbers JP18H04091, JP20K11666 and JP20H05794, Japan.
Y. Tamura—Partially supported by JSPS KAKENHI Grant Numbers JP21K21278, Japan.
X. Zhou—Partially supported by JSPS KAKENHI Grant Number JP19K11813, Japan.

C.-C. Lin et al. (Eds.): WALCOM 2023, LNCS 13973, pp. 279–290, 2023.
https://doi.org/10.1007/978-3-031-27051-2_24

by a step-by-step transformation. After the combinatorial reconfiguration framework was proposed by Ito et al. [14], it has been widely studied in the field of theoretical computer science and mathematics. (See, e.g., the surveys of van den Heuvel [13] and Nishimura [19]).

However, we sometimes face difficulties when applying the current framework to real-world systems. In the reconfiguration problem, we are required to have in advance a target (a more desirable) configuration. Sometimes it is hard to find a target configuration because there may exist exponentially many feasible configurations. Moreover, even if we successfully decide a target configuration, it may not be possible to reach it from the current configuration.

To resolve these situations, Ito et al. recently introduced the *optimization variant* of reconfiguration problems [15]. In this variant, we are given a single solution as a current configuration, and asked for the most desirable solution that can be transformed from the given one. The optimization variants of INDEPENDENT SET RECONFIGURATION [15,16] and DOMINATING SET RECONFIGURATION [1] have been studied, and very recently we applied the variant to (VERTEX-)COLORING RECONFIGURATION [22], which is one of the most studied reconfiguration problems [2–7,9,12,17,21].

1.1   Our Problem

We here define several notations and terminologies that are required to describe our problem. Let G be a graph with the vertex set $V(G)$ and the edge set $E(G)$. For an integer $k \geq 1$, let C be a *color set* consisting of k colors $1, 2, \ldots, k$. Recall that a *k-coloring* f of G is a mapping $f \colon V(G) \to C$ such that $f(v) \neq f(w)$ for every edge $vw \in E(G)$. For a *list function* $L \colon V(G) \to 2^C$, the k-coloring f is a *list L-coloring* of G if $f(v) \in L(v)$ holds for every vertex $v \in V(G)$. We call $L(v)$ a *list* of a vertex v.

In the (VERTEX-)COLORING RECONFIGURATION problem, given a graph G and two k-colorings f_0 and f_r of G, we are asked to determine whether there is a sequence $\langle f_0, f_1, \ldots, f_\ell \rangle$ of k-colorings of G such that $f_\ell = f_r$ and each f_i, $1 \leq i \leq \ell$, can be obtained from f_{i-1} by recoloring only a single vertex in G. Such a sequence is called *reconfiguration sequence* from f_0 to f_r. In the LIST (VERTEX-)COLORING RECONFIGURATION problem, we are given a list function L and two list L-colorings f_0 and f_r of an input graph G. Our task is to determine whether there is a reconfiguration sequence of list L-colorings between f_0 and f_r. COLORING RECONFIGURATION and LIST COLORING RECONFIGURATION are widely studied in the field of combinatorial reconfiguration [2–7,9–12,17,21]. See also the survey of Mynhardt and Nasserasr [18].

In this paper, we initiate the study of *optimization variant* of LIST COLORING RECONFIGURATION. In the OPT-LIST COLORING RECONFIGURATION problem, OPT-LCR for short, we are given a graph G, a list function L, a single list L-coloring f_0 of G and an integer p. Then we are asked to find a list L-coloring f_{sol} of G such that there exists a reconfiguration sequence of list L-colorings from f_0 to f_{sol} and f_{sol} uses at most p colors, that is, $|\{f_{\mathsf{sol}}(v) : v \in V(G)\}| \leq p$.

1.2   Related Results

Since LIST COLORING RECONFIGURATION generalizes COLORING RECONFIG-
URATION, all the hardness results of COLORING RECONFIGURATION are also
applicable to LIST COLORING RECONFIGURATION. It is known that COLORING
RECONFIGURATION is PSPACE-complete for bipartite graphs if the number k
of colors is any fixed constant of at least four [4]. Moreover, LIST COLORING
RECONFIGURATION remains PSPACE-complete for bipartite planar graphs even
if both the bandwidth of an input graph and k are bounded by some constant [21].
Hatanaka et al. proved that LIST COLORING RECONFIGURATION for threshold
graphs is PSPACE-complete [11]. On the positive side, LIST COLORING RECON-
FIGURATION is solvable in polynomial time for graphs of pathwidth at most one
(that is, caterpillars) [11]. Cereceda et al. [7] gave a polynomial-time algorithm
for COLORING RECONFIGURATION when $k \leq 3$, which is also applicable to LIST
COLORING RECONFIGURATION.

(LIST) COLORING RECONFIGURATION has also been studied from the viewpoint of parameterized complexity. COLORING RECONFIGURATION is W[1]-hard
when parameterized by ℓ, where ℓ is the upper bound of the length of the reconfiguration sequence [5]. On the other hand, COLORING RECONFIGURATION admits
an FPT algorithm parameterized by $k+\ell$ [5,17]. In LIST COLORING RECONFIGU-
RATION, the problem parameterized by vertex cover number vc is W[1]-hard [12],
while the problem is fixed-parameter tractable when parameterized by $k +$ mw,
where mw is the modular-width of an input graph [12].

The optimization variant of reconfiguration problems was recently proposed
by Ito et al. [15], and to the best of our knowledge, it has been applied to INDE-
PENDENT SET RECONFIGURATION [15,16], DOMINATING SET RECONFIGURA-
TION [1] and COLORING RECONFIGURATION [22]. In the optimization variant of
COLORING RECONFIGURATION, namely OPT-COLORING RECONFIGURATION,
we showed that for any $k \geq 4$ this problem is NP-hard for planar graphs with
degeneracy three and maximum degree four. On the other hand, we gave a
linear-time algorithm for OPT-COLORING RECONFIGURATION when $k \leq 3$ [22],
and linear-time algorithms for graphs with degeneracy two, chordal graphs and
cographs for any k [22].

1.3   Our Results

In this paper, we first observe that OPT-LCR remains NP-hard for empty graphs
even if every vertex of a given graph has a list of size two. Especially, OPT-LCR
on graphs of pathwidth pw is NP-complete for any pw $\in \{0, 1\}$. This incredi-
bly intractable result motivates us to study the problem when parameterized by
combinations of the number k of colors and structural graph parameters. We
prove that OPT-LCR is PSPACE-complete for bipartite graphs with bounded
bandwidth and pathwidth two, even if k is bounded by some constant. On the
positive side, we give an FPT algorithm parameterized by k for graphs of pathwidth one. We also design an FPT algorithm parameterized by $k+$vc, where vc is

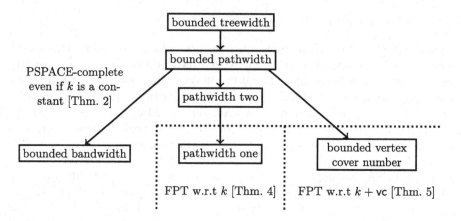

Fig. 1. Our results for OPT-LCR with respect to structural graph parameters, where k is the number of colors. Each arrow represents the inclusion relationship between two graph classes A and B: $A \to B$ means that B is a subclass of A.

vertex cover number of a given graph. Our results in this paper are summarized in Fig. 1. Proofs for the claims marked with ($*$) are omitted from this extended abstract.

2 Preliminaries

Let $G = (V, E)$ be a graph. We denote by $V(G)$ and $E(G)$ the vertex set and the edge set of G, respectively. We assume that all graphs in this paper are simple, undirected, and unweighted. For a vertex $v \in V$, we denote by $N(v)$ the neighborhood of v, that is, $N(v) = \{w : vw \in E\}$. For integers i and j with $i \leq j$, we write $[i, j]$ as the shorthand for the set $\{i, i + 1, \ldots, j\}$ of integers.

A list L-coloring f of a graph G is L-reachable from a list L-coloring f_0 of G if there is a sequence $\langle f_0, f_1, \ldots, f_\ell \rangle$ of list L-colorings of G such that $f_\ell = f$ and each f_i, $1 \leq i \leq \ell$, can be obtained from f_{i-1} by recoloring only a single vertex of G. Let $\#\mathsf{col}(f)$ be the number of colors used in f, that is, $\#\mathsf{col}(f) = |\{f(v) : v \in V(G)\}|$. Given a graph G, a list function L, an initial list L-coloring f_0 and an integer p, OPT-LCR is the problem of finding a list L-coloring f_{sol} such that f_{sol} is L-reachable from f_0 and $\#\mathsf{col}(f_{\mathsf{sol}}) \leq p$. We note that all algorithms in this paper compute the value $\#\mathsf{col}(f_{\mathsf{sol}})$ for simplicity, but one can find an actual solution f_{sol} with just a minor change in our algorithm.

2.1 Graph Parameters

For an n-vertex graph G, let ρ be an ordering of $V(G)$. For a vertex $v \in V(G)$, we denote by $\rho(v)$ an integer that represents the position of v in ρ. The *width* of an ordering ρ is defined as $\max\{|\rho(u) - \rho(v)| : uv \in E(G)\}$, and the *bandwidth* of G is the minimum width over all orderings of $V(G)$.

A *tree decomposition* of a graph G is a pair $\mathcal{T} = \langle T, \{X_t : t \in V(T)\}\rangle$, where T is a tree and $X_t \subseteq V(G)$ is called a *bag* for a node $t \in V(T)$, such that the following three conditions (i)–(iii) hold;

(i) $\bigcup_{t \in V(T)} X_t = V(G)$;
(ii) for every $uv \in E(G)$, there exists a node $t \in V(T)$ such that $u, v \in X_t$; and
(iii) for every $u \in V(G)$, the set $\{t \in V(T) : u \in X_t\}$ induces a connected subgraph of T.

The *width* of a tree decomposition $\mathcal{T} = \langle T, \{X_t : t \in V(T)\}\rangle$ is defined as $\max_{t \in V(T)} |X_t| - 1$, and the *treewidth* of a graph G is the minimum width over all possible tree decompositions of G. If T is a path, a tree decomposition $\mathcal{T}$ is also called a *path decomposition*. The *pathwidth* of a graph G is the minimum width over all possible path decompositions of G.

For a graph G, a subset $S \subseteq V(G)$ is a *vertex cover* of G if at least one of the endpoints of every edge $e \in E(G)$ is contained in S. It is easy to see that removing the vertices in S from G results in an *empty graph*, that is, an edgeless graph. The *vertex cover number* vc of G is the size of a minimum vertex cover of G.

Figure 1 also shows the relationship between treewidth, pathwidth, bandwidth and vertex cover number. If the vertex cover number of a graph G is bounded by a constant, the treewidth and the pathwidth of G are also bounded. The same property holds for bandwidth, but vertex cover number and bandwidth are incomparable.

3   NP-Hardness

We first observe that OPT-LCR is hard even for very restricted graphs.

Theorem 1 (∗). OPT-LCR *is NP-hard for empty graphs even if every vertex of a given graph has a list of size two.*

The pathwidth of empty graphs is zero and hence OPT-LCR is NP-hard even for graphs of pathwidth zero. On the other hand, LIST COLORING RECONFIGU-RATION for graphs of pathwidth at most one is solvable in polynomial time [10], which means that the solution for OPT-LCR can be verified in polynomial time if a given graph has pathwidth at most one. Therefore, the following corollary holds.

Corollary 1. OPT-LCR *on graphs of pathwidth* pw *is NP-complete for any* pw $\in \{0, 1\}$.

4   PSPACE-Completeness

Theorem 1 immediately implies that OPT-LCR is intractable even if treewidth, pathwidth, and bandwidth of a given graph are zero. Notice that, however, the number k of colors is not bounded by a constant in our reduction of Theorem 1.

This motivates us to analyze the complexity of OPT-LCR parameterized by treewidth, pathwidth, or bandwidth plus k. The following theorem states that OPT-LCR is quite hard even if both parameters are bounded by a constant.

Theorem 2. OPT-LCR *is PSPACE-complete for bipartite graphs of bounded bandwidth and pathwidth two, even if the number of colors is bounded by a constant.*

We first show that OPT-LCR is in PSPACE. For a yes-instance (G, L, f_0, p) of OPT-LCR and a sequence S from f_0 to a list L-coloring f_{sol} of G, one can verify in polynomial space whether S is a reconfiguration sequence and $\#\mathsf{col}(f_{sol}) \leq p$. Thus, OPT-LCR is in NPSPACE and hence in PSPACE by Savitch's theorem [20].

To prove the PSPACE-hardness of OPT-LCR, we observe the PSPACE-hardness of the H-WORD CHANGEABILITY problem defined as follows, and then reduce H-WORD CHANGEABILITY to OPT-LCR.

Let Σ be an *alphabet*, that is, a nonempty set of symbols. A *word* x over Σ is a sequence of symbols in Σ, and we denote by $|x|$ the number of symbols in x. Let $E \subseteq \Sigma^2$ be a binary relation over Σ. For a pair $H = (\Sigma, E)$, a word x over Σ is an H-*word* if every two consecutive symbols in x are in the binary relation E. Given an H-word x_0, a positive integer $q \leq |x_0|$ and a symbol $s \in \Sigma$, the H-WORD CHANGEABILITY problem asks whether there is a reconfiguration sequence $\langle x_0, x_1, \ldots, x_\ell \rangle$ of H-words such that x_i can be obtained from x_{i-1} by replacing a single symbol in x_{i-1} to another one for every $i \in [1, \ell]$, and a q-th symbol of x_ℓ is s.

The PSPACE-completeness of H-WORD CHANGEABILITY follows the result by Wrochna [21], which proves the PSPACE-completeness of the H-WORD REACHABILITY problem. In H-WORD REACHABILITY, given two H-words x_0 and x_ℓ, we are asked to determine whether there is a reconfiguration sequence of H-words between x_0 and x_ℓ.

Theorem 3 ([21]). *There is a pair $H = (\Sigma, E)$ for which H-WORD REACHABILITY is PSPACE-complete, even if the size of Σ is bounded by some constant.*

In the proof of Theorem 3, in fact, x_0 can be transformed into x_ℓ if and only if a second symbol of x_ℓ can be replaced with a special symbol. Therefore, we have the following proposition.

Proposition 1. *There is a pair $H = (\Sigma, E)$ for which H-WORD CHANGEABILITY is PSPACE-complete, even if $q = 2$ and the size of Σ is bounded by some constant.*

We construct an instance (G, L, f_0, p) of OPT-LCR from an instance $(x_0, 2, s)$ of H-WORD CHANGEABILITY, where $H = (\Sigma, E)$ such that Σ has a constant size. Our construction of (G, L, f_0, p) is based on the reduction from H-WORD REACHABILITY to LIST COLORING RECONFIGURATION by Wrochna [21].

We first construct a graph G. (See also Fig. 2.) A *chain of onions of width b and length n* is the graph with the vertex set $U \cup W$, where $U = \{u_1, \ldots, u_n\}$ and $W = \{w_i^j : i \in [1, n-1], j \in [1, b]\}$, and the edge set $\{u_i w_i^j, w_i^j u_{i+1} : i \in [1, n-1], j \in [1, b]\}$. A vertex c_i of a star T_i is the *center* if every vertex in $V(T_i) \backslash \{c_i\}$ is a leaf of T_i. To construct a graph G of OPT-LCR, we prepare a chain of onions of width $b = |\Sigma^2 \backslash E|$ and length $n = |x_0|$. We also add $|\Sigma| - 1$ stars $T_1, \ldots, T_{|\Sigma|-1}$, where each T_i, $1 \leq i \leq |\Sigma| - 1$, has the center c_i and $2|\Sigma| + 2$ leaves $\ell_i^1, \ldots, \ell_i^{2|\Sigma|+2}$. Then, for each $i \in [1, |\Sigma| - 1]$, we identify $\ell_i^{2|\Sigma|+2}$ with u_2 of the chain of onions. This completes the construction of G.

We next give a list function L as follows. Let Σ' be a disjoint copy of Σ and we write a' for the copy of $a \in \Sigma$. For a vertex $u_i \in U$ with $i \in [1, n]$ of a chain of onions, we let $L(u_i) = \Sigma$ if i is even, otherwise $L(u_i) = \Sigma'$. For each $(a_0, a_1) \in \Sigma^2 \backslash E$, we assign a distinct integer $j \in [1, b]$. Then, for a vertex $w_i^j \in W$ with $i \in [1, n-1]$ in a chain of onions, we let $L(w_i^j) = \{a_0, a_1'\}$ if i is even, otherwise $L(w_i^j) = \{a_0', a_1\}$. For the center c_i of a star T_i with $i \in [1, |\Sigma| - 1]$, we assign a distinct symbol $a_i \in \Sigma \backslash \{s\}$ and let $L(c_i) = \{a_i, z\}$, where z is the additional symbol that does not appear in Σ. For each leaf ℓ_i^j of a star T_i with $j \in [1, 2|\Sigma| + 1]$, we let $L(\ell_i^j) = \{z, z^j\}$, where z^j is the additional symbol that does not appear in Σ. As a consequence, the color set C is $\Sigma \cup \Sigma' \cup \{z, z^1, z^2, \ldots, z^{2|\Sigma|+1}\}$.

We then define an initial coloring f_0. If the i-th symbol of x_0 is a, then we let $f_0(u_i) = a$ if i is even, otherwise $f_0(u_i) = a'$. Suppose that i is even and $w_i^j \in W$ has a list $\{a_0, a_1'\}$. Since x_0 is an H-word, it holds that $f_0(u_i) \neq a_0$ or $f_0(u_{i+1}) \neq a_1'$; otherwise, $(a_0, a_1) \in E$, contradicting the construction of $L(w_i^j)$. Thus, we assign $f_0(w_i^j) = a_0$ if $f_0(u_i) \neq a_0$, otherwise $f_0(w_i^j) = a_1'$. Similarly, we assign an initial color to w_i^j when i is odd. For each star T_i with $i \in [1, |\Sigma| - 1]$, we then let $f_0(c_i) = z$ for the center c_i of T_i and $f_0(\ell_i^j) = z^j$ for a leaf ℓ_i^j of T_i for each $j \in [1, 2|\Sigma| + 1]$. It is easy to see that f_0 constructed as above is a list L-coloring of G with $\#\mathrm{col}(f_0) \geq 2|\Sigma| + 2$. Finally, we set $p = 2|\Sigma| + 1$.

Lemma 1. *Let (G, L, f_0, p) be an instance of OPT-LCR constructed from an instance $(x_0, 2, s)$ of H-WORD CHANGEABILITY as above, where $p = 2|\Sigma| + 1$. Then, $(x_0, 2, s)$ is a yes-instance of H-WORD CHANGEABILITY if and only if (G, L, f_0, p) is a yes-instance of OPT-LCR.*

Proof. We first prove the sufficiency. Suppose a reconfiguration sequence $S = \langle x_0, \ldots, x_{\mathrm{sol}} \rangle$ of H-words such that the second symbol of x_{sol} is s. Recall that each i-th symbol of x_0 corresponds to the color of $u_i \in U$ for f_0. If x_1 is obtained by replacing an i-th symbol $a_0 \in \Sigma$ of x_0 to $a_1 \in \Sigma$ and i is even, then we recolor $u_i \in U$ from a_0 to a_1. If there are vertices $w \in W$ adjacent to u_i such that $f_0(w) = a_1$, we recolor w to the other color of $L(w)$ before recoloring u_i. The recolorings of u_i and w are safe because $L(w)$ consists of a pair of symbols in $\Sigma^2 \backslash E$. Similarly, if i is odd, we recolor u_i from a_0' to a_1' and recolor $w \in W \cap N(u_i)$ if necessary. This yields a new list L-coloring of G corresponding to x_1. By repeating the above recoloring, eventually, we obtain a list L-coloring f of G such that $f(u_2) = s$. Recall that for the center c_i of every star T_i, $L(c_i)$

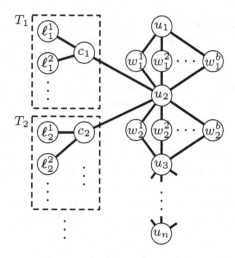

Fig. 2. The construction of a graph G for OPT-LCR. This graph consists of a chain of onions together with stars T_i which are surrounded by dotted rectangles

has no color corresponding to s. Then, we recolor the center c_i of each T_i from $z \in L(c_i)$ to $a_i \in L(c_i)$, and recolor all leaves of every T_i to z. Let $f_{\mathtt{sol}}$ be the list L-coloring of G obtained as above. The vertices in the chain of onions are assigned colors in $\Sigma \cup \Sigma'$ and the vertices in all stars are assigned colors in $\Sigma \cup \{z\}$ by $f_{\mathtt{sol}}$. Therefore, we have $\#\mathsf{col}(f_{\mathtt{sol}}) \leq |\Sigma \cup \Sigma' \cup \{z\}| \leq 2|\Sigma| + 1 = p$.

Conversely, we prove the necessity. Suppose a reconfiguration sequence $S = \langle f_0, \ldots, f_{\mathtt{sol}} \rangle$ of list L-colorings of G such that $\#\mathsf{col}(f_{\mathtt{sol}}) \leq p = 2|\Sigma| + 1$. Then, there is no center c_i of a star T_i for any $i \in [1, |\Sigma| - 1]$ such that $f_{\mathtt{sol}}(c_i) = z$; otherwise, there is a star T_i with $f_{\mathtt{sol}}(\ell_i^2) = z^j$ for any $j \in [1, 2|\Sigma| + 1]$ and hence $f_{\mathtt{sol}}$ uses colors $z, z^1, \ldots, z^{2|\Sigma|+1}$, which contradicts $\#\mathsf{col}(f_{\mathtt{sol}}) \leq p$. From the construction of $L(c_i)$ for the center c_i of T_i, we have $\bigcup_{i \in [1, |\Sigma| - 1]} f_{\mathtt{sol}}(c_i) = \Sigma \backslash \{s\}$ and hence $f_{\mathtt{sol}}(u_2) = s$. Since we can define the corresponding H-word for any list L-coloring of G, we obtain a reconfiguration sequence $S' = \langle x_0, \ldots, x_{\mathtt{sol}} \rangle$ of H-words from S such that the second symbol of $x_{\mathtt{sol}}$ is s. This completes the proof of Lemma 1.   □

We confirm the number k of colors and the property of G. The color set C is $\Sigma \cup \Sigma' \cup \{z, z^1, z^2, \ldots, z^{2|\Sigma|+1}\}$, and hence the number k of colors is $4|\Sigma| + 2$, which is a constant because the size of Σ is bounded by some constant. Clearly, G is bipartite. To show that the bandwidth of G is bounded by a constant, for each $i \in [1, n-1]$, consider the sub-orderings $\rho_i = \langle u_i, w_i^1, \ldots, w_i^b \rangle$ of vertices in the chain of onions of width b. In addition, let σ be an arbitrary ordering of $\bigcup_{i \in [1, |\Sigma| - 1]} V(T_i)$. Since each T_i has the center c_i and $2|\Sigma| + 1$ leaves, the length of σ is $\sum_{i \in [1, |\Sigma| - 1]} |V(T_i)| = (|\Sigma| - 1)(2|\Sigma| + 2)$. Then, consider the ordering $\rho = \langle \rho_1, \sigma, \rho_2, \ldots, \rho_{n-1}, u_n \rangle$. Since $b \leq |\Sigma|^2$, the width of ρ is

$$\max\{|\rho(u) - \rho(v)| : uv \in E(G)\} = \rho(u_2) - \rho(w_1^1)$$
$$= (1 + b + (|\Sigma| - 1)(2|\Sigma| + 2)) - 2$$
$$\leq 3|\Sigma|^2 - 3,$$

which is bounded by a constant. Finally, to show that the pathwidth of G is two, we give a sequence of bags with three vertices. For an integer $i \in [1, n - 1]$, let $\Pi_i = \langle A_i^1, A_i^2 \dots, A_i^b \rangle$ be the sequence of bags for the chain of onions of width b, where $A_i^j = \{u_i, w_i^j, u_{i+1}\}$. In addition, for an integer $i \in [1, |\Sigma| - 1]$, let $\Phi_i = \langle B_i^1, B_i^2 \dots, B_i^{2|\Sigma|+1} \rangle$ be the sequence of bags for stars, where $B_i^j = \{u_2, c_i, \ell_i^j\}$. Then, the pair $\langle P, \Pi \rangle$, where Π is the sequence $\langle \Pi_1, \Phi_1, \Phi_2, \dots, \Phi_{|\Sigma|-1}, \Pi_2, \dots, \Pi_{n-1} \rangle$ of bags and P is a path whose length equals the length of Π, forms the path decomposition of width two. This completes the proof of Theorem 2. □

5 Fixed Parameter Algorithms

5.1 Graphs of Pathwidth One

We give the following theorem.

Theorem 4 (∗). OPT-LCR *on graphs with pathwidth one is fixed-parameter tractable when parameterized by the number of colors.*

5.2 Vertex Cover Number

In this subsection, we show the following theorem.

Theorem 5. OPT-LCR *is fixed-parameter tractable when parameterized by* $k + \mathsf{vc}$.

Recall that a graph G with vertex cover number vc can be partitioned into a vertex cover S of size vc and an independent set I. It is known that a vertex cover S of size vc can be found in $O(1.2738^{\mathsf{vc}} + \mathsf{vc} \cdot n)$ time [8], and hence we suppose that S is given. To give an FPT algorithm for OPT-LCR, we again utilize the encoding graph. For a vertex subset $V' \subseteq V(G)$, we denote by $g[V']$ the restriction of a list L-coloring g of a graph G, that is, $g[V']$ is the mapping $V' \to L(v)$ such that $g[V'](v) = g(v)$ for each $v \in V'$. For simplicity, we write R instead of a reconfiguration graph R_G^L. For two nodes g, g' in R, we define $g \sim g'$ if there is a reconfiguration sequence $\langle g_0, g_1, \dots, g_\ell \rangle$ with $g_0 = g$ and $g_\ell = g'$ of list L-colorings such that $g[S] = g_i[S]$ for every $i \in [0, \ell]$. The relation $\sim$ is an equivalence relation, and we denote $V(\mathsf{R})/\sim$ the set of all equivalence classes in $V(\mathsf{R})$ with respect to $\sim$. The encoding graph H of R is obtained from R by contracting the nodes in each equivalence class in $V(\mathsf{R})/\sim$ to a single node.

Since the size of R depends on a super-polynomial function of $|V(G)|$ in general, we directly obtain H whose size is bounded by a function of $k + \mathsf{vc}$

without constructing R. Let $\widehat{H}'$ be an empty graph whose vertex set has one-to-one correspondence to a set of all list L-colorings of $G[S]$. For each two nodes $f, f' \in V(\widehat{H}')$, f and f' are joined by an edge if and only if the corresponding two colorings f and f' differ at only one vertex and $L(v)\backslash\{f(w), f'(w) : w \in N(v)\} \neq \emptyset$ holds for every vertex $v \in I$. Then, we define $\widehat{H}$ as the obtained graph. The following three lemmas ensure that $\widehat{H}$ constructed as above is equal to the encoding graph H of R.

Lemma 2 (∗). *For a node f of $\widehat{H}$, let R_f be a set of nodes g in R such that $f = g[S]$. Then, $g \sim g'$ for any two nodes $g, g' \in R_f$, and hence all nodes in R_f are contracted in a single node in the encoding graph H.*

Lemma 3 (∗). *Let f and f' be nodes of $\widehat{H}$. Then, f and f' are joined by an edge on $\widehat{H}$ if and only if there are colorings g, g' of G such that $f = g[S]$, $f' = g'[S]$ and g' is obtained from g by recoloring a single vertex in S.*

Lemma 4. *Let H be the encoding graph of R. Then, $H = \widehat{H}$.*

Proof. Let $\Phi(x)$ be the set of nodes in R that are contracted to a node x of the encoding graph H of R. Lemma 2 states that there is a one-to-one correspondence ϕ between $V(\widehat{H})$ and $V(H)$ such that $f = g[S]$ for every pair of a node $f \in V(\widehat{H})$ and a node $g \in \Phi(\phi(f))$ of R. Moreover, by Lemma 3, two nodes f and f' of $\widehat{H}$ are joined by an edge on $\widehat{H}$ if and only if there are nodes g, g' in R such that $g \in \Phi(\phi(f))$, $g' \in \Phi(\phi(f'))$ and g, g' are joined by an edge on R. This implies that for any two nodes $x, y \in V(\widehat{H})$, it holds that $xy \in E(\widehat{H})$ if and only if $\phi(x)\phi(y) \in E(H)$. Therefore, $\widehat{H}$ is indeed the encoding graph H of R. □

Suppose that we have obtained the encoding graph H of R by constructing $\widehat{H}$. For each node f in H, we determine whether there exists a list L-coloring g of G such that $f = g[S]$ and $\#\mathsf{col}(g) \leq p$. However, it is unlikely to be done in polynomial time of a given instance because a situation similar to Theorem 1 may arise. Instead, we guess a subset C' of the color set C such that $\{g(v) : v \in V(G)\} \subseteq C'$. For a node f in H and a subset $C' \subseteq C$, we say that C' is *compatible* with the list L-coloring f of $G[S]$ if $f(u) \in C'$ for every $u \in S$ and $C' \cap L(v)\backslash\{f(w) : w \in N(v)\} \neq \emptyset$ for every $v \in I$.

Lemma 5. *Let C' be a subset of the color set C and f be a list L-coloring of $G[S]$. Then, C' is compatible with f if and only if there exists a list L-coloring g of G such that $f = g[S]$ and $g(v) \in C'$ for every $v \in V(G)$.*

Proof. The necessity is clear, and hence we show the sufficiency. Since C' is compatible, we can construct a mapping h such that $h(v) \in C' \cap L(v)\backslash\{f(w) : w \in N(v)\}$ for every vertex $v \in I$. Clearly, $h(v) \neq f(w)$ holds for every pair of $v \in I$ and $w \in N(v)$. Thus, combined with f and h, we obtain a list L-coloring g of G with $f = g[S]$. Moreover, since $f(u) \in C'$ for each $u \in S$ and $h(v) \in C'$ for each $v \in I$, we have $g(v) \in C'$ for every $v \in V(G)$. □

Let H_0 be a connected component of H that contains a node $f_0[S]$. By Lemma 5, to solve OPT-LCR, it suffices to find a subset $C' \subseteq C$ of size at most p that is compatible with some node $f \in V(H_0)$.

Finally, we estimate the running time of our algorithm. Since the number of colors is bounded by k and the size of a vertex cover S is bounded by vc, obviously we have $|V(H)| \leq k^{\text{vc}}$. In addition, since the size of a list of each vertex in $V(G)$ is bounded by k, there are at most $k \cdot |V(H)| \leq k^{\text{vc}+1}$ pairs of nodes in H that can be joined by an edge. For such a pair of nodes f and f', we can decide in $O((k + \text{vc}) \cdot n)$ time whether $\{f(w), f'(w) : w \in N(v)\} \cap L(v) \neq \emptyset$ for every vertex $v \in I$. Thus, H can be constructed in $O(k^{\text{vc}+1}(k + \text{vc}) \cdot n)$ time. For each $C' \subseteq C$ and each node $f \in V(H_0)$, we determine in $O((k+\text{vc}) \cdot n)$ time whether C' is compatible with f and $|C'| \leq p$. Since there are 2^k subsets of C and $|V(H_0)| \leq k^{\text{vc}}$, we obtain the solution of a given instance in $O(2^k k^{\text{vc}}(k + \text{vc}) \cdot n)$ time. The total running time of our algorithm is $O(2^k k^{\text{vc}+1}(k + \text{vc}) \cdot n)$. This completes the proof of Theorem 5. $\square$

6 Conclusion

In this paper, we analyzed the parameterized complexity of OPT-LCR. We showed that OPT-LCR is NP-hard for empty graphs even if every vertex of a given graph has a list of size two. Furthermore, we proved the PSPACE-completeness for bipartite graphs with bounded bandwidth and pathwidth two, even if the number k of colors is bounded by some constant. We then gave an FPT algorithm parameterized by k for graphs of pathwidth one and designed an FPT algorithm parameterized by $k + \text{vc}$.

The PSPACE-completeness for graphs with pathwidth two immediately implies the PSPACE-completeness for graphs with treewidth two. It is interesting to settle the complexity of OPT-LCR on graphs of treewidth one, that is, forests. We note that the polynomial-time solvability of LIST COLORING RECONFIGURATION on forests is also still open.

References

1. Blanché, A., Mizuta, H., Ouvrard, P., Suzuki, A.: Decremental optimization of dominating sets under the reconfiguration framework. In: Gąsieniec, L., Klasing, R., Radzik, T. (eds.) IWOCA 2020. LNCS, vol. 12126, pp. 69–82. Springer, Cham (2020). https://doi.org/10.1007/978-3-030-48966-3_6
2. Bonamy, M., Bousquet, N.: Recoloring graphs via tree decompositions. Eur. J. Comb. **69**, 200–213 (2018)
3. Bonamy, M., Johnson, M., Lignos, I., Patel, V., Paulusma, D.: Reconfiguration graphs for vertex colourings of chordal and chordal bipartite graphs. J. Comb. Optim. **27**, 132–143 (2014)
4. Bonsma, P.S., Cereceda, L.: Finding paths between graph colourings: PSPACE-completeness and superpolynomial distances. Theoret. Comput. Sci. **410**(50), 5215–5226 (2009)

5. Bonsma, P., Mouawad, A.E., Nishimura, N., Raman, V.: The complexity of bounded length graph recoloring and CSP reconfiguration. In: Cygan, M., Heggernes, P. (eds.) IPEC 2014. LNCS, vol. 8894, pp. 110–121. Springer, Cham (2014). https://doi.org/10.1007/978-3-319-13524-3_10

6. Bonsma, P., Paulusma, D.: Using contracted solution graphs for solving reconfiguration problems. Acta Informatica **56**, 619–648 (2019)

7. Cereceda, L., van den Heuvel, J., Johnson, M.: Finding paths between 3-colourings. J. Graph Theor. **67**(1), 69–82 (2011)

8. Chen, J., Kanj, I.A., Xia, G.: Improved upper bounds for vertex cover. Theoret. Comput. Sci. **411**(40), 3736–3756 (2010)

9. Feghali, C., Johnson, M., Paulusma, D.: A reconfigurations analogue of brooks' theorem and its consequences. J. Graph Theor. **83**(4), 340–358 (2016)

10. Hatanaka, T., Ito, T., Zhou, X.: The list coloring reconfiguration problem for bounded pathwidth graphs. IEICE Trans. Fundam. Electron. Commun. Comput. Sci. **E98.A**(6), 1168–1178 (2015)

11. Hatanaka, T., Ito, T., Zhou, X.: Parameterized complexity of the list coloring reconfiguration problem with graph parameters. Theoret. Comput. Sci. **739**, 65–79 (2018)

12. Hatanaka, T., Ito, T., Zhou, X.: The coloring reconfiguration problem on specific graph classes. IEICE Trans. Fundam. Electron. Commun. Comput. Sci. **E102.D**(3), 423–429 (2019)

13. van den Heuvel, J.: The complexity of change. In: Blackburn, S.R., Gerke, S., Wildon, M. eds. Surveys in Combinatorics, London Mathematical Society Lecture Note Series, vol. 409, pp. 127–160. Cambridge University Press (2013)

14. Ito, T., et al.: On the complexity of reconfiguration problems. Theoret. Comput. Sci. **412**(12), 1054–1065 (2011)

15. Ito, T., Mizuta, H., Nishimura, N., Suzuki, A.: Incremental optimization of independent sets under the reconfiguration framework. In: Du, D.-Z., Duan, Z., Tian, C. (eds.) COCOON 2019. LNCS, vol. 11653, pp. 313–324. Springer, Cham (2019). https://doi.org/10.1007/978-3-030-26176-4_26

16. Ito, T., Mizuta, H., Nishimura, N., Suzuki, A.: Incremental optimization of independent sets under the reconfiguration framework. J. Comb. Optim. **43**(5), 1264–1279 (2022)

17. Johnson, M., Kratsch, D., Kratsch, S., Patel, V., Paulusma, D.: Finding shortest paths between graph colourings. Algorithmica **75**, 295–321 (2016)

18. Mynhardt, C.M., Nasserasr, S.: 50 Years of Combinatorics, Graph Theory, and Computing, chapter Reconfiguration of colourings and dominating sets in graphs, pp. 171–191. Chapman and Hall/CRC (2019)

19. Nishimura, N.: Introduction to reconfiguration. Algorithms **11**(4), 52 (2018)

20. Savitch, W.J.: Relationships between nondeterministic and deterministic tape complexities. J. Comput. Syst. Sci. **4**(2), 177–192 (1970)

21. Wrochna, M.: Reconfiguration in bounded bandwidth and tree-depth. J. Comput. Syst. Sci. **93**, 1–10 (2018)

22. Yanagisawa, Y., Suzuki, A., Tamura, Y., Zhou, X.: Decremental optimization of vertex-coloring under the reconfiguration framework. In: Chen, C.-Y., Hon, W.-K., Hung, L.-J., Lee, C.-W. (eds.) COCOON 2021. LNCS, vol. 13025, pp. 355–366. Springer, Cham (2021). https://doi.org/10.1007/978-3-030-89543-3_30

Parameterized Complexity of Path Set Packing

N. R. Aravind and Roopam Saxena[(⊠)]

Department of Computer Science and Engineering, IIT Hyderabad, Hyderabad, India
aravind@cse.iith.ac.in, cs18resch11004@iith.ac.in

Abstract. In PATH SET PACKING, the input is an undirected graph G, a collection $\mathcal{P}$ of simple paths in G, and a positive integer k. The problem is to decide whether there exist k edge-disjoint paths in $\mathcal{P}$. We study the parameterized complexity of PATH SET PACKING with respect to both natural and structural parameters. We show that the problem is $W[1]$-hard with respect to vertex cover plus the maximum length of a path in $\mathcal{P}$, and $W[1]$-hard with respect to pathwidth plus maximum degree plus solution size. These results answer an open question raised in [17]. On the positive side, we present an FPT algorithm parameterized by feedback vertex set plus maximum degree, and also provide an FPT algorithm parameterized by treewidth plus maximum degree plus maximum length of a path in $\mathcal{P}$.

Keywords: Path set packing · Set packing · Parameterized complexity · Fixed parameter tractability · Graph algorithms

1   Introduction

Xu and Zhang [17] introduced the PATH SET PACKING (PSP) problem and discussed its various applications, such as in software defined networks. The problem asks if for a given graph $G = (V, E)$ and a collection $\mathcal{P}$ of simple paths in G, there exists a set $S \subseteq \mathcal{P}$ of edge disjoint paths such that $|S| \geq k$. PSP is closely related to the well known SET PACKING problem. Xu and Zhang [17] showed that PSP is NP-complete even when the maximum length of the given paths is 3. Considering the optimization version of the problem, they showed that PSP is hard to approximate within $O(|E|^{\frac{1}{2}-\epsilon})$ unless $NP = ZPP$. They showed that PSP can be solved in polynomial time when the input graph is a tree and gave a parameterized algorithm with running time $O(|\mathcal{P}|^{tw(G)\Delta}|V|)$ where $tw(G)$ is treewidth of G and Δ is maximum degree. Further, they left open the question whether PSP is fixed parameter tractable with respect to treewidth of the input graph.

C.-C. Lin et al. (Eds.): WALCOM 2023, LNCS 13973, pp. 291–302, 2023.
https://doi.org/10.1007/978-3-031-27051-2_25

PATH SET PACKING:
Input: An instance $I = (G, \mathcal{P}, k)$, where $G = (V, E)$ is an undirected graph, $\mathcal{P}$ is a collection of simple paths in G, and $k \in \mathbb{N}$.
Output: YES if there is a set of k pairwise edge-disjoint paths in $\mathcal{P}$, NO otherwise.

1.1   Related Work

In SET PACKING we are given a list S of subsets of a universe U and it is asked if S has k pairwise disjoint sets. The SET PACKING problem is W[1]-hard when parameterized by solution size k [6]. For the maximum size of a set d, FPT algorithms for the combined parameter of k and d have been obtained [14,15]. Kernel of size $O(k^{d-1})$ [1] has also been obtained. Since PSP can be seen as a special case of SET PACKING, all the positive results obtained for SET PACKING are also applicable to PSP.

Path Set Packing (PSP) can also be seen as the problem of finding a Maximum Independent Set on the *conflict graph* obtained by considering each path as a vertex with two vertices being adjacent if the corresponding paths share an edge. When the input graph to a PSP instance is a grid graph, the corresponding conflict graphs are called EPG graphs [12]. It was shown in [12] that every graph is an EPG graph. This immediately implies the following, because of well-known results on MIS on general graphs.

Corollary 1. *PSP is W[1]-hard on Grid graphs when parameterized by solution size k.*

Corollary 2. *PSP doesn't admit a $2^{o(|\mathcal{P}|)}$ time exact algorithm, assuming ETH.*

Thus, it is natural to consider PSP with further or different restrictions on the input graph G. We mention some known results of this type.

1. When G is a tree, the conflict graph is called an EPT graph [11]. Recognizing EPT graphs is NP-Complete [10]; nevertheless MIS is solvable in polynomial time on the class of EPT graphs [16].
2. The class of B_k-EPG graphs was defined as graphs obtained as the edge intersection graph of paths on a grid, with the restriction that each path have at most k bends. MIS on B_1-EPG graphs is NP-hard [7]. In [3], the authors showed that for the class of B_1-EPG graphs, when the number of path shapes is restricted to three, MIS admits an FPT algorithm, while remanining W[1]-hard on B_2-EPG graphs.

1.2 Our Results

We studied PSP with respect to combination of both natural and structural parameters of the input graph G and obtained following results for PSP.

Theorem 1. *PSP is W[1]-hard when parameterized by vertex cover number of G + maximum length of a path in $\mathcal{P}$.*

Theorem 2. *PSP is W[1]-hard when parameterized by pathwidth of G + maximum degree of G + solution size.*

Theorem 3. *PSP admits an FPT algorithm when parameterized by feedback vertex number of G + maximum degree of G.*

Theorem 4. *PSP admits an FPT algorithm when parameterized by treewidth of G + maximum degree of G + maximum length of a path in $\mathcal{P}$.*

We note that the above positive results complement the hardness of PSP with respect to any subset of the parameters used in the respective algorithms. The hardness with respect to treewidth plus maximum path length, and hardness with respect to treewidth plus maximum degree are implied by (Theorem 1 and 2). The hardness with respect to feedback vertex number is implied by (Theorem 1). And for maximum degree plus maximum path length, we note that the reduction from maximum independent set to PSP given in [17] to prove the inapproximability of PSP also works to prove NP-hardness of PSP for bounded maximum degree in G and bounded maximum length of a path in $\mathcal{P}$ using the fact that independent set is NP-hard on bounded degree graphs [9].

2 Preliminaries

We use $[n]$ to denote the set $\{1, 2, ..., n\}$. For a sequence ρ with n elements, $set(\rho)$ denotes the set of all the elements of ρ, and for $j \in [n]$, $\rho[j]$ is the element of ρ at position j. All the graphs considered in this paper are simple and finite. We use standard graph notations and terminologies and refer the reader to [5]. A path P is simple if no vertex occurs more than once in it. We denote the set of all the edges and all the vertices of a path P by $E(P)$ and $V(P)$ respectively.

For a connected graph $G = (V, E)$, we say that a subset $S \subseteq V$ is a vertex cover if $V \backslash S$ is an independent set in G, and we say that S is a feedback vertex set if $G[V \backslash S]$ induces a forest. The minimum size of a vertex cover of G is called its vertex cover number and the minimum size of a feedback vertex set of G is called its feedback vertex number. For two disjoint sets $A, B \subseteq V$, $E(A, B)$ is the set of all the edges with one endpoint in A and another in B. For details on pathwidth and treewidth, we refer to [4] . For details on parameterized complexity and fixed parameter tractability (FPT) we refer to [4,6]. Informally, a $W[1]$-hard problem is unlikely to be fixed parameter tractable.

3 Hardness with Respect to Vertex Cover + Maximum Path Length

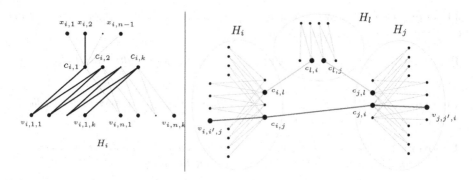

Fig. 1. Left: An example of vertex selection gadget H_i, the darkened edges forms a long path $P_{x_{i,2},v_{i,1}}$. Right: An example of inter gadget edges and darkened edges forms a short path $P_{v_{i,i',j},v_{j,j',i}}$ corresponding to an edge $v_{i,i'}v_{j,j'}$ in G.

In the k-MULTI COLORED CLIQUE (k-MCC) problem we are given a graph $G = (V, E)$, where V is partitioned into k disjoint sets $V_1, ..., V_k$, each of size n, and the question is if G has a clique $\mathcal{C}$ of size k such that $|\mathcal{C} \cap V_i| = 1$ for every $i \in [k]$. It is known that k-MCC is W[1]-hard parameterized by k [8].

We will give a parameterized reduction from k-MCC to PSP. Let $G = (V, E)$ and $\{V_1, ...V_k\}$ be an input of k-MCC. Let the vertices of V_i be labeled $v_{i,1}$ to $v_{i,n}$. We will construct an equivalent instance $(G' = (V', E'), \mathcal{P})$ of PSP (see Fig. 1 for overview). For every set V_i, we construct a vertex selection gadget H_i (an induced subgraph of G') as follows .

- Create a set $C_i = \{c_{i,1}, ...c_{i,k}\}$ of k vertices, a set $X_i = \{x_{i,1}, x_{i,2}, ..., x_{i,n-1}\}$ of $n - 1$ vertices, and connect every $x_{i,j}$ to $c_{i,1}$ we call these edges E_{C_i, X_i}.
- For every $v_{i,j} \in V_i$, create a vertex set $V_{i,j} = \{v_{i,j,1}, v_{i,j,2}, ..., v_{i,j,k}\}$ of k vertices. Connect $v_{i,j,l}$ to $c_{i,l}$ and $c_{i,l+1}$ where $l \in [k-1]$, and connect $v_{i,j,k}$ to $c_{i,k}$. We denote these edges by $E_{C_i, V_{i,j}}$.

Formally, $H_i = (\bigcup_{j=1}^n V_{i,j} \cup X_i \cup C_i, \bigcup_{j=1}^n E_{C_i, V_{i,j}} \cup E_{C_i, X_i})$. Further, let $C = \bigcup_{i=1}^k C_i$, we add the following edges in G'.

- For $1 \leq i < j \leq k$, we connect $c_{i,j} \in C_i$ to $c_{j,i} \in C_j$. We call these edges the inter gadget edges and denote them by E_C. Observe that there are $\binom{k}{2}$ inter gadget edges (Fig. 1).

The above completes the construction of $G' = (\bigcup_{i=1}^k V(H_i), \bigcup_{i=1}^k E(H_i) \cup E_C)$.

We now move on to the construction of the collection $\mathcal{P}$.

- Let $P_{x_{i,l},V_{i,j}} = (x_{i,l}, c_{i,1}, v_{i,j,1}, c_{i,2}, v_{i,j,2}, \ldots\ldots, c_{i,k}, v_{i,j,k})$, that is a path starting at $x_{i,l}$ and then alternates between a vertex in C_i and $V_{i,j}$ and ending at $v_{i,j,k}$. We call such a path a long path. For every $i \in [k]$, $l \in [n-1]$, and $j \in [n]$ we add $P_{x_{i,l},V_{i,j}}$ in $\mathcal{P}$. Observe that there are $n(n-1)$ long paths added from every H_i.

- For every edge $e = v_{i,i'}v_{j,j'} \in E$ where $v_{i,i'} \in V_i$ and $v_{j,j'} \in V_j$, w.l.o.g. assuming $i < j$, we add a path $P_{v_{i,i',j},v_{j,j',i}} = (v_{i,i',j}, c_{i,j}, c_{j,i}, v_{j,j',i})$ in $\mathcal{P}$ and call such a path, a short path. There are $|E|$ short paths added to $\mathcal{P}$.

The above completes the construction of instance $(G' = (V', E'), \mathcal{P})$ with $|\mathcal{P}| = |E| + kn(n-1)$.

The vertex set C forms a vertex cover for G' which is of size k^2 and the length of every long path is $2k + 1$. Further, the time taken for construction is $poly(|V|)$; the following concludes the correctness of the reduction and proof of Theorem 1.

Lemma 1 ($\star^1$). $(G = (V,E), \{V_1, \ldots V_k\})$ *is a yes instance of* k-*MCC if and only if* G' *has* $k(n-1) + \binom{k}{2}$ *edge disjoint paths in* $\mathcal{P}$.

4 Hardness with Respect to Pathwidth + Maximum Degree + Solution Size

We give a parameterized reduction from k-MCC to PSP. Let $G = (V,E)$ and $\{V_1, \ldots, V_k\}$ be the input for k-MCC, and let the vertices of set V_i be labeled $v_{i,1}$ to $v_{i,n}$. We will construct an equivalent instance $(G' = (V', E'), \mathcal{P})$ of PSP (see Fig. 2 for overview), the construction of G' is as follows.

- For every V_i, we construct a gadget (subgraph of G') which includes a vertex selection path P_i, a vertex set W_i, and k edge verification paths $P_{i,l}^e$ as follows.
  - Corresponding to V_i, we start with creating $n+1$ paths of $2k$ vertices each, one path for every vertex $v_{i,i'} \in V_i$ and an additional path. For every $i' \in [n+1]$, the i' path is $(v_{i,i',1}, u_{i,i',1}, v_{i,i',2}, u_{i,i',2}\ldots, v_{i,i',k}, u_{i,i',k})$. We now combine these $n+1$ paths into one path P_i by adding an edge between $u_{i,i',k}$ and $v_{i,i'+1,1}$ for every $i' \in [n]$.
  - We create n vertices $w_{i,1}$ to $w_{i,n}$ and call the set of these vertices W_i. For every $i' \in [n]$, we connect $w_{i,i'}$ to $v_{i,i',1}$ and $v_{i,i'+1,1}$.
  - We create k edge verification paths $P_{i,1}^e$ to $P_{i,k}^e$ with $n+1$ vertices each. The path $P_{i,j}^e$ is $(x_{i,1,j}, x_{i,2,j}, \ldots, x_{i,n,j}, c_{i,j})$.
  - For every $j \in [k]$, $i' \in [n]$, we connect $u_{i,i',j}$ to $x_{i,i',j}$. These edges connects the vertices of vertex selection path P_i to edge verification paths $P_{i,j}^e$.

---

[1] The proofs of statements marked with a $\star$ have been omitted.

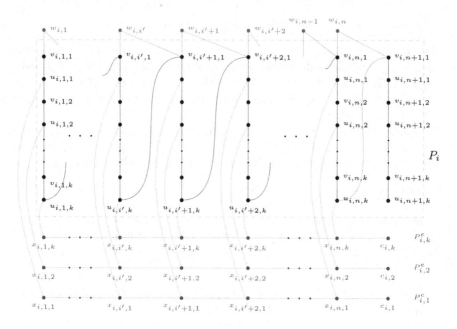

Fig. 2. An example of path P_i, edge verification paths $P_{i,1}^e$ $P_{i,2}^e$, and $P_{i,k}^e$, also the edges between vertices of vertex selection paths and edge verification paths.

- After constructing above mentioned gadgets for every vertex set in $\{V_1, .., V_k\}$, for $1 \le i < j \le k$, we connect $c_{i,j}$ to $c_{j,i}$. We call these edges the inter gadget edges. Observe that there are $\binom{k}{2}$ inter gadget edges.

The above completes the construction of G'. We now construct collection $\mathcal{P}$ of size $nk + |E|$ as follows.

- For every $i \in [k]$, from the subgraph of G' induced by $V(P_i) \cup W_i$, we will add n paths in the collection $\mathcal{P}$ as follows.
  - Add a path $l_{i,\bar{i'}} = (P_i(v_{i,1,1}, v_{i,i',1}), w_{i,i'}, P_i(v_{i,i'+1,1}, u_{i,n+1,k}))$ for every $i' \in [n]$, where $P_i(v_{i,1,1}, v_{i,i',1})$ is the path from vertex $v_{i,1,1}$ to $v_{i,i',1}$ in P_i (a unique path since P_i is a path). Intuitively, for every $i' \in [n]$ the $l_{i,\bar{i'}}$ contains all the edges of P_i except the edges which belong to subpath $P_i(v_{i,i',1}, v_{i,i'+1,1})$. We call these paths the long paths.
- For every edge $v_{i,i'}v_{j,j'} \in E$ where $i < j$, we add the path $s_{i,i',j,j'} = (v_{i,i',j}, u_{i,i',j}, P_{i,j}^e(x_{i,i',j}, c_{i,j}), P_{j,i}^e(c_{j,i}, x_{j,j',i}), u_{j,j',i}, v_{j,j',i})$ in $\mathcal{P}$, where $P_{i,j}^e(x_{i,i',j}, c_{i,j})$ is the path from $x_{i,i',j}$ to $c_{i,j}$ in $P_{i,j}^e$ (a unique path, since $P_{i,j}^e$ is a path). We note that every $s_{i,i',j,j'}$ contains exactly one inter gadget edge $c_{i,j}, c_{j,i}$. This finishes the construction of $\mathcal{P}$.

Observe that the construction of $(G', \mathcal{P})$ takes time polynomial in $|V|$. We now claim the bounds on pathwidth and maximum degree of G'.

Lemma 2 (⋆). *Pathwidth of G' is $O(k^2)$ and maximum degree of G' is $O(k)$.*

The following concludes the correctness of reduction and proof of Theorem 2.

Lemma 3 (⋆). *$G = (V, E)$ with partition V_1 to V_k is a yes instance of k-MCC if and only if $\mathcal{P}$ has $k + \binom{k}{2}$ pairwise edge disjoint paths.*

5 FPT Parameterized by Feedback Vertex Number + Maximum Degree

In this section, we will show that PSP is FPT parameterized by Γ (feedback vertex number) plus maximum degree Δ and prove Theorem 3. For a connected graph $G = (V, E)$, its feedback edge number, denoted by λ, is the minimum number of edges whose removal results in a tree and equals $|V| - |E| + 1$. Since the set of all edges incident on a feedback vertex set forms a feedback edge set, we have: $\lambda \leq \Gamma \cdot \Delta$. We will provide an algorithm which solves PSP in time $(\lambda \cdot \Delta)^{O(\lambda \cdot \Delta)} \cdot poly(|V| + |\mathcal{P}|)$. The approach used here is a non trivial adaptation of the approach given in [13].

5.1 Preliminaries: Defining Structures and Nice Solutions

Let $G = (V, E)$ be the input graph. We create 3 vertices z_1, z_2, z_3 and arbitrarily choose a vertex $v \in V$ and connect z_1, z_2, z_3, and v to each other forming a clique on 4 vertices in G. This modification increases the size of minimum feedback vertex set and maximum degree by only a constant, and is safe for our purposes. In this section we define structures (adapted from [13]) that we will need.

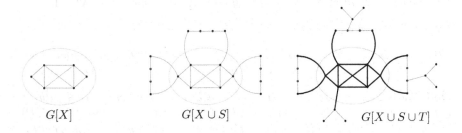

$G[X]$ $G[X \cup S]$ $G[X \cup S \cup T]$

Fig. 3. An example induced graphs $G[X], G[X \cup S]$, and $G[X \cup S \cup T]$. The darkened edges in rightmost graph represent core edges E_X.

Definition 1. *We define the vertex set T, S and X by the following process (refer Fig. 3).*

- *Initialize T as an empty set and $G' = (V', E')$ as a copy of G.*
- *While there is a vertex v in G' with degree $d_{G'}(v) = 1$, we set $T = T \cup \{v\}$ and $G' = G' - \{v\}$.*

– X is the set of all the vertices with degree at least 3 in $G[V \setminus T]$, and S is the set $V \setminus (T \cup X)$.

Observe that $G[T]$ is a forest, and every component of $G[T]$ is connected to $G[V \setminus T]$ by a single edge in G. $G[V \setminus T]$ has minimum degree 2. X is a non empty set as it contains at least z_1, z_2, z_3, and v as they form a clique. Every vertex in S has degree exactly 2 in $G[V \setminus T]$ and $G[S]$ is a union of paths (Fig. 3).

Definition 2. We define $E_X = E(G[X]) \cup E(X, S \cup T)$, i.e. all the edges in G with at least one endpoint in X. We will call these edges the core edges of G.

Observation 1 ($\star$)**.** Every component in $G[S \cup T]$ contains at most 1 component from $G[S]$.

Let $\mathcal{D}$ be the set of all the components of $G[S \cup T]$ which contain a component of $G[S]$. Further, let $\mathcal{T}$ be the set of all other components in $G[S \cup T]$, every component $C \in \mathcal{D} \cup \mathcal{T}$ is a tree, and the edges which connect C to $G[X]$, are called external edges of C, and these edges belong to E_X.

Observation 2 ($\star$)**.** Every component in $\mathcal{D}$ is connected to $G[X]$ by two edges in G (has 2 external edges), and every component in $\mathcal{T}$ is connected to $G[X]$ by one edge in G (has one external edge).

We call $\mathcal{D}$ the set of 2-external edge components, and $\mathcal{T}$ the set of 1-external edge components. To bound the size of X and $\mathcal{D}$, we recall the following from [13].

Proposition 1 ([13])**.** Let G be a connected graph of minimum degree at least two with cyclomatic number (feedback edge number) λ. Let X be the set of all the vertices of degree at least three in G. Then $|X| \leq 2\lambda - 2$ and if $X \neq \emptyset$, then the number of connected components of $G[V \setminus X]$ is at most $\lambda + |X| - 1$.

We get the following corollary.

Corollary 3. The size of vertex set $|X| = O(\lambda) = O(\Gamma \cdot \Delta)$, and $|E_X| = O(\lambda \cdot \Delta) = O(\Gamma \cdot \Delta^2)$. The size of component set $|\mathcal{D}| = O(\lambda) = O(\Gamma \cdot \Delta)$.

Definition 3. For a subgraph $H \subseteq G$, a path p in G is an internal path of H if $E(p) \subseteq E(H)$. Further, given a set P of paths in G, we define $INT(H, P) = \{p_i | \ p_i \in P \wedge E(p_i) \subseteq E(H)\}$, that is all the paths in P which are internal to H.

Definition 4. Let $(G, \mathcal{P}, k)$ be an instance of PSP, for a subgraph $H \subseteq G$, we define $OPT(H)$ as the maximum number of edge disjoint paths in $INT(H, \mathcal{P})$.

Lemma 4 ($\star$)**.** Let $(G, \mathcal{P}, k)$ be an instance of PSP. And let $M \subseteq \mathcal{P}$ be a path set packing of maximum size. Then there exists a path set packing $M' \subseteq \mathcal{P}$, such that $|M'| = |M|$ and the following holds

– for every component $D_i \in \mathcal{D}$, $OPT(D_i) \geq |M' \cap INT(D_i, \mathcal{P})| \geq OPT(D_i) - 1$, and
– for every component $T_i \in \mathcal{T}$, $|M' \cap INT(T_i, \mathcal{P})| = OPT(T_i)$.

*We call such M′ as a **nice solution**.*

While searching for a solution, we will only search for a nice solution of maximum size.

Proposition 2 ([16,17]). *PSP can be solved in time polynomial in $|V| + |\mathcal{P}|$ if the input graph is a tree.*

Corollary 4. *OPT(D_i) and OPT(T_i) can be computed in time polynomial in $|V| + |\mathcal{P}|$ for every $D_i \in \mathcal{D}$ and every $T_i \in \mathcal{T}$.*

5.2 Guessing and Extending the Solution

We first guess the number of internal paths that every $D_i \in \mathcal{D}$ will have in the solution. Formally, $f_d : \mathcal{D} \rightarrow \{0,1\}$ is a guessing that OPT(D_i) $- f_d(D_i)$ internal paths of D_i will be in the solution. For every $T_i \in \mathcal{T}$ we know that OPT(T_i) internal paths will be in the solution. Now, it is left to optimize the number of edge disjoint paths which are having one or more edges from E_X. Let $f_e : E_X \rightarrow [|E_X|] \cup \{0\}$. Let $E_i = \{e | e \in E_X \wedge f_e(e) = i\}$. Let there be l non-empty sets E_i except E_0; let them be E_1 to E_l. This is our second guess, where we are guessing the partition of E_X such that if a path p in the solution intersects with E_X, then $E(p) \cap E_X$ should be E_i for an $i \in [l]$. Further, all the edges of E_0 are guessed to be not part of any path in the solution. Thus, we are guessing that there will be l paths in the solution containing the edges from E_X.

Definition 5. *We say a path $p \in \mathcal{P}$ is of type E_i if $(E(p) \cap E_X) = E_i$.*

Definition 6. *We say that the pair (f_d, f_e) has a feasible solution, if there exists a set of edge disjoint paths $M \subseteq \mathcal{P}$ such that the following holds.*

1. *For every $i \in [l]$, there exists a path $p \in M$ such that p is of type E_i;*
2. *For every $D_i \in \mathcal{D}$, $|M \cap INT(D_i, M)| = OPT(D_i) - f_d(D_i)$;*
3. *For every $T_i \in \mathcal{T}$, $|M \cap INT(T_i, M)| = OPT(T_i)$.*

If a pair (f_d, f_e) has a feasible solution M, then $|M|$ will be equal to the sum of OPT(T_i) over every $T_i \in \mathcal{T}$, plus sum of OPT(D_i) $- f_d(D_i)$ over every $D_i \in \mathcal{D}$, plus l. Thus, taking maximum over the size of the feasible solution of every possible pair (f_d, f_e) which has a feasible solution will give us the maximum size of a path set packing as both f_d and f_e are exhaustive guesses, and we are searching for a nice solution (Lemma 4). There are at most $2^{O(\lambda)}$ distinct guesses f_d and at most $(\lambda \cdot \Delta)^{O(\lambda \cdot \Delta)}$ distinct guesses f_e. If we can verify whether a pair (f_d, f_e) has a feasible solution in time polynomial in $(|V| + |\mathcal{P}|)$, then we can bound the running time of the algorithm to $(\lambda \cdot \Delta)^{O(\lambda \cdot \Delta)} \cdot poly(|V| + |\mathcal{P}|)$. Given a pair (f_d, f_e), we will now discuss how to verify if (f_d, f_e) has a feasible solution.

Definition 7. *A set of paths P is (f_d, f_e)-compatible if the following holds.*

1. *For every pair $p, q \in P$, if $p \neq q$ then p and q are edge disjoint;*
2. *For every $p \in P$, there exists $i \in [l]$ such that p is of type E_i;*
3. *For every $D_i \in \mathcal{D}$, $OPT(D_i - E(P)) \geq OPT(D_i) - f_d(D_i)$;*
4. *For every $T_i \in \mathcal{T}$, $OPT(T_i - E(P)) = OPT(T_i)$.*

Observation 3 ($\star$). *Given a path set $P \subseteq \mathcal{P}$, in time polynomial in $(|V| + |P|)$ we can verify if P is (f_d, f_e)-compatible or not. Further, if P is (f_d, f_e)-compatible then every subset P' of P is also (f_d, f_e)-compatible.*

Lemma 5 ($\star$). *(f_d, f_e) has a feasible solution if and only if there exists a path set $P \subseteq \mathcal{P}$ such that $|P| = l$ and P is (f_d, f_e)-compatible.*

Due to above lemma, to verify if (f_d, f_e) has a feasible solution, it will suffice to verify if there exist an (f_d, f_e)-compatible path set $P \subseteq \mathcal{P}$ of size l.

Observation 4 ($\star$). *If there exist an E_i which contains exactly one external edge of three or more components in $\mathcal{D} \cup \mathcal{T}$, then no path can contain all the edges of E_i.*

If for any $i \in [l]$, E_i contains exactly one external edge of three or more components in $\mathcal{D} \cup \mathcal{T}$, then using Observation 4, we can conclude that (f_d, f_e) has no feasible solution. Thus we shall henceforth assume that no E_i where $i \in [l]$, contains exactly one external edge of three or more components in $\mathcal{D} \cup \mathcal{T}$.

Consider an auxiliary graph H with vertex set $\{E_1, ..., E_l\}$. In H, two vertices E_i and E_j are adjacent if and only if there exists a $D \in \mathcal{D}$ such that one of the external edges of D belongs to E_i and the other external edge belongs to E_j. Every D has two external edges, and thus, every D can cause at most one edge to be created in H. Combining this with Observation 4, we conclude that H has degree at most two. Thus, H is a union of paths and cycles. Let there be l' components in H which are labelled from 1 to l', let π_i be the the path formed by the component i, that is each π_i is a sequence of vertices (arbitrarily chose the first vertex of the path in case the component i is a cycle). We call π_i a type sequence. Let $\Pi = \{\pi_i | i \in [l']\}$.

Definition 8. *A sequence ρ of paths is a candidate for type sequence π_i, if $|\rho| = |\pi_i|$ and for every $j \in [|\rho|]$, $\rho[j]$ is of type $\pi_i[j]$.*

Lemma 6 ($\star$). *There exists $P \subseteq \mathcal{P}$ such that $|P| = l$ and P is (f_d, f_e)-compatible if and only if for every sequence $\pi_i \in \Pi$, there exists a candidate ρ such that $set(\rho)$ is (f_d, f_e)-compatible.*

Since there are at most $l' \leq |V|^2$ type sequences, it suffices to prove that given a type sequence $\pi_i \in \Pi$, in time polynomial in $(|\mathcal{P}| + |V|)$ we can verify if π_i has a candidate ρ_i such that $set(\rho_i)$ is (f_d, f_e)-compatible. Consider the following.

Lemma 7 ($\star$). *For every type sequence $\pi_i \in \Pi$, let ρ_i be a candidate of π_i. Then, $set(\rho_i)$ is (f_d, f_e)-compatible if and only if*

– *For every $j \in [|\pi_i|]$, $\{\rho_i[j], \rho_i[(j+1)mod|\pi_i|]\}$ is (f_d, f_e)-compatible.*

The above lemma helps us find a desired candidate for type sequence π_i in polynomial time, since we now only need to check if the paths of adjacent types in a sequence are (f_d, f_e)-compatible or not. If the size of π_i is ≤ 2, then we can verify if π_i has a candidate ρ_i such that $set(\rho_i)$ is (f_d, f_e)-compatible by checking for every distinct $p, q \in \mathcal{P}$ if (p, q) is a candidate of π_i as well as (f_d, f_e)-compatible or not, this will take time polynomial in $(|V| + |\mathcal{P}|)$. We now move on to the case when $|\pi_i| \geq 3$.

Given a type sequence π_i we create an auxiliary directed graph H_i as follows.

– For every type $\pi_i[j]$, create a vertex set $V_j = \{v_p | \ p \in \mathcal{P} \wedge p$ is of type $\pi_i[j] \wedge \{p\}$ is (f_d, f_e)-compatible $\}$. That is V_j contain vertices corresponding to every path of type $\pi_i[j]$ which is (f_d, f_e)-compatible.
– For every $j \in [|\pi_i|]$, let $j' = (j+1)mod|\pi_i|$, we add an arc (directed edge) from $v_p \in V_j$ to $v_q \in V_{j'}$ if and only if $\{p, q\}$ are (f_d, f_e)-compatible.

Lemma 8 ($\star$). *π_i has a candidate ρ_i such that $set(\rho_i)$ is (f_d, f_e)-compatible if and only if H_i has a cycle containing exactly one vertex from every V_j where $j \in [|\pi_i|]$.*

Lemma 9 ($\star$). *In time polynomial in $(|V| + |\mathcal{P}|)$, we can find a cycle containing exactly one vertex from every V_j in H_i or conclude that no such cycle exists.*

The above lemma finishes the proof of Theorem 3.

6 FPT When Combining Three Parameters

Given a pair $(G, \mathcal{P})$, let H be the conflict graph with vertex set $\mathcal{P}$. As noted earlier, solving PSP on $(G, \mathcal{P})$ is equivalent to finding a MIS in H. We can deduce the following fact about the structure of H.

Lemma 10 ($\star$). *If G has treewidth k and maximum degree Δ, and each path in $\mathcal{P}$ is of length at most r, then the treewidth of H is at most $(k+1)\Delta^r$.*

The Maximum Independent Set problem admits a $O(2^\tau)$ algorithm on graphs of treewidth τ [2]. Hence we obtain Theorem 4 as a corollary.

Acknowledgements. We thank anonymous reviewers of this and an earlier version of this paper for useful suggestions.

References

1. Abu-Khzam, F.N.: An improved kernelization algorithm for r-set packing. Inf. Process. Lett. **110**(16), 621–624 (2010)
2. Arnborg, S., Proskurowski, A.: Linear time algorithms for np-hard problems restricted to partial k-trees. Discret. Appl. Math. **23**(1), 11–24 (1989)

3. Bessy, S., Bougeret, M., Chaplick, S., Gonçalves, D., Paul, C.: On independent set in B1-EPG graphs. Discret. Appl. Math. **278**, 62–72 (2020)

4. Cygan, M., et al.: Parameterized Algorithms. Springer, Heidelberg (2015). https://doi.org/10.1007/978-3-319-21275-3

5. Diestel, R.: Graph Theory. Graduate Texts in Mathematics, vol. 173, 4th edn. Springer, Heidelberg (2012)

6. Downey, R.G., Fellows, M.R.: Parameterized Complexity. Monographs in Computer Science, Springer, New York (1999). https://doi.org/10.1007/978-1-4612-0515-9

7. Epstein, D., Golumbic, M.C., Morgenstern, G.: Approximation algorithms for B_1-EPG graphs. In: Dehne, F., Solis-Oba, R., Sack, J.-R. (eds.) WADS 2013. LNCS, vol. 8037, pp. 328–340. Springer, Heidelberg (2013). https://doi.org/10.1007/978-3-642-40104-6_29

8. Fellows, M.R., Hermelin, D., Rosamond, F.A., Vialette, S.: On the parameterized complexity of multiple-interval graph problems. Theor. Comput. Sci. **410**(1), 53–61 (2009)

9. Garey, M.R., Johnson, D.S.: The rectilinear steiner tree problem is np-complete. SIAM J. Appl. Math. **32**(4), 826–834 (1977)

10. Golumbic, M.C., Jamison, R.E.: Edge and vertex intersection of paths in a tree. Discret. Math. **55**(2), 151–159 (1985)

11. Golumbic, M.C., Jamison, R.E.: The edge intersection graphs of paths in a tree. J. Comb. Theory Ser. B **38**(1), 8–22 (1985)

12. Golumbic, M.C., Lipshteyn, M., Stern, M.: Edge intersection graphs of single bend paths on a grid. Networks **54**(3), 130–138 (2009)

13. Jansen, B.M.P.: On structural parameterizations of hitting set: hitting paths in graphs using 2-SAT. In: Mayr, E.W. (ed.) WG 2015. LNCS, vol. 9224, pp. 472–486. Springer, Heidelberg (2016). https://doi.org/10.1007/978-3-662-53174-7_33

14. Jia, W., Zhang, C., Chen, J.: An efficient parameterized algorithm for m-set packing. J. Algorithms **50**(1), 106–117 (2004)

15. Koutis, I.: A faster parameterized algorithm for set packing. Inf. Process. Lett. **94**(1), 7–9 (2005)

16. Tarjan, R.E.: Decomposition by clique separators. Discret. Math. **55**(2), 221–232 (1985)

17. Xu, C., Zhang, G.: The path set packing problem. In: Wang, L., Zhu, D. (eds.) COCOON 2018. LNCS, vol. 10976, pp. 305–315. Springer, Cham (2018). https://doi.org/10.1007/978-3-319-94776-1_26

Approximation Algorithm

Interweaving Real-Time Jobs with Energy Harvesting to Maximize Throughput

Baruch Schieber[(✉)], Bhargav Samineni, and Soroush Vahidi

New Jersey Institute of Technology, Newark, NJ, USA
{sbar,bs567,sv96}@njit.edu

Abstract. Motivated by baterryless IoT devices, we consider the following scheduling problem. The input includes n unit time jobs $\mathcal{J} = \{J_1, \ldots, J_n\}$, where each job J_i has a release time r_i, due date d_i, energy requirement e_i, and weight w_i. We consider time to be slotted; hence, all time related job values refer to slots. Let $T = \max_i \{d_i\}$. The input also includes an h_t value for every time slot t $(1 \leq t \leq T)$, which is the energy harvestable on that slot. Energy is harvested at time slots when no job is executed. The objective is to find a feasible schedule that maximizes the weight of the scheduled jobs. A schedule is feasible if for every job J_j in the schedule and its corresponding slot t_j, $t_j \neq t_{j'}$ if $j \neq j'$, $r_j \leq t_j \leq d_j$, and the available energy before t_j is at least e_j. To the best of our knowledge, we are the first to consider the theoretical aspects of this problem.

In this work we show the following. (1) A polynomial time algorithm when all jobs have identical r_i, d_i and w_i. (2) A $\frac{1}{2}$-approximation algorithm when all jobs have identical w_i but arbitrary r_i and d_i. (3) An FPTAS when all jobs have identical r_i and d_i but arbitrary w_i. (4) Reductions showing that all the variants of the problem in which at least one of the attributes r_i, d_i, or w_i are not identical for all jobs are NP-Hard.

Keywords: Scheduling · Approximation algorithms · NP-hardness · Throughput maximization · IoT

1 Introduction

The energy aware scheduling problem considered in this paper is defined as follows. Its input includes n jobs $\mathcal{J} = \{J_1, \ldots, J_n\}$, where each job J_i is associated with release time r_i, due date d_i, energy requirement e_i, and weight w_i. All jobs have equal (unit) processing time. We consider time to be slotted with unit length and hence all time related job values refer to slots. Let $T = \max_i \{d_i\}$. The input also includes an h_t value specified explicitly for every time slot t, $(1 \leq t \leq T)$, which is the energy harvestable on the slot. Energy can be harvested *only* at times when no job is executed. The energy available immediately before slot t is the energy harvested in slots $1, \ldots, t-1$ minus the energy consumed by the jobs scheduled in slots $1, \ldots t-1$. A schedule is *feasible* if it schedules no more than one job at a time, all executed jobs are scheduled between their release time and due date, and the energy available immediately before a job is executed is at least its energy requirement. The objective is to maximize throughput, namely, the weight of the jobs in a feasible schedule.

C.-C. Lin et al. (Eds.): WALCOM 2023, LNCS 13973, pp. 305–316, 2023.
https://doi.org/10.1007/978-3-031-27051-2_26

1.1   Motivation

Our problem is motivated by the proliferation of Internet of Things (IoT) devices, which are used for many applications such as sensor networks, control systems, and home and building automation, to name a few. One of the major challenges impacting the deployment of these devices is their power source. Most are powered by batteries, which are compact and lightweight options. However, the chemicals contained in these batteries pose a considerable risk to our environment [12]. Also, the scarcity of the materials needed for batteries makes them prohibitively expensive in some applications, especially on an industrial scale. Battery maintenance is another major issue, as the limited lifetime of batteries requires expensive and constant care to replace or recharge them. Consider, for example, IoT devices with humidity and temperature sensors that are stationed along an oil pipeline in a remote area and connected to its SCADA system [2]. Or, consider IoT devices for wildlife monitoring that are attached to animals and used to gather information like migration paths and population mortality [5]. Connecting these devices to the electric grid is not an option in some places, and using batteries is a logistical nightmare.

To overcome these issues, batteryless IoT devices have been proposed [14,17]. The energy used by these devices is directly harvested from environmental and renewable sources such as solar, wind, and radio-frequency (RF). In the simplest designs, the harvested energy output is connected directly to the load. However, this design is only appropriate when the harvested current and voltage match those required for a task's execution [13]. More sophisticated designs include either capacitors or super capacitors to store energy, and are *intermittent* systems [12,13] in which energy harvesting (charging) and task execution (discharging) are mutually exclusive to allow for a single control thread. This is captured in our model by interweaving job execution and energy harvesting.

A major challenge in the design of such an intermittent system is the variability of energy harvesting over time [16]. For example, consider a solar energy source. Certainly, this energy can only be harvested in the daytime; but even during the day the amount of harvestable energy varies based on the cloud cover and the sun angle and its predictability is challenging [8]. We consider the *offline* version of the problem, as proposed in [12], and assume that the energy harvesting profile over time is given as part of the input to our problem.

As noted in [12], intermittent systems are primarily used in monitoring and surveillance applications that collect data at a fixed rate and then process the data periodically. We model this by associating a release time and due date for each job (task). Due to the bare-bones design of these systems, all these tasks are pretty basic and require a minimal number of cycles. Thus, it can be assumed that all these tasks require equal processing time.

1.2   Our Results

We present both algorithms and hardness results for several variants of the problem. To the best of our knowledge, our work is the first theoretical analysis of this problem. Our objective is to find a feasible schedule that maximizes job throughput. When all the jobs have the same weight (i.e. the unweighted setting), this corresponds to finding a schedule that maximizes the number of jobs scheduled. We call this problem Energy Aware Scheduling (EAS). Otherwise, this corresponds to maximizing the weight of the jobs scheduled (i.e. the weighted setting). We call this problem Weighted Energy Aware Scheduling (WEAS).

In Sect. 2 we give an optimal polynomial time algorithm for EAS when all the jobs have identical release times and due dates. The dynamic programming algorithm is based on some properties of an optimal solution for this case. In the full version of the paper we give a more efficient algorithm (that is more involved) for the same problem that is based on additional properties of an optimal solution. In Sect. 3 we consider EAS when jobs have arbitrary release times and due dates and show that a simple greedy algorithm achieves a $\frac{1}{2}$-approximation. Interestingly, the proof of the approximation ratio of this simple algorithm is quite "tricky". In Sect. 4 we show an FPTAS for WEAS when all the jobs have identical release times and due dates. In Sect. 5.1 we prove that EAS is weakly NP-Hard whenever all jobs don't have both identical release times and identical due dates. We also show that WEAS is weakly NP-Hard in Sect. 5.2.

1.3   Prior Work

Our model is inspired by a similar model proposed by Islam and Nirjon [12]. While their model allows arbitrary processing times for both jobs and energy harvesting, we consider all processing times to be uniform. Additionally, they model jobs as belonging to a set of periodic tasks, while we do not enforce such a constraint. They give heuristic scheduling algorithms for the offline (in which the harvestable energy profile is part of the input) as well as the online (in which the harvestable energy profile is not known a priori) versions of the problem. They benchmarked their algorithms against other heuristics and in the offline case also compared their algorithm to the optimal solution computed using an IP solver.

Our problem is related to scheduling with *nonrenewable resources* in which jobs require a nonrenewable resource like energy or funding to be scheduled. In contrast to our problem, it is assumed that the replenishment of the resource is done instantly at predetermined times. As in our problem, jobs can be scheduled feasibly only if the amount of available resource when they are started is at least their resource requirement. The goal is to schedule these resource-consuming jobs to optimize various objectives. In [11] it was shown that computing a minimum makespan schedule of jobs with a single nonrenewable resource requirement with 2 replenishment times, arbitrary processing times, and identical release times on a single machine is (weakly) NP-Hard. The same paper also proved that in case the number of replenishment times of the resource is part of the input, the problem is strongly NP-Hard. [9] considered the same setting but with the objective of delay minimization. [6,10] (and references therein) also considered scheduling with nonrenewable resources.

Another related problem is *inventory constrained scheduling*, which considers the scheduling of two types of jobs: resource-producing and resource-consuming

jobs. A resource-consuming job cannot be scheduled unless its resource requirement is available. Unlike our problem where there is no need to schedule all the
resource-producing (energy harvesting) jobs, in this problem both the resource-
producing and resource-consuming jobs have to be scheduled. Several variants
of this problem were considered in [3,4,7]. These variants include the case of
jobs with equal processing times and the objective of minimizing the number of
tardy jobs, which is the complement of unweighted throughput maximization.

[1], and later [18], considered the "non-energy-aware" version of our problem and showed that minimizing the number of tardy jobs, which is equivalent
to maximizing the unweighted throughput, can be solved in polynomial time.
[19] identified special cases in which minimizing the weighted tardiness of the
"non-energy-aware" version of our problem is polynomial.

1.4   Preliminaries

As we mostly deal with integral values, we use the notation $[a, b]$, where $a, b \in \mathbb{Z}^+$
and $a \leq b$, to denote the set of integers between a and b inclusive. We assume that
time is slotted with unit length; that is, when referring to a job being scheduled
or energy being harvested at time slot $j \in \mathbb{Z}^+$, it means they are done on the
time interval $[j - 1, j)$. A job $J_i \in \mathcal{J} = \{J_1, \ldots, J_n\}$ can be executed at any slot
in $[r_i, d_i]$, assuming that the energy available before its execution time slot is at
least e_i.

We formally define a schedule S as a pair $(J(S), \pi_S)$ where $J(S) \subseteq \mathcal{J}$ is the
set of jobs scheduled by S and $\pi_S : J(S) \rightarrow [1, T]$ is a mapping that maps each
job $J_i \in J(S)$ to its execution time slot $t \in [1, T]$. We additionally define $E_S(t)$
to be the energy amount available immediately before slot t in schedule S, given
by the equation

$$E_S(t) = \sum_{\tau=1}^{t-1} h_\tau - \left(\sum_{J_i \in \{J' \in J(S) \,|\, \pi_S(J') < t\}} e_i + h_{\pi_S(J_i)} \right).$$

That is, it is the total energy harvestable from all time slots $\tau \in [1, t-1]$ minus
the energy consumed by the jobs scheduled up to slot t and the energy harvestable at the slots when these jobs are scheduled (since energy can be harvested only at slots in which no job is scheduled). A schedule S is feasible if π_S
is one-to-one, and for each job $J_i \in J(S)$ scheduled on time slot $t_i = \pi_S(J_i)$, we
have $E_S(t_i) \geq e_i$ and $t_i \in [r_i, d_i]$.

2   An Optimal Algorithm for **EAS** When All Jobs Have Identical Release Times and Due Dates

In this section, we consider EAS instances in which all jobs have identical release
times and due dates and present a polynomial time dynamic programming algorithm that produces an optimal schedule. From now on, we assume that the jobs

are sorted in non-decreasing order by their energy requirement ($e_1 \leq \ldots \leq e_n$). We consider all $r_i = 1$, though the algorithm can easily be adapted to cases where all $r_i = r$ for some $r > 1$. We begin with two claims.

Claim 1. *There exists an optimal schedule that schedules a prefix of the sequence* $J_1, \ldots, J_n$.

Proof. Suppose that the optimal algorithm schedules the m jobs $J_{i_1}, \ldots, J_{i_m}$, where $i_1 < \cdots < i_m$. Then, the schedule that replaces job J_{i_j} by J_j for $j \in [1, m]$ is also feasible since $e_j \leq e_{i_j}$. □

Claim 2. *There exists an optimal schedule in which the jobs are scheduled in non-decreasing order of their energy requirement.*

Proof. Consider a feasible schedule S in which a job J_i is scheduled before job J_j and $e_i > e_j$. To prove the claim it is sufficient to show that the schedule S' given by swapping J_i and J_j is also feasible. Let $t_i = \pi_S(J_i)$ and $t_j = \pi_S(J_j)$. The parts of schedule S' immediately before time slot t_i and after time slot t_j are feasible since for every $t \in [1, t_i - 1] \cup [t_j + 1, T]$, we have $E_{S'}(t) = E_S(t)$. Job J_j can be scheduled at slot t_i since $E_{S'}(t_i) = E_S(t_i) \geq e_i > e_j$. For every slot $t \in [t_i + 1, t_j - 1]$, we have $E_{S'}(t) = E_S(t) + e_i - e_j > E_S(t)$. Thus, the jobs scheduled in S' in these time slots are also feasible. Now consider slot t_j. Job J_j is scheduled in S at this slot, hence $E_S(t_j) \geq e_j$. It follows that $E_{S'}(t_j) = E_S(t_j) + e_i - e_j \geq e_i$, which implies J_i can be scheduled at slot t_j. Therefore, S' is a feasible schedule. Since the optimal schedule must also be feasible, the same swapping procedure can be applied. □

We apply these observations to obtain a dynamic programming algorithm to compute an optimal schedule O. Define the dynamic programming "table" as follows. For $i \in [1, n]$ and $t \in [1, T]$, let $A(i, t)$ be the maximum amount of available energy at the start of time slot $t + 1$, where the maximum is taken over all feasible schedules of jobs $J_1, \ldots, J_i$ on the time slots $[1, t]$. If such a feasible schedule does not exist, then $A(i, t) = -\infty$. Since the input size is $\Omega(n + T)$, the size of this table is polynomial. The maximum number of jobs that can be scheduled feasibly is given by the maximum m for which $A(m, T) \geq 0$. The respective optimal schedule can be computed by backtracking the intermediate values that contributed to $A(m, T)$. The computation of $A(i, t)$ is given in Algorithm 1, whose time complexity is $\mathcal{O}(nT)$. In the full version of the paper we describe a more efficient algorithm with time complexity $\mathcal{O}(n \log n + T)$ that is also optimal.

Theorem 1. *The maximum number of jobs that can be scheduled feasibly is given by the maximum m for which $A(m, T) \geq 0$ in the array $A(\cdot, \cdot)$ computed by Algorithm 1.*

Proof. Let m be maximum number of jobs that can be scheduled feasibly. For $i \in [1, m]$, let t_i be the time slot in which J_i is scheduled in such a schedule. It is easy to see that in this case $A(i, t_i) \geq 0$, for every $i \in [1, m]$. In the other direction, suppose that $A(i, t) \geq 0$. In this case, there exists a feasible schedule of jobs $J_1, \ldots, J_i$ on the time slots $[1, t]$. The schedule can be computed by backtracking the intermediate values that contributed to $A(i, t)$. □

Algorithm 1

Input: (1) n jobs $\{J_1, \ldots, J_n\}$, each with $r_i = 1$, $d_i = T$, and energy requirement e_i, and (2) h_t for each time slot $t \in [1, T]$

Output: $A(i, t)$, for $i \in [1, n]$ and $t \in [1, T]$

1: $A(\cdot, \cdot) \leftarrow -\infty$ ▷ *Initialize the table*
2: **for** $t = 2$ to T **do**
3:     **if** $\sum_{j=1}^{t-1} h_j \geq e_1$ **then**
4:        $A(1, t) \leftarrow \max\{A(1, t-1) + h_t, \sum_{j=1}^{t-1} h_j - e_1\}$
5: **for** $i = 2$ to n **do**
6:     **for** $t = i + 1$ to T **do**
7:        **if** $A(i-1, t-1) \geq e_i$ **then**
8:           $A(i, t) \leftarrow \max\{A(i, t-1) + h_t, A(i-1, t-1) - e_i\}$

---

3   A Greedy $\frac{1}{2}$-Approximation for EAS

This section considers the general case of EAS when jobs have arbitrary release times and due dates. This problem variant is NP-Hard as shown in Sect. 5.1. We present a $\frac{1}{2}$-approximation for this case that uses a greedy scheduling strategy.

Consider the following greedy approach to scheduling jobs. The algorithm works in iterations where in each iteration, either one job is scheduled or the algorithm stops. Let U be the set of unscheduled jobs and G the schedule constructed by the algorithm. Initially, $U = \{J_1, \ldots, J_n\}$ and $J(G) = \emptyset$. In iteration ℓ, the algorithm first checks for each job $J_i \in U$ whether there exists at least one time slot it can be feasibly scheduled in without impacting the feasibility of previously scheduled jobs. If it is not feasible to schedule any of the jobs, then the algorithm stops. Otherwise, for each job J_i that can be scheduled feasibly, find the time slot t_i that minimizes $Q = e_i + h_{t_i}$ over all its feasible time slots. The job scheduled in iteration ℓ is the job that minimizes Q over all feasible jobs that can be scheduled during this iteration. Denote this job as J_j, remove it from U, and add it to $J(G)$ with $\pi_G(J_j) = t_j$. The algorithm's pseudocode is given in Algorithm 2.

Let $O = (J(O), \pi_O)$ be an optimal schedule, where $|J(O)| = m \leq n$. Suppose that $|J(G)| = x$, which implies that the greedy algorithm stops after completing x iterations. Let $J_1^g, \ldots, J_x^g$ be the jobs scheduled by the greedy algorithm, where J_ℓ^g is scheduled in iteration ℓ. We prove the following lemma, which will later be used to prove the approximation ratio.

Lemma 1. *At the end of iteration ℓ of the greedy algorithm, for $1 \leq \ell \leq x$, there exists a feasible schedule S such that (i) $|J(S)| \geq \max\{\ell, m - \ell\}$, (ii) $\{J_1^g, \ldots, J_\ell^g\} \subseteq J(S)$ with $\pi_S(J_i^g) = \pi_G(J_i^g)$, for $1 \leq i \leq \ell$, and (iii) $J(S) \backslash \{J_1^g, \ldots, J_\ell^g\} \subseteq J(O)$.*

Proof. We prove the lemma by induction. For the induction base we add a "dummy" iteration 0 before the actual start of the greedy algorithm. The claim holds for $\ell = 0$ since at the beginning of the greedy algorithm the schedule O

Algorithm 2

    Input: (1) n jobs $\{J_1, \ldots, J_n\}$, each with release time r_i, due date d_i, and energy requirement e_i, and (2) h_t for each time slot $t \in [1, T]$

    Output: A feasible schedule G

1: $U \leftarrow \{J_1, \ldots, J_n\}$
2: $J(G) \leftarrow \emptyset, \pi_G(\cdot) \leftarrow \emptyset$
3: $G \leftarrow (J(G), \pi_G)$
4: $Q_{\min} \leftarrow 1$ ▷ *The tentative minimum energy value*
5: **while** $U \neq \emptyset$ **and** $Q_{\min} > 0$ **do**
6:     $E \leftarrow 0$ ▷ *Tracks the energy available*
7:     $Q_{\min} \leftarrow 0$
8:     **for** $t = 1$ **to** T **do**
9:         **if** a job J_i is already scheduled at slot t **then**
10:             $E \leftarrow E - e_i$
11:             ▷ *Tentatively scheduled job causes already scheduled jobs to be infeasible*
12:             **if** $E < 0$ **then**
13:                 $E \leftarrow E + Q_{\min}$
14:                 $Q_{\min} \leftarrow 0$
15:         **else if** $\{J_i \in U \mid t \in [r_i, d_i]\} = \emptyset$ **then**
16:             $E \leftarrow E + h_t$
17:         **else**
18:             Let J_k be the job that minimizes e_k over jobs in $\{J_i \in U \mid t \in [r_i, d_i]\}$
19:             **if** $E + Q_{\min} \geq e_k$ **and** $(Q_{\min} = 0$ **or** $e_k + h_t < Q_{\min})$ **then**
20:                 $E \leftarrow E + Q_{\min} - e_k$ ▷ *J_k becomes the tentatively scheduled job*
21:                 $Q_{\min} \leftarrow e_k + h_t$
22:                 $j \leftarrow k$
23:                 $t_j \leftarrow t$
24:             **else**
25:                 $E \leftarrow E + h_t$
26:     ▷ *Schedule the job with index j at slot t_j if it is feasible*
27:     **if** $Q_{\min} > 0$ **then**
28:         $J(G) \leftarrow J(G) \cup \{J_j\}$
29:         $\pi_G(J_j) \leftarrow t_j$
30:         $U \leftarrow U \setminus \{J_j\}$

---

is feasible and $|J(O)| = m$. Consider the end of iteration ℓ, for $\ell \geq 1$. By the inductive hypothesis, at the start of iteration ℓ, there exists a feasible schedule S of at least $\max\{\ell - 1, m - \ell + 1\}$ jobs that schedules the jobs $J_1^g, \ldots, J_{\ell-1}^g$ at time slots $\pi_G(J_1^g), \ldots, \pi_G(J_{\ell-1}^g)$, respectively, and the remaining jobs belong to $J(O)$. Suppose that S schedules job J_ℓ^g on time slot $t_\ell = \pi_G(J_\ell^g)$. In this case S satisfies the conditions of the lemma also for ℓ, since it schedules at least $\max\{\ell, m - \ell\}$ jobs, including the jobs $J_1^g, \ldots, J_\ell^g$ at slots $\pi_G(J_1^g), \ldots, \pi_G(J_\ell^g)$, and the remaining jobs belong to $J(O)$.

From now on assume that S has not scheduled job J_ℓ^g at time slot t_ℓ. We show how to obtain a schedule S' that satisfies the conditions of the lemma for ℓ. We start with schedule S and modify it as follows. First, we schedule the job J_ℓ^g at slot t_ℓ (in case $J_\ell^g \in J(S)$ this would just change the execution time slot of J_ℓ^g). If S already scheduled another job at slot t_ℓ, then this job is discarded.

Otherwise (that is, if S has not scheduled another job at slot t_ℓ), then the job in $J(S)\backslash\{J_1^g, \ldots, J_\ell^g\}$ that S schedules earliest is discarded.

Before showing that schedule S' is feasible, we show that it satisfies the conditions of the lemma. Clearly, S' schedules the jobs $J_1^g, \ldots, J_\ell^g$ at time slots $\pi_G(J_1^g), \ldots, \pi_G(J_\ell^g)$, and the remaining jobs in $J(S')$ belong to $J(O)$. Also, since exactly one job is discarded from $J(S)$, $|J(S')| \geq \max\{\ell, m - \ell\}$.

Schedule S' is feasible if and only if for all $J_i \in J(S')$, $E_{S'}(\pi_{S'}(J_i)) \geq e_i$. This is clearly the case for all jobs in $J(S')$ that are scheduled at slots $[1, t_\ell - 1]$ since schedule S is feasible. Next, consider job J_ℓ^g scheduled in S' at slot t_ℓ and the rest of the jobs in S' that are scheduled after this time slot. We distinguish between three cases.

Case 1: S schedules another job $J_j \in J(S)\backslash\{J_1^g, \ldots, J_\ell^g\}$ at time slot t_ℓ. In this case $E_{S'}(t_\ell) = E_S(t_\ell) \geq e_j$. However, since the greedy algorithm preferred to schedule job $J_i = J_\ell^g$ at slot t_ℓ while J_j was also feasible at the same time, we must have $e_i \leq e_j$ and thus $J_i = J_\ell^g$ is feasible at slot t_ℓ. Since $e_i + h_{t_\ell} \leq e_j + h_{t_\ell}$, we have that for all $t \in [t_\ell + 1, T]$, $E_{S'}(t) \geq E_S(t)$, and thus the jobs in S' scheduled after time slot t_ℓ are also feasible.

In the remaining two cases, S has not scheduled another job at slot t_ℓ. In these cases, the job in $J(S)\backslash\{J_1^g, \ldots, J_\ell^g\}$ that S schedules earliest is discarded. Let this job be denoted by J_j and the time slot it was scheduled in by $t_j = \pi_S(J_j)$.

Case 2: $t_j < t_\ell$. Again, since the greedy algorithm preferred to schedule $J_i = J_\ell^g$ at time slot t_ℓ while J_j was also feasible at time slot t_j, we must have $e_i + h_{t_\ell} \leq e_j + h_{t_j}$. This implies that $E_{S'}(t_\ell) \geq E_S(t_\ell) + e_j + h_{t_j} \geq e_i$ and thus $J_i = J_\ell^g$ is feasible at slot t_ℓ. It also implies that for all $t \in [t_\ell + 1, T]$, $E_{S'}(t) = E_S(t) + e_j + h_{t_j} - e_i - h_{t_\ell} \geq E_S(t)$, and thus the jobs in S' scheduled after time slot t_ℓ are also feasible.

Case 3: $t_j > t_\ell$. Again, since the greedy algorithm preferred to schedule $J_i = J_\ell^g$ at time slot t_ℓ while J_j was also feasible at time slot t_j, we must have $e_i + h_{t_\ell} \leq e_j + h_{t_j}$. This implies that for all $t \in [t_j + 1, T]$, $E_{S'}(t) \geq E_S(t)$, and thus the jobs in S' scheduled after time slot t_j are feasible. Since J_j is the earliest job in $J(S)\backslash\{J_1^g, \ldots, J_\ell^g\}$, all the jobs in $J(S')$ scheduled before t_j are in $\{J_1^g, \ldots, J_\ell^g\}$. Since the greedy schedule is guaranteed to be feasible, the schedule of these jobs in S' is also feasible.

Therefore, the schedule S' generated from modifying S is always feasible. As S' was already shown to satisfy the constraints of the lemma, we have the proof of the inductive step. □

Theorem 2. *The greedy algorithm yields a $\frac{1}{2}$-approximation of the optimal solution.*

Proof. Consider any $1 \leq \ell \leq \lceil \frac{m}{2} \rceil$. Lemma 1 implies that at the start of iteration ℓ there exists a feasible schedule of at least $m - (\ell - 1) \geq \ell$ jobs that schedules the jobs $J_1^g, \ldots, J_{\ell-1}^g$ at times $\pi_G(J_1^g), \ldots, \pi_G(J_{\ell-1}^g)$. Thus, there exists at least one job that can be feasibly scheduled in iteration ℓ. It follows that the greedy algorithm completes at least $\lceil \frac{m}{2} \rceil$ iterations, which implies that $2x \geq m$. □

4   An FPTAS for **WEAS** When All Jobs Have Identical Release Times and Due Dates

We now consider WEAS in the special case of jobs with identical release times and due dates. In the full version of the paper we show that this variant of the problem is NP-Hard. We present a fully polynomial time approximation scheme (FPTAS) that for any constant ε finds a feasible schedule that is a $(1 - \varepsilon)$-approximation to the maximum weight of the scheduled jobs in any feasible schedule. Due to space constraints the description is given in the full version of the paper.

5   Hardness Results

5.1   Unweighted Setting

In EAS we consider the cases of arbitrary release times and identical due dates, identical release times and arbitrary due dates, and arbitrary release times and arbitrary due dates and show they are all (weakly) NP-Hard. We mainly use a reduction from the $k-$Sum problem (i.e. the parameterized version of Subset Sum whose hardness is shown in [15]).

Theorem 3. *EAS when jobs have arbitrary release times and identical due dates is (weakly) NP-Hard.*

Proof. A reduction is given from $k-$Sum. An instance of this problem consists of a set of positive integers $\mathcal{A} = \{\alpha_1, \dots, \alpha_n\}$, a target value $0 < \beta < S = \sum_{i=1}^{n} \alpha_i$, and an integer $k < n$. The objective is to decide whether there exists a subset $A \subseteq \mathcal{A}$ such that $|A| = k$ and $\sum_{\alpha \in A} \alpha = \beta$. To simplify notation, we assume that $\mathcal{A}$ is sorted in non-increasing order ($\alpha_1 \geq \dots \geq \alpha_n$) and assume without loss of generality that $S > 2$ and $n > 2$. Construct a corresponding instance of EAS with the following:

- Time slots $[1, 2n - k + 2]$ and a threshold value of n jobs
- $\mathcal{J} = \{J_1, \dots, J_n\}$ where each job $J_i \in \mathcal{J}$ has release time, due date, and energy requirement given by $r_i = i+1$, $d_i = 2n-k+2$, and $e_i = S^2 n^2 + \alpha_i (Sn)$, respectively
- Energy harvesting values defined by

$$
h_t = \begin{cases}
k \left(S^2 n^2 \right) + \beta \left(Sn \right) & \text{if } t = 1 \\
S - \alpha_{t-1} & \text{if } t \in [2, n+1] \\
(n - k)(S^2 n^2) + (S - \beta)\left(Sn \right) - S(n - k - 1) - \beta & \text{if } t = n+2 \\
0 & \text{otherwise}
\end{cases}
$$

We claim that any feasible schedule that schedules all n jobs must assign exactly k jobs to time slots in $\tau_1 = [2, n + 1]$ and the remaining $n-k$ jobs to time slots in $\tau_2 = [n + 3, 2n - k + 2]$. To see this, note that energy must always be harvested at the first time slot since no jobs are yet released. Moreover, the total amount of energy that can be harvested on τ_1 is $\sum_{t=2}^{n+1}(S - \alpha_{t-1}) = S(n-1)$. Thus, the total amount of energy that can be harvested on slots $[1, n+1]$ is

$$k\left(S^2 n^2\right) + \beta\left(Sn\right) + S(n-1) < k\left(S^2 n^2\right) + Sn(S+1) < (k+1)(S^2 n^2),$$

which is strictly less than the energy required to execute more than k jobs. This implies that no more than k jobs can be scheduled in τ_1, and that energy must be harvested at time slot $n + 2$ in order to schedule more jobs. This leaves exactly $n - k$ time slots in τ_2 until the due date to schedule the remaining $n - k$ jobs. Additionally, we note that by the construction of the energy harvesting profile, it is optimal to schedule jobs in τ_1 immediately at their release time as the later a job is scheduled, the more energy is lost from not harvesting energy at that time slot. We proceed with the assumption that jobs in τ_1 are scheduled in this way.

We now show that the total energy requirement of the k jobs scheduled in τ_1 is exactly $\gamma_1 = k\left(S^2 n^2\right) + \beta\left(Sn\right)$. Assume for the sake of contradiction that this is not the case; that is, the total energy requirement of the k jobs is $R = k\left(S^2 n^2\right) + \beta'\left(Sn\right)$ where $\beta' \neq \beta$. If $\beta' > \beta$, then by integrality it must be that $\beta' \geq \beta + 1$, which implies $R \geq \gamma_1 + Sn$. However, as shown before, the total amount of energy harvestable on τ_1 is $S(n-1)$, so it is not possible to feasibly schedule the k jobs on τ_1. Now consider the case when $\beta' < \beta$. The energy harvested on τ_1 is $S(n-k) - (S - \beta')$ and $(\beta - \beta')(Sn)$ energy is leftover, so at the start of time slot $n+2$, there is at most $\gamma_2 = (\beta - \beta')(Sn) + S(n-k-1) + \beta'$ energy. Since we must harvest energy at slot $n + 2$, we get that at the start of slot $n + 3$, the energy available is

$$\gamma_3 = (n - k)(S^2 n^2) + (S - \beta)(Sn) - S(n - k - 1) - \beta + \gamma_2$$
$$= (n - k)(S^2 n^2) + (S - \beta')(Sn) - \beta + \beta'.$$

However, the energy requirement of the remaining $n - k$ jobs is $(n - k)(S^2 n^2) + (S - \beta')(Sn)$, so it is not possible to feasibly schedule the remaining jobs in τ_2. Hence, there is a contradiction so $\beta' = \beta$ and $R = \gamma_1$.

Therefore, a "yes" instance of this problem implies that the corresponding k−Sum instance is also a "yes" instance since the first k jobs scheduled correspond to k positive integers that sum to β. Conversely, a "yes" instance of k−Sum implies that the corresponding EAS instance is a "yes" instance since we can schedule the jobs that correspond to elements of A at time slots in τ_1 at their release time and the remaining $n - k$ jobs at time slots in τ_2. Thus, EAS when jobs have arbitrary release times and identical due dates is NP-Hard. □

A similar type of reduction from k−Sum can also be used to show the hardness of EAS with identical release times and arbitrary due dates. Full details of the proof of the following theorem are given in the full version of the paper.

Theorem 4. *EAS when jobs have identical release times and arbitrary due dates is (weakly) NP-Hard.*

Since jobs having arbitrary release times and identical due dates (or identical release times and arbitrary due dates) is a special case of both being arbitrary, we get the following as an immediate consequence of Theorem 3 (or Theorem 4).

Theorem 5. *EAS when jobs have arbitrary release times and due dates is* *(weakly)* NP-Hard.

5.2 Weighted Setting

In WEAS it can be shown that the problem is NP-Hard even when all the jobs have identical release time and due dates. A proof of the following theorem (which is a straightforward reduction from Knapsack) is given in the full version of the paper.

Theorem 6. *WEAS when jobs have identical release times and due dates is* *(weakly)* NP-Hard.

Since the case of jobs having identical release times and due dates is a special case of either one or both of them being arbitrary, we get the following as an immediate consequence.

Theorem 7. *WEAS is (weakly)* NP-Hard.

6 Conclusions and Open Problems

We conclude with a brief summary of our results and open problems. We presented three algorithms: (1) an optimal polynomial time algorithm for EAS with identical release times and due dates (Sect. 2), (2) a greedy $\frac{1}{2}$-approximation algorithm for EAS with arbitrary release times and due dates (Sect. 3), and (3) an FPTAS for WEAS in the case of identical release times and due dates (Sect. 4).

It would be interesting to see if there exists a PTAS or a better constant factor approximation for EAS with arbitrary release times and due dates, or if special cases of EAS where only one of them is arbitrary admit better approximation ratios. Another natural direction to consider is the extension of the greedy approach to WEAS with arbitrary release times and due dates.

In Sect. 5, we study the hardness of both EAS and WEAS and give nontrivial reductions from the $k-$Sum problem to show that except for the case of identical release times and due dates, EAS is weakly NP-Hard (Sect. 5.1). It is open whether EAS admits an FPTAS or whether there is a reduction from a strongly NP-Hard problem to EAS.

One could also consider expanding our model. A natural extension is to consider the case of arbitrary processing times for jobs. Another is to consider online versions of our problems. This may include either considering an online energy harvesting profile as considered in [12], or considering a model in which both jobs and the harvesting profile are revealed in an online manner.

References

1. Baptiste, P.: Polynomial time algorithms for minimizing the weighted number of late jobs on a single machine with equal processing times. J. Sched. **2**(6), 245–252 (1999)

2. Boyer, S.A.: SCADA: Supervisory Control and Data Acquisition. International Society of Automation, 4th edition (2010)

3. Briskorn, D., Choi, B.-C., Lee, K., Leung, J., Pinedo, M.: Inventory constrained scheduling on a single machine. Manuskripte aus den Instituten für Betriebswirtschaftslehre der Universität Kiel 640, Christian-Albrechts-Universität zu Kiel, Institut für Betriebswirtschaftslehre (2008)

4. Briskorn, D., Choi, B.-C., Lee, K., Leung, J., Pinedo, M.: Complexity of single machine scheduling subject to nonnegative inventory constraints. Eur. J. Oper. Res. **207**(2), 605–619 (2010)

5. Bäumker, E., Schüle, F., Woias, P.: Development of a batteryless VHF-beacon and tracker for mammals. In: Journal of Physics: Conference Series, vol. 1052, p. 012005 (2018)

6. Caruso, A., Chessa, S., Escolar, S., del Toro, X., López, J.C.: A dynamic programming algorithm for high-level task scheduling in energy harvesting IoT. IEEE Internet Things J. **5**(3), 2234–2248 (2018)

7. Davari, M., Ranjbar, M., De Causmaecker, P., Leus, R.: Minimizing makespan on a single machine with release dates and inventory constraints. Eur. J. Oper. Res. **286**(1), 115–128 (2020)

8. Faceira, J., Afonso, P., Salgado, P.: Prediction of solar radiation using artificial neural networks. In: Moreira, A.P., Matos, A., Veiga, G. (eds.) CONTROLO'2014. LNEE, vol. 321, pp. 397–406. Springer, Cham (2015). https://doi.org/10.1007/978-3-319-10380-8_38

9. Gafarov, E.R., Lazarev, A.A., Werner, F.: Single machine scheduling problems with financial resource constraints: some complexity results and properties. Math. Soc. Sci. **62**(1), 7–13 (2011)

10. Grigoriev, A., Holthuijsen, M., van de Klundert, J.: Basic scheduling problems with raw material constraints. Nav. Res. Logist. **52**(6), 527–535 (2005)

11. Györgyi, P., Kis, T.: Approximation schemes for single machine scheduling with non-renewable resource constraints. J. Sched. **17**, 135–144 (2014)

12. Islam, B., Nirjon, S.: Scheduling computational and energy harvesting tasks in deadline-aware intermittent systems. In: 2020 IEEE Real-Time and Embedded Technology and Applications Symposium (RTAS), pp. 95–109. IEEE (2020)

13. Lucia, B., Balaji, V., Colin, A., Maeng, K., Ruppel, E.: Intermittent computing: challenges and opportunities. In: 2nd Summit on Advances in Programming Languages (SNAPL 2017) (2017)

14. Merrett, G.V.: Invited: energy harvesting and transient computing: a paradigm shift for embedded systems? In: 53rd ACM/EDAC/IEEE Design Automation Conference (DAC), pp. 1–2 (2016)

15. Pătraşcu, M., Williams, R.: On the possibility of faster SAT algorithms. In: Proceedings of the Twenty-First Annual ACM-SIAM Symposium on Discrete Algorithms (SODA), pages 1065–1075. SIAM (2010)

16. Shaikh, F.K., Zeadally, S.: Energy harvesting in wireless sensor networks: a comprehensive review. Renew. Sustain. Energy Rev. **55**, 1041–1054 (2016)

17. Sudevalayam, S., Kulkarni, P.: Energy harvesting sensor nodes: survey and implications. IEEE Commun. Surv. Tutorials **13**(3), 443–461 (2010)

18. Vakhania, N.: Branch less, cut more and minimize the number of late equal-length jobs on identical machines. Theoret. Comput. Sci. **465**, 49–60 (2012)

19. van den Akker, J.M., Diepen, G., Hoogeveen, J.A.H.: Minimizing total weighted tardiness on a single machine with release dates and equal-length jobs. J. Sched. **13**, 561–576 (2010)

Recognizing When a Preference System is Close to Admitting a Master List

Ildikó Schlotter[1,2(✉)]

[1] Centre for Economic and Regional Studies, Budapest, Hungary
`schlotter.ildiko@krtk.hu`
[2] Budapest University of Technology and Economics, Budapest, Hungary

Abstract. A preference system $\mathcal{I}$ is an undirected graph where vertices have preferences over their neighbors, and $\mathcal{I}$ admits a master list if all preferences can be derived from a single ordering over all vertices. We study the problem of deciding whether a given preference system $\mathcal{I}$ is *close to* admitting a master list based on three different distance measures. We determine the computational complexity of the following questions: can $\mathcal{I}$ be modified by (i) k swaps in the preferences, (ii) k edge deletions, or (iii) k vertex deletions so that the resulting instance admits a master list? We investigate these problems in detail from the viewpoint of parameterized complexity and of approximation. We also present two applications related to stable and popular matchings.

1 Introduction

A preference system models a set of agents as an undirected graph where agents are vertices, and each agent has preferences over its neighbors. Preference systems are a fundamental concept in the area of matching under preferences which, originating in the seminal work of Gale and Shapley [16] on stable matchings, is a prominent research field in the intersection of algorithm design and computational social choice that has steadily gained attention over the last two decades.

Preference systems may admit a master list, that is, a global ranking over all agents from which agents derive their preferences. Master lists arise naturally in many practical scenarios such as P2P networks [26], job markets [21], and student housing assignments [30]. Consequently, master lists and its generalizations have been the focus of research in several papers [7,9,11,21,22,24,29].

In this work we aim to investigate the computational complexity of recognizing preference systems that are close to admitting a master list. Such instances may arise as a result of noise in the data set, or in scenarios where a global ranking of agents is used in general, with the exception of a few anomalies.

Our Contribution. We introduce three measures to describe the distance of a given preference system $\mathcal{I}$ from the class of preference systems admitting a

Supported by the Hungarian Academy of Sciences under its Momentum Programme (LP2021-2) and the Hungarian Scientific Research Fund (OTKA grants K128611 and K124171).

C.-C. Lin et al. (Eds.): WALCOM 2023, LNCS 13973, pp. 317–329, 2023.
https://doi.org/10.1007/978-3-031-27051-2_27

master list. The first measure, $\Delta^{\mathrm{swap}}(\mathcal{I})$ is based on the swap distance between agents' preferences, while the measures $\Delta^{\mathrm{edge}}(\mathcal{I})$ and $\Delta^{\mathrm{vert}}(\mathcal{I})$ are based on classic graph operations, the deletion of edges or vertices; precise definitions follow in Sect. 2. We study in detail the complexity of computing these values for a given preference system $\mathcal{I}$. After proving that computing any of these three measures is NP-hard, we apply the framework of parameterized complexity and of approximation algorithms to gain a more fine-grained insight.

In addition to the problems of computing $\Delta^{\mathrm{swap}}(\mathcal{I})$, $\Delta^{\mathrm{edge}}(\mathcal{I})$, and $\Delta^{\mathrm{vert}}(\mathcal{I})$, we briefly look at two applications. First, we show that if a strict preference system $\mathcal{I}$ is close to admitting a master list, then we can bound the number of stable matchings as a function of the given distance measure. This yields an efficient way to solve a wide range of stable matching problems in instances that are close to admitting a master list. Second, we consider an optimization problem over popular matchings where the task is to find a maximum-utility popular matching while keeping the number (or cost) of blocking edges low. We prove that this notoriously hard problem can be efficiently solved if preferences are close to admitting a master list. In both of these applications, the running time of the obtained algorithms heavily depends on the distance measure used.

Related Work. Master lists have been extensively studied in the context of stable matchings [7,11,21,22]. Various models have been introduced in the literature to generalize master lists, and capture preferences that are similar to each other in some sense. Closest to our work might be the paper by Bredereck et al. [7] who examine the complexity of multidimensional stable matching problems on instances that are close to admitting a master list. Abraham et al. investigated a setting where agent pairs are ranked globally [1]. Bhatnagar et al. [4] examined three restrictions on preference systems—the k-attribute, the k-range, and the k-list models—that aim to capture similarities among preferences; these models have been studied subsequently by several researchers [9,24,29].

Restricted preference profiles have been also examined in the broader context of computational social choice; see the survey by Elkind et al. [13]. In election systems, computing the Kemeny score [23] for a multiset of votes (where each vote is a total linear order over a set of candidates) is analogous to computing the value $\Delta^{\mathrm{swap}}(\mathcal{I})$ for a preference system $\mathcal{I}$, although there are some differences between these two problems. Besides the extensive literature on the complexity of Kemeny voting (see e.g. [3,15]), our work also relates to the problem of computing certain distance measures between elections [5]. Some of the distance measures we use were considered by Gupta et al. in their paper on committee selection [18].

2   Preliminaries

We assume that the reader is familiar with basic concepts in graph theory, classic and parameterized computational complexity, and approximation theory. For directed and undirected graphs, we will use the notation of the book by Bang-Jensen and Gutin [2], unless otherwise stated. For more on complexity and approximations, we refer the reader to corresponding books [12,17,33]. We provide all definitions and notations we use (apart from those defined below), as well as all formal proofs in the full version of our paper [31].

Preference Systems. A *preference system* is a pair $\mathcal{I} = (G, \preceq)$ where G is an undirected graph and $\preceq = \{\preceq_v : v \in V(G)\}$ where $\preceq_v$ is a weak or a strict order over $N_G(v)$ for each vertex $v \in V(G)$, indicating the *preferences* of v. For some $v \in V(G)$ and $a, b \in N_G(v)$, we say that v *prefers* b to a, denoted by $a \prec_v b$, if $a \not\succeq_v b$. We write $a \sim_v b$, if $a \preceq_v b$ and $b \preceq_v a$. A *tie* in v's preferences is a maximal set $T \subseteq N_G(v)$ such that $t \sim_v t'$ for each t and t' in T. If each tie has size 1, then $\mathcal{I}$ is a *strict preference system*, and we may denote it by $(G, \prec)$.

Deletions and Swaps. For a set X of edges or vertices in G, let $\mathcal{I} - X$ denote the preference system whose underlying graph is $G - X$ and where the preferences of each vertex $v \in V(G - X)$ is the restriction of $\preceq_v$ to $N_{G-X}(v)$. We may refer to $\mathcal{I} - X$ as a *sub-instance* of $\mathcal{I}$.

If vertex v has strict preferences $\prec_v$ in $\mathcal{I}$, then a *swap* is a triple $(a, b; v)$ with $a, b \in N_G(v)$, and it is *admissible* if a and b are consecutive[1] in v's preferences. *Performing* an admissible swap $(a, b; v)$ in $\mathcal{I}$ means switching a and b in v's preferences; the resulting preference system is denoted by $\mathcal{I} \lhd (a, b; v)$. For a set S of swaps, $\mathcal{I} \lhd S$ denotes the preference system obtained by performing the swaps in S in $\mathcal{I}$ in an arbitrary order as long as each swap is admissible (if this is not possible, $\mathcal{I} \lhd S$ is undefined). For non-strict preferences, similar notions will be discussed in Sect. 3.

Master Lists. A weak or strict order $\preceq^{ml}$ over $V(G)$ is a *master list* for $(G, \preceq)$, if for each $v \in V(G)$, the preferences of v are *consistent* with $\preceq^{ml}$, that is, $\preceq_v$ is the restriction of $\preceq^{ml}$ to $N_G(v)$. We will denote by $\mathcal{F}_{ML}$ the family of those preference systems that admit a master list. Notice that $\mathcal{F}_{ML}$ is closed under taking subgraphs: if we delete a vertex or an edge from a preference system in $\mathcal{F}_{ML}$, the remainder still admits a master list.

3   Problem Definition and Initial Results

Let us first introduce the notion of a *preference digraph*, a directed graph associated with a given preference system, which can be exploited to obtain a useful characterization of preference systems that admit a master list. We then proceed with defining our measures for describing the distance from $\mathcal{F}_{ML}$.

Characterization of $\mathcal{F}_{ML}$ Through the Preference Digraph. With a strict preference system $\mathcal{I} = (G, \prec)$ where $G = (V, E)$, we associate an arc-labelled directed graph $D_\mathcal{I}$ that we call the *preference digraph* of $\mathcal{I}$. We let $D_\mathcal{I}$ have the same set of vertices as G, and we define the arcs in $D_\mathcal{I}$ by adding an arc (a, b) labelled with v whenever $a \prec_v b$ holds for some vertices a, b and v in V. Note that several parallel arcs may point from a to b in $D_\mathcal{I}$, each having a different label, so we have $|V(D_\mathcal{I})| = |V|$ but $|A(D_\mathcal{I})| = O(|V||E|)$. Observation 1 immediately follows from the fact that acyclic digraphs admit a topological order.

---

[1] Vertices a and b are consecutive in v's preferences, if either $a \prec_v b$ but there is no vertex c with $a \prec_v c \prec_v b$, or $b \prec_v a$ but there is no vertex c with $b \prec_v c \prec_v a$.

Observation 1. *A strict preference system* $(G, \prec)$ *admits a master list if and only if the preference digraph of* G *is acyclic.*

For a preference system $\mathcal{I} = (G, \preceq)$ with $G = (V, E)$ that is not necessarily strict we extend the concept of the preference digraph of $\mathcal{I}$ as follows. Again, we let $D_{\mathcal{I}}$ have V as its vertex set, but now we add two types of arcs to $D_{\mathcal{I}}$: for any v in V and $a, b \in N_G(V)$ with $a \neq b$ we add a *strict arc* (a, b) with label v whenever $a \prec_v b$, and we add a pair of *tied arcs* (a, b) and (b, a), both with label v, whenever $a \sim_v b$. Note that this way we indeed generalize our definition above for the preference digraph of strict preference systems. We will call a cycle of $D_{\mathcal{I}}$ that contains a strict arc a *strict cycle*. The following lemma is a straightforward generalization of Observation 1.

Lemma 2. *A preference system* $(G, \preceq)$ *admits a master list if and only if no cycle of the preference digraph of* G *is strict.*

Measuring the Distance from $\mathcal{F}_{\mathrm{ML}}$**.** Let us now define our three measures to describe the distance of a given strict preference system $\mathcal{I} = (G, \prec)$ from the family $\mathcal{F}_{\mathrm{ML}}$ of preference systems that admit a master list:

- $\Delta^{\mathrm{swap}}(\mathcal{I}) = \min\{|S| : S \text{ is a set of swaps in } \mathcal{I} \text{ such that } \mathcal{I} \lhd S \in \mathcal{F}_{\mathrm{ML}}\}$;
- $\Delta^{\mathrm{edge}}(\mathcal{I}) = \min\{|S| : S \subseteq E(G), \mathcal{I} - S \in \mathcal{F}_{\mathrm{ML}}\}$;
- $\Delta^{\mathrm{vert}}(\mathcal{I}) = \min\{|S| : S \subseteq V(G), \mathcal{I} - S \in \mathcal{F}_{\mathrm{ML}}\}$.

The measures $\Delta^{\mathrm{edge}}(\mathcal{I})$ and $\Delta^{\mathrm{vert}}(\mathcal{I})$ can be easily extended for preference systems that are not necessarily strict, since the above definitions are well-defined for any preference system $(G, \preceq)$.

Extending the measure $\Delta^{\mathrm{swap}}(\mathcal{I})$ for non-strict preference systems is, however, not entirely straightforward. If there are ties in the preferences of some vertex v, how can we define an admissible swap? In this paper we use the following definition for swap distance, which seems to be standard in the literature [6,8]. Let $\preceq_u$ and $\preceq_v$ be weak orders. If they are not defined over the same sets, then the *swap distance* of $\preceq_u$ and $\preceq_v$, denoted by $\Delta(\preceq_u, \preceq_v)$ is ∞, otherwise

$$\Delta(\preceq_u, \preceq_v) = |\{\{a, b\} : a \prec_u b \text{ but } b \preceq_v a\}| + |\{\{a, b\} : a \sim_u b \text{ but } a \not\prec_v b\}|.$$

For two preferences systems $\mathcal{I} = (G, \preceq)$ and $\mathcal{I}' = (G', \preceq')$ with $G = (V, E)$ and $G' = (V', E')$, we let their swap distance, denoted by $\Delta(\mathcal{I}, \mathcal{I}')$, be ∞ if they are not defined over the same vertex set; otherwise (that is, if $V = V'$) we let $\Delta(\mathcal{I}, \mathcal{I}') = \sum_{v \in V} \Delta(\preceq_v, \preceq'_v)$. Using this, we can define

$$\Delta^{\mathrm{swap}}(\mathcal{I}) = \min\{\Delta(\mathcal{I}, \mathcal{I}') : \mathcal{I}' \in \mathcal{F}_{\mathrm{ML}}\}.$$

The following lemma follows easily from the definitions.

Lemma 3. $\Delta^{\mathrm{swap}}(\mathcal{I}) \geq \Delta^{\mathrm{edge}}(\mathcal{I}) \geq \Delta^{\mathrm{vert}}(\mathcal{I})$ *for any preference system* $\mathcal{I}$.

Let MASTER LIST BY SWAPS (or MLS for short) be the problem whose input is a preference system $\mathcal{I}$ and an integer k, and the task is to decide whether $\Delta^{\mathrm{swap}}(\mathcal{I}) \leq k$. We define the MASTER LIST BY EDGE DELETION (or MLED) and the MASTER LIST BY VERTEX DELETION (or MLVD) problems analogously.

4   Computing the Distance from Admitting a Master List

Let us now present our main results on recognizing when a given preference list is close to admitting a master list. We investigate the classical and parameterized complexity of each of our problems MLS, MLED, and MLVD. In Sect. 4.1 we consider strict preference systems, and then extend our results for weakly ordered preferences in Sect. 4.2.

4.1   Strict Preferences

We show that computing the distance from $\mathcal{F}_{\mathrm{ML}}$ is NP-hard for each of our three distance measures. However, when viewed from the perspective of approximation or of parameterized complexity, intrinsic differences between MLS, MLED, and MLVD will surface.

We start with Theorem 4 showing that we cannot expect a polynomial-time algorithm for MLS or for MLED and even a polynomial-time approximation is unlikely to exist already for bipartite graphs, assuming the so-called *Unique Games Conjecture* [25], a standard assumption in complexity theory. The proof of Theorem 4 relies on a connection between MLS, MLED, and the FEEDBACK ARC SET problem which, given a directed graph D and an integer k, asks whether there exists a set of at most k arcs in D whose deletion from D yields an acyclic graph. Interestingly, the connection of this problem to MLS and to MLED can be used both ways: on the one hand, it serves as the basis of our reduction for proving computational hardness, and on the other hand, we will be able to apply already existing algorithms for FEEDBACK ARC SET in our quest for solving MLS and MLED.

Theorem 4. *MLS and MLED are both NP-hard, and assuming the Unique Games Conjecture they are NP-hard to approximate by any constant factor in polynomial time. All of these hold even if the input graph is bipartite with all vertices on one side having degree 2, and preferences are strict.*

Thanks to Lemma 1 below, for any strict preference system $\mathcal{I}$ we can decide whether $\Delta^{\mathrm{swap}}(\mathcal{I}) \leq k$ for some $k \in \mathbb{N}$ by applying the FPT algorithm of Lokshtanov et al. [27] for FEEDBACK ARC SET on the preference digraph $D_{\mathcal{I}}$ and parameter k. Their algorithm runs in time $O(k!4^k k^6(n+m))$ on an input graph with n vertices and m arcs [27]. If $G = (V, E)$ is the graph underlying $\mathcal{I}$, then $D_{\mathcal{I}}$ has $|V|$ vertices and $O(|V| \cdot |E|)$ arcs, implying a running time of $O(k!4^k k^6 |V| \cdot |E|)$.

Lemma 1. *For a strict preference system $\mathcal{I}$, $\Delta^{\mathrm{swap}}(\mathcal{I}) \leq k$ if and only if the preference digraph of $\mathcal{I}$ admits a feedback arc set of size at most k.*

Corollary 5. *If preferences are strict, then MLS is fixed-parameter tractable with parameter k, and can be solved in time $O(k!4^k k^6 |V| \cdot |E|)$.*

We remark that MLS for strict preferences can be formulated as a variant of the KEMENY SCORE problem with incomplete votes as studied by Betzler et al. [3]. Their results also imply that MLS is FPT with parameter k, though the running time we obtain in Corollary 5 is better than the one stated in [3, Theorem 10].

Algorithm 1. Obtaining a 2-approximation for MLED on input $(\mathcal{I}, k)$ with strict preferences

---

1: Construct the graph $H_\mathcal{I}$.
2: Let F be a solution for FEEDBACK ARC SET on input $(H_\mathcal{I}, k)$.
3: Ensure that each arc in F is incident to some vertex in V by replacing all arcs of F entering some a_v^- with (a_v^-, a).
4: Return $S_F = \{\{a, v\} \in E : (a, a_v^+) \in F \text{ or } (a_v^-, a) \in F\}$.

In contrast to MLS, the MLED problem is W[1]-hard with k as the parameter; the reduction is from MULTICOLORED CLIQUE [14].

Theorem 6. *MLED is W[1]-hard with parameter k, even for strict preferences.*

Although Theorem 6 provides strong evidence that there is no FPT algorithm for MLED with parameter k, and by Theorem 4 we cannot hope for a polynomial-time approximation algorithm for MLED either, our next result shows that combining these two approaches yields a way to deal with the computational hardness of the problem. Namely, Theorem 7 provides a 2-approximation for MLED whose running time is FPT with parameter k. This result again relies heavily on the connection between MLED and FEEDBACK ARC SET.

Theorem 7. *There exists an algorithm that achieves a 2-approximation for MLED if preferences are strict, and runs in FPT time with parameter k.*

2-Approximation FPT Algorithm for MLED (Strict Preferences). Let the strict preference system $\mathcal{I} = (G, \prec)$ with underlying graph $G = (V, E)$ and $k \in \mathbb{N}$ be our input for MLED. See Algorithm 1 for a formal description.

First, we construct a directed graph $H_\mathcal{I}$ by setting

$$V(H_\mathcal{I}) = V \cup \{a_v^-, a_v^+ : \{a, v\} \in E\},$$
$$A(H_\mathcal{I}) = \{(a_c^+, b_c^-) : a, b, c \in V, a \prec_c b\} \cup \{(a_v^-, a), (a, a_v^+) : v \in V, a \in N_G(v)\}.$$

Our approximation factor relies on the property if $H_\mathcal{I}$ that, roughly speaking, the effect of deleting an edge from G can be achieved by deleting two arcs from $H_\mathcal{I}$.

Next, we compute a minimum feedback arc set F in $H_\mathcal{I}$ using the algorithm by Lokshtanov et al. [27]. We may assume that F only contains arcs incident to some vertex in V, as we can replace any arc (a_c^+, b_c^-) with the sole are leaving b_c^-, namely (b_c^-, b), since all cycles containing (a_c^+, b_c^-) must also go through (b_c^-, b).

Finally, we return the set $S_F = \{\{a, v\} \in E : (a, a_v^+) \in F \text{ or } (a_v^-, a) \in F\}$.

Note that $H_\mathcal{I}$ has $|V| + 2|E|$ vertices and at most $|V| \cdot |E| + 4|E|$ arcs. The total running time of Algorithm 1 is therefore $O(k! 4^k k^6 |V| \cdot |E|)$ which is indeed FPT with parameter k.

Contrasting our positive results for MLS and MLED, a reduction from the classic HITTING SET problem shows that MLVD is computationally hard both in the classic and in the parameterized sense, and cannot be approximated by any FPT algorithm, as stated by Theorem 8.

Theorem 8. *MLVD is NP-hard and W[2]-hard with parameter k. Furthermore, no FPT algorithm with k as the parameter can achieve an $f(k)$-approximation* *for MLVD for any computable function f, unless* $\mathsf{FPT} = \mathsf{W}[1]$. *All of these hold even if the input graph is bipartite and preferences are strict.*

4.2 Weakly Ordered Preferences

Let us now consider preference systems that are not necessarily strict. The hardness results of Sect. 4.1 trivially hold for weakly ordered preferences, so we will focus on extending the algorithmic results of the previous section.

Lemma 2. *For any preference system* $\mathcal{I} = (G, \preceq)$, $\Delta^{\mathrm{swap}}(\mathcal{I}) \leq k$ *if and only if there exists a set of at most k arcs in the preference digraph $D_{\mathcal{I}}$ of $\mathcal{I}$ that hits every strict cycle of $D_{\mathcal{I}}$.*

Thanks to Lemma 2, we can reduce MLS to a generalization of the FEEDBACK ARC SET problem where, instead of searching for a feedback arc set, the task is to seek an arc set that only hits certain *relevant* cycles. In the SUBSET FEEDBACK ARC SET (or SFAS) problem the input is a directed graph D, a vertex set $W \subseteq V(D)$ and an integer k, and the task is to find a set of at most k arcs in D that hits all *relevant* cycles in D, where a cycle is relevant if it goes through some vertex of W.

To solve SFAS, we apply an FPT algorithm by Chitnis et al. [10] for the vertex variant of SFAS, the DIRECTED SUBSET FEEDBACK VERTEX SET (or DSFVS) problem that, given a directed graph D, a set $W \subseteq V(D)$ and a parameter $k \in \mathbb{N}$, asks for a set of at most k *vertices* that hits all relevant cycles in D. Applying a simple, well-known reduction from SFAS to DSFVS, we can use the algorithm by Chitnis et al. [10] to obtain an FPT algorithm for MLS with parameter k.

Theorem 9. *MLS is fixed-parameter tractable with parameter k, even if preferences are weak orders.*

Next we extend Theorem 7 for weak orders, by reducing MLED to SFAS.

Theorem 10. *There exists an algorithm that achieves a 2-approximation for MLED, and runs in FPT time with parameter k.*

2-Approximation FPT Algorithm for MLED. Let the preference system $\mathcal{I}$ with underlying graph $G = (V, E)$ and $k \in \mathbb{N}$ be our input for MLED. For each $v \in V$, let T_v be the set family containing every tie that appears in the preferences of v. See Algorithm 2 for a formal description.

First, we construct a directed graph $H_{\mathcal{I}}$ with $V(H_{\mathcal{I}}) = V \cup T \cup U \cup Z$ where

$$T = \{t : v \in V, t \in T_v\},$$
$$U = \{a_v^-, a_v^+ : \{a, v\} \in E\},$$
$$Z = \{z_{(a,b,v)} : a \prec_v b \text{ for some } a, b, v \in V\},$$

and with arc set $A(H_{\mathcal{I}}) = A_T \cup A_U \cup A_Z$ where

$$A_T = \{(t, a_v^-), (a_v^+, t) : v \in V, t \in T_v, a \in t\}$$
$$A_U = \{(a_v^-, a), (a, a_v^+) : v \in V, a \in N_G(v)\}$$
$$A_Z = \{(a_v^+, z_{(a,b,v)}), (z_{(a,b,v)}, b_v^-) : z_{(a,b,v)} \in Z\}.$$

Algorithm 2. Obtaining a 2-approximation for MLED on input $(\mathcal{I}, k)$

1: Construct the graph $H_\mathcal{I}$.
2: Let F be a solution for SUBSET FEEDBACK ARC SET on input $(H_\mathcal{I}, Z, k)$.
3: Ensure $F \subseteq A_U$ by replacing all arcs pointing to some $a_v^- \in U$ with (a_v^-, a) and all arcs leaving some $a_v^+ \in U$ with (a, a_v^+).
4: Return $S_F = \{\{a, v\} \in E : (a, a_v^+) \in F \text{ or } (a_v^-, a) \in F\}$.

Next, we solve the SUBSET FEEDBACK ARC SET problem $(H_\mathcal{I}, Z, k)$ by applying the above reduction from SFAS to DSFVS and then using the algorithm of Chitnis et al. [10]; let F be the solution obtained for $(H_\mathcal{I}, Z, k)$. Observe that w.l.o.g. we may assume that F only contains arcs of A_U. Indeed we can replace any arc $f \in F$ in $A_T \cup A_Z$ by an appropriately chosen arc $f' \in A_U$: note that f either points to some $a_v^- \in U$ or it leaves some $a_v^+ \in U$; in the former case we set $f' = (a_v^-, a)$, while in the latter case we set $f' = (a, a_v^+)$. Then any cycle containing f must also contain f', so we can safely replace f with f', as $F \setminus \{f\} \cup \{f'\}$ still hits all relevant cycles. Hence, we will assume $F \subseteq A_U$.

Finally, we return the set $S_F = \{\{a, v\} \in E : (a, a_v^+) \in F \text{ or } (a, a_v^+) \in F\}$.

It is clear that the above algorithm runs in FPT time with parameter k.

5   Applications

In this section we consider two examples related to stable and popular matchings where we can efficiently solve computationally hard optimization problems on preference systems that are close to admitting a master list; see the book [28] for the definition of stability and popularity.

5.1   Optimization over Stable Matchings

One of the most appealing property of the distances defined in Sect. 3 is that whenever the distance of a *strict* (but not necessarily bipartite) preference system from admitting a master list is small, we obtain an upper bound on the number of stable matchings contained in the given preference system. Therefore, strict preference systems that are close to admitting a master list are easy to handle, as we can efficiently enumerate their stable matchings, as Lemmas 11 and 13 show.

Lemma 11. *Given a strict preference system $\mathcal{I} = (G, \prec)$ with $G = (V, E)$ and a set $S \subseteq E$ of edges such that $\mathcal{I} - S \in \mathcal{F}_{\mathrm{ML}}$, the number of stable matchings in $\mathcal{I}$ is at most $2^{|S|}$, and it is possible to enumerate all of them in time $2^{|S|} \cdot O(|E|)$.*

Corollary 12. *In a strict preference system $\mathcal{I}$, the number of stable matchings is at most $2^{\Delta^{\mathrm{swap}}(\mathcal{I})}$.*

Observe that although the number of stable matchings may grow exponentially as a function of the distance Δ^{edge} or Δ^{swap}, this growth does not depend on the size of the instance. By contrast, this is not the case for the distance Δ^{vert}.

Lemma 13. *Given a strict preference system $\mathcal{I} = (G, \prec)$ with $G = (V, E)$ and a set $S \subseteq V$ of vertices such that $\mathcal{I} - S \in \mathcal{F}_{ML}$, the number of stable matchings in $\mathcal{I}$ is at most $|V|^{|S|}$, and it is possible to enumerate all stable matchings of $\mathcal{I}$ in time $|V|^{|S|} \cdot O(|E|)$.*

There exists an algorithm by Gusfield and Irving [19, 20] that outputs the set $\mathcal{S}(\mathcal{I})$ of stable matchings in a preference system $\mathcal{I}$ over a graph $G = (V, E)$ in $O(|\mathcal{S}(\mathcal{I})| \cdot |E|)$ time after $O(|V| \cdot |E| \log |V|)$ preprocessing time. As a consequence, the bounds of Lemma 11, Corollary 12, and Lemma 13 on $|\mathcal{S}(\mathcal{I})|$ directly yield a way to handle computationally hard problems on any preference system $\mathcal{I}$ where $\Delta^{swap}(\mathcal{I})$, $\Delta^{edge}(\mathcal{I})$, or $\Delta^{vert}(\mathcal{I})$ has small value, even without the need to determine a set S of edges or vertices for which $\mathcal{I} - S \in \mathcal{F}_{ML}$ or a set S of swaps for which $I \lhd S \in \mathcal{F}_{ML}$. Thus, we immediately have the following result, even without having to use our results in Sect. 4. For the definitions of the NP-hard problems mentioned as an example in Theorem 14, see the book [28].

Theorem 14. *Let $\mathcal{I}$ be a strict (but not necessarily bipartite) preference system, and Q any optimization problem where the task is to maximize or minimize some function f over $\mathcal{S}(\mathcal{I})$ such that $f(M)$ can be computed in polynomial time for any matching $M \in \mathcal{S}(\mathcal{I})$. Then Q can be solved*

(i) in FPT time with parameter $\Delta^{edge}(\mathcal{I})$ or $\Delta^{swap}(\mathcal{I})$;
(ii) in polynomial time if $\Delta^{vert}(\mathcal{I})$ is constant.

In particular, these results hold for SEX-EQUAL STABLE MATCHING, BALANCED STABLE MATCHING, (GENERALIZED) MEDIAN STABLE MATCHING[2], EGALITARIAN STABLE ROOMMATES, *and* MAXIMUM-WEIGHT STABLE ROOMMATES.

We remark that the bounds in Lemmas 11 and 13 are tight in the following sense. For any $k, n \in \mathbb{N}$ with $n \geq k$, there exist strict preference systems $\mathcal{I}_k$ and $\mathcal{J}_{k,n}$ such that (i) $\Delta^{edge}(\mathcal{I}_k) = k$ and $\mathcal{I}_k$ admits 2^k stable matchings, and (ii) $\Delta^{vert}(\mathcal{J}_{k,n}) = k$, the number of vertices in $\mathcal{J}_{k,n}$ is $2n$, and $\mathcal{J}_k$ admits $\binom{n}{k}$ stable matchings. See the full paper [31] for the details of these constructions.

5.2 Maximum-Utility Popular Matchings with Instability Costs

We now turn our attention to the MAX-UTILITY POPULAR MATCHING WITH INSTABILITY COSTS problem, studied in [32]: given a strict preference system $\mathcal{I} = (G, \prec)$, a utility function $\omega : E(G) \to \mathbb{N}$, a cost function $c : E(G) \to \mathbb{N}$, an objective value $t \in \mathbb{N}$ and a budget $\beta \in \mathbb{N}$, the task is to find a popular matching in $\mathcal{I}$ whose utility is at least t and whose blocking edges have total cost at most β. Our aim is to investigate whether we can solve this problem efficiently for instances that are close to admitting a master list.

Note that in general this problem is computationally hard even if the given preference system is strict, bipartite, admits a master list, and the cost and

---

[2] Although the problem of finding a (generalized) median matching is not an optimization problem over $\mathcal{S}(\mathcal{I})$, it is clear that it can be solved in $|\mathcal{S}(\mathcal{I})| \cdot O(|\mathcal{I}|)$ time.

utility functions are very simple. Namely, given a strict, bipartite preference system $(G, \prec) \in \mathcal{F}_{\mathrm{ML}}$ for which a stable matching has size $|V(G)|/2 - 1$, it is NP-hard and W[1]-hard with parameter β to find a complete popular matching (i.e., one that is larger than a stable matching) that admits at most β blocking edges [32]. Nevertheless, if the total cost β of the blocking edges that we allow is a constant and each edge has cost at least 1, then MAX-UTILITY POPULAR MATCHING WITH INSTABILITY COSTS can be solved in polynomial time for bipartite, strict preference systems that admit a master list (in fact, it suffices to assume that the preferences of all vertices on one side of the bipartite input graph are consistent with a master list), representing an island of tractability for this otherwise extremely hard problem [32]. Therefore, it is natural to ask whether we can extend this result for strict preferences systems that are close to admitting a master list. Theorem 15 answers this question affirmatively.

Theorem 15. *Let $\mathcal{I}$ be a strict (but not necessarily bipartite) preference system with $G = (V, E)$. Then an instance $(\mathcal{I}, \omega, c, t, \beta)$ of* MAX-UTILITY POPULAR MATCHING WITH INSTABILITY COSTS *where $c(e) \geq 1$ for all edges $e \in E$, and β is constant can be solved*

(i) in FPT time with parameter $\Delta^{\mathrm{edge}}(\mathcal{I})$ or $\Delta^{\mathrm{swap}}(\mathcal{I})$;
(ii) in polynomial time if $\Delta^{\mathrm{vert}}(\mathcal{I})$ is constant.

We apply the same approach as in Sect. 5.1, with a crucial difference: for the algorithms proving Theorem 15 we will need to determine a set of edges or vertices whose deletion yields an instance in $\mathcal{F}_{\mathrm{ML}}$. Using such a set, we then apply Lemma 16 or 17 below; these are generalizations of Lemmas 11 and 13 for the case when we allow a fixed set of edges to block the desired matching.

Lemma 16. *Given a strict preference system $\mathcal{I} = (G, \prec)$ with $G = (V, E)$ and edge sets $B \subseteq E$ and $S \subseteq E$ such that $\mathcal{I} - S \in \mathcal{F}_{\mathrm{ML}}$, the number of matchings M for which $B = \mathrm{bp}(M)$ is at most $2^{|S|}$, and it is possible to enumerate them in time $2^{|S|} \cdot O(|E|)$.*

Lemma 17. *Given a strict preference system $\mathcal{I} = (G, \prec)$ with $G = (V, E)$, an edge set $B \subseteq E$, and a vertex set $S \subseteq V$ such that $\mathcal{I} - S \in \mathcal{F}_{\mathrm{ML}}$, the number of matchings M for which $B = \mathrm{bp}(M)$ is at most $|V|^{|S|}$, and it is possible to enumerate them in time $|V|^{|S|} \cdot O(|E|)$.*

The algorithms proving Theorem 15 start with searching for the set of blocking edges using brute force: recall that our budget β is constant, and since each edge has cost at least 1, we know that $|\mathrm{bp}(M)| \leq \beta$ for our desired popular matching M. Thus, there are only polynomially many sets to consider as the set B of blocking edges.

Next, to prove statement (ii) of Theorem 15, we again use brute force to find a set S of $\Delta^{\mathrm{vert}}(\mathcal{I})$ vertices such that $\mathcal{I} - S \in \mathcal{F}_{\mathrm{ML}}$. Thus having the sets S and B at hand, we can apply Lemma 17. For statement (i) however, we need to find a set S of edges such that $\mathcal{I} - S \in \mathcal{F}_{\mathrm{ML}}$ in FPT time with $\Delta^{\mathrm{edge}}(\mathcal{I})$ or $\Delta^{\mathrm{swap}}(\mathcal{I})$ as parameter. Notice that it suffices to use Theorem 7 to obtain an edge set S of size at most $2\Delta^{\mathrm{edge}}(\mathcal{I})$, and then we can apply Lemma 16. For a more detailed description of these algorithms and their correctness, see the full paper [31].

Table 1. Summary of our results on MLS, MLED, and MLVD. Results marked by the sign † assume the Unique Games Conjecture.

Problem	Parameterized complexity	Approximation
MLS	FPT wrt k (Corollary 5, Theorem 9)	Constant-factor approx. is NP-hard (Theorem 4)[†]
MLED	W[1]-hard wrt k (Theorem 6)	Constant-factor approx. is NP-hard (Theorem 4)[†] 2-approx. FPT alg wrt k (Theorems 7, 10)
MLVD	W[2]-hard wrt k (Theorem 8)	$f(k)$-approx. is W[1]-hard wrt k (Theorem 8)

6   Summary and Further Research

We summarize our main results on MLS, MLED, and MLVD in Table 1. Interestingly, all our hardness results hold for strict preference systems, and we were able to extend all our positive results for preference systems with weak orders.

There are a few questions left open in the paper. We gave asymptotically tight bounds on the maximum number of stable matchings in a strict preference system $\mathcal{I}$ as a function of $\Delta^{\mathrm{edge}}(\mathcal{I})$ and $\Delta^{\mathrm{vert}}(\mathcal{I})$, but we were not able to do the same for $\Delta^{\mathrm{swap}}(\mathcal{I})$. Another question is whether the approximation factor of our 2-approximation FPT algorithm for MLED can be improved.

A possible direction of future research would be to identify further problems that can be solved efficiently on preference systems that are close to admitting a master list. Also, it would be interesting to see how these measures vary in different real-world scenarios, and to find those practical applications where preference profiles are usually close to admitting a master list.

References

1. Abraham, D.J., Levavi, A., Manlove, D.F., O'Malley, G.: The stable roommates problem with globally-ranked pairs. Internet Math. **5**, 493–515 (2008)
2. Bang-Jensen, J., Gutin, G.Z.: Digraphs: Theory, Algorithms and Applications. Springer Monographs in Mathematics, 2nd edn. Springer, London (2009). https://doi.org/10.1007/978-1-84800-998-1
3. Betzler, N., Fellows, M.R., Guo, J., Niedermeier, R., Rosamond, F.A.: Fixed-parameter algorithms for Kemeny rankings. Theor. Comput. Sci. **410**(45), 4554–4570 (2009)
4. Bhatnagar, N., Greenberg, S., Randall, D.: Sampling stable marriages: why spouse-swapping won't work. In: SODA 2008, pp. 1223–1232. SIAM (2008)
5. Boehmer, N., Faliszewski, P., Niedermeier, R., Szufa, S., Wąs, T.: Understanding distance measures among elections. In: IJCAI-22, pp. 102–108 (2022)
6. Bredereck, R., Chen, J., Knop, D., Luo, J., Niedermeier, R.: Adapting stable matchings to evolving preferences. In: AAAI 2020, pp. 1830–1837 (2020)
7. Bredereck, R., Heeger, K., Knop, D., Niedermeier, R.: Multidimensional stable roommates with master list. In: Chen, X., Gravin, N., Hoefer, M., Mehta, R. (eds.) WINE 2020. LNCS, vol. 12495, pp. 59–73. Springer, Cham (2020). https://doi.org/10.1007/978-3-030-64946-3_5

8. Chen, J., Skowron, P., Sorge, M.: Matchings under preferences: strength of stability and tradeoffs. ACM Trans. Econ. Comput. **9**(4), 20:1–20:55 (2021)
9. Cheng, C.T., Rosenbaum, W.: Stable matchings with restricted preferences: structure and complexity. In: EC 2021, pp. 319–339 (2021)
10. Chitnis, R., Cygan, M., Hajiaghayi, M., Marx, D.: Directed subset feedback vertex set is fixed-parameter tractable. ACM Trans. Algorithms **11**(4), 28:1–28:28 (2015)
11. Chowdhury, R.: A simple matching domain with indifferences and a master list. Rev. Econ. Des. (2022). https://doi.org/10.1007/s10058-022-00292-9
12. Cygan, M., et al.: Parameterized Algorithms. Springer, Cham (2015). https://doi.org/10.1007/978-3-319-21275-3
13. Elkind, E., Lackner, M., Peters, D.: Structured preferences. In: Endriss, U. (ed.) Trends in Computational Social Choice, chap. 10, pp. 187–207. AI Access (2017)
14. Fellows, M.R., Hermelin, D., Rosamond, F.A., Vialette, S.: On the parameterized complexity of multiple-interval graph problems. Theor. Comput. Sci. **410**, 53–61 (2009)
15. Fitzsimmons, Z., Hemaspaandra, E.: Kemeny consensus complexity. In: IJCAI-21, pp. 196–202 (2021)
16. Gale, D., Shapley, L.S.: College admissions and the stability of marriage. Amer. Math. Monthly **69**(1), 9–15 (1962)
17. Garey, M.R., Johnson, D.S.: Computers and Intractability: A Guide to the Theory of NP-Completeness. W. H. Freeman & Co., New York (1979)
18. Gupta, S., Jain, P., Saurabh, S.: Well-structured committees. In: IJCAI-20, pp. 189–195 (2020)
19. Gusfield, D.: The structure of the stable roommate problem: efficient representation and enumeration of all stable assignments. SIAM J. Comput. **17**(4), 742–769 (1988)
20. Gusfield, D., Irving, R.W.: The Stable Marriage Problem: Structure and Algorithms. Foundations of Computing Series, MIT Press, Cambridge (1989)
21. Irving, R.W., Manlove, D.F., Scott, S.: The stable marriage problem with master preference lists. Discret. Appl. Math. **156**(15), 2959–2977 (2008)
22. Kamiyama, N.: Many-to-many stable matchings with ties, master preference lists, and matroid constraints. In: AAMAS 2019, pp. 583–591 (2019)
23. Kemeny, J.: Mathematics without numbers. Daedalus **88**, 577–591 (1959)
24. Khanchandani, P., Wattenhofer, R.: Distributed stable matching with similar preference lists. In: OPODIS 2016. LIPIcs, vol. 70, pp. 12:1–12:16. Schloss Dagstuhl - Leibniz-Zentrum für Informatik (2016)
25. Khot, S.: On the power of unique 2-prover 1-round games. In: STOC 2002, pp. 767–775. Association for Computing Machinery (2002)
26. Lebedev, D., Mathieu, F., Viennot, L., Gai, A.T., Reynier, J., de Montgolfier, F.: On using matching theory to understand P2P network design. In: INOC 2007 (2007)
27. Lokshtanov, D., Ramanujan, M.S., Saurabh, S.: When recursion is better than iteration: a linear-time algorithm for acyclicity with few error vertices. In: SODA 2018, pp. 1916–1933 (2018)
28. Manlove, D.F.: Algorithmics of Matching Under Preferences, Series on Theoretical Computer Science, vol. 2. World Scientific, Singapore (2013)
29. Meeks, K., Rastegari, B.: Solving hard stable matching problems involving groups of similar agents. Theor. Comput. Sci. **844**, 171–194 (2020)

30. Perach, N., Polak, J., Rothblum, U.G.: A stable matching model with an entrance criterion applied to the assignment of students to dormitories at the Technion. Int. J. Game Theory **36**, 519–535 (2008)
31. Schlotter, I.: Recognizing when a preference system is close to admitting a master list. CoRR abs/2212.03521 (2022). arXiv:2212.03521 [cs.CC]
32. Schlotter, I., Cseh, Á.: Maximum-utility popular matchings with bounded instability. CoRR abs/2205.02189 (2022). arXiv:2205.02189 [cs.DM]
33. Vazirani, V.V.: Approximation Algorithms. Springer, Heidelberg (2001)

Groups Burning: Analyzing Spreading Processes in Community-Based Networks

Gennaro Cordasco[1(✉)], Luisa Gargano[2], and Adele A. Rescigno[2]

[1] Department of Psychology, University of Campania "L.Vanvitelli", Caserta, Italy
gennaro.cordasco@unicampania.it
[2] Department of Computer Science, University of Salerno, Fisciano, Italy
{lgargano,arescigno}@unisa.it

Abstract. Graph burning is a deterministic, discrete-time process that can be used to model how influence or contagion spreads in a graph. In the graph burning process, each node starts as dormant, and becomes informed/burned over time; when a node is burned, it remains burned until the end of the process. In each round, one can burn a new node (source of fire) in the network. Once a node is burned in round t, in round $t + 1$, each of its dormant neighbors becomes burned. The process ends when all nodes are burned; the goal is to minimize the number of rounds. We study a variation of graph burning in order to model spreading processes in community-based networks. With respect to a specific piece of information, a community is *satisfied* when this information reaches at least a prescribed number of its members. Specifically, we consider the problem of identifying a minimum length sequence of nodes that, according to a graph burning process, allows to satisfy all the communities of the network. We investigate this NP-hard problem from an approximation point of view, showing both a lower bound and a matching upper bound. We also investigate the case when the number of communities is constant and show how to solve the problem with a constant approximation factor. Moreover, we consider the problem of maximizing the number of satisfied groups, given a budget k on the number of rounds.

Keywords: Burning number · Spreading processes · Group influence

1 Introduction

Spreading processes in complex networks have recently gained a great deal of attention. There are many situations where members of a network may influence their neighbors' behavior and decisions, by swaying their opinions, by suggesting what products to buy, but also acting in many social problems as public health awareness, financial inclusion, and more. A fundamental aspect to understand and control the spreading dynamics is the identification of spreaders that can diffuse information within the network in the least possible amount of time.

Small and marginalized groups within a larger community are those who need the most from attention, information and assistance. It is important, then, to ensure that each group receives an appropriate amount of information and resources, so to respect the diverse composition of the communities.

To address the above issue, in this paper we consider the problem of ensuring the right amount of informed representatives for each group in the network within fast spreading processes.

In our model the spreading process reflects the *burning process* in a graph. Bonato [5] introduced the notion of graph burning as a simplified model for the spread of memes in a network. Imagine someone trying to optimize the spread of a meme, hitting key actors in the network with the meme in a given priority sequence. To recap graph burning, nodes start off as dormant, and become informed/burned over time. If a dormant node is neighboring an informed one, it becomes informed too. One can burn/inform a new node anywhere in the network in each step. We can then see the process to proceed in sequential discrete steps, where one node is selected at each time-step t as a *source* of fire and burns all of its neighbors at the next time step $t + 1$. The nodes that are burned, can burn their neighbors at the next time step. That is, the information can pass from a node who have been informed in the previous time step. Furthermore, an informed node remains informed or burned throughout the process.

The analysis of the burning process enables evaluating the robustness of a network with respect to misinformation strategies (diffusion of negative behaviors) and on the other hand, it allows optimizing the impact of positive strategies.

Problems Definitions. We model the network as an undirected graph $G = (V, E)$, where V is the set of individuals and the set of edges E represents the relationships among members of the network, i.e., $(u, v) \in E$ if individuals u and v can directly communicate. We denote by $n = |V|$ the number of individuals in the network. For any $u, v \in V$, we denote by $d(u, v)$ the distance between u and v in G. We denote by $N_d[v] = \{w \mid w \in V, 0 \le d(v, w) \le d\}$ the set of all nodes having distance at most d from $v \in V$; we call $N_d[v]$ the *neighborhood of radius d around v (d-neighborhood of v)*. We denote by $N[v] = N_1[v]$ the closed neighborhood of v, that is, the set composed by v and its neighbors. Note also that $N_0[v] = \{v\}$.

The Graph Burning problem [5] is defined as follows.

GRAPH BURNING
Input: A graph $G = (V, E)$.
Output: A sequence $(v_1, v_2, \ldots, v_k)$, with $v_i \in V$ for each[1] $i \in [k]$, of minimum length k such that $\bigcup_{i=1}^{k} N_{k-i}[v_i] = V$.

If $(v_1, v_2, \ldots, v_k)$ is a burning sequence for the given graph G, then a source v_i, where $i \in [k]$, will burn only all the nodes within distance $k - i$ from v_i by the end of the k-th step. Each node $v \in V$ must be either a source or burned from at least one of the sources by the k-th round.

In our setting, we are given a family of pairwise disjoint subsets $V_1, \ldots, V_\omega$, referred to as *groups*, of the node set V and we are interested in the minimum

---

[1] For any integer a, we denote by [a] the set $\{1, 2, \ldots, a\}$.

time needed to reach at least a given number of individuals in each group. In particular, a positive integer p_j is assigned to each group V_j. The value p_j represents the minimum amount of nodes of group V_j that have to be informed (burned) during the spreading process.

BURNING WITH GROUPS (BG)
Input: $G = (V, E)$, a group family $\Pi = \{V_1, V_2, \ldots, V_\omega\}$ (a collection of pairwise disjoint subsets of V), and a vector $R = (p_1, p_2, \ldots, p_\omega)$ of requirements for each group.
Output: A sequence $(v_1, v_2, \ldots, v_k)$, where $v_i \in V$, of minimum length k such that

$$\left| \left(\bigcup_{i=1}^{k} N_{k-i}[v_i] \right) \cap V_j \right| \geq p_j, \text{ for each } j \in [\omega].$$

We notice that BG is a generalization of the GRAPH BURNING problem. Indeed, starting from BG, the GRAPH BURNING problem is obtained considering each node as a single group (i.e. $w = n$) and fixing all the requirements equal to 1, or alternatively, considering all the nodes in a single group (i.e., $w = 1$) and fixing its requirement to n.

We will also consider the following corresponding maximization problem.

BURNING MAX GROUPS (BMG)
Input: $G = (V, E)$, a group family $\Pi = \{V_1, V_2, \ldots, V_\omega\}$ (a collection of pairwise disjoint subsets of V), a vector $R = (p_1, p_2, \ldots, p_\omega)$ of requirements for each group, and an integer k.
Output: A sequence $(v_1, v_2, \ldots, v_k)$, where $v_i \in V$, and a set of integers $J \subseteq [\omega]$ such that

$$\left| \left(\bigcup_{i=1}^{k} N_{k-i}[v_i] \right) \cap V_j \right| \geq p_j, \text{ for each } j \in J \text{ and } |J| \text{ is maximized.}$$

Given a sequence $(v_1, v_2, \ldots, v_k)$, in the following we will say that the group V_j, for $j \in [\omega]$, is *satisfied* iff $\left| \left(\bigcup_{i=1}^{k} N_{k-i}[v_i] \right) \cap V_j \right| \geq p_j$. In words, in the BG problem, the goal is to find the sequence of the minimum length that satisfies all the groups, while in the BMG problem, given a budget of at most k sources, the goal is to maximize the number of satisfied groups.

2   Related Works

The problem of finding a source set of minimum size which, according to a spreading process, is able to spread a piece of information to the whole (or a fixed fraction of the) network, has its roots in the area of the spread of information in Social Networks.

The spread of viral information across a network naturally suggests many interesting optimization problems like Influence Maximization and Target Set

Selection (see [8,21] and references quoted therein). The first authors to study the spread of information in networks from an algorithmic point of view were Kempe et al. [24] who proposed two diffusion models named Linear Threshold and Independent Cascade. Thereafter, a series of papers that isolated interesting variants of the problem were proposed [1,4,9–15,17–19,22,26]. Being the problem hard in general, several heuristics have been proposed [16,20,27].

All the considered models are applied on a static snapshot of the network and asks for the identification of a set of initial adapter which will be in charge to trigger the diffusion of the information. In a real setting, on the other hand, the network is dynamic and algorithms should be able to adapt to network changes.

Recently, the classical Influence Maximization problem has been revised with the aim of fetching the top influential users in social networks under a group influence perspective [28]. The assumption is that the interaction within a group is high and this favors the rapid dissemination of information inside it. The goal becames reaching all the groups or maximizing the number of reached groups.

Closely related to our work is the COLORFUL K-CENTER PROBLEM, which has been studied in [3,23].

COLORFUL K-CENTER PROBLEM
Input: A set P of n points in a metric space, and an integer k, a partition $\{P_1, P_2, \ldots, P_c\}$ of P into c color classes, and a coverage requirement $0 \leq t_j \leq |P_j|$ for each color class $j \in [c]$.
Output: Find the smallest radius ρ such that using k balls of radius ρ, centered at points of P, we can simultaneously cover at least t_j points from each class P_j with $j \in [c]$.

Apart from the fact that the problem is defined in a metric space, the main difference with our BG problem is the fact that the number of sources/centers is fixed to k, while the radius of the neighbourhood (that is, the value to be minimized) is equal for each source, while in the BG problem the number of sources varies as well as the radius of the neighbourhood, because it depends on the position of the source in the output burning sequence.

3   Burning with Groups: The General Case

3.1   Hardness Results for BURNING WITH GROUPS

We show that BG cannot be approximated in polynomial time within a factor $c \log n$, for some constant $c < 1$, unless NP has quasi polynomial time (i.e., $NP \subset TIME(n^{O(\text{polylog } n)})$, where $TIME(t)$ denote the class of problems that admit a deterministic algorithm that runs in time t). To this aim, we provide an approximation preserving reduction from SET COVER.

SET COVER (SC)
Input: A universe $\mathcal{U} = \{1, 2, \ldots, n\}$ and a collection $\mathcal{S}$ of m subsets of $\mathcal{U}$, whose union equals the universe.
Output: A collection $\mathcal{C} \subseteq \mathcal{S}$ of minimum size such that $\bigcup_{C \in \mathcal{C}} C = \mathcal{U}$.

The following result has been proved in [2].

Theorem 1 [2]. SET COVER *cannot be approximated, in polynomial time, within a factor* $(\log n)/48$, *unless NP has quasi polynomial time.*

Remark 1. We remark that in the instance of SET COVER produced by the reduction in the proof of Theorem 1, the number of subsets (m) and the size of the universe (n) are polynomially related (i.e., $m \approx n^a$, for some constant $a > 0$).

Theorem 2. BG *cannot be approximated, in polynomial time, to a factor* $c \log n$ *where* $c < 1$ *is a certain constant, unless NP has quasi polynomial time, even if all the requirements are equal to 1.*

Proof. We give an approximation preserving reduction from SET COVER.

The theorem will follow by Theorem 1 and Remark 1, since the construction below provides a graph $G = (V, E)$ having $|V| = O(n \times m)$ nodes.

To our aim, given a SC instance $\langle \mathcal{U}, \mathcal{S} \rangle$, we construct an instance $\langle G, \Pi, R \rangle$ of BG. Let $|\mathcal{U}| = n$ and $\mathcal{S} = \{C_1, \ldots, C_m\}$. We build the graph $G = (V, E)$ where V is partitioned into the disjoint sets $V_0, V_1, \ldots, V_{n+2}$ (i.e., the group family is $\Pi = \{V_0, V_1, \ldots, V_{n+2}\}$) and where all the group requirements are fixed to 1 (i.e., $R = (1, 1, \ldots, 1)$). Formally, $G = (V, E)$ is build as follows:

- For any $C_i \in \mathcal{S}$, if $C_i = \{a_1, a_2, \ldots, a_r\}$ we add to G a star S_i of $r + 1$ nodes $\{u_i, v_{i,1}, v_{i,2}, \ldots, v_{i,r}\}$ and edges $\{(u_i, v_{i,j}) \mid j \in [r]\}$. The center node u_i is assigned to the group V_0 and the leaf node $v_{i,j}$ is assigned to V_{a_j}, for $j \in [r]$.
- Then, we add a node u_{m+1}. We assign u_{m+1} to V_{n+1}.
- Finally, for each $i \in [m]$, we add a path P_i of length $2n + 1$ connecting the node u_i to u_{i+1}; all the nodes of the these m paths are assigned to V_{n+2}.

The correctness of the reduction is implied by the following claim.

Claim 1. The instance $\langle \mathcal{U}, \mathcal{S} \rangle$ of SC admits a solution of size k if and only if the instance $\langle G, \Pi, R \rangle$ of BG admits a solution of size $k + 1$.   □

3.2   Approximation Algorithms for BURNING WITH GROUPS

We show that BG can be approximated to a factor $\log n + 1$. To this aim, we first define a novel maximization problem called MAXIMUM MULTI-COVERAGE BURNING (MMCB) and show that it admits a 2-approximation. We will then use the 2 approximation for MMCB to obtain a $\log n + 1$ approximation for BG.

MAXIMUM MULTI-COVERAGE BURNING (MMCB)
Input: $G = (V, E)$, a group family $\Pi = \{V_1, V_2, \ldots, V_\omega\}$ (a collection of pairwise disjoint subsets of V), a vector $L = (\ell_1, \ell_2, \ldots, \ell_\omega)$ of ω non-negative values and an integer k.
Output: A sequence $S = (v_1, v_2, \ldots, v_k)$, where $v_i \in V$, such that

$$\sum_{j=1}^{\omega} \min\left\{ \left| \left(\bigcup_{i=1}^{k} N_{k-i}[v_i] \right) \bigcap V_j \right|, \ell_j \right\} \text{ is maximum.} \tag{1}$$

Algorithm 1. MMCB Algorithm(G, Π, L, k)

1: $S = ()$
2: **for** $i = 1$ to k **do**
3:    $v = argmax_{u \in V} f_i(u \mid S)$
4:    $S(i) = v$
5: **end for**
6: **return** S

Let $S = (v_1, \ldots, v_h)$ be a sequence of $0 \leq h \leq k$ nodes. We define $N_{S,k} = \bigcup_{i=1}^{h} N_{k-i}[v_i]$ as the set of all the nodes burned by S by the k-th round and

$$f(S,k) = \sum_{j=1}^{\omega} \min \left\{ \left| \left(\bigcup_{i=1}^{h} N_{k-i}[v_i] \right) \cap V_j \right|, \ell_j \right\} = \sum_{j=1}^{\omega} \min \{|N_{S,k} \cap V_j|, \ell_j\}. \quad (2)$$

When S is an empty sequence, we have $N_{S,k} = \emptyset$ and $f(S,k) = 0$. In the following, we will use N_S and $f(S)$, instead of $N_{S,k}$ and $f(S,k)$, whenever the value of k is clear from the context.

Notice that MMCB asks for a sequence S of size k that maximizes $f(S)$.

Moreover, let $S = (v_1, \ldots, v_h)$ be a sequence of $0 \leq h \leq k$ nodes, let $i \in [k]$ and let $v \in V$ be a node, we define

$$f_i(v \mid S) = \sum_{j=1}^{\omega} \min\{|(N_{k-i}[v] - N_S) \cap V_j|, \max\{0, \ell_j - |N_S \cap V_j|\}\}, \quad (3)$$

the gain that is provided by burning the neighborhood of radius $k - i$ around v, assuming that nodes in N_S are already burned.

By Eq. (3) we can easily observe that given two sequences S and S' such that $N_S \subseteq N_{S'}$ and a node $v \in V$ we have

$$f_i(v \mid S) \geq f_i(v \mid S') \text{ for each } i \in [k]. \quad (4)$$

Given the instance $\langle G, \Pi, L, k \rangle$ of MMCB, the algorithm MMCB proceeds by iteratively adding nodes to the sequence S. At each iteration i, for $i \in [k]$, the node v to be added to the sequence is greedily chosen to give with its $(k - i)$-neighborhood the maximum contribution to the sum in (1), see line 3.

The following properties will be useful to prove the approximation factor.

Lemma 1. *Let* $S = (v_1, v_2, \ldots, v_{h-1})$ *be a sequence of* $h - 1$ *nodes and let* $S \perp v = (v_1, v_2, \ldots, v_{h-1}, v)$ *be the sequence obtained by queuing* v *at* S. *We have*

$$f(S \perp v) - f(S) = f_h(v \mid S). \quad (5)$$

Let $S = (v_1, v_2, \ldots, v_k)$ be the solution provided by the algorithm MMCB on the instance $\langle G, \Pi, L, k \rangle$ and let $O = (u_1, u_2, \ldots, u_k)$ be an optimal solution for the MMCB problem on the same instance $\langle G, \Pi, L, k \rangle$.

Denote by $S_i = (v_1, \ldots, v_i)$ the sequence constructed by the algorithm MMCB by the end of the i-th step, for $i \in [k]$. We denote by $S_0 = ()$ the empty sequence and recall that $f(S_0) = 0$. Let v_i be the node selected at step i.

Lemma 2. *For $i \in [k]$, $f_i(v_i \mid S_{i-1}) \geq f_i(u_i \mid S)$.*

Lemma 3. $\sum_{i=1}^{k} f_i(u_i \mid S) \geq f(O) - f(S)$.

Theorem 3. *MMCB admits a 2-approximation algorithm.*

Proof. We have that,

$$
\begin{aligned}
f(S) &= (f(S) - f(S_{k-1})) + (f(S_{k-1}) - f(S_{k-2})) + \ldots + (f(S_1) - f(S_0)) \\
&= f_k(v_k \mid S_{k-1}) + f_{k-1}(v_{k-1} \mid S_{k-2}) + \ldots + f_1(v_1 \mid S_0) \qquad \text{by Lemma 1} \\
&= \sum_{i=1}^{k} f_i(v_i \mid S_{i-1}) \\
&\geq \sum_{i=1}^{k} f_i(u_i \mid S) \qquad\qquad \text{by Lemma 2} \\
&\geq f(O) - f(S). \qquad\qquad \text{by Lemma 3} \qquad\qquad\qquad (6)
\end{aligned}
$$

Inequality (6) implies $f(S) \geq f(O)/2$. $\qquad\qquad\qquad\qquad\qquad\qquad\qquad$ □

We show now how the 2-approximation bound for MMCB can be used to obtain a $\log n + 1$ approximation for BG.

Consider an instance $\langle G, \Pi, L, k \rangle$ of MMCB. Let $\ell = \sum_{j=1}^{\omega} \ell_j$. By Definition (2), we have that, for any sequence S of size k: (i) $f(S) \leq \ell$ and (ii) $f(S) = \ell$ if and only if $|N_S \cap V_j| \geq \ell_j$, for $j \in [\omega]$. Hence, whenever $f(S) = \ell$, the sequence S also satisfies the BG instance $\langle G, \Pi, R = L \rangle$.

Similarly, let $\langle G, \Pi, R \rangle$ be an instance of BG, let $r = \sum_{j=1}^{\omega} p_j$ and let k^* be the size of the smallest sequence S^* that satisfies all the requirements (i.e., the optimal value for the given instance). Using the same sequence S^* we get $f(S^*, k^*) = r$ for the instance $\langle G, \Pi, R, k^* \rangle$ of MMCB and we have: (i) the optimum value for the instance $\langle G, \Pi, R, k^* \rangle$ of MMCB is r and (ii) if the optimum value for $\langle G, \Pi, R, k \rangle$ of MMCB is r then $k \geq k^*$.

Let $\langle G, \Pi, R \rangle$ be an instance of BG. For each $k = 1, 2, \ldots, n$, we execute the Algorithm 2 and take the smallest set obtained as the solution of the problem. Theorem 4 shows that, exploiting Algorithm 2 and the above properties, one can obtain the desired approximation factor for BG.

Theorem 4. *BG can be approximated to $\log n + 1$.*

Proof. For a given k the obtained sequence $S = S^1 \bot S^2 \bot \ldots \bot S^{t_k}$ has length $k \times t_k$. We show now that by choosing the value k such that $k \times t_k$ is minimum one can get the desired approximation factor.

Let k^* denote the value of an optimal solution $S_{BG} = (v_1, \ldots, v_{k^*})$ for BG instance $\langle G, \Pi, R \rangle$. Clearly $f(S_{BG}, k^*) = r$. Hence, when the Algorithm 2 is executed with $k = k^*$, we know, by Theorem 3 and the above properties, that the greedy MMCB algorithm will compute a sequence S^1 such that $f(S^1, k^*) \geq r/2$. By iterating the MMCB algorithm t times, we get the sequences $S =$

Algorithm 2. BG Algorithm($G = (V, E), \Pi, R, k$)

1: $h = 1$, $\Pi^1 = \Pi$, $R^1 = R$, $V^1 = V$
2: $S^1 = MMCB(G, \Pi^1, R^1, k)$
3: **while** $\left(\sum_{i=1}^{h} f(S^i) < r \right)$ **do**
4:     $h = h + 1$
5:     $V^h = V^{h-1} - N_{S^{h-1}}$ ▷ V^h is the set of nodes not burned by $S^1, S^2, \ldots, S^{h-1}$
6:     $\Pi^h = \{V_1^h, \ldots, V_\omega^h\}$ ▷ $V_j^h = V^h \cap V_j$, for $j \in [\omega]$
7:     $R^h = (p_1^h, \ldots, p_\omega^h)$ ▷ $p_j^h = \max\{0, p_j^{h-1} - |N_{S^{h-1}} \cap V_j|\}$, for $j \in [\omega]$
8:     $S^h = MMCB(G, \Pi^h, R^h, k)$
9: **end while**
10: **return** $S = S^1 \perp S^2 \perp \ldots \perp S^h$ ▷ the concatenation of $S^1, S^2, \ldots, S^h$

$S^1 \perp S^2 \perp \ldots \perp S^t$ (that is S is the concatenation of $S^1, S^2, \ldots, S^t$) providing a solution for the instance $\langle G, \Pi, R, k^* \times t \rangle$ of MMCB of value at least

$$f(S, k^* \times t) \geq \sum_{i=1}^{t} f(S^i, k^*) \geq r \left(\frac{1}{2} + \frac{1}{4} + \ldots + \frac{1}{2^t} \right) = r \left(1 - \frac{1}{2^t} \right).$$

Hence, for some $t \leq \log r + 1$, the algorithm will find a sequence S such that $f(S, k^* \times t) \geq r$ and the corresponding burning time is

$$|S| = k^* \times t \leq k^* \times (\log r + 1) \leq k^* \times (\log n + 1). \qquad \square$$

4   Burning with $O(1)$ Groups

In this section we assume that $\omega = O(1)$ and obtain a constant approximation factor for the BG problem. Our solution goes through the following problem:

SQUARE DOMINATION WITH GROUPS (SDG)
Input: $G = (V, E)$, a group family $\Pi = \{V_1, V_2, \ldots, V_\omega\}$ (a collection of pairwise disjoint subsets of V), and a vector $R = (p_1, p_2, \ldots, p_\omega)$ of requirements for each group.
Output: A set $\{v_1, v_2, \ldots, v_k\} \subseteq V$ of minimum size k such that
$$\left| \left(\bigcup_{i=1}^{k} N_k[v_i] \right) \cap V_j \right| \geq p_j, \quad \text{for each } j \in [\omega].$$

SDG differs from BG because in SDG each source covers a neighbourhood of the same radius k and consequently the order of nodes in the solution does not matter, while in BG the radius of the neighbourhood depends on the position of the source in the sequence.

We notice that given any sequence S, solution for BG on $\langle G, \Pi, R \rangle$, then the set containing the nodes in S is a solution for SDG for the same instance. Moreover, from any solution $\{v_1, v_2, \ldots, v_k\}$ to SDG on $\langle G, \Pi, R \rangle$ we can get a solution for BG, of size $2k$, for the same instance as any sequence $(v_1, \ldots, v_k, u_1, \ldots, u_k)$ where $u_1, \ldots, u_k$ are k arbitrary chosen nodes. Hence,

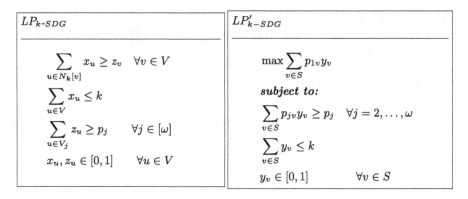

Fig. 1. (left) The natural LP relaxation for the k-SDG problem; (right) A simplified LP relaxation for the same problem.

denoted by O_B and O_D the sizes of the optimal solution of the BG problem and the SDG problem, respectively, we have $O_D \leq O_B \leq 2O_D$.

In the following, we will show how to obtain, from this observation, a polynomial time algorithm for the BG problem, whose solution is upper bounded by $3O_B + \omega - 1$.

We assume that $O_D > \omega$, otherwise $O_D = O(1)$ and we are able to find the optimal solution for the SDG problem in polynomial time by simply enumerating over all subsets of V of size O_D, which results in an exact algorithm having running time $|V|^{O(O_D)} = |V|^{O(1)}$.

We are going to use a result in [3], where the authors describe a pseudo-approximation algorithm for the COLORFUL K-CENTER PROBLEM (with two colors). They show how to find a solution of radius at most $2r^*$ using at most $k + 1$ centers (i.e., sources), where r^* is the optimum radius for the considered problem.

Let k-SDG be the decision version of SDG for a given integer k, that is, the problem asking if there exists a subset $S \subseteq V$, with $|S| = k$ such that $\left| \left(\bigcup_{v \in S} N_k[v] \right) \cap V_j \right| \geq p_j$, for each $j \in [\omega]$.

We use the solution of the following natural LP relaxation for the k-SDG problem, that we name $LP_{k\text{-}SDG}$ (Fig. 1 (left)), to have a simplified LP version (Fig. 1 (right)). This will allow to find a 2-approximation for the SDG problem.

Given a fractional solution (x, z) to $LP_{k\text{-}SDG}$, the variable x_u, for $u \in V$, represents the fraction of node u that is used as a source and z_v, for $v \in V$, represents the fraction of coverage that node v receives by the other (fractional) sources nodes (namely, nodes u at distance at most k from v with $x_u > 0$).

Following [3], we present an algorithm that, given a feasible solution (x, z) to $LP_{k\text{-}SDG}$, finds a clustering of the nodes of V and a subset S of cluster centers (nodes with $z_v > 0$), that we can use to write a simplified version of $LP_{k\text{-}SDG}$.

Let S and $\{C_v \mid v \in S\}$ be the sets returned by Algorithm 3. For any $v \in S$ and $j \in [\omega]$, let $P_{jv} = C_v \cap V_j$ be the set of nodes of group V_j in the cluster C_v. Fix $p_{jv} = |P_{jv}|$. By using S and the values p_{jv}, for each $v \in S$ and $j \in [\omega]$, we define the linear program $LP'_{k\text{-}SDG}$ (Fig. 1 (right)).

Algorithm 3. Clustering Algorithm(G, k, x, z)

1: $S = \emptyset, \quad V' = V$
2: **while** $V' \neq \emptyset$ and $\max_{v \in V'} z_v > 0$ **do**
3: $\quad v = argmax_{v' \in V'} z_{v'}$
4: $\quad S = S \cup \{v\}$
5: $\quad y_v = \min\{1, \sum_{u \in N_k[v]} x_u\}$
6: $\quad C_v = N_{2k}[v] \cap V'$
7: $\quad V' = V' - C_v$
8: **end while**
9: **return** $S, \{C_v \mid v \in S\}, y$

The variable y_v in LP'_{k-SDG} represents the fraction of node v that is used as a source to cover p_{jv} nodes of group V_j. The following Lemma shows that the vector y returned by Algorithm 3 is a feasible solution to LP'_{k-SDG}.

Lemma 4. *Given a feasible solution (x, z) to $LP_{k\text{-}SDG}$ and the sets $S \subseteq V$ and C_v, for $v \in S$ returned by Algorithm 1, the following proprieties hold:*
(i)   The clusters C_v, for $v \in S$ are pairwise disjoint.
(ii) y is a feasible solution to $LP'_{k\text{-}SDG}$.

As $LP'_{k\text{-}SDG}$ has only ω non-trivial constraints, any extreme point will have at most ω variables attaining strictly fractional values [25]. So by the second constraint in $LP'_{k\text{-}SDG}$ at most $k - 1 + \omega$ variables of any feasible y are non-zero.

So, we choose to round up the fractional variables y_v to 1 since the coverage of each group V_i for $i \in [\omega]$ can only increase, and set $S' = \{v \in S \mid y_v > 0\}$. By said above, we get $|S'| = k - 1 + \omega$. Recalling that for each node $v \in S'$ it holds $C_v \subseteq N_{2k[v]}$ and that $k > \omega$ we have that $|S'| = k - 1 + \omega < 2k - 1$ and S' is a solution for the k-SDG problem.

Recalling by the definition of the k-SDG problem that $k \leq diam(G)$, we can repeat all the above procedure for $k = 1, \cdots, diam(G)$, and stop at the smallest value of k for which a solution S' is possible. Now, we notice that, if O_D is the size of the optimal solution of the SDG problem, then $k \leq O_D$, since an optimal solution S_{O_D} for the SDG problem satisfies all the constraints of $LP_{k\text{-}SDG}$ whenever $k \geq O_D$. Hence the following theorem follows.

Theorem 5. *For $\omega = O(1)$, there exists a polynomial time algorithm that finds a 2-approximation for the SDG problem.*

By observing that the above strategy enables to identify a set of $k - 1 + \omega$ sources such that their $2k$-neighborhoods satisfy all the requirements, we have that one can obtain a sequence S, solution of the BG problem, by tacking all the above sources in any order followed by other arbitrary chosen $2k$ nodes. Hence, $|S| = 3k + \omega - 1$ and recalling that $k \leq O_D \leq O_B$ the following result holds.

Theorem 6. *For $\omega = O(1)$, there exists a polynomial time algorithm to find a solution for the BG problem whose size is upper bounded by $3O_B + \omega - 1$.*

Recalling that the GRAPH BURNING problem is a special case of BG in which all the nodes form a single group (i.e., $\omega = 1$) with requirement n, the above result generalizes the 3-approximation for the GRAPH BURNING [6].

5 Burning Max Groups with Unitary Requirements

When all the requirements are equal to one, an instance of BURNING MAX GROUPS (BMG) can be seen as an instance of the following problem [7].

MCG
Input: $\mathcal{S} = \{S_1, S_2, \ldots, S_m\}$ partitioned into ℓ subsets $G_1, G_2, \ldots, G_\ell$ of a ground set $\mathcal{X}$, a global bound k and ℓ subset bounds k_i, for each G_i.
Output: $H \subseteq \mathcal{S}$ such that $|H| \leq k$ and $|H \cap G_i| \leq k_i$ for $i \in [\ell]$, and $|\bigcup_{S \in H} S|$ is maximized.

Indeed given an instance $\langle G, \Pi = \{V_1, V_2, \ldots, V_\omega\}, R, k \rangle$ of the BMG problem, where all the requirements are fixed to 1 (i.e., $R = (1, \ldots, 1)$), we can set $\mathcal{X} = \{1, 2, \ldots, \omega\}$ and $\ell = k$. For each $i \in [k]$, and for each $v \in V$ let $S_{i,v} = \{j \mid N_{k-i}[v] \cap V_j \neq \emptyset\}$, we set $\mathcal{S} = \{S_{i,v} \mid v \in V, i \in [k]\}$ and for each $i \in [k]$, $G_i = \{S_{i,v} \mid v \in V\}$. Finally set $k_i = 1$, for each $i \in [k]$. Since MCG admits a 2-approximation algorithm [7], we have the following results.

Theorem 7. *There exists a polynomial time algorithm to find a 2-approximation for the Burning Max Groups problem with unitary requirements.*

Open Problem: Can the above result be extended to the BMG problem in the general case?

By considering each node as a separate subset, the BMG problem becomes

MAX BURNING (MB)
Input: $G = (V, E)$ and a integer k.
Output: A sequence of nodes $(v_1, v_2, \ldots, v_k)$, such that maximizes $\left| \bigcup_{i=1}^{k} N_{k-i}[v_i] \right|$.

Theorem 8. *There exists a polynomial time algorithm to find a 2-approximation for the MAX BURNING problem.*

References

1. Ackerman, E., Ben-Zwi, O., Wolfovitz, G.: Combinatorial model and bounds for target set selection. Theor. Comp. Sci. **411**(44–46), 4017–4022 (2010)
2. Arora, S., Lund, C.: Hardness of approximation. In: Hochbaum, Ed. D. (ed.) Approximation Algorithms for NP-Hard Problems, pp. 399–446 PWS Publishers (1995)

3. Bandyapadhyay, S., Inamdar, T., Pai, S., Varadarajan, K.: A constant approximation for colorful k-center. arXiv:1907.08906v1 (2019)
4. Ben-Zwi, O., Hermelin, D., Lokshtanov, D., Newman, I.: Treewidth governs the
   complexity of target set selection. Discrete Optim. **8**(1), 87–96 (2011)
5. Bonato, A.: A survey of graph burning. Contr. Discret. Math. **16**, 185–197 (2021)
6. Bonato, A., Kamali, S.: Approximation algorithms for graph burning. In: Gopal,
   T.V., Watada, J. (eds.) TAMC 2019. LNCS, vol. 11436, pp. 74–92. Springer, Cham
   (2019). https://doi.org/10.1007/978-3-030-14812-6_6
7. Chekuri, C., Kumar, A.: Maximum coverage problem with group budget constraints and applications. In: Jansen, K., Khanna, S., Rolim, J.D.P., Ron, D.
   (eds.) APPROX/RANDOM -2004. LNCS, vol. 3122, pp. 72–83. Springer, Heidelberg (2004). https://doi.org/10.1007/978-3-540-27821-4_7
8. Chen, W., Lakshmanan, L.V.S., Castillo, C.: Information and influence propagation in social networks. In: Synthesis Lectures on Data Management, vol. 5.4
   (2013)
9. Chopin, M., Nichterlein, A., Niedermeier, R., Weller, M.: Constant thresholds can
   make target set selection tractable. Theory Comput. Syst. **55**(1), 61–83 (2014).
   https://doi.org/10.1007/s00224-013-9499-3
10. Cicalese, F., Cordasco, G., Gargano, L., Milanic, M., Vaccaro, U.: Latency-bounded
    target set selection in social networks. Theor. Comp. Sci. **535**, 1–15 (2014)
11. Coja-Oghlan, A., Feige, U., Krivelevich, M., Reichman, D.: Contagious sets in
    expanders. In: Proceedings of SODA, pp. 1953–1987 (2015)
12. Cordasco, G., Gargano, L., Rescigno, A.A.: Influence propagation over large scale
    social networks. In: Proceedings of ASONAM, pp. 1531–1538 (2015)
13. Cicalese, F., Cordasco, G., Gargano, L., Milanic, M., Peters, J., Vaccaro, U.: Spread
    of influence in weighted networks under time and budget constraints. Theor. Comp.
    Sci. **586**, 40–58 (2015)
14. Cordasco, G., Gargano, L., Rescigno, A.A.: On finding small sets that influence
    large networks. Soc. Netw. Anal. Min. **6**, 1–20 (2016)
15. Cordasco, G., Gargano, L., Rescigno, A.A., Vaccaro, U.: Evangelism in social net-
    works: algorithms and complexity. Networks **71**(4), 346–357 (2018)
16. Cordasco, G., Gargano, L., Mecchia, M., Rescigno, A.A., Vaccaro, U.: Discovering
    small target sets in social networks: a fast and effective algorithm. Algorithmica
    80, 1804–1833 (2018). https://doi.org/10.1007/s00453-017-0390-5
17. Cordasco, G., Gargano, L., Rescigno, A.A.: Active influence spreading in social
    networks. Theor. Comp. Sci. **764**, 15–29 (2019)
18. Cordasco, G., et al.: Whom to befriend to influence people. Theor. Comp. Sci. **810**,
    26–42 (2020)
19. Cordasco, G., Gargano, L., Peters, J.G., Rescigno, A.A., Vaccaro, U.: Fast and
    frugal targeting with incentives. Theor. Comp. Sci. **812**, 62–79 (2020)
20. Dinh, T.N., Zhang, H., Nguyen, D.T., Thai, M.T.: Cost-effective viral marketing
    for time-critical campaigns in large-scale social networks. IEEE Trans. Net. **22**,
    2001–2011 (2014)
21. Easley, D., Kleinberg, J.: Networks, Crowds, and Markets: Reasoning About a
    Highly Connected world. Cambridge University Press, Cambridge (2010)
22. Gargano, L., Hell, P., Peters, J.G., Vaccaro, U.: Influence diffusion in social net-
    works under time window constraints. Theor. Comp. Sci. **584**(C), 53–66 (2015)
23. Jia, X., Sheth, K., Svensson, O.: Fair colorful k-center clustering.
    arXiv:2007.04059v1 (2020)
24. Kempe, D., Kleinberg, J., Tardos, E.: Maximizing the spread of influence through
    a social network. In: Proceedings of KDD 2003 (2003)

25. Lev, B., Soyster, A.L.: Integer programming with bounded variables via canonical separation. J. Oper. Res. Soc. **29**(5), 477–488 (1978)
26. Reddy, T.V.T., Rangan, C.P.: Variants of spreading messages. J. Graph Algorithms Appl. **15**(5), 683–699 (2011)
27. Shakarian, P., Eyre, S., Paulo, D.: A scalable heuristic for viral marketing under the tipping model. Soc. Netw. Anal. Min. **3**(4), 1225–1248 (2013). https://doi.org/10.1007/s13278-013-0135-7
28. Zhu, J., Ghosh, S., Wu, W.: Group influence maximization problem in social networks. IEEE Trans. Comput. Soc. Syst. **6**(6), 1156–1164 (2019)

Roman k-Domination: Hardness, Approximation and Parameterized Results

A. Mohanapriya[1(✉)], P. Renjith[2], and N. Sadagopan[1]

[1] Indian Institute of Information Technology, Design and Manufacturing,
Kancheepuram, Chennai, India
{coe19d003,sadagopan}@iiitdm.ac.in
[2] National Institute of Technology, Calicut, Kozhikode, India
renjith@nitc.ac.in

Abstract. We investigate the computational complexity of finding a minimum Roman k-dominating function (RKDF) on split graphs. We prove that RKDF on split graphs is NP-complete on $K_{1,2k+3}$-free split graphs. We also show that finding RKDF on star-convex bipartite graphs and comb-convex bipartite graphs are NP-complete. Further, we also show that finding RKDF on bipartite chain graphs is polynomial-time solvable, which is a non-trivial subclass of comb-convex bipartite graphs. On the parameterized front, we show that finding RKDF on split graphs is in W[1]-hard when the parameter is the solution size. From the approximation perspective, we show that there is no constant factor approximation algorithm for RKDF.

Keywords: Roman k-dominating function · Split graphs · Bipartite graphs

1 Introduction

The set D is a dominating set of G if every vertex in $V(G) \setminus D$ is adjacent to at least one vertex in D. The dominating set problem and its variants are of fundamental interest in graph theoretic research. Some of the variants of Domination which are studied from function perspectives are: $\{k\}$-dominating function [1], k-rainbow domination function [2], Roman domination [3], global Roman domination [4], double Roman domination [5], Roman $\{k\}$-domination [6] and Roman k-domination [7]. In this paper we analyze the computational complexity of Roman k-domination for every fixed $k \in \mathbb{Z}^+$. Let G be a simple graph with vertex set $V(G)$. A Roman dominating function on a graph G is a vertex labeling $f : V(G) \to \{0, 1, 2\}$ such that every vertex with label 0 has a neighbor with label 2. The weight $w(f)$ of an RDF f is $\Sigma_{v \in V(G)} f(v)$. The Roman domination number $\gamma_R(G)$ is the minimum weight of an RDF of G. A Roman k-dominating function (RKDF) on G is a function $f : V(G) \to \{0, 1, 2\}$ such that every vertex u for which $f(u) = 0$ is adjacent to at least k vertices

$v_1, v_2, \ldots, v_k$ with $f(v_i) = 2$ for $i = 1, 2, \ldots, k$. The weight of a Roman k-dominating function is the value $w(f) = \Sigma_{u \in V(G)} f(u)$. The minimum weight of a Roman k-dominating function on a graph G is called the Roman k-domination number $\gamma_{kR}(G)$ of G. Note that the Roman 1-domination number $\gamma_{1R}(G)$ is the usual Roman domination number $\gamma_R(G)$.

On the complexity front, Roman domination is polynomial-time solvable on interval graphs, cographs, and AT-free graphs [8], and Roman domination is NP-complete on bipartite graphs [9,10]. Global Roman Domination is NP-complete for bipartite graphs and chordal graphs [4], and double Roman Domination is NP-complete for bipartite graphs and chordal graphs [11]. Finding a minimum weight RKDF is called the minimum weight Roman k-domination problem (RKDP). RKDP is known to be polynomial-time solvable for cactus graphs [7].

To the best of our knowledge, RKDF has not been studied for a subclass of chordal graphs and a subclass of bipartite graphs. We make the first attempt in identifying the complexity of the Roman k-domination problem on a subclass of chordal graphs and for a subclass of bipartite graphs. In this paper, we focus on the computational complexity of the Roman k-domination problem on split graphs and some subclass of bipartite graphs such as star-convex bipartite graphs and comb-convex bipartite graphs.

Our Results: In this paper, we consider connected undirected unweighted simple graphs.

1. We show that deciding the Roman k-domination problem is NP-complete for split graphs, star-convex bipartite graphs, and comb-convex bipartite graphs. We show that for a positive integer k, the Roman k-domination problem is NP-complete for $K_{1,2k+3}$-free split graphs. Further, we prove that finding a minimum weight Roman k-dominating function on bipartite chain graphs is polynomial-time solvable.
2. As far as approximation algorithms are concerned, we show that there is no constant factor polynomial-time approximation algorithm for RKDF unless $P = NP$.
3. On the parameterized complexity front, we show that for split graphs finding RKDF is W[2]-hard with the parameter being the weight of the Roman k-domination function.

Graph Preliminaries:
For a graph G, $V(G)$ denotes the vertex set and $E(G)$ represents the edge set. For a set $S \subseteq V(G)$, $G[S]$ denotes the subgraph of G induced on the vertex set S. The open neighborhood of a vertex v is $N_G(v) = \{u \mid \{u, v\} \in E(G)\}$ and the closed neighborhood of v is $N_G[v] = \{v\} \cup N_G(v)$. For $S \subseteq V(G)$, $N_G(S) = \bigcup_{v \in S} N_G(v)$. The degree of vertex v is $d_G(v) = |N_G(v)|$. A split graph is a graph G in which $V(G)$ can be partitioned into two sets; a clique K and an independent set I. A split graph is written as $G = (K \cup I, E)$, where K is a maximal clique and I is an independent set. In a split graph, for each vertex u in K, $N_G^I(u) = N_G(u) \cap I$, $d_G^I(u) = |N_G^I(u)|$, and for each vertex v in I, $N_G^K(v) =$

$N(v) \cap K$, $d_G^K(v) = |N_G^K(v)|$. For each vertex u in K, $N_G^I[u] = N_G(u) \cap I \cup \{u\}$, and for each vertex v in I, $N_G^K[v] = N(v) \cap K \cup \{v\}$. For a split graph G, $\Delta_G^I = \max\{d_G^I(u)\}$, $u \in K$ and $\Delta_G^K = \max\{d_G^K(v)\}$, $v \in I$. A bipartite graph G with $V(G) = (X, Y)$ is called tree-convex if there exists a tree T on X such that, for each y in Y, the neighbors of y induce a subtree in T. When T is a star (comb), G is called star- (comb-) convex bipartite graph. If G by its definitions contains more than one partition, then each partition is non-empty.

Roadmap: In Sect. 2, we present classical complexity results and bounds for RKDF on split graphs. In Sect. 3, we present classical complexity results for RKDF on star-convex bipartite graphs and comb-convex bipartite graphs, and RKDF is polynomial-time solvable on bipartite chain graphs. We present parameterized complexity results and approximation results in Sect. 4, and Sect. 5, respectively. We present some observations about Roman k-domination,

Claim 1. *For a graph G, if there exists a vertex v whose degree is less than k, then RKDF having minimum weight has $f(v) \neq 0$.*

Proof. We know that if a vertex v such that $f(v) = 0$ means that it has u_i, $1 \leq i \leq k$ neighbours such that $f(u_i) = 2$. Since the degree of v is less than k, there can not be k vertices adjacent to v. Hence RKDF having minimum weight has $f(v) \neq 0$.

2   Roman k-Domination on Split Graphs

We analyze the classical complexity of Roman k-domination on split graphs for any $k \geq 1$.

2.1   Roman k-Domination on Split Graphs Is NP-Complete

In this section, we show that the decision version of the Roman k-domination problem is NP-complete for split graphs by giving a polynomial-time reduction from a well-known NP-complete problem, Exact-3-Cover (X3C) [12], which is defined as follows:

EXACT-3-COVER $(X, \mathcal{C})$

Instance: A finite set X with $|X| = 3q$ and a collection $\mathcal{C} = \{C_1, C_2, \ldots, C_m\}$ of 3-element subsets of X.

Question: Is there a subcollection $\mathcal{C}' \subseteq \mathcal{C}$ such that for every $x \in X$, x belongs to exactly one member of $\mathcal{C}'$ (that is, $\mathcal{C}'$ partitions X) ?

The decision version of Roman k-domination problem is defined below:

The Roman k-domination problem

Instance: A graph G, $\alpha \in \mathbb{N}$

Question: Does there exist a function $f : V(G) \rightarrow \{0, 1, 2\}$ such that for every vertex u for which $f(u) = 0$ is adjacent to at least k vertices $v_1, v_2, \ldots, v_k$ for which $f(v_i) = 2, 1 \leq i \leq k$ and $\Sigma_{u \in V(G)} f(u) \leq \alpha$?

We can observe that Roman 1-domination is not a special case of Roman 2-domination. For instance, assume that we have G as $K_{1,10}$ and we ask "does there exist a Roman 1-dominating function (RKDF) of weight at most 2?" The answer is yes. But for the same G as $K_{1,n}$ and we ask "does there exist a Roman 2-dominating function (RKDF) of weight at most 2?" The answer is no.

From this, we can observe that YES instance of Roman 1-domination becomes NO instance of Roman 2-domination. Therefore, the computational complexity of Roman k-domination for each k needs to be analyzed separately. The computational complexity of Roman k-domination can vary for each k. Thus we present a reduction wherein the reduction instances use parameter k as part of the construction.

Theorem 1. *For split graphs, the Roman k-domination problem (RKDP) is NP-complete.*

Proof. **RKDP is in NP:** Given a graph G, an integer α and a function f, whether f is a RKDF of size at most α can be checked in polynomial time. Hence RKDP is in NP.

RKDP is NP-Hard: X3C can be reduced to RKDP on split graphs in polynomial time using the following reduction. We map an instance $(X, \mathcal{C})$ of X3C, where $X = \{x_1, \ldots, x_{3q}\}$ and $\mathcal{C} = \{c_1, \ldots, c_m\}$, to the corresponding instance (G, α) of RKDP as follows: $V(G) = V_1 \cup V_2 \cup V_3$, $V_1 = \{c_{ij} \mid c_i \in \mathcal{C},\ 1 \le i \le m,\ 1 \le j \le k\}$, $V_2 = \{x_i \mid x_i \in X\}$, $V_3 = \{y_{ij} \mid 1 \le i \le m,\ 1 \le j \le 2k-2\}$, and $E(G) = \{\{c_{ij}, x_l\} \mid x_l \in c_i,\ c_{ij} \in V_1,\ x_l \in V_2,\ 1 \le i \le m,\ 1 \le j \le k,\ 1 \le l \le 3q\} \cup \{\{y_{ij}, c_{il}\} \mid 1 \le i \le m,\ 1 \le j \le 2k-2,\ 1 \le l \le k\} \cup \{\{w, z\} \mid w, z \in V_1\}$. Note that G is a split graph with V_1 as a clique and $V_2 \cup V_3$ as an independent set. Using the following claims, we establish that this reduction is a solution preserving reduction. That is Exact-3-Cover $(X, \mathcal{C})$ if and only if G has a RKDF with weight at most $\alpha = (2kq) + ((2k-2)(m-q))$. An illustration for X3C to Roman 4-domination on split graphs is shown in Fig. 1. In Fig. 1, For $X = \{x_1, x_2, x_3, x_4, x_5, x_6\}$ and $\mathcal{C} = \{c_1 = \{x_1, x_2, x_3\}, c_2 = \{x_2, x_3, x_4\}, c_3 = \{x_4, x_5, x_6\}\}$, the corresponding graph G with $\alpha = 2qk + ((2k-2)(m-q)) = 6$. For $\mathcal{C}' = \{c_1, c_3\}$ which is a solution to X3C we obtain the corresponding Roman 4-domination function as follows: $f(c_{11}) = \ldots = f(c_{14}) = f(c_{31}) = \ldots = f(c_{34}) = 2$, $f(y_{11}) = \ldots = f(y_{16}) = f(y_{31}) = \ldots = f(y_{36}) = 0$, $f(c_{21}) = \ldots = f(c_{24}) = f(x_1) = \ldots = f(x_6) = 0$, $f(y_{21}) = \ldots = f(y_{26}) = 1$.

Proof. Necessity: Suppose $\mathcal{C}'$ is a solution for X3C with $|\mathcal{C}'| = q$, then the corresponding function $f : V \rightarrow \{0, 1, 2\}$ in G is defined as follows.
For $c_i \in \mathcal{C}'$:
$f(v) = 2$, if $v \in V_1$, $v = c_{ij}$, $1 \le j \le k$,
$f(v) = 0$, if $v \in V_2$,
$f(v) = 0$, if $v \in V_3$, $v \in N_G^I(c_{ij})$
For $c_i \notin \mathcal{C}'$ $(c_i \in \mathcal{C} - \mathcal{C}')$:
$f(v) = 1$, if $v \in V_3$, $v \in N_G^I(c_{ij})$, $1 \le j \le k$,
$f(v) = 0$, if $v \in V_1$, $v = c_{ij}$, $1 \le j \le k$

Observe that there are kq vertices in V_1 having label 2, and there are $(2k - 2)(m - q)$ vertices in V_3 having label 1. Therefore, f is a RKDF with weight $\alpha = (2kq) + ((2k - 2)(m - q))$.

Sufficiency: Suppose that G has a RKDF f with weight at most $\alpha = (2kq) + ((2k - 2)(m - q))$.

To complete the proof of sufficiency, we present a series of claims using the structural understanding of G and f.

Claim 1A. If G has a RKDF f with weight at most α, then for each $x \in V_2$, $f(x) = 0$.

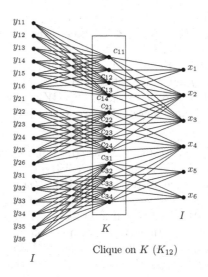

Fig. 1. Reduction: An instance of X3C to RKDP in split graphs

Proof. Suppose there exists a vertex $x \in V_2$ such that $f(x) \neq 0$. Let $x_1, \ldots, x_p \subseteq V_2$ such that $f(x_i) \neq 0$, $1 \leq i \leq p$. Clearly $|V_2 - \{x_1, \ldots, x_p\}| = 3q - p$. Let $V_2' = V_2 - \{x_1, \ldots, x_p\}$. Since f is a RKDF, each $a \in V_2'$ with $f(a) = 0$ is adjacent to at least k vertices say $\{w_1, \ldots, w_k\}$ with $f(w_i) = 2$, $1 \leq i \leq k$. By our construction for any $w \in V_1$, $|N(w) \cap V_2| = 3$. Hence there are at least $(\lceil \frac{3q-p}{3} \rceil)k$ vertices, say $Z = \{z_1, \ldots, z_{(\lceil \frac{3q-p}{3} \rceil)k}\}$ in V_1 such that for each $z \in Z$, $f(z) = 2$.

Case 1: For each i, $1 \leq i \leq \lceil \frac{3q-p}{3} \rceil, j \neq l, 1 \leq j \leq k, 1 \leq l \leq k, f(c_{ij}) = f(c_{il}) = 2$.

Let $r = \lceil \frac{3q-p}{3} \rceil$. Without loss of generality we relabel Z as $Z = \{z_1 = c_{11}, \ldots, z_k = c_{1k}, z_{k+1} = c_{21}, \ldots, z_{2k} = c_{2k}, \ldots, z_{rk-k+1} = c_{r1}, \ldots, z_{rk} = c_{rk}\}$ such that for each $z \in Z$, $f(z) = 2$.

Case 1.1: For each $c \in (V_1 \setminus Z)$ such that $f(c) = 0$, for each $y \in V_3$, $y \in N(z)$ where $z \in Z$ such that $f(y) = 0$ and for each $a \in V_3$, $a \in N(c)$ where

$c \in (V_1 \setminus Z)$ such that $f(a) = 1$. Then the weight of RKDF of G is at least $2k(\lceil \frac{3q-p}{3} \rceil) + p + (2k-2)(m - (\lceil \frac{3q-p}{3} \rceil)) = p + 2km - 2m + 2(\lceil \frac{3q-p}{3} \rceil)$, which is greater than α, a contradiction.

Note that if we assume other values for each $r \in (V_1 \setminus Z)$ such that $f(r) = 1$ or $f(r) = 2$, for each $y \in V_3$, $y \in N(z)$ where $z \in Z$ such that $f(y) = 1$ or $f(y) = 2$ and for each $a \in V_3$, $a \in N(w)$ where $w \in (V_1 \setminus Z)$ such that $f(a) = 2$, we arrive at a contradiction.

Case 2: For some i, $1 \leq i \leq \lceil \frac{3q-p}{3} \rceil$, $j \neq l$, $1 \leq j \leq k$, $1 \leq l \leq k$, $f(c_{ij}) = 2$ and $f(c_{il}) \neq 2$.

Let $c_1, c_2, \ldots, c_t \subseteq V_1$ such that for each c_i, $1 \leq i \leq t$, $j \neq l$, $1 \leq j \leq k$, $1 \leq l \leq k$, $f(c_{ij}) = 2$ and $f(c_{il}) = 2$. we know that $t < r$. For some $1 \leq i \leq t < r$, $S = \{c_{ij} \mid c_{ij} \in V_1, 1 \leq i \leq t, 1 \leq j \leq k$ such that $f(c_{ij}) = 2\}$. Let $W = \{c_{ij} \mid c_{ij} \in V_1, 1 \leq j \leq k$, there exist c_{ij}, c_{il} such that $f(c_{ij}) = 2$ and $f(c_{il}) \neq 2$ for some i, $1 \leq i \leq \lceil \frac{3q-p}{3} \rceil$, $j \neq l$, $1 \leq j \leq k$, $1 \leq l \leq k\}$. Note that $W \cap S = \emptyset$. Let $W_0 = \{w \mid w \in W, f(w) = 0\}$, $W_1 = \{w \mid w \in W, f(w) = 1\}$, $W_2 = \{w \mid w \in W, f(w) = 2\}$. Let $U = W \cup S$.

Note that for each vertex $y \in V_3$, $y \in N(W)$, $f(y) \neq 0$ because by our construction $d_G(y) = k$ and there exist two vertices in $a, b \in N(y)$ such that $f(a) \neq 2$, $f(b) = 2$.

Case 2.1: For each $c \in (V_1 \setminus U)$ such that $f(c) = 0$, for each $y \in V_3$, $y \in N(z)$, where $z \in Z$ such that $f(y) = 0$, for each $g \in V_3$, $g \in N(w)$ where $w \in (V_1 \setminus Z)$ such that $f(g) = 1$, for each vertex in $w \in W_0$, $f(w) = 0$, for each vertex in $l \in W_1$, $f(l) = 1$, and for each vertex in $d \in W_2$, $f(d) = 2$.

Then the weight of RKDF of G is at least $2k(t) + |W_1| + 2|W_2| + (2k-2)(m-t)$ where $t < r$, which is greater than α, a contradiction.

Note that if we assume other values for each $c \in (V_1 \setminus U)$ such that $f(c) = 1$ or $f(c) = 2$, for each $y \in N(z)$, $y \in V_3$ where $z \in Z$ such that $f(y) = 1$ or $f(y) = 2$ and for each $g \in N(w)$ and $g \in V_3$ where $w \in V_1 \setminus Z$ such that $f(g) = 2$, For W any other combination, for each vertex in $w \in W_0$, $f(w) = 0$, for each vertex in $l \in W_1$, $f(l) = 1$, and for each vertex in $d \in W_2$, $f(d) = 2$, we obtain a contradiction.

Therefore there does not exist $\{x_1, \ldots, x_p\} \subseteq V_2$ such that $f(x_i) \neq 0$, $1 \leq i \leq p$. Thus for each $x \in V_2$, $f(x) = 0$.

Claim 1B. Given f, without loss of generality $Z = \{c_{11}, \ldots, c_{1k}, \ldots, c_{q1}, \ldots, c_{qk}\}$ such that for each $z \in Z$, $f(z) = 2$.

Proof. By our construction, for any $w \in V_1$, $|N(w) \cap V_2| = 3$, then at least qk say $Z = \{c_{11}, \ldots, c_{1k}, \ldots, c_{q1}, \ldots, c_{qk}\}$ such that for each $z \in Z$, $f(z) = 2$. Suppose that there exists a $c_i \in Z$, $1 \leq i \leq q$ such that $f(c_{ij}) \neq 2$, $1 \leq j < k$. Since for each $x_i \in V_2$, $f(x_i) = 0$, there exist $U = \{u_l \in (V_1 \setminus Z)\}$ such that $f(u_l) = 2$, $1 \leq l \leq j$. Note that the vertices in U need not be consecutive. Let $W = \{w \mid w \in (V_1 \setminus U)\}$.

Case 1: For each $w \in W$, $f(w) = 0$, for each $y \in V_3$, there exists c_i, $1 \leq i \leq m$, $f(c_{ij}) \neq 2$, $f(c_{il}) = 2$, $1 \leq j \leq k$, $1 \leq l \leq k$, $j \neq l$ such that $f(y) = 1$, for each

$y \in V_3$, there does not exists c_i, $1 \leq i \leq m$, $f(c_{ij}) \neq 2$, $f(c_{il}) = 2$, $1 \leq j \leq k$, $1 \leq l \leq k$, $j \neq l$ such that $f(y) = 0$. Then the weight of RKDF of G is at least $2kq + ((2k - 2)(m - t))$ where $t < q$, which is greater than α, a contradiction. (Note that t refers to the number of consecutive vertices $c_i \in V_1$, $1 \leq i \leq t$ such that $f(c_i) = 2$)

Note that when we assume other values for each $w \in W$, $f(w) = 0$, for each $y \in V_3$, there exists c_i, $1 \leq i \leq m$, $f(c_{ij}) \neq 2$, $f(c_{il}) = 2$, $1 \leq j \leq k$, $1 \leq l \leq k$, $j \neq l$ such that $f(y) = 2$, for each $a \in V_3$, there does not exists c_i, $1 \leq i \leq m$, $f(c_{ij}) \neq 2$, $f(c_{il}) = 2$, $1 \leq j \leq k$, $1 \leq l \leq k$, $j \neq l$ such that $f(a) = 1$, $f(a) = 2$, we still arrive at a contradiction. Therefore the vertices in Z are consecutive and $z \in Z$, $f(z) = 2$.

Claim 1C. If G has a RKDF f with weight at most α, then $w \in (V_1 \setminus Z)$, $Z = \{c_{11}, \ldots, c_{1k}, \ldots, c_{q1}, \ldots, c_{qk}\}$ such that for each $z \in Z$, $f(z) = 2$, $f(w) = 0$.

Proof. Case 1: Suppose that there exist at least k vertices such that $U = \{u_1, \ldots, u_k\} \subseteq (V_1 \setminus Z)$ such that $f(u_i) = 2$, $1 \leq i \leq k$.

Recall that $y \in N(w)$ such that $w \in (V_1 \setminus (Z \cup U))$, $f(y) \neq 0$.

Case 1.1: Let $t = q + 1$, $U = \{c_{t1}, \ldots, c_{tk}\}$ such that for each $u \in U$, $f(u) = 1$, $U \cap Z = \emptyset$, for each $w \in (V_1 \setminus (Z \cup U))$, $f(w) = 0$, for each $y \in N(w)$ such that $y \in V_3$, $w \in Z$, $f(y) = 0$, for each $y \in N(w)$ such that $y \in V_3$, $w \in U$, $f(y) = 0$ and for each $y \in V_3$, $y \in N(w)$, $w \in (V_1 \setminus (Z \cup U))$, $f(y) = 1$. Then the weight of RKDF of G is $2kq + k + ((2k - 2)(m - q - 1))$, which is greater than α, a contradiction. Note that if we assume other values for each $u \in U$, $f(u) = 2$, $U \cap Z = \emptyset$, for each $w \in (V_1 \setminus (Z \cup U))$, $f(w) = 0$, for each $y \in N(w)$ such that $y \in V_3$, $w \in Z$, $f(y) = 1$ or $f(y) = 2$, for each $y \in N(w)$ such that $y \in V_3$, $w \in U$, $f(y) = 1$ or $f(y) = 2$ and for each $y \in V_3$, $y \in N(w)$, $w \in (V_1 \setminus (Z \cup U))$, $f(y) = 2$, we still arrive at a contradiction.

Case 2: Suppose that there exist at least one vertex $u \in (V_1 \setminus Z)$ such that $f(u) = 2$.

Recall that $y \in N(v)$ such that $v \in (V_1 \setminus (Z \cup \{u\}))$, $f(y) \neq 0$.

Case 2.1: For each $v \in (V_1 \setminus (Z \cup \{u\}))$, $f(v) = 0$, for each $y \in V_3$, $y \in N(z)$, $z \in Z$, $f(y) = 0$ and for each $y \in V_3$, $y \in N(w)$, $w \in (V_1 \setminus Z)$, $f(y) = 1$. Then the weight of RKDF of G is $2kq + 2 + ((2k - 2)(m - q))$, which is greater than α, a contradiction. Note that if we assume other values for vertices in Case 2.1, we still arrive at a contradiction.

Case 3: Suppose that there exists a vertex such that $w \in (V_1 \setminus Z)$ such that $f(w) = 1$.

Case 3.1: For each $v \in (V_1 \setminus (Z \cup \{u\}))$, $f(v) = 0$, for each $y \in V_3$, $y \in N(w)$, $w \in Z$, $f(y) = 0$ and for each $a \in V_3$, $a \in N(w)$, $a \in (V_1 \setminus Z)$, $f(a) = 1$. Then the weight of RKDF of G is $2kq + 1 + ((2k - 2)(m - q))$, which is greater than α, a contradiction. Note that if we assume other values for vertices in Case 3.1, we still arrive at a contradiction. Therefore for any vertex $w \in (V_1 \setminus Z)$ such that $f(w) = 0$. Thus $\Sigma_{u \in V_1} f(u) = 2kq$.

Claim 1D. If G has a RKDF f with weight at most α, then $u \in N(w)$, $w \in (V_1 \setminus Z)$, $Z = \{c_{11}, \ldots, c_{1k}, \ldots, c_{q1}, \ldots, c_{qk}\}$, $f(u) = 1$, and $v \in N(z)$, $z \in Z$, $Z = \{c_{11}, \ldots, c_{1k}, \ldots, c_{q1}, \ldots, c_{qk}\}$, $f(v) = 0$.

Proof. Case 1: There exists a vertex $y \in V_3$, $y \in N(z)$, $z \in Z$, $f(y) = 1$ and for each $a \in V_3$, $a \in N(w)$, $w \in (V_1 \setminus Z)$, $f(a) = 1$ (Recall that $f(a) = 1$ or $f(a) = 2$).

Then the weight of RKDF of G is $2kq + ((2k-2)(m-q)) + 1$, which is greater than α, a contradiction. If we assume other values for vertices in Case 1, we still arrive at a contradiction. Therefore $y \in V_3$, $y \in N(z)$, $z \in Z$, $f(y) = 0$. We know that $a \in V_3$, $a \in N(w)$, $w \in (V_1 \setminus Z)$, $Z = \{c_{11}, \ldots, c_{1k}, \ldots, c_{q1}, \ldots, c_{qk}\}$, $f(a) \neq 0$.

Case 2: There exist a vertex $y \in V_3$, $y \in N(w)$, $w \in (V_1 \setminus Z)$, $f(y) = 2$, for each $a \in N(c)$, $a \in V_3$, $c \in (V_1 \setminus (Z \cup \{c\}))$, $f(a) = 1$. Then the weight of RKDF of G is $2kq + ((2k-2)(m-q-1)) + 2$, which is greater than α, a contradiction. Therefore $y \in V_3$, $y \in N(w)$, $w \in (V_1 \setminus Z)$, $f(y) = 1$. Thus $\Sigma_{y \in V_3} f(y) = (2k-2)(m-q)$.

Since each $c_{ij} \in V_1$ has exactly three neighbors in X, clearly there exists kq vertices of $c_{ij} \in V_1$ with weight 2 such that $(\bigcup_{f(c_{ij})=2} N_G(c_{ij}) \cap X) = X$. Consequently, $\mathcal{C}' = \{c_i \mid f(c_{ij}) = 2\}$ is an exact-3-cover of $\mathcal{C}$.

Therefore, the Roman k-domination problem is NP-complete on split graphs.   □

3   Roman k-Domination on Some Subclasses of Bipartite Graphs

In this section, we show that finding RKDF on star-convex bipartite graphs and comb-convex bipartite graphs are NP-complete. Further, we show that finding RKDF on bipartite chain graphs is polynomial-time solvable.

3.1   Roman k-Domination on Star-Convex Bipartite Graphs

In this section, we show that the decision version of the Roman k-domination problem is NP-complete for star-convex bipartite graphs by giving a polynomial time reduction from Exact-3-Cover (X3C).

Theorem 2. *For star-convex bipartite graphs, the Roman k-domination problem is NP-complete.*

The proof is omitted due to space constraint.

3.2   Roman k-Domination on Comb-Convex Bipartite Graphs

In this section, we show that the decision version of the Roman k-domination problem is NP-complete for comb-convex bipartite graphs by giving a polynomial time reduction from Exact-3-Cover (X3C).

Theorem 3. *For comb-convex bipartite graphs, the Roman k-domination problem is NP-complete.*

The proof is omitted due to space constraint.

3.3   Roman k-Domination on Bipartite Chain Graphs

It is known from Theorem 1, that for P_5-free graphs finding RKDF is NP-complete. A natural question to ask is the complexity of RKDF on P_5-free bipartite graphs. In this section, we present a polynomial-time algorithm for finding RKDF on bipartite chain graphs. A bipartite graph $G(X, Y, E)$ is called a bipartite chain graph if and only if the neighborhood of the vertices in X can be linearly ordered with respect to inclusion (subset or equal) and the neighborhood of the vertices in Y can also be linearly ordered with respect to inclusion (subset or equal).

We show that for any fixed $k \in \mathbb{Z}^+$, finding the minimum Roman k-domination function is polynomial-time solvable for bipartite chain graphs. Let $G(X, Y, E)$ be a bipartite chain graph such that $X = (x_1, \ldots, x_n)$, $Y = (y_1, \ldots, y_n)$ such that $d_G(x_1) \geq d_G(x_2) \geq \ldots \geq d_G(x_n)$ and $d_G(y_1) \geq d_G(y_2) \geq \ldots d_G(y_m)$. It is known that $N_G(x_1) \supseteq N_G(x_2) \supseteq \ldots \supseteq N_G(x_n)$ and $N_G(y_1) \supseteq N_G(y_2) \supseteq \ldots \supseteq N_G(y_m)$. Let l be the number of vertices in G whose degree is less than k.

Theorem 4. *For $G(X, Y, E)$ be a bipartite chain graph such that $X = (x_1, \ldots, x_n)$, $Y = (y_1, \ldots, y_n)$ such that $d_G(x_1) \geq d_G(x_2) \geq \ldots \geq d_G(x_n)$ and $d_G(y_1) \geq d_G(y_2) \geq \ldots d_G(y_m)$, then*

$$\gamma_{kR}(G) = \begin{cases} 4k + l & \text{if } G \text{ has at least } 3k \text{ vertices in } X \text{ of degree at least } k \\ & \text{and } G \text{ has at least } 3k \text{ vertices in } Y \text{ of degree at least } k \\ 2k + l + r & \text{if } G \text{ has at least } 2k \text{ vertices in } X \text{ of degree at least } k \\ & \text{and } G \text{ has at least } k \text{ vertices in } Y \text{ of degree at least } k, \\ & \text{where } r \text{ represents number of vertices in } y_{k+1}, \ldots, y_m \\ & \text{having degree at least} k \text{ or vice versa} \\ n + m & \text{otherwise.} \end{cases}$$

Proof. Case 1: If G has at least $3k$ vertices in X of degree at least k and G has at least $3k$ vertices in Y of degree at least k, then $\gamma_{kR} = 4k + l$. Since $x_1, \ldots, x_{3k}, y_1, \ldots, y_{3k}$ has degree at least k, then $Z = \{x_1, \ldots, x_k, y_1, \ldots, y_k\}$ such that $z \in Z$, $f(z) = 2$. There are k vertices in $Z \cap X$, there are $2k$ vertices $y_i \in Y$, $k + 1 \leq i \leq m$ such that $d_G(y_i) \geq k$, y_i is adjacent to $Z \cap X$. Hence $f(y_i) = 0$, $k + 1 \leq i \leq m$. Similarly, there are k vertices in $Z \cap Y$, there are $2k$ vertices $x_i \in X$, $k + 1 \leq i \leq n$ such that $d_G(x_i) \geq k$, x_i is adjacent to $Z \cap Y$. Hence $f(x_i) = 0$, $k + 1 \leq i \leq m$. The vertices whose degree is strictly less than k do not have k neighbors labeled 2, hence they are labeled 1.

Case 2: If G has at least $2k$ vertices in X of degree at least k and G has at least k vertices in Y of degree at least k or vice versa, then $\gamma_{kR} = 2k + l + r$.

Without loss of generality assume that the number of vertices of degree k in X is more than the number of vertices of degree k in Y. It is known that $Z = \{x_1, \ldots, x_{2k}, y_1, \ldots, y_k\}$ are of degree at least k. Since the number of vertices in Y having degree at least k is at least $2k$ and at most $3k - 1$, and the number of vertices in X having degree at least K is at least k and at most $3k-1$, k vertices in Y say $y_1, \ldots, y_k$ such that $f(y_i) = 2$, $1 \leq i \leq k$. Therefore for all $x \in X$, $d_G(x) \geq k$, $f(x) = 0$. Since the number of $y_i \in Y$, $k+1 \leq i \leq m$ having degree at least k is at most $2k-1$, then all $y_i \in Y$, $k+1 \leq i \leq m$ having degree at least k are labeled 1. The vertices whose degree is strictly less than k do not have k neighbors labeled 2, hence they are labeled 1. The number of vertices in y having label 2 are $2k$, and the rest of the vertices in y having label 1 and degree at least k are r. Hence the minimum Roman k-domination function of G is $\gamma_{kR}(G) = 2k + l + r$.

Case 3: The number of vertices of degree at least k in X is at most $2k-1$ or the number of vertices of degree at least k in Y is at most $k-1$ or vice versa. Then all the vertices in $V(G)$ are labeled 1.   □

4   W-Hardness Results for Roman Domination on Split Graphs

In this section, we shall analyze the complexity of the parameterized version of Roman domination problem. By Roman domination, we mean RKDF when $k = 1$.

The parameterized version of Roman domination problem (PRD) with solution size k as the parameter is defined below:

PRD (G, k)

Instance: A graph G. **Parameter:** A positive integer α.

Question: Does there exist a function $f : V(G) \to \{0, 1, 2\}$ such that for every vertex u for which $f(u) = 0$ is adjacent to at least one vertex v for which $f(v) = 2$ and $\Sigma_{u \in V(G)} f(u) \leq \alpha$?

Theorem 5. *PRD is W[2]-hard on split graphs.*

Proof. We prove this by giving a reduction from the dominating set problem in general graphs. Given an instance of (G, k) of dominating set, we construct a corresponding instance of split graph $(H, \alpha = 2k)$ as follows; $V(H) = W \cup X \cup Y$, $W = \{w_i \mid u_i \in V\} \cup X = \{x_i \mid u_i \in V\} \cup Y = \{y_i \mid u_i \in V\}$, and $E(H) = \{\{w_i, x_j\}, \{w_i, y_j\}, \{x_i, w_j\}, \{y_i, w_j\} \mid \{u_i, u_j\} \in E(G)\} \cup \{\{w_i, x_i\}, \{w_i, y_i\} \mid 1 \leq i \leq n\} \cup \{\{w_i, w_j\} \mid 1 \leq i < j \leq n\}$. We can observe that H is a split graph with $H[W]$ as a clique and $H[X \cup Y]$ as an independent set.

We show that G has a dominating set D of size k if and only if H has a RKDF with weight $\alpha = 2k$. Clearly for each vertex $u_i \in D$, $f(w_i) = 2$, for the vertices $u_j \notin D$, $w_j \in W$, $f(w_j) = 0$ and for each vertex $z \in X \cup Y$, $f(z) = 0$. We obtain a RKDF of weight $2k$ in H. Conversely, by our construction, we know that a vertex $w_i \in W$ such that $|N_G^I(w_i)| \geq 4$, whereas for the corresponding $x_i \in X$, $|N(x_i)| \geq 2$ and $y_i \in Y$, $|N(y_i)| \geq 2$. Suppose that a subset of vertices from $X \cup Y$ has weight 2. Then $w \in W$, $f(w) = 0$. Further, $2n - k$ vertices in $X \cup Y$ should have weight at least 1. Thus $\alpha \geq 2k + n + n - k = 2n + k$ which is a contradiction that $\alpha = 2k$. Similarly, if a subset of vertices from X has weight 2 or a subset of vertices from Y has weight 2, we obtain a contradiction for α. Thus, a set of k vertices in W have the corresponding k vertices of G is in D.

This proves that D is a dominating set of size k for G and establishes the theorem. □

Corollary 1. *PRD is W[2]-hard on chordal graphs.*

Given the ordering of vertices in bipartite chain graphs. Since we visit each vertex exactly once, we can find a minimum Roman k-dominating set in linear time.

5   Approximation Hardness for RKDF

In this section, we prove that split graphs do not admit any constant-time approximation algorithm for MRKDF. To show the hardness result for finding a minimum weight Roman k-dominating function (MRKDP), we provide an approximation ratio preserving reduction from the MIN SET COVER problem. Minimum set cover problem: Let X be any non-empty set and C be a family of subsets of X. For the set system (X, C), a set $C' \subseteq C$ is called a cover of X, if every element of X belongs to at least one element of C. The MIN SET COVER problem is to find a minimum cardinality cover of X for a given set system (X, C). The following result is proved in [13].

Theorem 6 [13]. *The MIN SET COVER problem for the input instance (X, C) does not admit a $(1 - \epsilon) \ln |X|$-approximation algorithm for any $\epsilon > 0$ unless $NP \subseteq DTIME(|X|^{O(\log \log |X|)})$. Furthermore, this inapproximability result holds for the case when the size of the input collection C is no more than the size of the set X.*

Theorem 7. *For a fixed integer ρ, MRKDP on split graphs does not admit a ρ-approximation algorithm, unless $P = NP$.*

Proof. In order to prove the theorem, we propose the following approximation-ratio preserving reduction. Let $X = \{x_1, x_2, \ldots, x_p\}$ and $C = \{c_1, c_2, \ldots, c_q\}$ be an instance of the MIN SET COVER problem. From this, with similar arguments as in Theorem 1, we construct an instance $G(K, I, E)$ of MRKDP for split graphs. We state the following:

Claim 1: MIN SET COVER instance (X, C) has a cover of cardinality m if and only if G has a RKDF of size $(2mk) + ((2k - 2)(q - m))$.

Proof. The proof is similar to the claim that is Exact-3-cover $(X, \mathcal{C})$ if and only if G has a RKDF with weight at most $\alpha = (2kq) + ((2k - 2)(m - q))$ of Theorem 1, and details are omitted due to space constraint.

If α is MRKDF of G and C' is a minimum set cover of X for the set system (X, C), then $\alpha = (2mk) + ((2k - 2)(q - m))$. Suppose that MRKDF can be approximated within a ratio of ρ, where ρ is a fixed integer, by using some approximation algorithm A, that runs in polynomial time. Then the algorithm SET-COVER-APPROX presented in Algorithm 1 constructs a solution for the MIN SET COVER problem.

Algorithm 1. *SET-COVER-APPROX(X, C)*

1: **Input**: A set X and a collection C of subsets of X.
2: **Output**: A cover C^* of X.
3: Construct the graph G
4: Compute a RKDF g on G by using algorithm A
5: Construct a cover C^* of X from the weight of RKDF of G which is α (as illustrated in the proof of Claim 1 in Theorem 7)
6: return C^*

Clearly, SET-COVER-APPROX runs in polynomial time. Let R be the RKDF of G obtained from Algorithm A and let R^* be the optimal RKDF of G. Since RKDF algorithm A admits ρ-approximation algorithm, $\frac{R}{R^*} = \rho$. This proves that the algorithm SET-COVER-APPROX approximates the set cover of X within ratio ρ', where $1 \leq \rho' \leq \rho$. By Theorem 6, it is known that set cover does not admit a $(1 - \epsilon) \ln |X|$-approximation algorithm for any $\epsilon > 0$ unless $NP \subseteq DTIME(|X|^{O(\log \log |X|)})$. However, it contradicts Theorem 6 which says MIN SET COVER for any arbitrary instance (X, C) does not admit a $(1 - \epsilon) \ln |X|$-approximation algorithm for any $\epsilon > 0$ unless NP $\subseteq$ DTIME $(|X|^{O(\log \log |X|)})$.

$\square$

Concluding Remarks and Further Research: In this paper, we have presented NP-Hardness result for split graphs, star-convex and comb-convex bipartite graphs. We have presented a polynomial-time algorithm for RKDF on bipartite chain graphs. We have also presented approximation and parameterized complexity results for RKDF. The natural direction for further research is to explore these graph classes in other variants of the Roman dominating function such as the global Roman dominating function and double Roman dominating function.

References

1. Domke, G.S., Hedetniemi, S.T., Laskar, R.C., Fricke, G.: Relationships between integer and fractional parameters of graphs. In: Graph Theory, Combinatorics, and Applications, Proceedings of the Sixth Quadrennial International Conference on the Theory and Applications of Graphs, vol. 1, pp. 371–387 (1991)
2. Brešar, B., Henning, M.A., Rall, D.F.: Rainbow domination in graphs. Taiwanese J. Math. **12**(1), 213–225 (2008)
3. Cockayne, E.J., Dreyer, P.A., Jr., Hedetniemi, S.M., Hedetniemi, S.T.: Roman domination in graphs. Discrete Math. **278**(1–3), 11–22 (2004)
4. Panda, B.S., Goyal, P.: Hardness results of global roman domination in graphs. In: Mudgal, A., Subramanian, C.R. (eds.) CALDAM 2021. LNCS, vol. 12601, pp. 500–511. Springer, Cham (2021). https://doi.org/10.1007/978-3-030-67899-9_39

5. Beeler, R.A., Haynes, T.W., Hedetniemi, S.T.: Double roman domination. Discrete Appl. Math. **211**, 23–29 (2016)

6. Wang, C.X., Yang, Y., Wang, H.J., Xu, S.J.: Roman {k}-domination in trees and complexity results for some classes of graphs. J. Comb. Optim., pp. 1–13 (2021)

7. Kammerling, K., Volkmann, L.: Roman k-domination in graphs. J. Korean Math. Soc. **46**(6), 1309–1318 (2009)

8. Liedloff, M., Kloks, T., Liu, J., Peng, S.-L.: Roman domination over some graph classes. In: Kratsch, D. (ed.) WG 2005. LNCS, vol. 3787, pp. 103–114. Springer, Heidelberg (2005). https://doi.org/10.1007/11604686_10

9. Liu, C.H., Chang, G.J.: Roman domination on strongly chordal graphs. J. Comb. Optim. **26**(3), 608–619 (2013)

10. Padamutham, C., Palagiri, V.S.R.: Algorithmic aspects of roman domination in graphs. J. Appl. Math. Comput. **64**(1), 89–102 (2020). https://doi.org/10.1007/s12190-020-01345-4

11. Ahangar, H.A., Chellali, M., Sheikholeslami, S.M.: On the double roman domination in graphs. Discrete Appl. Math. **232**, 1–7 (2017)

12. Garey, M.R., Johnson, D.S.: Computers and Intractability, vol. 174. Freeman San Francisco, San Francisco (1979)

13. Feige, U.: A threshold of ln n for approximating set cover. J. ACM (JACM) **45**(4), 634–652 (1998)

Parameterized Complexity

On the Parameterized Complexity of Compact Set Packing

Ameet Gadekar[(⊠)]

Aalto University, Espoo, Finland
`ameet.gadekar@aalto.fi`

Abstract. The SET PACKING problem is, given a collection of sets $\mathcal{S}$ over a ground set $\mathcal{U}$, to find a maximum collection of sets that are pairwise disjoint. The problem is among the most fundamental NP-hard optimization problems that have been studied extensively in various computational regimes. The focus of this work is on parameterized complexity, PARAMETERIZED SET PACKING (PSP): Given $r \in \mathbb{N}$, is there a collection $\mathcal{S}' \subseteq \mathcal{S} : |\mathcal{S}'| = r$ such that the sets in $\mathcal{S}'$ are pairwise disjoint? Unfortunately, the problem is not fixed parameter tractable unless W[1] = FPT, and, in fact, an "enumerative" running time of $|\mathcal{S}|^{\Omega(r)}$ is required unless the exponential time hypothesis (ETH) fails. This paper is a quest for tractable instances of SET PACKING from parameterized complexity perspectives. We say that the input $(\mathcal{U}, \mathcal{S})$ is "compact" if $|\mathcal{U}| = f(r) \cdot \Theta(\text{poly}(\log |\mathcal{S}|))$, for some $f(r) \geq r$. In the COMPACT PSP problem, we are given a compact instance of PSP. In this direction, we present a "dichotomy" result of PSP: When $|\mathcal{U}| = f(r) \cdot o(\log |\mathcal{S}|)$, PSP is in FPT, while for $|\mathcal{U}| = r \cdot \Theta(\log(|\mathcal{S}|))$, the problem is W[1]-hard; moreover, assuming ETH, COMPACT PSP does not admit $|\mathcal{S}|^{o(r/\log r)}$ time algorithm even when $|\mathcal{U}| = r \cdot \Theta(\log(|\mathcal{S}|))$. Although certain results in the literature imply hardness of compact versions of related problems such as SET r-COVERING and EXACT r-COVERING, these constructions fail to extend to COMPACT PSP. A novel contribution of our work is the identification and construction of a gadget, which we call Compatible Intersecting Set System pair, that is crucial in obtaining the hardness result for COMPACT PSP. Finally, our framework can be extended to obtain improved running time lower bounds for COMPACT r-VECTORSUM.

1 Introduction

Given a graph $G = (V, E)$, the problem of finding a maximum-size subset of disjoint edges (matching) is tractable, but its generalization to hypergraphs, even when the edge length is 3, is NP-hard. This general problem is known as the **Hypergraph Matching** problem. The hyper-graph $H = (W, F)$ can be equivalently viewed as a set system $(\mathcal{U}, \mathcal{S})$, where the universe (or the ground set) $\mathcal{U}$ corresponds to the vertex set W and $\mathcal{S}$ corresponds to the collection of hyperedges F. Then finding a maximum matching in H is equivalent to finding maximum number of pairwise disjoint sets (packing) in $\mathcal{S}$. Hence the **Hypergraph Matching** problem is also known

© The Author(s), under exclusive license to Springer Nature Switzerland AG 2023
C.-C. Lin et al. (Eds.): WALCOM 2023, LNCS 13973, pp. 359–370, 2023.
https://doi.org/10.1007/978-3-031-27051-2_30

as the SET PACKING problem, which is a fundamental problem in combinatorial optimization with numerous applications [24]. While this problem captures many classical combinatorial problems such as maximum independent set (or maximum clique), k-dimensional matching and also, some graph packing problems [9,14], this generalization also makes it intractable in several regimes. One computational regime in which SET PACKING has been explored extensively is approximation algorithms. Since SET PACKING generalizes the maximum independent set problem [1], it inherits the inapproximability of the latter problem [15]. This immediately implies that the trivial approximation of picking simply one set in the packing is roughly the best to hope for. Furthermore, approximations in terms of $|\mathcal{U}|$ are also not hopeful since the result also implies inapproximability bound of $|\mathcal{U}|^{1/2-\epsilon}$, which is matched by [13]. To combat these intractabilities, various restrictions of SET PACKING have been studied. Particularly, a restriction where the size of the sets in $\mathcal{S}$ is bounded by some integer k, which is known as k-Set Packing, is also a well-studied problem. However, k-Set Packing captures the independent set problem in bounded degree graphs, which again is a notoriously hard problem to approximate beyond the "trivial" bound [2,3]. While [16] improves the lower bound for k-Set Packing to $\Omega(k/\ln k)$, the best known approximation is $(k+1+\epsilon)/3$ [8,10], yielding a logarithmic gap between the bounds. Besides approximation algorithms, SET PACKING has also been studied from the parameterized complexity perspectives (with the standard parameter on the size of an optimal packing solution). In this problem, known as the PARAMETERIZED SET PACKING (PSP) problem, we are given an instance $(\mathcal{U}, \mathcal{S}, r)$ and the task is to decide if there exists a packing of size r. Unfortunately, even PSP remains intractable and is, actually, W[1]-complete [12]. In fact, *Exponential Time Hypothesis* (ETH) implies that the trivial *enumerative algorithm* running in $O^*(|\mathcal{S}|^r)$ time to find an r-packing is asymptotically our best hope [11]. The algorithmic outlook for PSP worsens further due to [7], which rules out $o(r)$-FPT-approximation algorithm assuming the *Gap Exponential Time Hypothesis* (Gap-ETH). Assuming a weaker hypothesis of FPT $\neq$ W[1], very recently [22] showed that there is no FPT algorithm for PSP problem that finds a packing of size $r/r^{1/H(r)}$, for any increasing function $H(\cdot)$, when given a promise that there is an r-packing in the instance. Thus, the flurry of these negative results make it likely that SET PACKING problem is intractable in all computational regimes.

In this paper, we consider PSP on *compact* instances. We say that an instance $(\mathcal{U}, \mathcal{S}, r)$ of PSP is compact if $|\mathcal{U}| = f(r) \cdot \Theta(\text{poly}(\log |\mathcal{S}|))$, for some function $f(r) \geq r$, that is, the universe is relatively small compared to the number of sets.[1] Besides the algorithmic motivation, compact instances have recently been used as an "intermediate step" to prove FPT inapproximability results of the (non-compact) classical problems (see, e.g., [4,18] where the compact instances were used in proving FPT-inapproximability of the k-EvenSet and Dominating Set).

---

[1] In fact there is another way to define compactness: when $|\mathcal{S}| = f(r) \cdot \Theta(\text{poly}(\log |\mathcal{U}|))$. However in this case, the *enumerative algorithm* running in time $O^*(|\mathcal{S}|^r)$ is already fixed parameter tractable [6]. Thus, the interesting case is when the universe is compact, which is the case we will be focusing on.

We hope that studying COMPACT PSP would lead to some ideas that would be useful in proving tight FPT inapproximability of PSP (that is, to weaken the Gap-ETH assumption used in [7]).

1.1 Our Results

Our main result is the following dichotomy of PARAMETERIZED SET PACKING.

Theorem 1 (Dichotomy). *The following dichotomy holds for* PSP.

○ *If* $|\mathcal{U}| = f(r) \cdot o(\log |\mathcal{S}|)$, *for any* f, *then* PSP *is in* FPT.
○ PSP *remains* W[1]*-hard even when* $|\mathcal{U}| = r \cdot \Theta(\log |\mathcal{S}|)$.

The algorithmic result follows from well-known dynamic programming based algorithms [5,11] that run in time $O^*(2^{|\mathcal{U}|})$, and observing that this running time is fixed parameter tractable [6] when $|\mathcal{U}| = f(r)o(\log |\mathcal{S}|)$. The main contribution of our work is the W[1]-hardness of PSP even when $|\mathcal{U}| = r \cdot \Theta(\log |\mathcal{S}|)$. Towards this, we show an FPT-reduction from SUBGRAPH ISOMORPHISM (SGI) to COMPACT PSP. The hardness result follows since SGI is W[1]-hard. In fact, our hardness result can be strengthened assuming Exponential Time Hypothesis (ETH) [11] to obtain the following result.

Theorem 2. COMPACT PSP *requires time* $|\mathcal{S}|^{\Omega(r/\log r)}$ *even when* $|\mathcal{U}| = r \cdot \Theta(\log |\mathcal{S}|)$, *unless ETH fails.*

The result of Theorem 2 follows from the ETH-hardness result of SGI due to [20], and from the fact that the hardness reduction of Theorem 1 is parameter preserving up to a multiplicative constant. Note that since PSP can be trivially solved by enumeration in time $O^*(|\mathcal{S}|^r)$, the above result says that, even for the compact instances this is essentially our best hope, up to a log factor in the exponent. An interesting consequence of the dichotomy theorem coupled with Theorem 2 is the fact that, as soon as instances get asymptotically smaller, not only we beat the enumerative algorithm, but we actually obtain an FPT algorithm. We would like to remark that the universe size in Theorem 2 is tight (up-to $\log r$ factor) since having $|\mathcal{U}| = o(r/\log r) \cdot \Theta(\log |\mathcal{S}|)$ would already allow $|\mathcal{S}|^{o(r/\log r)}$ time algorithm. Further, note that for W[1]-hardness, it is sufficient to have $|\mathcal{U}| = f(r)\Theta(\log |\mathcal{S}|)$, for some f, since we can add $f(r) - r$ new sets each with a unique dummy element and inflate the parameter to $f(r)$. However, this is not true for ETH based running time lower bounds as such inflation fail to transfer the lower bounds asymptotically.

Finally, we extend our construction framework (Theorem 3) to improve the running time lower bound (matching the trivial upper bound up to a log factor in the exponent) for the compact version of r-VECTORSUM: Given a collection $\mathcal{C}$ of N vectors in $\mathbb{F}_2^d$, and a target vector $b \in \mathbb{F}_2^d$, r-VECTORSUM asks if there are r vectors in $\mathcal{C}$ that sum to b. COMPACT r-VECTORSUM is defined when $d = f(r) \cdot \Theta(\mathsf{poly}(\log N))$, for some $f(r) \geq r$.

Theorem 3. COMPACT r-VECTORSUM *requires time* $N^{\Omega(r/\log r)}$, *even when* $d = r \cdot \Theta(\log N)$, *unless ETH fails.*

The present bound of [4] rules out $N^{o(\sqrt{r})}$ time under ETH. Due to space constraints, the proof of this theorem is deferred to the full version.

1.2 Our Contributions and Comparison to Existing Work

In this section, we compare our contribution with existing works to highlight its significance. To our best knowledge, the compact version of combinatorial problems has not previously been formalized and investigated. However, several existing reductions already imply the hardness of compact version of some of the combinatorial problems. Here we review and compare the related results.

Our Contribution. As far as we know, there are no results showing W[1]-hardness of COMPACT PSP, and hence the corresponding dichotomy (Theorem 1). The key contribution of this paper is to show the hardness result for COMPACT PSP. On the way, we also show an ETH-based almost tight running time lower bound for COMPACT PSP, with tight (up-to $\log r$ factor) universe size $|\mathcal{U}| = r \cdot \Theta(\log |\mathcal{S}|)$. Interestingly, we show both of these results with a single FPT reduction. In addition, we extend our framework to improve the running time lower bounds for COMPACT r-VECTORSUM.

Next we survey some known hardness results for SET r-COVERING in the compact regime and argue their limitations in extending them to PSP. In particular, [17, Lemma 25] shows a reduction from SGI to a variant of SET r-COVERING called COMPACT EXACT, where we want to find an r-packing that is also a covering (in fact they show hardness for COMPACT SET. But a closer inspection of their construction shows that the intended set cover is also a packing). The high level idea of the construction is similar to ours: first assign each vertex of G a logarithm length binary pattern vector. Then, create two kinds of sets: V-sets that capture the mapping of the vertices and E-sets that capture the mapping of edges. The idea is to use the pattern vectors to create these sets so that there is an isomorphic copy of H in G if and only if there are $|V_H|$ many V-sets and $|E_H|$ many E-sets covering the universe exactly once. However, if we consider the Soundness (No case) proof of this reduction, then it crucially relies on the fact that no candidate solution can cover the entire universe exactly once. In fact, it is quite easy to find r sets that are mutually disjoint but do not form a cover. Therefore, it fails to yield hardness for COMPACT PSP. The heart of our construction lies in ensuring that in the No case, any r sets intersect. To this end, we construct a combinatorial gadget called Compatible Intersecting Set System (ISS) pair. This gadget is a pair of set systems $(\mathcal{A}, \mathcal{B})$ over a universe U that guarantees two properties: First, every pair of sets within each set system intersects, and second, for any set $a \in \mathcal{A}$, there exists $b \in \mathcal{B}$ such that a intersects every set in $\mathcal{B}$ except b. Further, we present a simple greedy algorithm that finds such compatible ISS pair $(\mathcal{A}, \mathcal{B})$ over a universe of size N, each having roughly $2^{\Omega(N)}$ sets. Note that this gadget, which we use to build our compact hard instance, also has a "compact" universe. While, on the other hand, [23] shows COMPACT SET is W[1]-hard using a reduction from k-CLIQUE to SET r-COVERING with $r = \Theta(k^2)$ and $|\mathcal{U}| = r^{3/2} \cdot \Theta(\log |\mathcal{S}|)$, but does not yield a tight ETH- based running time lower bound. In contrast, [21] shows such tight

ETH lower bound for COMPACT SET: requiring time $|\mathcal{S}|^{\Omega(r)}$, which can be easily modified to obtain a similar running time lower bound for COMPACT EXACT (by reducing from 1-IN-3-SAT, instead from 3-SAT).

1.3   Overview of Techniques

In this section we sketch the main ideas of our hardness proof of Theorem 1. To this end, we present a reduction from SGI, which asks, given a graph G on n vertices and another graph H with k edges, if there is a subgraph of G isomorphic (not necessarily induced) to H, with parameter k. The reduction produces an instance $\mathcal{I} = (\mathcal{U}, \mathcal{S}, r)$ of PSP in FPT time such that $r = \Theta(k)$ and $|\mathcal{U}| = \Theta(r \log |\mathcal{S}|)$. We remark that the classical reduction given in [12] also has parameter $r = \Theta(k)$, but $|\mathcal{U}|$ is linear in the size of G, which is the size of $|\mathcal{S}|$. Below, we attempt a reduction to construct a compact instance that, despite falling short of its goal, illustrates some of the key ideas of the actual hardness proof. This failed attempt also highlights the crucial properties of the gadget that are necessary for the correct reduction.

Our reduction constructs the instance $\mathcal{I} = (\mathcal{U}, \mathcal{S}, r)$ of PSP using a special set system gadget – which we call the Intersecting Set System (ISS) gadget. A set system $\mathcal{A} = (U_A, S_A)$ with M sets over N elements is called an (M, N)-Intersecting set system, if every pair $s^i, s^j \in S_A$ intersects (i.e., s^i, s^j has a non-empty intersection). We show how to efficiently construct an (M, N)-ISS $\mathcal{A} = (U_A, S_A)$ with $M = 2^{N-1}$. Let $U_A = \{1, 2, \cdots, N+1\}$. Then, for every subset $s \subseteq \{2, 3, \cdots, N+1\}$, add the set $s' := \{1\} \cup s$ to S_A. Note that $\mathcal{A}$ has a compact universe since $|U_A| = \log_2 M + 1 = \log_2 |S_A| + 1$, which is crucial in constructing a compact instance of PSP. We are now ready to present the reduction using this compact ISS gadget. Let the given instance of SGI be $\mathcal{J} = (G = (V_G, E_G), H = (V_H, E_H), k)$. Let $\ell := |V_H|$, $n := |V_G|$ and $m := |E_G|$. Further, let $V(G) = \{1, \cdots, n\}$. Note that $\ell \leq 2k$, since isomorphic sub-graph in G to H is not necessarily induced. Let $\mathcal{A} = (U_A, S_A)$ be the (M, N)-ISS gadget specified above with $N = \lceil \log n \rceil + 1$. Since $M \geq n$, assume $S_A = \{s^\alpha\}_{\alpha \in V(G)}$ by arbitrarily labeling sets in S_A and ignoring the sets $s^\alpha, \alpha > n$. We construct an instance $\mathcal{I} = (\mathcal{U}, \mathcal{S}, r)$ of COMPACT PSP as follows. For every $v \in V_H$, and $w \in N_H(v)$, let $\mathcal{A}_{v,w} = (U_{v,w}, S_{v,w})$ be a distinct copy of ISS $\mathcal{A}$ (that is, the universes $\{U_{v,w}\}_{w \in N_H(v)}$ are disjoint) with the same labeling of sets as that of S_A. Note that $\mathcal{A}_{v,w}$ and $\mathcal{A}_{w,v}$ are distinct copies of $\mathcal{A}$. Let $U_v := \cup_{w \in N(v)} U_{v,w}$. The universe $\mathcal{U}$ in $\mathcal{I}$ is defined as $\mathcal{U} = \cup_{v \in V_H} U_v$. Now for $\mathcal{S}$, we will construct two types of sets, that we call V-sets and E-sets. For every $\alpha \in V(G)$ and $v \in V_H$, add the set $S_{\alpha \mapsto v} := \cup_{w \in N(v)} s^\alpha_{v,w}$ to $\mathcal{S}$. These sets are referred as V-sets. For each edge $(\alpha, \beta) \in E_G$ and each edge $(v, w) \in E_H$, add the set $S_{(\alpha,\beta) \mapsto (v,w)} := \bar{s}^\alpha_{v,w} \cup \bar{s}^\beta_{w,v}$ to $\mathcal{S}$. These sets are called E-sets. Finally, setting the parameter $r = \ell + k$, concludes the construction of PSP instance $\mathcal{I} = (\mathcal{U}, \mathcal{S}, r)$. First, note that for the base ISS gadget $\mathcal{A} = (U_A, S_A)$, we have that $|U_A| = \Theta(\log n)$. Hence, $|\mathcal{U}| = \sum_{i \in [\ell]} \sum_{j \in [d(v_i)]} |U_A| = \Theta(k \log n)$, where as $|\mathcal{S}| = \Theta(mk + n\ell) = \Theta(n^2 k)$. Since $r = \Theta(k)$, we have $|\mathcal{U}| = \Theta(r \log |\mathcal{S}|)$, yielding a COMPACT PSP instance.

To illustrate the main ideas, we analyze the completeness and discuss how the soundness fails. In the completeness case, let us assume that there exists an injection $\phi : V_H \to V_G$ which specifies the isomorphic subgraph in G. Consider $T = \{S_{\phi(v) \mapsto v}\}_{v \in V_H} \bigcup \{S_{(\phi(v), \phi(w)) \mapsto (v,w)}\}_{(v,w) \in E_H}$, and note that $T \subseteq S$ due to ϕ. Notice that we have chosen $\ell + k$ sets from S. To see that T forms a packing, fix $v, w \in V_H$ such that $w \in N(v)$. Let $T \mid_{U_{v,w}}$ be the restriction of T on $U_{v,w}$ (Formally, $T \mid_{U_{v,w}} := \{t \cap U_{v,w} : t \in T\}$). Then note that $T \mid_{U_{v,w}}$ forms a packing since $T \mid_{U_{v,w}} = \{s_{v,w}^{\phi(v)}, \bar{s}_{v,w}^{\phi(v)}\}$.

For soundness, we show the proof for a simpler case. Suppose $T \subseteq S$ with $|T| = r$ is a packing with at most one V-set from each vertex of G. Further, assume $|T_V| = \ell$ and $|T_E| = k$, where T_V and T_E denote the V-sets and E-sets of T respectively. Finally, we also assume that T covers $\mathcal{U}$. Let $V_H = \{v_1, \cdots, v_\ell\}$. Relabel the sets in T_V as $T_V = \{T_V^i : \exists S_{\alpha \mapsto v_i} \in T_V, \text{ for some } \alpha \in V_G, v_i \in V_H\}$. Now consider $V' := \{\alpha \mid T_V^i = S_{\alpha \mapsto v_i}\} \subseteq V_G$, and relabel the vertices of V' as $V' = \{\alpha_1', \cdots, \alpha_\ell'\}$, where $\alpha_i' = \alpha$ such that $T_V^i = S_{\alpha \mapsto v_i}$. We claim that $G[V']$ is isomorphic to H with injection $\phi : V_H \to V_G$ given by $\phi(v_i) = \alpha_i'$. To this end, we show $(v_i, v_j) \in E_H \implies (\phi(v_i), \phi(v_j)) \in E_{G[V']}$. Note that since $\mathcal{A}_{v_i, v_j}$ is an ISS, $T_V \mid_{U_{v_i, v_j}} = T_V^i \mid_{U_{v_i, v_j}} = s_{v_i, v_j}^{\alpha_i'}$. Further, combining the fact that T is a packing covering $\mathcal{U}$ with the fact $|T_E| = k$, we have that $T_E \mid_{U_{v_i, v_j}} = \bar{s}_{v_i, v_j}^{\alpha_i'}$. Similarly, it holds that $T_V \mid_{U_{v_j, v_i}} = T_V^j \mid_{U_{v_j, v_i}} = s_{v_j, v_i}^{\alpha_j'}$, and hence $T_E \mid_{U_{v_j, v_i}} = \bar{s}_{v_j, v_i}^{\alpha_j'}$. But this implies that $S_{(\alpha_i', \alpha_j') \mapsto (v_i, v_j)} \in T_E$, which means $(\phi(v_i), \phi(v_j)) = (\alpha_i', \alpha_j') \in E_{G[V']}$, as desired. However, for the general case, we would require a gadget that enforces all the above assumptions in any candidate packing.

1.4   Open Problems

An interesting problem is to show COMPACT PSP also needs time $|S|^{\Omega(r)}$ even when $|\mathcal{U}| = r \cdot \Theta(\log |S|)$, assuming ETH. Another interesting direction is FPT approximating COMPACT PSP: Given a promise that there is an r-packing, is it possible to find a packing of size $\omega(1)$ in FPT time? Note that for the general PSP problem, there is no $o(r)$ FPT-approximation, assuming Gap-ETH. However, recent results [19, 22] use a weaker assumption of W[1] $\neq$ FPT but also obtain weaker FPT-inapproximibility. It is also interesting to show such hardness of approximation for COMPACT PSP.

2   Preliminaries

2.1   Notations

For $q \in \mathbb{N}$, denote by $[q]$, the set $\{1, \cdots, q\}$. For a finite set $[q]$ and $i \in [q]$, we overload '$+$' operator and denote by $i+1$ as the (cyclic) successor of i in $[q]$. Thus, the successor of q is 1 in $[q]$. All the logs are in base 2. For a graph $G = (V, E)$ and a vertex $v \in V$, denote by $N(v)$, the set of vertices adjacent to v. Further, $d(v)$ denotes the degree of v, i.e., $d(v) := |N(v)|$. For a finite universe U and $s \subseteq U$, denote by $\bar{s}$ as the complement of s under U, i.e., $\bar{s} := U \setminus s$. Similarly, for a family of sets $S = \{s_1, \cdots, s_M\}$ over U, we denote by $\mathsf{comp}(S) = \{\bar{s}_1, \cdots, \bar{s}_M\}$. Further, for a subset $s \subseteq U$ and a sub-universe $U' \subseteq U$, denote by $s \mid_{U'}$ as the restriction of s on sub-universe U', i.e., $s \mid_{U'} := s \cap U'$. Similarly, for a family of sets $S = \{s_1, \cdots, s_M\}$ over U, denote by $S \mid_{U'}$ as the restriction of every set of S on U', i.e., $S \mid_{U'} := \{s_1 \mid_{U'}, \cdots, s_M \mid_{U'}\}$. For a set system $A = (U_A, S_A)$, we denote the complement set system by $\bar{A} = (U_A, \mathsf{comp}(S_A))$. For $s, t \subseteq U$, we say s and t intersects if $s \cap t \neq \emptyset$.

For a background on parameterized complexity, please refer to [11,12].

2.2   Problem Definitions

Definition 1 (PARAMETERIZED SET PACKING (PSP)). *Given a collection of sets $S = \{S_1, \ldots, S_m\}$ over an universe $\mathcal{U} = \{e_1, \ldots, e_n\}$ and an integer r, the PSP problem asks if there is a collection of sets $S' \subseteq S$ such that $|S'| = r$ and, $S_i \cap S_j = \emptyset$ for every $S_i \neq S_j \in S'$. An instance of PSP is denoted as $(\mathcal{U}, S, r)$.*

COMPACT PSP is defined when the instances have $|\mathcal{U}| = f(r) \cdot \Theta(\mathsf{poly}(\log|S|))$, for some function $f(r) \geq r$.

Given two graphs $G = (V_G, E_G)$ and $H = (V_H, E_H)$, a homomorphism from H to G is a map $\phi : V_H \rightarrow V_G$ such that if $(v_i, v_j) \in E_H$ then $(\phi(v_i), \phi(v_j)) \in E_G$.

Definition 2 (SUBGRAPH ISOMORPHISM (SGI)). *Given a graph $G = (V_G, E_G)$ and a smaller graph $H = (V_H, E_H)$ with $|E_H| = k$, the SGI problem asks if there is an injective homomorphism from H to G. An instance of SGI is denoted as $(G = (V_G, E_G), H = (V_H, E_H), k)$.*

The parameterized version of SGI has parameter $\kappa = |E_H| = k$. Without loss of generality, we assume $|V_H| \leq 2k$, and every vertex of H has degree at most k.

3   Dichotomy of PSP

In this section we prove the hardness part of the dichotomy theorem (Theorem 1). First in Sect. 3.1, we identify the gadget and its associated properties that are crucial for the reduction. Then, using this gadget, in Sect. 3.2 we show an FPT-reduction from SGI to COMPACT PSP.

3.1   Compatible Intersecting Set System Pair

A set system $\mathcal{A} = (U_A, S_A)$ is called an (M, N)-Intersecting set system (ISS), if it contains M sets over N elements such that every pair $s, t \in S_A$ intersects.

Definition 3 (Compatible ISS pair). *Given two ISS $\mathcal{A} = (U, S_A)$ and $\mathcal{B} = (U, S_B)$ on a universe U, we say that $(\mathcal{A}, \mathcal{B})$ is a compatible ISS pair if there exists an efficiently computible bijection $f : S_A \rightarrow S_B$ such that*

- *(Complement partition) $\forall s \in S_A$, s and $f(s)$ forms a partition of U, and*
- *(Complement exchange) $\forall s \in S_A$, $\mathcal{A}_s := (U, (S_A \setminus \{s\}) \cup \{f(s)\})$ is an ISS.*

Since f is as bijection, we have $|S_A| = |S_B|$, and $\forall t \in S_B$, the set system $\mathcal{B}_t := (U, (S_B \setminus \{t\}) \cup \{f^{-1}(t)\})$ is also an ISS. Also, for $(s,t) \in (S_A, S_B)$ if $s \cup t = U$, then $t = f(s)$. The following lemma, whose proof is deferred to the full version, efficiently constructs a compatible (M, N)-ISS pair, which is a key ingredient in our hardness proof. The idea of the construction is simple: for every $N/2$-sized subset s of $[N]$, add s to $\mathcal{A}$ and $\bar{s}$ to $\mathcal{B}$.

Lemma 1. *For even $N \geq 2$, we can compute a compatible (M, N)-ISS pair $(\mathcal{A}, \mathcal{B})$ with $M \geq 2^{N/2-1}$ in time polynomial in M and N. Further, $\mathcal{B} = \bar{\mathcal{A}}$.*

3.2   Hardness of COMPACT PSP

Our hardness result follows from the following FPT-reduction from SGI that yields compact instances of PSP and the fact that SGI is W[1]-hard.

Theorem 4. *There is an FPT-reduction that, for every instance $\mathcal{I} = (G = (V_G, E_G), H = (V_H, E_H), k)$ of SGI with $|V_G| = n$ and $|E_G| = m$, computes $\mu = k!$ instances $\mathcal{J}_p = (\mathcal{U}_p, \mathcal{S}_p, r), p \in [\mu]$ of PSP with $|\mathcal{U}_p| = \Theta(k \log n)$, $|\mathcal{S}_p| = \Theta(n^2 k + mk)$, and $r = \Theta(k)$, such that there is a subgraph of G isomorphic to H if and only if there is an r-packing in at least one of the instances $\{\mathcal{J}_p\}_{p \in [\mu]}$.*

Proof. The construction follows the approach outlined in Sect. 1.3. Let $V_G = \{1, \cdots, n\}$. Let $(\mathcal{A}, \bar{\mathcal{A}})$ be the compatible (M, N)-ISS pair given by Lemma 1, for $N = 2\lceil \log(n+1) \rceil + 2$. We call $\mathcal{A} = (U_A, S_A)$ and $\bar{\mathcal{A}} = (U_A, \text{comp}(S_A))$ as the base ISS gadgets. Further, assume an arbitrary ordering on $S_A = \{s^1, \cdots, s^M\}$. Since $M \geq 2^{N/2-1} > n$, every $\alpha \in V_G$ can be identified by the set $s^\alpha \in S_A$ corresponding to the index $\alpha \in [M]$. For each ordering $p : V_H \to [\ell]$, create an instance $\mathcal{J}_p = (\mathcal{U}_p, \mathcal{S}_p, r)$ of COMPACT PSP as follows. Rename the vertices of V_H as $\{v_1, \cdots, v_\ell\}$ with $v_i := v \in V_H$ such that $p(v) = i$. For each $v_i \in V_H$, create a collection $\mathcal{C}_{v_i}$ of $d(v_i) + 1$ many different copies of base ISS gadget $\mathcal{A}$ (i.e., each has its own distinct universe) as: $\mathcal{C}_{v_i} := \{\mathcal{A}_{v_i,0}, \{\mathcal{A}_{v_i,w}\}_{w \in N(v_i)}\}$, where $\mathcal{A}_{v_i,0} = (U_{v_i,0}, S_{v_i,0})$ and $\mathcal{A}_{v_i,w} = (U_{v_i,w}, S_{v_i,w})$. Let $U_{v_i} = \cup_{w \in N(v)} U_{v_i,w}$. For each $\mathcal{C}_{v_i}$, let $U_{\mathcal{C}_i} = U_{v_i,0} \cup U_{v_i}$. Define the universe $\mathcal{U}_p$ of $\mathcal{J}_p$ as $\mathcal{U}_p = \bigcup_{i \in [\ell]} U_{\mathcal{C}_i}$.

The sets in $\mathcal{S}_p$ are of two types: V-sets and E-sets as defined below. For $\alpha \in V_G$ and $v \in V_H$, denote by $S_v^\alpha = \cup_{w \in N(v)} s_{v,w}^\alpha$. Recall that for $\alpha \in V_G$ and $(v,w) \in E_H$, the set $s_{v,w}^\alpha$ is the αth set in $S_{v,w}$ of ISS $\mathcal{A}_{v,w} = (U_{v,w}, S_{v,w})$.

V-sets: For each $\alpha \in V_G$, for each $v_i \in \{v_1, \cdots, v_{\ell-1}\}$, and for each $\beta \in V_G, \beta > \alpha$, add a set $S_{\alpha \mapsto v_i, \beta}$ to $\mathcal{S}_p$ such that

$$S_{\alpha \mapsto v_i, \beta} := s_{v_i,0}^\alpha \bigcup S_{v_i}^\alpha \bigcup \bar{s}_{v_{i+1},0}^\beta$$

Further, for each $\alpha \in V_G$, and for each $\beta \in V_G, \beta < \alpha$, add a set $S_{\alpha \mapsto v_\ell, \beta}$ to $\mathcal{S}_p$,

$$S_{\alpha \mapsto v_\ell, \beta} := s_{v_\ell,0}^\alpha \bigcup S_{v_\ell}^\alpha \bigcup \bar{s}_{v_1,0}^\beta$$

E-sets: For each edge $(\alpha, \beta) \in E_G$ and each edge $(v_i, v_j) \in E_H$, add a set $S_{(\alpha,\beta) \mapsto (v_i,v_j)}$ to $\mathcal{S}_p$ such that

$$S_{(\alpha,\beta) \mapsto (v_i,v_j)} := \bar{s}_{v_i,v_j}^\alpha \bigcup \bar{s}_{v_j,v_i}^\beta$$

Parameter: Set $r := k + \ell$.

This concludes the construction. Before we prove its correctness, we note the size of the constructed instance $\mathcal{J}_p$. First, $r = \Theta(k)$, since $\ell \leq 2k$. Then, $|\mathcal{U}_p| = \sum_{i=1}^{\ell} |U_{\mathcal{C}_i}| = \sum_{i=1}^{\ell} (d(v_i)+1)N = \Theta(k \log n)$, and $|\mathcal{S}_p| = \Theta(n^2\ell + mk) = \Theta(n^2 k)$.

Yes case. Suppose there is a subgraph $G' = (V_{G'}, E_{G'})$ of G that is isomorphic to H with injection $\phi : V_H \to V_{G'}$. Let $V_{G'} = \{\alpha_1, \alpha_2, \cdots, \alpha_\ell\} \subseteq [n]$ such that $\alpha_1 < \alpha_2 < \cdots < \alpha_\ell$. Relabel the vertices of H as $\{v_1, \cdots, v_\ell\}$, where $v_i := \phi^{-1}(\alpha_i), i \in [\ell]$. Now, consider the ordering p of V_H such that $p(v_i) = i$, for $i \in [\ell]$, and fix the corresponding instance $\mathcal{J}_p = (\mathcal{U}_p, \mathcal{S}_p, r)$. Consider the following collection of V-sets and E-sets: $T_V := \bigcup_{i \in [\ell]} S_{\alpha_i \mapsto v_i, \alpha_{i+1}}$ and $T_E := \bigcup_{(v_i, v_j) \in E_H} S_{(\alpha_i, \alpha_j) \mapsto (v_i, v_j)}$. Let $T = T_V \cup T_E$. Note that for this choice of p, we have $T_V \subseteq \mathcal{S}_p$ due to construction, and $T_E \subseteq \mathcal{S}_p$ due to ϕ, and hence $T \subseteq \mathcal{S}_p$. Further, $|T| = |T_V| + |T_E| = \ell + k = r$, as required. Now, we claim that T forms a packing in $\mathcal{J}_p$. Towards this goal, note that it is sufficient to show that the sets in $T |_{U_{\mathcal{C}_i}}$ are mutually disjoint, for all $i \in [\ell]$. To this end, it is sufficient to show that both $T |_{U_{v_i, 0}}$ and $T |_{U_{v_i}}$ are packing, for all $i \in [\ell]$. Fix $i \in [\ell]$, and consider the following cases:

1. $T |_{U_{v_i, 0}}$: Since $T_E |_{U_{v_i, 0}} = \emptyset$ by construction, we focus on $T_V |_{U_{v_i, 0}}$. But, $T_V |_{U_{v_i, 0}}$ is a packing since,

$$T_V |_{U_{v_i, 0}} = \begin{cases} \{S_{\alpha_{i-1} \mapsto v_{i-1}, \alpha_i} |_{U_{v_i, 0}}, S_{\alpha_i \mapsto v_i, \alpha_{i+1}} |_{U_{v_i, 0}}\} = \{\bar{s}_{v_i, 0}^{\alpha_i}, s_{v_i, 0}^{\alpha_i}\}, \text{if } i \neq 1 \\ \{S_{\alpha_\ell \mapsto v_\ell, \alpha_1} |_{U_{v_1, 0}}, S_{\alpha_1 \mapsto v_1, \alpha_2} |_{U_{v_1, 0}}\} = \{\bar{s}_{v_1, 0}^{\alpha_1}, s_{v_1, 0}^{\alpha_1}\}, \text{ if } i = 1. \end{cases}$$

2. $T |_{U_{v_i}}$: It is sufficient to show that $T |_{U_{v_i, v_j}}$ is a packing, $\forall v_j \in N(v_i)$. But this follows since, $\forall v_j \in N(v_i)$,

$$T |_{U_{v_i, v_j}} = \{T_V |_{U_{v_i, v_j}}, T_E |_{U_{v_i, v_j}}\}$$
$$= \{S_{\alpha_i \mapsto v_i, \alpha_{i+1}} |_{U_{v_i, v_j}}, S_{(\alpha_i, \alpha_j) \mapsto (v_i, v_j)} |_{U_{v_i, v_j}}\} = \{s_{v_i, v_j}^{\alpha_i}, \bar{s}_{v_i, v_j}^{\alpha_i}\}.$$

No case. Suppose there is an r-packing $T \subseteq \mathcal{S}_p$ in some instance $\mathcal{J}_p, p \in [\mu]$, then we show that there is a subgraph G_T of G that is isomorphic to H. First note that $p \in [\mu]$ gives a labeling $\{v_1, \cdots, v_\ell\}$ of V_H such that $v_i = p^{-1}(i)$, for $i \in [\ell]$. Next, partition T into T_V and T_E, such that T_V and T_E correspond to the V-sets and E-sets of T respectively. This can be easily done since $t \in T$ is a V-set if and only if $t |_{U_{v_i, 0}} = s_{v_i, 0}^{\alpha} \in \mathcal{A}_{v_i, 0}$, for some $\alpha \in V_G, v_i \in V_H$. Let $U_0 = \{U_{v_i, 0}\}_{v_i \in V_H}$ and $U_1 = \{U_{v_i}\}_{v_i \in V_H}$. We claim the following.

Lemma 2. $|T_V| = \ell$ and $|T_E| = k$.

Proof. Note that for $t \in T_V$, we have $t |_{U_0} = \{s_{v_i, 0}^{\alpha}, \bar{s}_{v_{i+1}, 0}^{\beta}\}$, for some $\alpha, \beta \in V_G$ and $v_i, v_{i+1} \in V_H$. Hence, it follows that $|t |_{U_0}| = N$. Since $|U_0| = \ell N$ and T_V is a packing, we have $|T_V| \leq \ell$. For bounding $|T_E|$, consider $t \in T_E$, and note that $t |_{U_1} = \{\bar{s}_{v_i, v_j}^{\alpha}, \bar{s}_{v_j, v_i}^{\beta}\}$, for some $(\alpha, \beta) \in E_G$ and $(v_i, v_j) \in E_H$. But also note that we have $\bar{s}_{v_i, v_j}^{\alpha} \in \bar{\mathcal{A}}_{v_i, v_j}$ and $\bar{s}_{v_j, v_i}^{\beta} \in \bar{\mathcal{A}}_{v_j, v_i}$. Hence, by the virtue of T_E being a packing and using the facts that U_1 is the union of universes of $2k$ many base ISS $\{\bar{\mathcal{A}}_{v,w}\}_{v \in V_H, w \in N(v)}$, and each $t \in T_E$ contains sets from two of such ISS, it follows $|T_E| \leq k$. Finally, $|T| = r = \ell + k$ implies $|T_V| = \ell$ and $|T_E| = k$.   □

For $i \in [\ell]$, as $\mathcal{A}_{v_i,0} = (U_{v_i,0}, S_{v_i,0})$ is an ISS, we can relabel the sets in T_V as $T_V = \{T_V^1, \cdots, T_V^\ell\}$, where $T_V^i := t \in T_V$ such that $t \mid_{U_0} \ni s_{v_i,0}^\alpha$, for some $s_{v_i,0}^\alpha \in S_{v_i,0}$. The following lemma is our key ingredient.

Lemma 3. *T covers the whole universe $\mathcal{U}_p$.*

Proof. Since $\mathcal{U}_p = U_0 \cup U_1$, we will show that $T \mid_{U_j}$ covers U_j, for $j = \{0, 1\}$. For U_0, note that $T \mid_{U_0} = T_V \mid_{U_0}$ by construction. For $T_V^i \in T_V$, we have $|T_V^i \mid_{U_0}| = N$ due to complement partition axiom of $(\mathcal{A}_{v_i,0}, \bar{\mathcal{A}}_{v_i,0})$. Since T_V forms a packing, we have that $|\cup_{i \in [\ell]} T_V^i \mid_{U_0}| = \ell N = |U_0|$, as desired. Next, we have $|U_1| = 2kN$. Consider $T_V^i \in T_V$ and notice $|T_V^i \mid_{U_1}| = \frac{N}{2} d(v_i)$ since $T_V^i \mid_{U_1} = S_{v_i}^\alpha$, for some $\alpha \in V_G$. Since T_V forms a packing, we have $|\cup_{i \in [\ell]} T_V^i \mid_{U_1}| = \sum_{i=1}^\ell |T_V^i \mid_{U_1}| = kN$. Now consider $t = S_{(\alpha,\beta) \mapsto (v_i,v_j)} \in T_E$, for some $(\alpha, \beta) \in E_G$ and $(v_i, v_j) \in E_H$. Since, $t \mid_{U_1} = \{\bar{s}_{v_i,v_j}^\alpha, \bar{s}_{v_j,v_i}^\beta\}$, we have $|t \mid_{U_1}| = N$. As T_E forms a packing, we have $|\cup_{t \in T_E} t \mid_{U_1}| = \sum_{t \in T_E} |t \mid_{U_1}| = kN$. Finally, T being a packing, we have $|\cup_{\tau \in T} \tau \mid_{U_1}| = |\cup_{i \in [\ell]} T_V^i \mid_{U_1}| + |\cup_{t \in T_E} t \mid_{U_1}| = 2kN = |U_1|$ as desired.   □

Let $\alpha_i = \alpha \in V_G$ such that $T_V^i \mid_{U_{v_i,0}} \ni s_{v_i,0}^\alpha$, for $i \in [\ell]$, and let $V_T = \{\alpha_i\}_{i \in [\ell]}$.

Lemma 4. *For each vertex $\alpha \in V_G$, there is at most one V-set $S_{\alpha \mapsto v_i, \beta}$ in T_V, for some $v_i \in V_H$ and $\beta \in V_G$.*

Proof. It is sufficient to show $\alpha_i < \alpha_{i+1}$, for $i \in [\ell-1]$. Fix such i and consider the universe $U_{v_{i+1},0}$ of $\mathcal{A}_{v_{i+1},0}$. Then, note that only T_V^i and T_V^{i+1} contain elements of $U_{v_{i+1},0}$. Let $T_V^i = S_{\alpha_i \mapsto v_i, \beta}$ for $\beta > \alpha_i$, and let $T_V^{i+1} = S_{\alpha_{i+1} \mapsto v_{i+1}, \gamma}$, for $\gamma > \alpha_{i+1}$. As T covers $U_{v_i,0}$ (Lemma 3), and using the complement partition property of the compatible ISS pair $(\mathcal{A}_{v_{i+1},0}, \bar{\mathcal{A}}_{v_{i+1},0})$, we have that $\alpha_{i+1} = \beta > \alpha_i$.   □

Lemma 5. *For every edge $(\alpha, \beta) \in E_G$, there is at most one E-set $S_{(\alpha,\beta) \mapsto (v_i,v_j)}$ in T_E, for some $(v_i, v_j) \in E_H$.*

Proof. Suppose there are two sets $S_{(\alpha,\beta) \mapsto (v_i,v_j)}, S_{(\alpha,\beta) \mapsto (v_i',v_j')} \in T_E$, for some $(\alpha, \beta) \in E_G$. Without loss of generality assume $v_i \neq v_i'$. Then, we will show that $S_{\alpha \mapsto v_i, \gamma}, S_{\alpha \mapsto v_i', \delta} \in T_V$, for some $\gamma, \delta \in V_G$, contradicting Lemma 4. Since $S_{(\alpha,\beta) \mapsto (v_i,v_j)}, S_{(\alpha,\beta) \mapsto (v_i',v_j')} \in T_E$, it holds that $T_E \mid_{U_{v_i,v_j}} = \bar{s}_{v_i,v_j}^\alpha$, and $T_E \mid_{U_{v_i',v_j'}} = \bar{s}_{v_i',v_j'}^\alpha$. As T covers $\mathcal{U}_p$, in particular, T covers U_{v_i,v_j}, it must be that $T_V \mid_{U_{v_i,v_j}} = s_{v_i,v_j}^\alpha$ as $(\mathcal{A}_{v_i,v_j}, \bar{\mathcal{A}}_{v_i,v_j})$ is a compatible ISS pair. By similar reasoning for $U_{v_i',v_j'}$, it must be that $T_V \mid_{U_{v_i',v_j'}} = s_{v_i',v_j'}^\alpha$. This implies that $T_V \mid_{U_{v_i}} = S_{v_i}^\alpha$ and $T_V \mid_{U_{v_i'}} = S_{v_i'}^\alpha$. Thus, $S_{\alpha \mapsto v_i, \gamma}, S_{\alpha \mapsto v_i', \delta} \in T_V$ for $v_i \neq v_i'$, for some $\gamma, \delta \in V_G$.   □

Let $G_T = G[V_T] = (V_T, E_T)$, be the induced subgraph of G on V_T. Note that from the above lemmas, $|V_T| = \ell$ and $|E_T| = k$. To finish the proof, we claim that G_T is isomorphic to H with the injective homomorphism $\phi : V_H \to V_T$ given by $\phi(v_i) = \alpha_i$, for $i \in [\ell]$. To this end, we will show that for any $(v_i, v_j) \in E_H$, it holds that $(\phi(v_i), \phi(v_j)) = (\alpha_i, \alpha_j) \in E_T$. Consider the universe U_{v_i, v_j}, and note that $T_V^i \mid_{U_{v_i, v_j}} = s_{v_i, v_j}^{\alpha_i}$. As T covers U_{v_i, v_j}, it holds that $T_E \mid_{U_{v_i, v_j}} = \bar{s}_{v_i, v_j}^{\alpha_i}$ since $(\mathcal{A}_{v_i, v_j}, \bar{\mathcal{A}}_{v_i, v_j})$ is a compatible ISS pair. Hence $S_{(\alpha_i, \beta) \mapsto (v_i, v_j)} \in T_E$, for some $(\alpha_i, \beta) \in E_G$. This implies that $T_E \mid_{U_{v_j, v_i}} = \bar{s}_{v_j, v_i}^{\beta}$. By similar arguments for U_{v_j, v_i}, we have that $\beta = \alpha_j$ as $T_v^j \mid_{U_{v_j, v_i}} = s_{v_j, v_i}^{\alpha_j}$. Hence $(\alpha_i, \alpha_j) = (\phi(v_i), \phi(v_j)) \in E_G$. □

Acknowledgments. This work has been partially supported by European Research Council (ERC) under the European Union's Horizon 2020 research and innovation programme (grant agreement No. 759557). I thank Parinya Chalermsook for the informative discussions about the results in the paper, and for providing guidance on writing this paper. I also thank anonymous reviewers for their valuable suggestions on improving the readability of the paper.

References

1. Ausiello, G., D'Atri, A., Protasi, M.: Structure preserving reductions among convex optimization problems. J. Comput. Syst. Sci. **21**(1), 136–153 (1980)
2. Austrin, P., Khot, S., Safra, M.: Inapproximability of vertex cover and independent set in bounded degree graphs. In: 2009 24th Annual IEEE Conference on Computational Complexity, pp. 74–80. IEEE (2009)
3. Bansal, N., Gupta, A., Guruganesh, G.: On the lovász theta function for independent sets in sparse graphs. SIAM J. Comput. **47**(3), 1039–1055 (2018)
4. Bhattacharyya, A., Gadekar, A., Ghoshal, S., Saket, R.: On the hardness of learning sparse parities. In: Sankowski, P., Zaroliagis, C. (eds.) 24th Annual European Symposium on Algorithms (ESA 2016). Leibniz International Proceedings in Informatics (LIPIcs), vol. 57, pp. 1–17. Schloss Dagstuhl-Leibniz-Zentrum fuer Informatik, Dagstuhl, Germany (2016). https://doi.org/10.4230/LIPIcs.ESA.2016.11, http://drops.dagstuhl.de/opus/volltexte/2016/6362
5. Björklund, A., Husfeldt, T., Koivisto, M.: Set partitioning via inclusion-exclusion. SIAM J. Comput. **39**(2), 546–563 (2009)
6. Cai, L., Juedes, D.: Subexponential parameterized algorithms collapse the w-hierarchy. In: Orejas, F., Spirakis, P.G., van Leeuwen, J. (eds.) ICALP 2001. LNCS, vol. 2076, pp. 273–284. Springer, Heidelberg (2001). https://doi.org/10.1007/3-540-48224-5_23
7. Chalermsook, P., et al.: From Gap-ETH to FPT-inapproximability: clique, dominating set, and more. In: 2017 IEEE 58th Annual Symposium on Foundations of Computer Science (FOCS), pp. 743–754. IEEE (2017)
8. Chan, Y.H., Lau, L.C.: On linear and semidefinite programming relaxations for hypergraph matching. Math. program. **135**(1–2), 123–148 (2012)
9. Chataigner, F., Manić, G., Wakabayashi, Y., Yuster, R.: Approximation algorithms and hardness results for the clique packing problem. Discrete Appl. Math. **157**(7), 1396–1406 (2009)
10. Cygan, M.: Improved approximation for 3-dimensional matching via bounded pathwidth local search. In: 2013 IEEE 54th Annual Symposium on Foundations of Computer Science, pp. 509–518. IEEE (2013)

11. Cygan, M., et al.: Parameterized Algorithms, 1st edn. Springer, Cham (2015). https://doi.org/10.1007/978-3-319-21275-3
12. Downey, R.G., Fellows, M.R.: Parameterized Complexity. Springer, Cham (2012)
13. Halldórsson, M.M., Kratochvíl, J., Telle, J.A.: Independent sets with domination constraints. Discret. Appl. Math. **99**(1–3), 39–54 (2000)
14. Hassin, R., Rubinstein, S.: An approximation algorithm for maximum triangle packing. Discrete Appl. Math. **154**(6), 971–979 (2006)
15. Hastad, J.: Clique is hard to approximate within n1-. In: Proceedings of the 37th Annual Symposium on Foundations of Computer Science FOCS 1996, p. 627. IEEE Computer Society, USA (1996)
16. Hazan, E., Safra, S., Schwartz, O.: On the complexity of approximating k-set packing. Comput. Complex. **15**(1), 20–39 (2006). https://doi.org/10.1007/s00037-006-0205-6
17. Jones, M., Lokshtanov, D., Ramanujan, M.S., Saurabh, S., Suchý, O.: Parameterized complexity of directed Steiner tree on sparse graphs. In: Bodlaender, H.L., Italiano, G.F. (eds.) ESA 2013. LNCS, vol. 8125, pp. 671–682. Springer, Heidelberg (2013). https://doi.org/10.1007/978-3-642-40450-4_57
18. Lin, B.: A Simple gap-producing reduction for the parameterized set cover problem. In: Baier, C., Chatzigiannakis, I., Flocchini, P., Leonardi, S. (eds.) 46th International Colloquium on Automata, Languages, and Programming (ICALP 2019). Leibniz International Proceedings in Informatics (LIPIcs), vol. 132, pp. 81:1–81:15. Schloss Dagstuhl-Leibniz-Zentrum fuer Informatik, Dagstuhl, Germany (2019). https://doi.org/10.4230/LIPIcs.ICALP.2019.81, http://drops.dagstuhl.de/opus/volltexte/2019/10657
19. Lin, B.: Constant approximating k-clique is w[1]-hard. In: Khuller, S., Williams, V.V. (eds.) STOC 2021: 53rd Annual ACM SIGACT Symposium on Theory of Computing, Virtual Event, Italy, 21–25 June 2021, pp. 1749–1756. ACM (2021). https://doi.org/10.1145/3406325.3451016
20. Marx, D.: Can you beat treewidth? In: 48th Annual IEEE Symposium on Foundations of Computer Science (FOCS 2007), pp. 169–179. IEEE (2007)
21. Pătraşcu, M., Williams, R.: On the possibility of faster SAT algorithms. In: Proceedings of the Twenty-First Annual ACM-SIAM Symposium on Discrete Algorithms, pp. 1065–1075. SIAM (2010)
22. Karthik, C.S., Khot, S.: Almost polynomial factor inapproximability for parameterized k-clique (2021)
23. Karthik, C.S., Laekhanukit, B., Manurangsi, P.: On the parameterized complexity of approximating dominating set. In: Proceedings of the 50th Annual ACM SIGACT Symposium on Theory of Computing, pp. 1283–1296. STOC 2018, Association for Computing Machinery, New York, NY, USA (2018). https://doi.org/10.1145/3188745.3188896
24. Vemuganti, R.: Applications of set covering, set packing and set partitioning models: a survey. In: Du, DZ., Pardalos, P.M. (eds.) Handbook of Combinatorial Optimization, pp. 573–746. Springer, Boston (1998). https://doi.org/10.1007/978-1-4613-0303-9_9

Structural Parameterization of Cluster Deletion

Giuseppe F. Italiano[1], Athanasios L. Konstantinidis[1]($\boxtimes$),
and Charis Papadopoulos[2]

[1] LUISS University, Rome, Italy
{gitaliano,akonstantinidis}@luiss.it
[2] Department of Mathematics, University of Ioannina, Ioannina, Greece
charis@uoi.gr

Abstract. In the WEIGHTED CLUSTER DELETION problem we are given
a graph with non-negative integral edge weights and the task is to determine, for a target value k, if there is a set of edges of total weight at
most k such that its removal results in a disjoint union of cliques. It is
well-known that the problem is FPT parameterized by k, the total weight
of edge deletions. In scenarios in which the solution size is large, naturally one needs to drop the constraint on the solution size. Here we study
WEIGHTED CLUSTER DELETION where there is no bound on the size of the
solution, but the parameter captures structural properties of the input
graph. Our main contribution is to classify the parameterized complexity of WEIGHTED CLUSTER DELETION with three structural parameters,
namely, vertex cover, twin cover and neighborhood diversity. We show
that the problem is FPT when parameterized by the vertex cover, whereas
it becomes paraNP-hard when parameterized by the twin cover or the
neighborhood diversity. To illustrate the applicability of our FPT result,
we use it in order to show that the unweighted variant of the problem,
CLUSTER DELETION, is FPT parameterized by the twin cover. This is the
first algorithm with single-exponential running time parameterized by the
twin cover. Interestingly, we are able to achieve an FPT result parameterized by the neighborhood diversity that involves an ILP formulation. In
fact, our results generalize the parameterized setting by the solution size,
as we deduce that both parameters, twin cover and neighborhood diversity, are linearly bounded by the number of edge deletions.

Keywords: Cluster deletion problem · Twin cover · Neighborhood
diversity

G. F. Italiano—Partially supported by MUR, the Italian Ministry for University and
Research, under PRIN Project AHeAD (Efficient Algorithms for HArnessing Net-
worked Data).
C. Papadopoulos—Supported by the Hellenic Foundation for Research and Innovation
(H.F.R.I.) under the "First Call for H.F.R.I. Research Projects to support Faculty
members and Researchers and the procurement of high-cost research grant", Project
FANTA (eFficient Algorithms for NeTwork Analysis), number HFRI-FM17-431.

C.-C. Lin et al. (Eds.): WALCOM 2023, LNCS 13973, pp. 371–383, 2023.
https://doi.org/10.1007/978-3-031-27051-2_31

1 Introduction

Highly-connected parts of complex systems reveal clustering data that are important in numerous application fields such as computational biology [2] and machine learning [1, 25]. In graph-theoretic terms, those dense homogeneous sets are often identified as cliques. A core algorithmic theme that has received considerably interest is to modify a given graph as little as possible in order to reveal disjoint cliques. In the CLUSTER DELETION problem we seek to delete the minimum number of edges of a given graph such that the resulting graph is a vertex-disjoint union of cliques (cluster graph). Here we also consider its natural variant with weights on the edges of the graph, named WEIGHTED CLUSTER DELETION: each edge has an associated non-negative weight and the goal is to minimize the sum of the weights of the removed edges. It is known that the problems are NP-hard on general graphs [26] and settling their complexity status even on restricted settings has attracted several researchers.

With regards to parameterized complexity, the general result by Cai [6] shows that (WEIGHTED) CLUSTER DELETION is FPT parameterized by the number of deleted edges. Considering the same parameter, it is known that the problem admits a linear kernel [7] and recently, Cao et al. [8] devised a different, still linear, kernel. In particular, both variations of the problem admit several fast FPT algorithms parameterized by the solution size [3, 8, 27]. However, as with the principle of parameterized complexity, such algorithms are efficient whenever the considered parameter is rather small. Combined with the light of lower bounds refuting the existence of subexponential FPT algorithms [20], it seems reasonable to study different distance measures. If the remaining edges inside the cliques are used as a parameter then the unweighted variant of the problem has shown to be FPT and does not admit a polynomial kernel [18]. In contrast, by considering either the maximum degree or the diameter of the graph as a parameter of the problem, paraNP-hardness results occur. More specifically, Komusiewicz and Uhlmann [20] showed that CLUSTER DELETION is NP-hard on graphs of maximum degree 4, but it is polynomial-time solvable on graphs with maximum degree 3. Further, CLUSTER DELETION is NP-hard on P_5-free graphs [5, 23], although there is a polynomial algorithm that computes an optimal solution on P_4-free graph [16]. Interestingly, CLUSTER DELETION is FPT when parameterized by the size of a minimum cluster vertex deletion set [21].

Naturally, the weighted variant of the problem may behave differently than the unweighted on the same class of graphs. For instance, WEIGHTED CLUSTER DELETION is NP-hard even on P_4-free graphs and split graphs [5]. Apart from some restricted subclasses of chordal graphs for which the problem can be solved in polynomial time [5], the weighted variant of the problem has received less interest when parameterized by distance measures other than the solution size. Our focus is to complement existing results and analyze both variations of the problem under graph structural parameters. To capture the powerfulness of such parameters, we consider generalizations of the vertex cover number. These type of parameterizations proved to be successful in a wide range of problems [4, 14, 15, 24]. Here we exploit their impact towards the (WEIGHTED) CLUSTER DELETION problem.

Our Results. We consider parameterizations of the problem with respect to structural properties of the given graph. As (WEIGHTED) CLUSTER DELETION admits fast algorithms parameterized by the solution size, it is natural to consider variations of the vertex cover number such as the twin cover and the neighborhood diversity. We note that both notions constitute generalizations of the vertex cover number [15, 24], though there is no relation among them.

We first show that both parameters are linearly upper-bounded in the number of edge deletions required to obtain a cluster graph. Thus we explore further venues to attack CLUSTER DELETION, since the unweighted and weighted variations of the problem were already known to admit fast parameterized algorithms by the solution size [6, 8, 17, 27].

As an initial point, we establish that WEIGHTED CLUSTER DELETION is paraNP-hard when parameterized by the twin cover or the neighborhood diversity. This is achieved through an interesting reduction from a terminal cut problem with a small number of terminals.

Theorem 1.1. WEIGHTED CLUSTER DELETION *is NP-hard on graphs with twin cover number at most three and graphs with neighborhood diversity at most two.*

Based on this negative result, we also consider the more restrictive vertex cover number as a structural parameter. The vertex cover number is unrelated to the solution size. However, with our next algorithm, vertex cover can be considered as one of the few parameters for which the weighted variant admits a positive result. Our technique relies on carefully applying a dynamic programming approach that handles vertices that lie outside the vertex cover. We note that CLUSTER DELETION is expressible in monadic second order logic (MSO_2) as explicitly given in [22]. Thus WEIGHTED CLUSTER DELETION is FPT when parameterized by treewidth [9]. However, we are not aware if such an approach leads to a single-exponential running time, as we deduce for the vertex cover.

Theorem 1.2. WEIGHTED CLUSTER DELETION *can be solved in $2^{O(\mathsf{vc})} \cdot O(n^2)$ time, where vc is the vertex cover number of the input graph.*

To illustrate the wider applicability of the algorithm given in Theorem 1.2, we turn our attention to the unweighted variant of CLUSTER DELETION. Twin cover introduced by Ganian [15] generalizes vertex cover in the sense that vertices outside the cover set form an independent set or a true twin class. Our approach for CLUSTER DELETION relies on carefully contracting true twin classes that lie outside the cover set. It turns out that this process results in an edge-weighted graph of bounded vertex cover. Then we apply the algorithm given in Theorem 1.2 for the WEIGHTED CLUSTER DELETION problem.

Theorem 1.3. CLUSTER DELETION *can be solved in $2^{O(\mathsf{tc})} \cdot O(n^2)$ time, where tc is the twin cover number of the input graph.*

It should be noted that the FPT membership given in Theorem 1.3, can be obtained with a different approach, though with a worse running time. In particular, one could use the *cluster vertex deletion* number (also known as *distance to cluster*) which stands for the number of deleted vertices that is required to obtain a cluster graph. Doucha and Kratochvíl [12] showed that for any graph G, the cluster vertex deletion number of G is at most the twin cover number of G. Combined with the fact that CLUSTER DELETION is FPT parameterized by the cluster vertex deletion number [21], we get an algorithm with running time $2^{O(\text{tc} \log \text{tc})} \cdot n^{O(1)}$. Thus Theorem 1.3 reveals the first FPT algorithm with single-exponential running time. Moreover, we believe that our algorithm is interesting on its own because it exploits further connections between the two variations of the problem.

Regarding the neighborhood diversity which was introduced by Lampis [24], we use a completely different approach. This notion is based on true twin and false twin classes of vertices. We use integer linear programming (ILP) as a subroutine in our FPT algorithm. In particular, we translate part of our problem as an instance of choosing sufficient maximum cliques in an auxiliary graph of bounded size. As the size is bounded, we show that the formulation to an ILP problem with bounded number of variables is feasible.

Theorem 1.4. CLUSTER DELETION *can be solved in* $2^{2^{O(\text{nd})}} \cdot n^{O(1)}$ *time, where* nd *is the neighborhood diversity of the input graph.*

2   Preliminaries

All graphs considered here are simple and undirected. For $S \subseteq V$, $N(S) = \bigcup_{v \in S} N(v) \setminus S$ and $N[S] = N(S) \cup S$. For $X \subseteq V(G)$, the subgraph of G *induced* by X, $G[X]$, has vertex set X, and for each vertex pair u, v from X, uv is an edge of $G[X]$ if and only if $u \neq v$ and uv is an edge of G. For $R \subseteq E(G)$, $G \setminus R$ denotes the graph $(V(G), E(G) \setminus R)$, that is a subgraph of G and for $S \subseteq V(G)$, $G - S$ denotes the graph $G[V(G) - S]$, that is an induced subgraph of G. For two disjoint sets of vertices A and B, we write $E(A, B)$ to denote the edges that have one endpoint in A and one endpoint in B. A *matching* in G is a set of edges having no common endpoint. A *cluster graph* is a graph in which every connected component is a clique. *Contracting* a set of vertices S is the operation of substituting the vertices of S by a new vertex w with $N(w) = N(S)$. We next formalize CLUSTER DELETION, as a decision problem.

CLUSTER DELETION

Input: A graph $G = (V, E)$ and a non-negative integer k.
Task: Decide whether there is $E' \subseteq E(G)$ such that $G \setminus E'$ is a cluster graph and $|E'| \leq k$.

In the optimization setting, the task of CLUSTER DELETION is to turn the input graph G into a cluster graph by deleting the minimum number of edges. We describe a solution of CLUSTER DELETION in two equivalent ways: either a solution is given as a set of edges $E' \subseteq E(G)$ such that $G \setminus E'$ is a cluster graph, or it is given as a vertex partition $S = \{C_1, \ldots, C_t\}$ of $V(G)$ such that each $G[C_i]$ is a clique. The equivalence follows because the connected components of the cluster graph $G \setminus E'$ correspond to the induced subgraphs $G[C_i]$, so that the set $E' = \bigcup E(C_i, C_j)$ where $C_i, C_j \in S$ forms the required set of deleted edges.

Let $S = \{C_1, \ldots, C_t\}$ be a solution of CLUSTER DELETION such that each $G[C_i]$ is a clique. In such terms, the problem can be viewed as a vertex partition problem into $C_1, \ldots, C_t$. Each C_i is called *cluster*. Edgeless clusters, i.e., clusters containing exactly one vertex, are called *trivial clusters*. An *optimal solution S* for CLUSTER DELETION is a clique partition of G such that the number of edges $E(G) \setminus E(S)$ is minimum, where $E(S)$ stands for the set of edges $\bigcup E(C_i)$.

For the edge-weighted variation, every edge of the input graph admits a cost of deletion (represented by a weight function w) and the task is to perform the minimum sum of weights deletions. Given a subset E' of edges, we let $w(E') = \sum_{e \in E'} w(e)$, for the ease of notation. Hereafter we assume that the edge weights are positive integers. The reason of not considering negative weights comes from the fact that any graph can be completed into a clique with arbitrary edge-weights, so that WEIGHTED CLUSTER DELETION becomes trivially difficult even on graphs with sufficiently large cliques.

WEIGHTED CLUSTER DELETION
Input: A graph $G = (V, E)$, a weight function $w : E(G) \to \mathbb{Z}^+$, and a non-negative integer k.
Task: Decide whether there is $E' \subseteq E(G)$ such that $G \setminus E'$ is a cluster graph and $w(E') \leq k$.

Notice that if all edge weights are equal to one then the two variants of the problem coincide. Moreover, positive results propagate from WEIGHTED CLUS-TER DELETION towards CLUSTER DELETION, whereas negative results propagate in the reverse order. However, a notable difference among the two problems is the aspect of computing a minimum solution. Indeed, the formal description of the edge-weighted problem is not sufficient to find a minimum weight solution. Despite this fact, we point out that our positive results concerning both problems are able to compute an optimal solution with an additional polynomial factor on the stated running times.

Next we provide some useful properties concerning twin vertices. Two adjacent vertices u and v are called *true twins* if $N[u] = N[v]$, whereas two non-adjacent vertices x and y are called *false twins* if $N(x) = N(y)$. A *true twin class* of G is a maximal set of vertices that are pairwise true twins. Note that the set of true twin classes of G constitutes a partition of $V(G)$. We denote by $\mathcal{T}(G) = \{T_1, \ldots, T_r\}$ the true twin classes of G, so that each set of vertices T_i forms a true twin class in G. Observe that the partition $\mathcal{T}(G) = \{T_1, \ldots, T_r\}$ of $V(G)$ into classes of true twins can be constructed in linear time.

Lemma 2.1 ([5]). *Let x and y be true twin vertices in G. Then, in any optimal solution for CLUSTER DELETION x and y belong to the same cluster.*

It is not difficult to extend Lemma 2.1 for a set of true twin vertices.

Observation 2.1. *The vertices of a true twin class of G belong to the same cluster in any optimal solution for CLUSTER DELETION.*

We should point out that the above characterization does not hold for the edge-weighted variant of the problem, even if we relax the restriction to certain (rather than *any*) optimal solutions. However the following result holds.

Lemma 2.2. *Let X be a true twin class of G such that all edges incident to the vertices of X have the same positive weight. Then the vertices of X belong to the same cluster in any optimal solution for* WEIGHTED CLUSTER DELETION.

Graph Parameters. A *vertex cover* of G is a set of vertices that includes at least one endpoint of every edge of the graph. The *vertex cover number*, denoted by vc(G), is the size of a minimum cardinality vertex cover in G. Notice that a set of vertices X is a vertex cover if and only if $V(G) \setminus X$ is an independent set. Unfortunately, vertex cover is a rather restrictive graph parameter and, for that reason the following generalizations have been proposed. The twin cover of a graph has been introduced by Ganian [15] as follows.

Definition 2.1. *A set of vertices $X \subseteq V(G)$ is a* twin cover *of G if for every edge $uv \in E(G)$ either one of the following holds: (i) $u \in X$ or $v \in X$, (ii) u and v are true twin vertices. Then G has twin cover number* tc *if* tc *is the minimum possible size of a twin cover of G.*

It is known that if a minimum twin cover in G has size at most k, then it is possible to find a twin cover of size k in time $O(|E| + k|V| + 1.2738^k)$ [15].

Another generalization of vertex cover is the neighborhood diversity which has been defined by Lampis [24]. Two vertices x, y of G have the *same type* if $N(x) \setminus \{y\} = N(y) \setminus \{x\}$. The relation of having the same type is an equivalence. In particular, two vertices x and y have the same type if and only if x and y are either true twin or false twin vertices.

Definition 2.2. *A graph $G = (V, E)$ has neighborhood diversity at most* nd, *if there exists a partition of $V(G)$ into at most* nd *sets, such that all vertices in each set have the same type.*

Observe that the vertices of a given type not only have the same (closed) neighborhood in G, but also form either a clique or an independent set in G. Moreover, it is useful to consider a *type graph* H of a graph G, in which every node V_i of H is a type class of G and two such nodes V_i, V_j are adjacent in H if and only if $uv \in E(G)$ for $u \in V_i$ and $v \in V_j$. There exists an algorithm which runs in polynomial time and given a graph $G = (V, E)$ finds a minimum partition of $V(G)$ into neighborhood types [24].

Notice here that there are graphs that have bounded twin cover and unbounded neighborhood diversity, and vice versa (see for e.g., [15]). Moreover, twin cover and neighborhood diversity are incomparable with treewidth (tw) but more restrictive than cliquewidth (cw) [15, 24].

We now relate the above mentioned parameters with the number k of deleted edges for CLUSTER DELETION. We consider connected graphs, because any solution for CLUSTER DELETION or WEIGHTED CLUSTER DELETION of a disconnected graph G is obtained by the union of partial solutions on each connected component of G. Regarding vc there are simple examples for which $k = O(n)$ and vc $= O(1)$, whereas other examples exist to show the opposite situation: a star graph is typical example for the former case and a graph consisting of two vertex-disjoint cliques with an additional edge is an example for the latter case. Thus k and vc are unrelated. Notice that this comes in contrast to similar relations with respect to the *cluster vertex deletion* number (also known as *distance to cluster*) which stands for the number of deleted vertices that is required to obtain a cluster graph. It is not difficult to see that the cluster vertex deletion number is at most $2k$. Let us now show that nd and tc are both linearly upper-bounded in k.

Proposition 2.1. *Let G be a connected graph and let H be a cluster subgraph of G with $k = |E(G) \setminus E(H)|$. Then, $\mathsf{nd}(G) \leq 3k + 1$ and $\mathsf{tc}(G) \leq 2k$.*

3   Algorithmic Results for WEIGHTED CLUSTER DELETION

In this section, we present our results on WEIGHTED CLUSTER DELETION when the parameter is the twin cover (tc) or neighborhood diversity (nd) or vertex cover (vc) of the given graph. We begin with the hardness result for the more general parameters tc and nd and then provide an efficient algorithm for the restricted parameter vc.

We obtain our result from the k-MULTIWAY CUT problem: given a graph $G = (V, E)$, a set $T = \{t_1, \ldots, t_k\} \subseteq V(G)$ of k terminals, and a non-negative integer ℓ, the task is to find a set of edges $F' \subseteq E(G)$ such that $|F'| \leq \ell$ and each terminal belongs to a separate connected component in $G \setminus F'$. It is known that k-MULTIWAY CUT problem remains NP-hard even if the number of terminals is three (i.e., $k = 3$) [11].

Theorem 3.1. WEIGHTED CLUSTER DELETION *is NP-hard on graphs with twin cover number at most three and graphs with neighborhood diversity at most two.*

Proof. We give a polynomial-time reduction to WEIGHTED CLUSTER DELETION on graphs with twin cover 3 or neighborhood diversity 2 from the NP-hard problem 3-MULTIWAY CUT. Let $(G = (V, E), T = \{t_1, t_2, t_3\}, \ell)$ be an instance of 3-MULTIWAY CUT where $|V(G)| = n$ and $|E(G)| = m$. We assume that G is connected and $G[T]$ is edgeless, because we can restrict to the connected components of G and any edge between terminals belongs to a solution. Starting from G, we construct a graph H by adding all necessary edges so that (i) $V(G) \setminus T$ is a clique and (ii) every vertex t of T is adjacent to every vertex of $V(G) \setminus T$. Observe that H has the same vertex set with G and contains all edges of the complete graph except the edges of the triangle among the three terminals. We assign the following edge-weight function for the edges of H: if $e \in E(G)$ then $\mathsf{w}(e) = n^2$; otherwise, $\mathsf{w}(e) = 1$.

Notice that H can be constructed in polynomial time and contains n vertices and $\frac{n(n-1)}{2} - 3$ edges. Since the vertices of $V(G) \setminus \{t_1, t_2, t_3\}$ form a clique and the neighborhood of t_1, t_2 and t_3 is exactly this clique, the vertices of the clique are true twins. Thus, by the definition of twin cover, we get that $\mathsf{tc}(H) = 3$, with a twin cover set $\{t_1, t_2, t_3\}$. Moreover, the vertices of H can be partitioned into a clique $V(G)$ and an independent set T such that every vertex of the independent set T is adjacent to every vertex of the clique $V(G)$. Hence H not only has a bounded twin cover, but it also admits a neighborhood diversity exactly 2, since one type class is the clique $V(G) \setminus T$ and the other type class is the independent set T. Let $W = \ell \cdot n^2 + p$, where $p = \frac{n(n-1)}{2} - m - 3$. We prove that 3-MULTIWAY CUT has a solution $F' \subseteq E(G)$ with $|F'| \leq \ell$ if and only if WEIGHTED CLUSTER DELETION has a solution $E' \subseteq E(H)$ with $\mathsf{w}(E') \leq W$.   □

3.1 Vertex Cover

Here we provide an FPT algorithm for the WEIGHTED CLUSTER DELETION problem parameterized by the vertex cover. As a consequence, notice that CLUSTER DELETION can be solved within the same running time.

Theorem 3.2. WEIGHTED CLUSTER DELETION *can be solved in* $2^{O(\mathsf{vc})} \cdot O(n^2)$ *time, where* vc *is the vertex cover number of the input graph.*

Proof. Let (G, k) be an instance of WEIGHTED CLUSTER DELETION and let X be a vertex cover of $G = (V, E)$ of size vc. The vertices of $V \setminus X$ form an independent set I. Let $S = \{C_1, \ldots, C_r\}$ be a solution for WEIGHTED CLUSTER DELETION. Observe that any cluster C_i contains at most one vertex from I. That is, $|C_i \cap I| \leq 1$, for every $1 \leq i \leq r$. Since $|X| = \mathsf{vc}$, there are at most vc vertices of I that belong to non-trivial clusters. We design an FPT algorithm that computes a solution by applying a dynamic programming scheme. For a set of vertices Y, we assign its total edge-weight $\mathsf{w}(Y)$ as $-\infty$ whenever $G[Y]$ is not a cluster graph and $\mathsf{w}(Y)$ is the sum of the edge-weights in $G[Y]$, otherwise.

We number the vertices of I in an arbitrary order $I = \{1, \ldots, |I|\}$. For technical reasons, we extend I by adding a vertex u in G that is non-adjacent to any vertex, so that $I' = I \cup \{u\}$. Let G be the resulting graph and let $I' = \{0, 1, \ldots, |I|\}$, assuming that u is numbered with zero. For the dynamic programming, we construct a table T as follows: given a subset X' of X and an integer $j \in \{0, 1, \ldots, |I|\}$, $T[X', j]$ denotes the total edge-weights of a maximum edge-weighted cluster subgraph of $G[X' \cup \{0, \ldots, j\}]$. Clearly $T[X, |I|]$ is the desired value for our problem. As a base case, observe that $T[\emptyset, j] = 0$ for any j. For the recurrence, we have the following equation:

$$T[X', j] = \max_{\emptyset \neq Y' \subseteq X'} \begin{cases} T[X' \setminus Y', 0] + \mathsf{w}(Y'), & \text{if } j = 0 \\ T[X' \setminus Y', j - 1] + \mathsf{w}(Y' \cup \{j\}), & \text{otherwise.} \end{cases} \tag{1}$$

□

4 An Application on Twin Cover

In this section, we illustrate how Theorem 3.2 can be applied in the more relaxed variation with no edge weights but in a more powerful setting. We consider twin cover number as a parameter for CLUSTER DELETION and we show that CLUSTER DELETION can be solved in FPT time under this parameterization. For doing so, we apply a natural process related to the true twin vertices resulting in an edge-weighted graph of bounded vertex cover.

Even though we deal with the unweighted variant, we consider edge-weighted graphs in a natural way. If there is no weight function defined on the edges of a graph G, we assign a weight to each edge of G equal to one and assume that G is an edge-weighted graph. A true twin class T of an edge-weighted graph G, is called *1-class* if all edges incident to the vertices of T have weight one.

Definition 4.1 (T-contraction). *Given a 1-class T of G, we define the T-contraction of G as the edge-weighted graph H obtained from G by contracting T into a single vertex v_T s.t. all edges incident to v_T in H have weight $|T|$.*

Observe that after a T-contraction, H has $|V(G)| - |T| + 1$ vertices, whereas the total value of the new edge weights in H is at most $|E(G)|$.

Lemma 4.1. *Let T be a 1-class of a graph G and let H be the T-contraction of G. There is a solution $E_1 \subseteq E(G)$ for WEIGHTED CLUSTER DELETION on G with $w(E_1) \leq k$ if and only if there is a solution $E_2 \subseteq E(H)$ for WEIGHTED CLUSTER DELETION on H with $w(E_2) \leq k$.*

Theorem 4.1. CLUSTER DELETION *can be solved in $2^{O(\text{tc})} \cdot O(n^2)$ time, where* tc *is the twin cover number of the input graph.*

Proof. Let (G, k) be an instance of CLUSTER DELETION and let X be a twin cover of $G = (V, E)$ of size $\text{tc} = |X|$. By the definition of twin cover, the vertices of $V(G) \setminus X$ induce a disjoint union of cliques in G. In particular, observe that every connected component of $G - X$ is a clique and forms a 1-class in G. Based on this fact, it is not difficult to prove the following. Let $Y_1, \ldots, Y_p$ be the connected components of $G - X$ such that $|Y_i| \geq 2$, for every $1 \leq i \leq p$. Any Y_i-contraction results in an edge-weighted graph H_i in which the connected components of $H_i - X$ with at least one edge are $\{Y_1, \ldots, Y_p\} \setminus Y_i$, and every $Y' \in \{Y_1, \ldots, Y_p\} \setminus Y_i$ is a 1-class.

We apply a Y_i-contraction in an arbitrary order with respect to $Y_1, \ldots, Y_p$. The resulting graph H has vertex cover at most $|X|$: every connected component of $H - X$ has size one implying that the vertices of $H - X$ form an independent set. Therefore we can apply the WEIGHTED CLUSTER DELETION algorithm on the edge-weighted graph H given in Theorem 3.2.   □

5   CLUSTER DELETION and Neighborhood Diversity

In this section, we provide an FPT algorithm for CLUSTER DELETION when parameterized by neighborhood diversity. We will use integer linear programming as a subroutine of our main result. In particular, we translate part of our problem to an instance of the p-VARIABLE INTEGER LINEAR PROGRAMMING FEASIBILITY problem: given an $m \times p$ matrix A over $\mathbb{Z}$ and a vector $b \in \mathbb{Z}^m$, decide whether there is a vector $x \in \mathbb{Z}^p$ such that $Ax \leq b$. Lenstra [19] showed that the above problem is FPT parameterized by p, while Frank and Tardos [13] showed that this algorithm can be made to run also in polynomial space. We will make use of these results, that we formally state next[1].

Theorem 5.1 ([13,19]). p-Variable Integer Linear Programming Feasibility *can be solved using* $O(p^{2.5p+o(p)} \cdot L)$ *arithmetic operations and space polynomial in* L, *where* L *is the number of bits in the input.*

Before giving the details of our FPT algorithm for Cluster Deletion when parameterized by neighborhood diversity, we describe the basic idea how to compute a solution for Cluster Deletion by using the type graph. Let G be a graph and let H be its type graph of size nd. Moreover, let $\{V_1, \ldots, V_{nd}\}$ be the type classes of G, or equivalently, the nodes of H. In the beginning, we find all possible cliques (not necessarily maximal) of H by taking all the subsets of nodes of H that form cliques. For every clique H_i of H, it is possible to find a maximum clique in G that is induced by the vertices of the nodes which belong to H_i. Now, our task is to choose some cliques H_i of H and for each H_i to choose a specific number of maximum cliques contained in G. Those maximum cliques will be considered as clusters for the problem.

Theorem 5.2. Cluster Deletion *can be solved in* $2^{2^{O(nd)}} \cdot n^{O(1)}$ *time, where* nd *is the neighborhood diversity of the input graph.*

Proof. Let (G, k) be an instance of Cluster Deletion and let H be the type graph of G of size nd. By the definition of neighborhood diversity, the vertices of G can be partitioned in nd type classes. Let $\{V_1, \ldots, V_{nd}\}$ be the type classes of G. For ease of notation, we let $G_i = G[V_i]$. In the forthcoming arguments, we assume that $S = \{S_1, \ldots, S_r\}$ is the set of clusters of an optimal solution for Cluster Deletion on (G, k).

Since the vertices of a type class that forms a clique are true twin vertices, they belong to exactly one cluster of the solution S by Lemma 2.2. On the other hand, the vertices of a type class of G that forms an independent set belong to different clusters by the definition of cluster. Hence, for any cluster $S_j \in S$ that contains a vertex of V_i, we have the properties: (i) $V_i \subseteq S_j$, if G_i is a clique and (ii) $|V_i \cap S_j| = 1$, if G_i is an independent set. Based on this, we define the following quantities:

$$\|V_i\|_F = \begin{cases} |V_i| & \text{if } G_i = \text{clique} \\ 1 & \text{if } G_i = \text{independent set} \end{cases} \qquad \|V_i\|_T = \begin{cases} 1 & \text{if } G_i = \text{clique} \\ |V_i| & \text{if } G_i = \text{independent set} \end{cases}$$

Now we focus on the type graph H of G. We consider all subgraphs of H that form a clique in H. Since every type class of G is a node in H, there are at most 2nd cliques in H. Let $\mathcal{H} = \{H_1, \ldots, H_q\}$ be the set of cliques in H, where

---

[1]   In general the applicability of Theorem 5.1 has revealed close connections between FPT and ILP (see for e.g. [10]).

$q = |\mathcal{H}| \leq 2^{\text{nd}}$. Given a clique $H_j \in \mathcal{H}$, we denote by $G[H_j]$ the graph induced by the vertices of G that belong to the nodes of H_j. Formally, the set of vertices of $G[H_j]$ is exactly $\{u \in V_i \mid V_i \in V(H_j)\}$.

For every clique $H_j \in \mathcal{H}$, we define the following values: n_j is the number of vertices of a maximum clique in $G[H_j]$, m_j is the number of edges of a maximum clique in $G[H_j]$, $t_j = \min \|V_i\|_T$, for any $V_i \in V(H_j)$.

Claim 5.1. *For any $H_j \in \mathcal{H}$, a maximum clique S_j in $G[H_j]$ fulfils properties (i) and (ii) and contains the following number of vertices: $n_j = \sum_{V_i \in V(H_j)} \|V_i\|_F$.*

Thus, by Claim 5.1 we can compute in polynomial time both numbers n_j and m_j for every clique H_j. Similarly, t_j can be computed in time linear in the size of H_j. Now consider a maximum clique of $G[H_j]$ for a clique H_j. Observe that $G[H_j]$ may contain more than one maximum cliques that are vertex-disjoint. We capture this property by a non-negative integer y_j assigned to $H_j \in \mathcal{H}$, which describes some number of vertex-disjoint maximum cliques that belong to $G[H_j]$.

Given the set of cliques $\mathcal{H}$, for every node V_i of the type graph H we define $M(i)$ as the cliques of $\mathcal{H}$ that contain V_i. Formally, we have $M(i) = \{H_j \in \mathcal{H} \mid V_i \in V(H_j)\}$. Based on the number of cliques $q = |\mathcal{H}|$, we define a sequence $X = (x_1, \ldots, x_q)$ of non-negative integers and we say that X is *valid vector* of $\mathcal{H}$ if the following conditions hold: $0 \leq x_j \leq t_j$ and $\sum_{H_j \in M(i)} x_j = \|V_i\|_T$, for every $V_i \in V(H)$.

Furthermore, for a valid vector X of $\mathcal{H}$, we let $|E(X)|$ be the total number of edges if we choose x_j number of vertex-disjoint maximum cliques that belong to $G[H_j]$ for all $x_j \in X$. That is, $|E(X)| = \sum_{x_j \in X} x_j \cdot m_j$. With the next claim, our task is translated into finding a valid vector with an appropriate cost.

Claim 5.2 *There is a solution $E' \subseteq E(G)$ for (G, k) if and only if there is a valid vector X of $\mathcal{H}$ with $|E(G)| - |E(X)| \leq k$.*

Regarding the running time, the set $\mathcal{H}$ can be computed in $2^{\text{nd}} \cdot n^{O(1)}$ time by taking all possible subset of nodes of H and checking if each subset induces a clique in H. Additionally, for every V_i the set $M(i)$ can be found in $|\mathcal{H}| \cdot n^{O(1)}$ time by traversing all cliques of $\mathcal{H}$. Moreover, the ILP has at most 2nd number of variables, where the value of any variable is bounded by n. Thus by applying Theorem 5.1 the ILP can be solved in $2^{2^{O(\text{nd})}} \cdot n^{O(1)}$ time. Then we can compute a solution for CLUSTER DELETION from a valid vector in the same running time, as described in Claim 5.1. Therefore the overall running time is bounded by the time needed to solve the ILP system as it contains 2nd number of variables. □

References

1. Bansal, N., Blum, A., Chawla, S.: Correlation clustering. Mach. Learn. **56**, 89–113 (2004). https://doi.org/10.1023/b:mach.0000033116.57574.95
2. Ben-Dor, A., Shamir, R., Yakhini, Z.: Clustering gene expression patterns. J. Comput. Biol. **6**, 281–297 (1999)

3. Böcker, S., Briesemeister, S., Bui, Q.B.A., Truß, A.: Going weighted: parameterized algorithms for cluster editing. Theor. Comput. Sci. **410**, 5467–5480 (2009)
4. Bonnet, É., Sikora, F.: The graph motif problem parameterized by the structure of the input graph. Discrete Appl. Math. **231**, 78–94 (2017)
5. Bonomo, F., Durán, G., Valencia-Pabon, M.: Complexity of the cluster deletion problem on subclasses of chordal graphs. Theor. Comput. Sci. **600**, 59–69 (2015)
6. Cai, L.: Fixed-parameter tractability of graph modification problems for hereditary properties. Inf. Process. Lett. **58**, 171–176 (1996)
7. Cao, Y., Chen, J.: Cluster editing: kernelization based on edge cuts. Algorithmica **64**(1), 152–169 (2012). https://doi.org/10.1007/s00453-011-9595-1
8. Cao, Y., Ke, Y.: Improved kernels for edge modification problems. In: Proceedings of IPEC 2021, pp. 1–14 (2021)
9. Courcelle, B.: The monadic second-order logic of graphs I: recognizable sets of finite graphs. Inf. Comput. **85**, 12–75 (1990)
10. Cygan, M., et al.: Parameterized Algorithms. Springer, Cham (2015). https://doi.org/10.1007/978-3-319-21275-3
11. Dahlhaus, E., Johnson, D.S., Papadimitriou, C.H., Seymour, P.D., Yannakakis, M.: The complexity of multiterminal cuts. SIAM J. Comput. **23**, 864–894 (1994)
12. Doucha, M., Kratochvíl, J.: Cluster vertex deletion: a parameterization between vertex cover and clique-width. In: Proceedings of MFCS 2012, vol. 7464, pp. 348–359 (2012). https://doi.org/10.1007/978-3-642-32589-2
13. Frank, A., Tardos, É.: An application of simultaneous Diophantine approximation in combinatorial optimization. Combinatorica **7**, 49–65 (1987). https://doi.org/10.1007/BF02579200
14. Ganian, R.: Twin-cover: beyond vertex cover in parameterized algorithmics. In: Marx, D., Rossmanith, P. (eds.) IPEC 2011. LNCS, vol. 7112, pp. 259–271. Springer, Heidelberg (2012). https://doi.org/10.1007/978-3-642-28050-4_21
15. Ganian, R.: Improving vertex cover as a graph parameter. Discrete Math. Theor. Comput. Sci. **17**(2), 77–100 (2015)
16. Gao, Y., Hare, D.R., Nastos, J.: The cluster deletion problem for cographs. Discrete Math. **313**, 2763–2771 (2013)
17. Gramm, J., Guo, J., Hüffner, F., Niedermeier, R.: Graph-modeled data clustering: fixed-parameter algorithms for clique generation. In: Petreschi, R., Persiano, G., Silvestri, R. (eds.) CIAC 2003. LNCS, vol. 2653, pp. 108–119. Springer, Heidelberg (2003). https://doi.org/10.1007/3-540-44849-7_17
18. Grüttemeier, N., Komusiewicz, C.: On the relation of strong triadic closure and cluster deletion. Algorithmica **82**(4), 853–880 (2019). https://doi.org/10.1007/s00453-019-00617-1
19. Lenstra, J.H.W.: Integer programming with a fixed number of variables. Math. Oper. Res. **8**, 538–548 (1983)
20. Komusiewicz, C., Uhlmann, J.: Cluster editing with locally bounded modifications. Discrete Appl. Math. **160**, 2259–2270 (2012)
21. Komusiewicz, C., Uhlmann, J.: Alternative parameterizations for cluster editing. In: Černá, I., et al. (eds.) SOFSEM 2011. LNCS, vol. 6543, pp. 344–355. Springer, Heidelberg (2011). https://doi.org/10.1007/978-3-642-18381-2_29
22. Konstantinidis, A.L., Papadopoulos, C.: Maximizing the strong triadic closure in split graphs and proper interval graphs. Discrete Appl. Math. **285**, 79–95 (2020)
23. Konstantinidis, A.L., Papadopoulos, C.: Cluster deletion on interval graphs and split related graphs. Algorithmica **83**(7), 2018–2046 (2021). https://doi.org/10.1007/s00453-021-00817-8

24. Lampis, M.: Algorithmic meta-theorems for restrictions of treewidth. Algorithmica **64**(1), 19–37 (2012). https://doi.org/10.1007/s00453-011-9554-x
25. Li, P., Puleo, G.J., Milenkovic, O.: Motif and hypergraph correlation clustering. IEEE Trans. Inf. Theor. **66**, 3065–3078 (2020)
26. Shamir, R., Sharan, R., Tsur, D.: Cluster graph modification problems. Discrete Appl. Math. **144**, 173–182 (2004)
27. Tsur, D.: Cluster deletion revisited. Inf. Process. Lett. **173**, 106171 (2022)

Parity Permutation Pattern Matching

Virginia Ardévol Martínez[1]([✉]) [ID], Florian Sikora[1] [ID], and Stéphane Vialette[2] [ID]

[1] Université Paris-Dauphine, PSL University, CNRS, LAMSADE,
75016 Paris, France
{virginia.ardevol-martinez,florian.sikora}@dauphine.fr
[2] LIGM, CNRS, Univ Gustave Eiffel, 77454 Marne-la-Vallée, France
stephane.vialette@univ-eiffel.fr

Abstract. Given two permutations, a pattern σ and a text π, PARITY PERMUTATION PATTERN MATCHING asks whether there exists a parity and order preserving embedding of σ into π. While it is known that PERMUTATION PATTERN MATCHING is in FPT, we show that adding the parity constraint to the problem makes it W[1]-hard, even for alternating permutations or for 4321-avoiding patterns. However, it remains in FPT if the text avoids a fixed permutation, thanks to a recent meta-theorem on twin-width. On the other hand, as for the classical version, PARITY PERMUTATION PATTERN MATCHING remains polynomial-time solvable when both permutations are separable, or if both are 321-avoiding, but NP-hard if the pattern is 321-avoiding and the text is 4321-avoiding.

Keywords: PERMUTATION PATTERN MATCHING · Fixed parameter tractability · Parameterized hardness · NP-hardness

1 Introduction

Permutations are one of the most fundamental objects in discrete mathematics, and in concrete, deciding if a permutation contains another permutation as a pattern is one of the most natural decision problems related to them. More precisely, in the well-known problem PERMUTATION PATTERN MATCHING (PPM), given two permutations σ and π, the task is to determine if σ is a pattern of π, or equivalently, if π contains a subsequence which is order-isomorphic to σ. For example, if $\pi = 3\,1\,5\,4\,2$, it contains $\sigma = 2\,3\,1$, as $3\,5\,2$ is a subsequence of π with the same relative order as σ, but π does not contain $\sigma = 1\,2\,3$, as there are no 3 increasing elements in π. In the latter case, we say that π *avoids* $1\,2\,3$. The notion of avoidance allows to define classes of permutations as sets of permutations that avoid certain patterns, for example, 321-avoiding permutations, or $(2413, 3142)$-avoiding permutations, which are known as *separable permutations*.

PPM was proven to be NP-complete by Bose, Buss, and Lubiw in 1998 [6]. This motivated the search for exact exponential time algorithms [1,4,8,13]. However, some special cases, such as LONGEST INCREASING SUBSEQUENCE, or the cases where both σ and π are separable or 321-avoiding, are known to be polynomial time solvable [2,6,10,16]. In fact, it was shown in [17] that PPM is

C.-C. Lin et al. (Eds.): WALCOM 2023, LNCS 13973, pp. 384–395, 2023.
https://doi.org/10.1007/978-3-031-27051-2_32

always polynomial-time solvable if the pattern avoids any fixed permutation $\tau \in \{1, 12, 21, 132, 231, 312, 213\}$, and NP-complete otherwise. This result was then extended in [18].

Its parameterized complexity was open for a long time, with a series of partial results, but a breakthrough result of Guillemot and Marx showed that it is fixed parameter tractable when parameterized by the size of the pattern σ, using a new *width measure* structure theory of permutations [15]. They showed that the problem can be solved in time $2^{\mathcal{O}(k^2 \log k)} n$, and later on, Fox improved the running time of the algorithm by removing a factor $\log k$ from the exponent [12].

This led to the question of whether a graph-theoretic generalization of their permutation parameter could exist, that was answered positively in [5], by introducing the notion of *twin-width*, which has proven huge success recently. They showed that graphs of bounded twin-width define a very natural class with respect to computational complexity, as FO model checking becomes linear in them.

Pattern matching for permutations, together with its many variants, has been widely studied in the literature (the best general reference is [19], see also [7]). Here we introduce a natural variation of PPM, which we call PARITY PERMUTATION PATTERN MATCHING, and that incorporates the additional constraint that the elements of σ have to map to elements of π with the same parity, i.e., even (resp. odd) elements of σ have to be mapped to even (resp. odd) elements of π. For one thing, pattern avoidance with additional constraints [3,9], including parity restrictions [14,20], has emerged as a promising research area. For another, PARITY PERMUTATION PATTERN MATCHING aims at providing concrete use cases of the 2-colored extension of PPM introduced in [16]. We show that, surprisingly, it does not fit into the twin-width framework, and this increases the complexity of the problem, as it becomes W[1]-hard parameterized by the length of the pattern.

In fact, the approach used by Guillemot and Marx [15] to prove that PPM is FPT is based on a result that states that given a permutation π, there exists a polynomial time algorithm that either finds an $r \times r$-grid of π or determines that the permutation has bounded width (and returns the merge sequence of the decomposition, which is used to solve the PPM problem in FPT time). This win-win approach works because, if π contains an $r \times r$-grid, it's not hard to see that it contains every possible pattern σ. However, this cannot be generalized to PARITY PPM, as here we have no information on the parity of the elements of the grid, and thus, it is not guaranteed that every pattern maps via a parity respecting embedding into the grid.

Structure of the Paper. The paper is organized as follows. Section 2 briefly introduces the necessary concepts and definitions. In Sect. 3, we study the parameterized complexity of PARITY PPM, showing that it is harder than PPM in general, but that it remains in FPT for some cases, namely when the *twin-width* of the host permutation is bounded. Finally, in Sect. 4, we show that concerning the classical P vs NP questions, PARITY PPM is similar to PPM. A summary of the complexity of the problems is given in Table 1.

Table 1. Summary of known results (for PPM) and our results (for PARITY PPM).

	PPM	PARITY PPM
General case	NP-hard, FPT	W[1]-hard
Separable permutations	P	P
321-av σ and 321-av π	P	P
321-av σ and 4321-av π	NP-hard	NP-hard
4321-av σ	FPT	W[1]-hard
Alternating π and σ	FPT	W[1]-hard
π is fixed pattern avoiding	FPT	FPT

Due to space constraints, some proofs (marked with ($\star$)) are deferred to the full version of this paper.

2 Preliminaries

Let $[n] = \{1, \ldots, n\}$. A permutation of length n is a bijection $f : [n] \longrightarrow [n]$. Given two permutations $\sigma \in S_k$ and $\pi \in S_n$, we say that π (the text, or the host) *contains* σ (the pattern) if there is an embedding from σ into π, i.e., an injective function f such that for every pair of elements x and y of σ, their images $f(x)$ and $f(y)$ of π are in the same relative order as x and y. Otherwise, we say that π *avoids* σ. If π contains σ, we write $\sigma \le \pi$.

A *permutation class* is a set $\mathcal{C}$ of permutations such that for every permutation $\pi \in \mathcal{C}$, every pattern of π is also contained in $\mathcal{C}$. Every permutation class can be defined by the minimal set of permutations that do not lie inside it, and we define this as $\mathcal{C} = \mathrm{Av}(B)$, where B is the minimal set of avoided permutations.

In this manner, we can define the class $\mathrm{Av}(4321)$, which is the set of permutations that avoid 4321, $\mathrm{Av}(321)$, which is the set of permutations that avoid 321, and $\mathrm{Av}(2413, 3142)$, i.e., the class of permutations that avoid both 2413 and 3142. As we mentioned in the introduction, the latter is known as the class of *separable permutations*, and it can also be characterized as the set of permutations that have a separating tree. In other words, a permutation is separable if there exists an ordered binary tree $\mathcal{T}$ in which the elements of the permutation appear in the leaves and such that the descendants of a tree node form a contiguous subset of these elements.

Furthermore, we define the set of *alternating permutations* as the set of permutations $\sigma \in S_n$ such that $\sigma_1 > \sigma_2 < \sigma_3 > \ldots$.

The problem of determining whether a fixed pattern is contained in a permutation has been well studied in the literature, and it is referred to as PERMUTATION PATTERN MATCHING. Here, we study a natural variation of PPM, PARITY PERMUTATION PATTERN MATCHING, which we define formally below.

Definition 1. *Given two permutations, a pattern $\sigma \in S_k$ and a text $\pi \in S_n$, the problem* PERMUTATION PATTERN MATCHING *asks whether π contains σ.*

Definition 2. *An injective function f from σ to π is a parity respecting embedding if for all elements x and y of σ, $f(x)$ and $f(y)$ are in the same relative order as x and y, and for every element x of σ, $f(x)$ has the same parity as x.*

We say that an occurrence of a pattern σ in a permutation π respects parity if there is a parity respecting embedding of σ into π. Furthermore, if there is an occurrence of σ in π which respects parity, we say that π parity contains σ, and we write $\sigma \leq_P \pi$. Otherwise, we say that π parity avoids σ.

Definition 3. PARITY PPM *is the problem of determining whether given a pattern σ and a text π, there exists a parity respecting embedding of σ into π.*

As a remark, note that if instead of considering the problem PPM with the constraint that even (resp. odd) elements have to map to even (resp. odd) elements, we require that elements in even (resp. odd) indices (positions) map to elements in even (resp. odd) indices, the problem is equivalent. Indeed, σ parity avoids π if and only if σ^{-1} parity index avoids π^{-1}.

For example, the parity+order preserving embedding of $\sigma = 2413$ into $\pi = 4276315$ yields the parity-index+order-preserving embedding of $\sigma^{-1} = 3142$ into $\pi^{-1} = 6251743$ (occurrences are depicted with bold integers).

To see this, assume that there is a parity respecting embedding of σ into π. Denote by $P_\sigma(i)$ the position in σ of the element with value i and by f the parity respecting injective map between σ and π associated to the embedding. Since $\sigma^{-1} = P_\sigma(1) \ldots P_\sigma(k)$, and f respects parity, if $f(i) = j$, both i and j have the same parity, and thus, the indices in the inverses will also have the same parity (by definition, odd elements are placed in odd indices in the inverses, and vice versa). Furthermore, since f is an embedding, for $i < j$, $\sigma_i < \sigma_j$ if and only if $f(\sigma_i) < f(\sigma_j)$. Thus, $P(\sigma_i)$ is to the left of $P(\sigma_j)$ in both σ^{-1} and π^{-1}, and by assumption, we also have $i < j$, so f induces a parity index respecting embedding between the inverses.

In this paper, we focus mainly on the parameterized complexity of the above-mentioned problem. Parameterized complexity allows the classification of NP-hard problems on a finer scale than in the classical setting. Fixed parameter tractable (FPT) algorithms are those with running time $O(f(k) \cdot poly(n))$, where n is the size of the input and f is a computable function that depends only on some well-chosen parameter k. On the other hand, problems for which we believe that there does not exist an algorithm with that running time belong to the W-hierarchy. We refer to [11] for more background on the topic.

3   Parameterized Complexity

We already saw in the introduction that PPM is in FPT in general, and why the win-win approach of Guillemot and Marx for the parameterized algorithm for PPM doesn't work for PARITY PPM. We show that this intuition is indeed true, proving that the problem is W[1]-hard. In fact, we prove something stronger, which is that PARITY PPM is W[1]-hard even when restricted to alternating permutations or when the pattern is 4321-avoiding. Note that both results are independent from each other, as alternating permutations and 4321-avoiding permutations are not comparable, but they both imply the W[1]-hardness of the general case.

However, the twin-width framework (on which the parameterized algorithm of Guillemot and Marx is an initial step) will be useful to prove that PARITY PPM remains in FPT when the text avoids a fixed pattern.

3.1   Parameterized Hardness for Alternating Permutations

Theorem 4. PARITY PPM *is* W[1]-*hard parameterized by the length* k *of the pattern, even for alternating permutations* σ *and* π.

Proof. We reduce from k-CLIQUE in general graphs, which is known to be W[1]-hard parameterized by the size of the clique k [11]. Given as input a graph G and a parameter k, k-CLIQUE asks whether G contains a clique of size k. For our reduction, given a graph $G = (V, E)$, with $|V| = n$ and $|E| = m$, and a parameter k, we construct a permutation σ that depends only on the parameter k, and a permutation π, that depends on G, such that there exists a clique of size k in the graph G if and only if there is a parity respecting embedding of σ into π.

Construction. We explain the construction of π for a general graph G. The high-level idea is to construct different gadgets to represent the vertices and the edges of the graph, and to somehow *link* each edge gadget to the corresponding vertex gadgets, that is, we *link* the gadget associated to edge (u, v) with the gadgets associated to vertices u and v by placing elements of value greater than the minimum element of each vertex gadget and smaller than the maximum element of each vertex gadget between the elements of the edge gadget.

We define the following gadgets (see also Fig. 1):

- A *vertex gadget* $\pi[V]$, which is a direct sum of n decreasing permutations, all order-isomorphic to 21 and composed of odd elements. It contains $2n$ elements and starts at element $8m + 3$.
- An *edge gadget* $\pi[E]$, which is a direct sum of m permutations, all order-isomorphic to 435261 and formed by odd elements. It contains $6m$ elements and starts at element 3.
- The *separator gadget* is composed of the four even integers $4(n + 3m) + 4$, $8m + 2$, $4(n + 3m) + 2$ and 2 (and hence is order-isomorphic to 4231). The *separator gadget* lies between the *vertex gadget* and the *edge gadget*.
- Let Even be the $2(n + 3m) - 2$ even integers between 4 and $4(n + 3m)$ that do not appear in the *vertex gadget*, the *separator gadget* or the *edge gadget*. The *even garbage gadget* is the alternating sequence composed of the even integers of Even. It is constructed recursively from left to right as follows: place (and remove from Even) the maximum of Even, place (and remove from Even) the minimum of Even and recurse. It is placed to the right of the edge gadget.

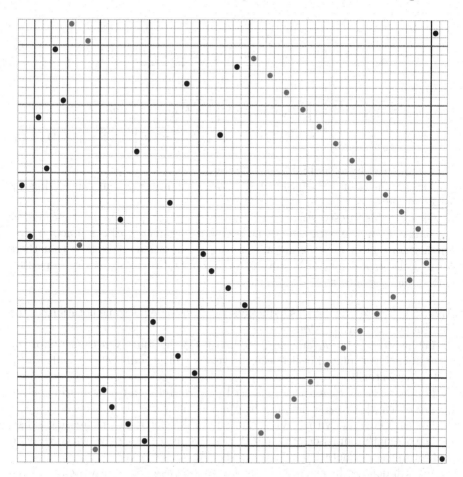

Fig. 1. Illustration of the construction introduced in the proof of Theorem 4. The permutation in the figure corresponds to a clique of size 3 with vertices v_1, v_2, v_3 and edges (v_1, v_2), (v_1, v_3), (v_2, v_3). The odd elements are represented in black while even elements are colored in red. Furthermore, blue lines delimit the vertex boxes. (Color figure online)

- Let Odd be the two odd elements $4(n + 3m) + 3$ and 1 that do not appear in the *vertex gadget*, the *separator gadget* or the *edge gadget*. The *odd garbage gadget* is the decreasing sequence composed of the two odd integers of Odd. It is constructed as the *even garbage gadget* and placed directly to its right.

Formally, define,

$$\forall v_i \in V, \ \pi[v_i] = \boxed{8m + 2 + 2\sum_{j<i}(deg(v_j) + 2) + 2\sum_i (deg(v_i)) + 3}$$
$$\boxed{8m + 2 + 2\sum_{j<i}(deg(v_j) + 2) + 1} \quad (1)$$

$$\forall e_k = (i,j) \in E, \ \pi[e_k] = \boxed{6k+1} \ \boxed{6k-1} \ \boxed{8m+2+2\sum_{j'<i} (deg(v'_j)+2) + 2\sum_{(i,j')/j'<j} (1) + 3}$$

$$\boxed{6k-3} \ \boxed{8m+2+2\sum_{j'<j}(deg(v'_j)+2)+2\sum_{(i',j)/i'<i}(1)+3} \ \boxed{6k-5} \tag{2}$$

$$\pi = \boxed{\pi[v_1]} \ldots \boxed{\pi[v_n]} \ \boxed{4(n+3m)+4} \ \boxed{4(n+3m)+2} \ \boxed{8m+2} \ \boxed{2}$$

$$\boxed{\pi[e_1]}, \ldots \boxed{\pi[e_m]}, \boxed{EVEN} \ \boxed{ODD} \tag{3}$$

(boxes are used for readability purposes only).

The permutation σ is constructed as the permutation π but considering K_k as the graph G.

Clearly, this construction can be carried out in polynomial time and σ depends only on the parameter k, i.e., the new parameter $|\sigma|$ is a function of k. Furthermore, both σ and π are alternating permutations. We claim that there exists a clique of size k in the graph G if and only if there is a parity respecting embedding of σ into π.

Notation. Before proving this reduction, we need to define some notation for the elements of the permutations.

Let us denote by w_i and w'_i ($i \in \{1,2,3,4\}$) the four even elements of the separator gadget, placed in between the vertex gadget and the edge gadget of σ and π, respectively.

For each vertex v_i, with $i \in \{1,\ldots,n\}$, we will refer to the decreasing subsequence of length two associated to it, $\sigma[v_i]$, as the vertex box associated to v_i.

For each edge e_i, with $i \in \{1,\ldots,m\}$, we will refer to the decreasing subsequence of length four associated to it (i.e., the elements of $\sigma[e_i]$ which correspond to 4321 in the permutation 435261) as the edge box of e_i.

We will denote the elements of the vertex box associated to vertex v_i as $v_{i,1}$ and $v_{i,2}$, from left to right (i.e., $v_{i,1} > v_{i,2}$), and the elements of the edge box associated to edge e_i as $e_{i,1}$, $e_{i,2}$, $e_{i,3}$ and $e_{i,4}$, again from left to right. On the other hand, for each edge, we denote the two elements placed in between $e_{i,2}$ and $e_{i,3}$, and between $e_{i,3}$ and $e_{i,4}$, as $h_{i,1}$ and $h_{i,2}$, respectively, where here $h_{i,1} < h_{i,2}$ (these are the elements that correspond to the subsequence 56 in $\sigma[e_i]$).

Finally, the even elements to the right of the edge gadget placed below w_2 are referred to as $w_{i,1}$, $w_{i,2}$, $w_{i,3}$ and $w_{i,4}$, for every edge $i \in \{1,\ldots,m\}$, where $w_{i,1}$ is the element $e_{i,4}+1$, $w_{i,2}$ is $e_{i,3}+1$, $w_{i,3}$ is $e_{i,2}+1$, and $w_{i,4}$ is $e_{i,1}+1$. Note that $w_{i,4}$ is not defined for the last edge. On the other hand, the even elements to the right of the edge gadget placed above w_2 are denoted as $x_{i,t}$, for every vertex $i \in \{1,...,n\}$ and every edge incident to v_i, $t \in \{1,...,m_i\}$ ($x_{i,t} = h_{x,y}+1$ for some pair x,y).

Furthermore, we denote by $x_{i,0}$ and x_{i,m_i+1} the even elements in the extremes such that $x_{i,0} = v_{i,2}+1$ and $x_{i,m_i+1} = v_{i,1}+1$. Again, note that x_{n,m_n+1} is not defined.

For the elements of π, we follow an analogous notation denoting the elements by $v'_{i,1}$, $e'_{i,1}$, etc.

Direct Implication

Claim 5. $(\star)$ *If there exists a clique of size k in the graph G, then there is a parity respecting embedding of σ into π.*

Reverse Implication. Suppose now that there exists a parity respecting embedding between σ and π and let f be the associated injective mapping. We want to show that we have enough structure in the permutations to infer that there must be a clique of size k in the graph G. In order to do so, we will prove the following sequence of claims that will restrict the map f.

Claim 6. *Any parity respecting embedding f from σ to π must map w_i to w'_i, for $i \in \{1, 2, 3, 4\}$.*

Proof of Claim. Since the pattern matching needs to respect parity, f must map the w_i's to even elements of π. Towards a contradiction, assume first that $f(w_i) \neq w'_j$, $i, j \in \{1, 2, 3, 4\}$. That means that $f(w_i) = w'_{i',j}$ or $x'_{i',t}$, for some indices i', j or i', t. But then, the odd elements to the right of w_i in σ cannot map to elements to the right of $f(w_i)$ in π (as there would be at most 2 odd elements to the right of $f(w_i)$ and there are strictly more than 2 odd elements to the right of w_i), so f cannot be an embedding of σ into π. Finally, since both the w_i's and the w'_i's form 4231 subsequences, it is clear that there exists a unique way to embed the w_i's into the w'_i's, which is mapping each w_i to its corresponding w'_i, for every $i \in \{1, 2, 3, 4\}$. Thus, if $f(w_i) \neq w'_i$, f cannot be an embedding. ◁

Claim 7. *All the elements to the left (resp. to the right) of the w_i's in σ map to elements to the left (resp. to the right) of w'_i's in π. Similarly, the elements above (resp. below) w_2 in σ map to elements above (resp. below) w'_2 in π.*

Proof of Claim. This is a direct corollary of Claim 6. ◁

Claim 8. *Any parity respecting embedding f from σ to π must map vertex blocks of σ to vertex blocks of π.*

Proof of Claim. By Claim 7, since elements to the left of w_2 in σ map to elements to the left of w'_2 in π, we have that $f(v_{i,j}) = v'_{i',j'}$, for $i \in \{1, ..., k\}$, $i' \in \{1, ..., n\}$ and $j, j' \in \{1, 2\}$. Assume that $f(v_{i,1}) = v'_{i',j'}$ and $f(v_{i,2}) = v'_{i'',j''}$, with $i' \neq i''$. Since $v_{i,1}$ is to the left of $v_{i,2}$, it means that f must map $v_{i,2}$ to an element placed to the right of $f(v_{i,1}) = v'_{i',j'}$. But $v_{i,1} > v_{i,2}$ and every element which is to the right of $v'_{i',j'}$ and which does not belong to the vertex block of $v'_{i'}$, is greater than $v'_{i',j'}$. Thus, if $i'' \neq i'$, then f would not be an embedding. ◁

Claim 9. *Any parity respecting embedding f from σ to π must map edge blocks of σ to edge blocks of π.*

Proof of Claim. Again, we have that $f(e_{i,j}) = e'_{i',j'}$ for some pair i', j', and since the structure of the gadget has the same properties as the vertex gadget, we can use the same argument as in the proof of Claim 8. ◁

Claim 10. *Any parity respecting embedding f from σ to π must map $h_{i,j}$ to $h'_{i',j}$, where $e'_{i'}$ is the edge associated to the edge block where $e_{i,1}$ maps to.*

Proof of Claim. By Claim 7, we have that necessarily, $f(e_{i,j}) = e'_{i',j}$, for $i \in \{1, ..., l\}, i' \in \{1, ..., m\}$ and $j \in \{1, 2, 3, 4\}$.

First, since $f(e_{i,2}) = e'_{i',2}$ and $f(e_{i,3}) = e'_{i',3}$, and f is an embedding, the fact that $h_{i,1}$ is in between $e_{i,2}$ and $e_{i,3}$ implies that it must map to an element between $e'_{i',2}$ and $e'_{i',3}$. Similarly, $h_{i,2}$ must map to an element in between $e'_{i',3}$ and $e'_{i',4}$. Since edge blocks map to edge blocks, there is at most one element that satisfies each of these conditions. And these elements are $h'_{i',1}$ and $h'_{i',2}$, respectively.   ◁

Claim 11. *All the even elements to the right of the edge gadgets in σ must map to even elements to the right of the edge gadgets in π.*

Proof of Claim. This follows from Claim 6. Since $f(w_i) = w'_i$ for $i \in \{1, 2, 3, 4\}$ and f has to respect parity, the rest of the even elements cannot map anywhere else.   ◁

Now, suppose that there is a parity respecting embedding f of σ into π and assume, towards a contradiction, that G does not contain a clique of size k. Since there is no clique of size k, it means that we cannot have $l = \binom{k}{2}$ edges between the k vertices of G which are in the image of f (that is, the vertices associated to the images of the k vertex boxes of σ).

We know that the k vertex blocks of σ map to k vertex blocks in π and the $\binom{k}{2}$ edge blocks of σ map to $\binom{k}{2}$ edge blocks of π. Since G does not contain a clique, one of the k vertices corresponding to the k vertex blocks in the image of f will have degree strictly smaller than $k - 1$ when we restrict G to the k selected vertices. Let i' be the vertex with degree strictly smaller than $k - 1$ and suppose it is the image of vertex block i in σ. Then, there are two possible cases. The first case is that in the image of f, between the values $f(v_{i,1})$ and $f(v_{i,2})$, there are less than k odd elements (these elements are necessarily of the form $h'_{i,j}$). Since in between $v_{i,1}$ and $v_{i,2}$ in σ there are k odd elements of the form $h_{i,j}$, this would imply that f cannot be a parity respecting embedding. The second possibility is that in between the values $f(v_{i,1})$ and $f(v_{i,2})$ there are k odd elements (which again are necessarily of the form $h'_{i,j}$) but one of them is not in between $f(e_{l,2})$ and $f(e_{l,3})$, or $f(e_{l,3})$ and $f(e_{l,4})$, for some $l \in \{1, ..., m\}$. This would also contradict the fact that f is a parity respecting isomorphism, as all the $h_{i,j}$ in σ are between some pair $e_{l,2}, e_{l,3}$, or $e_{l,3}, e_{l,4}$ (with respect to the x-axis). Therefore, if there is a parity respecting embedding of σ into π, it must map the k vertex boxes of σ into k vertex boxes of π associated to k vertices that form a clique in G.   □

Corollary 12. $(\star)$ *Given a pattern $\sigma \in S_k$ and a text $\pi \in S_n$, PARITY PPM cannot be solved in time $f(k) \cdot n^{o(\sqrt{k})}$ for any computable function f, under the Exponential Time Hypothesis (ETH).*

Note that reducing from SUBGRAPH ISOMORPHISM instead of k-CLIQUE in the proof of Theorem 4 to get a better lower bound under the ETH is not trivial since there is a notion of order of the pattern in PARITY PPM (i.e., two isomorphic subgraphs can result in different permutations depending on the ordering of their vertices).

3.2   Parameterized Hardness for 4321-Avoiding Patterns

In this subsection, we complement the previous hardness result by showing that the problem remains hard for patterns belonging to the class of 4321-avoiding permutations. Our proof uses a colored version of PPM defined in [16], proven W[1]-hard parameterized by $k = |\sigma|$ in [16].

Definition 13. 2-COLORED 2IPP *(2 INCREASING PERMUTATION PATTERN) consists on, given a 321-avoiding permutation $\sigma \in S_k$ and an arbitrary permutation π such that both σ and π are 2-colored permutations, finding a color-preserving embedding of σ into π.*

Theorem 14. ($\star$) PARITY PPM *is W[1]-hard parameterized by the length k of the pattern, even if the pattern is 4321-avoiding.*

3.3   Parameterized Algorithm for Fixed Pattern Avoiding Text

In the previous subsection, we showed that restricting the pattern does not necessarily reduce the complexity of the problem. However, we now see that restricting the text allows us to use the twin-width meta-theorem [5] to have a positive result. In fact, to see that PARITY PPM is FPT if the text avoids a fixed pattern x, it suffices to show that we can describe the problem using first-order (FO) logic, i.e., that we can express it as a formula which uses quantified variables over non-logical objects, and sentences (formulas without free variables) that contain the variables. Indeed, adding unary relations to mark the odd and even values preserves bounded twin-width, and therefore FPT tractability. The result follows from [5]:

Lemma 15 ([5]). *FO model checking is FPT on every hereditary proper subclass of permutation graphs.*

This implies that FO model checking is FPT in the class of permutations avoiding a fixed pattern. Here, FO model checking refers to the problem of, given a first-order sentence ϕ of FO and a finite model $\mathcal{M}$ of FO (which specifies the domain of disclosure of the variables), deciding whether $\mathcal{M}$ satisfies ϕ, i.e., whether there exists an assignment of the variables which respects the domain imposed by $\mathcal{M}$ and that satisfies ϕ. Therefore, we can state the following theorem:

Theorem 16. ($\star$)PARITY PPM *is in FPT if the text π avoids a fixed permutation.*

4 Classical Complexity

Even though PARITY PPM is harder than PPM from the parameterized point of view, we will show that this is not the case concerning its classical complexity.

4.1 Hardness

A nice quite recent result showed that PPM remains NP-hard, even if the pattern is 321-avoiding and the text is 4321-avoiding [17]. In the following, we show that it remains true for PARITY PPM.

Theorem 17. (⋆)PARITY PPM *is* NP-*hard, even if* σ *is a* 321-*avoiding permutation and* π *is a* 4321-*avoiding permutation.*

4.2 Polynomial-Time Solvable Cases

For some specific cases of PERMUTATION PATTERN MATCHING, polynomial time algorithms that solve the problem exactly have been proposed. Here, we show that some of these algorithms can be adapted to solve the problem PARITY PERMUTATION PATTERN MATCHING while still running in polynomial time.

Theorem 18. (⋆)*Let* σ *be a permutation in* S_k *and* π *be a permutation in* S_n. PARITY PPM *can be solved in polynomial time in the following cases:*

1. *If both permutations are separable. In particular, if both permutations are (231, 213)-avoiding, it can be solved in linear time.*
2. *If both permutations are 321-avoiding.*

Acknowledgements. Thanks to Édouard Bonnet and Eun Jung Kim for pointing out the link with the twin-width framework, and to the reviewers for their useful comments.

References

1. Ahal, S., Rabinovich, Y.: On complexity of the subpattern problem. SIAM J. Discrete Math. **22**(2), 629–649 (2008)
2. Albert, M.H., Lackner, M., Lackner, M., Vatter, V.: The complexity of pattern matching for 321-avoiding and skew-merged permutations. Discrete Math. Theor. Comput. Sci. **18**(2) (2016)
3. Alexandersson, P., Fufa, S.A., Getachew, F., Qiu, D.: Pattern-avoidance and fuss-catalan numbers. arXiv preprint. arXiv:2201.08168 (2022)
4. Berendsohn, B.A., Kozma, L., Marx, D.: Finding and counting permutations via CSPs. Algorithmica **83**(8), 2552–2577 (2021). https://doi.org/10.1007/s00453-021-00812-z
5. Bonnet, É., Kim, E.J., Thomassé, S., Watrigant, R.: Twin-width I: tractable FO model checking. J. ACM **69**(1), 3:1–3:46 (2022)
6. Bose, P., Buss, J.F., Lubiw, A.: Pattern matching for permutations. Inf. Process. Lett. **65**(5), 277–283 (1998)

7. Bruner, M.L., Lackner, M.: The computational landscape of permutation patterns. Pure Mathematics and Applications: Special Issue for the Permutation Patterns 2012 Conference, vol. 24, no. 2, pp. 83–101 (2013)

8. Bruner, M., Lackner, M.: A fast algorithm for permutation pattern matching based on alternating runs. Algorithmica **75**(1), 84–117 (2016). https://doi.org/10.1007/s00453-015-0013-y

9. Bulteau, L., Fertin, G., Jugé, V., Vialette, S.: Permutation pattern matching for doubly partially ordered patterns. In: Proceedings of the CPM. LIPIcs, vol. 223, pp. 21:1–21:17 (2022)

10. Cormen, T.H., Leiserson, C.E., Rivest, R.L., Stein, C.: Introduction to Algorithms. MIT Press, Cambridge (2022)

11. Cygan, M., et al.: Parameterized Algorithms. Springer, Cham (2015)

12. Fox, J.: Stanley-wilf limits are typically exponential. arXiv preprint. arXiv:1310.8378 (2013)

13. Gawrychowski, P., Rzepecki, M.: Faster exponential algorithm for permutation pattern matching. In: 5th SOSA@SODA 2022, pp. 279–284. SIAM (2022)

14. Gil, J.B., Tomasko, J.A.: Restricted Grassmannian permutations. Enumerative Comb. Appl. **2**(4), #S4PP6 (2021)

15. Guillemot, S., Marx, D.: Finding small patterns in permutations in linear time. In: Proceedings of the SODA, pp. 82–101. SIAM (2014)

16. Guillemot, S., Vialette, S.: Pattern matching for 321-avoiding permutations. In: Dong, Y., Du, D.-Z., Ibarra, O. (eds.) ISAAC 2009. LNCS, vol. 5878, pp. 1064–1073. Springer, Heidelberg (2009). https://doi.org/10.1007/978-3-642-10631-6_107

17. Jelínek, V., Kynčl, J.: Hardness of permutation pattern matching. In: Proceedings of the SODA, pp. 378–396. SIAM (2017)

18. Jelínek, V., Opler, M., Pekárek, J.: Griddings of permutations and hardness of pattern matching. In: Proceedings of the MFCS. LIPIcs, vol. 202, pp. 65:1–65:22 (2021)

19. Kitaev, S.: Patterns in Permutations and Words. Monographs in Theoretical Computer Science. An EATCS Series, Springer, Berlin (2011). https://doi.org/10.1007/978-3-642-17333-2

20. Tanimoto, S.: Combinatorics of the group of parity alternating permutations. Adv. Appl. Math. **44**(3), 225–230 (2010)

Author Index

C.-C. Lin et al. (Eds.): WALCOM 2023, LNCS 13973, pp. 397–398, 2023.
https://doi.org/10.1007/978-3-031-27051-2

Printed in the United States
by Baker & Taylor Publisher Services